CHEMISTRY FOR ENVIRONMENTAL ENGINEERING

Third Edition

KU-318-674

Clair N. Sawyer

Environmental Consultant
Formerly Vice President and Director of Research
Metcalf and Eddy, Engineers, Boston
and Professor of Sanitary Chemistry
Massachusetts Institute of Technology

Perry L. McCarty

Silas H. Palmer Professor of Environmental Engineering
Stanford University

McGraw-Hill Book Company

New York   St. Louis   San Francisco   Auckland   Bogotá   Düsseldorf
Johannesburg   London   Madrid   Mexico   Montreal   New Delhi   Panama
Paris   São Paulo   Singapore   Sydney   Tokyo   Toronto

CHEMISTRY FOR ENVIRONMENTAL ENGINEERING

1 2 3 4 5 6 7 8 9 0   D O D O   7 8 3 2 1 0 9 8

Library of Congress Cataloging in Publication Data

Sawyer, Clair N
    Chemistry for environmental engineering.

    (McGraw-Hill series in water resources and environmental engineering)
    First-2d ed. published under title: Chemistry for sanitary engineers.
    Includes index.
    1. Environmental chemistry. 2. Sanitary engineering. I. McCarty, Perry L., joint author. II. Title. QD31.2.S28 1978   628'.01'54   77-24619
ISBN 0-07-054971-0

This book was set in Times Roman by Bi-Comp, Incorporated.
The editors were B. J. Clark and J. W. Maisel;
the production supervisor was Dennis J. Conroy.
New drawings were done by ECL Art Associates, Inc.
R. R. Donnelley & Sons Company was printer and binder.

To
Orphelia and Martha
for their patience and encouragement

CONTENTS

PREFACE

Education in environmental engineering is usually conducted at the graduate level, and up to the present time has drawn mainly on students with a civil engineering background. In general, civil engineering training does not prepare a student well in chemistry and biology. Since a knowledge of these sciences is vital to the environmental engineer, the graduate program must be designed to correct this deficiency. In recent years students from other engineering disciplines and from the natural sciences have been attracted to this field. Some have a deficiency in chemistry and biology similar to that of the civil engineer and need an exposure to general concepts of importance.

This book is written to serve as a textbook for a first course in chemistry for environmental engineering students with one year of college-level chemistry. Environmental engineers need a wide background in chemistry, and in recognition of this need, the book summarizes the important aspects from various areas of chemistry. This treatment should help orient the students' thinking, aid them in their choice of areas for advanced study, and help them develop a better concept of what they should expect to derive from further study.

The purpose of this book is twofold: (1) It attempts to bring into focus those aspects of chemistry which are particularly valuable to environmental engineering practice, and (2) it attempts to lay a groundwork of understanding in the area of specialized quantitative analysis, commonly referred to as water and wastewater analysis, that will serve the student as a basis in all the common phases of environmental engineering practice and research.

Substantial changes continue to occur in the emphasis of courses for environmental engineers. The trend is toward a more fundamental understanding of the chemical phenomena causing changes in the quality of natural waters and of water undergoing treatment. In addition, emphasis is being directed toward more advanced methods of analysis, which are required to help solve many of the complex modern problems facing the environmental engineer.

This new edition is organized to better meet the current needs. Part One is concerned solely with fundamentals of chemistry for engineers. It includes chapters on general chemistry, organic chemistry, physical chemistry, colloid chemistry, biochemistry, and nuclear chemistry, and a new chapter on equilibrium chemistry. Each gives emphasis to environmental engineering applications. Many schools offer a lecture course on chemical principles, and it is felt that the revisions in Part One make the text more suitable for this purpose.

All chapters concerned with analytical measurements are now together in Part Two of the book. The first several chapters here contain general information on quantitative, qualitative, and instrumental methods of analysis, which can be used as background material for the subsequent chapters concerned with water and wastewater analyses of particular value to environmental engineers. These chapters are written to stress the basic chemistry of each analysis and to point out their significance in environmental engineering practice. They should be particularly useful when employed with "Standard Methods for the Examination of Water and Wastewater," published by the American Public Health Association, which gives the details for carrying out each analytical determination. Part Two is considered to be most useful as lecture material to be utilized along with a laboratory course on water and wastewater analysis.

Problems are included at the end of most chapters to stress fundamentals and to increase the usefulness of this book as a classroom text. Example problems are also given throughout the text to help increase the students' understanding of the principles outlined. In the first portion of the text, where the emphasis is on chemical fundamentals, answers are included after many of the problems to allow students to evaluate independently their understanding of the principles emphasized.

In order to meet the requirements of a textbook, brevity has been an important consideration throughout. For those who believe that we have been too brief, we can only beg their indulgence and recommend that they seek further information in standard references on the subject. References of particular value are listed at the end of each chapter in the first portion of the book.

We wish to continue to express our thanks to Dr. Wertheim for his permission to use certain material from the book by Wertheim and Jeskey entitled "Introductory Organic Chemistry." Special thanks are also due to Dr. Alonzo W. Lawrence of Koppers Company, Inc., and Dr. James J. Bisogni, Jr., of Cornell University for their many helpful suggestions for reorganizing this book, and to Dr. George A. Parks of Stanford University for his comments on Chapter 5. In addition, we greatly appreciate the several suggestions made by Dr. Irvine W. Wei, Northeastern University, a reviewer selected by the publishers.

Clair N. Sawyer
Perry L. McCarty

PART

ONE

FUNDAMENTALS OF CHEMISTRY FOR ENVIRONMENTAL ENGINEERING

CHAPTER
ONE

INTRODUCTION

The important role that environmental and public health engineers have played in providing us with pure and adequate water supplies, facilities for wastewater and refuse disposal, safe recreational areas, and a healthy environment within our homes and places of employment is not generally appreciated by the public at large. Those who have experienced living in the underprivileged areas of the world usually return home with a new sense of respect for the guardians of the public health. Among these guardians are the engineers who are in the front lines of defense, employing their knowledge of science and engineering to erect barriers against the ever-present onslaught of diseases and plagues, the most terrible of the "Four Horsemen of the Apocalypse."

For many years the attention of environmental engineering was devoted largely to the development of safe water supplies and the sanitary disposal of human wastes. Because of the success in controlling the spread of enteric diseases through the application of engineering principles, a new concept of the potentialities of preventive medicine was born. Expanding populations with resultant increased industrial operations, power production and use of motor-driven vehicles, plus new industries based upon new technology have intensified old problems and created new ones in the fields of water supply, waste disposal, and air pollution. Many of these have offered a real challenge to environmental engineers, and the profession as a whole has been ready to accept the challenge.

Over the years, intensification of old problems and the introduction of new ones have led to basic changes in the philosophy of environmental engineering practice. Originally the major objectives were to produce hygienically safe water supplies and to dispose of wastes in a manner that would prevent the development of nuisance conditions. Many other factors concerned with aesthetics, economics, recreation, and other elements of better living are important considerations and have become part of the responsibilities of the modern environmental engineer.

1-1 WATER

Water is one of the materials required to sustain life and has long been suspected of being the source of many of the illnesses of man. It was not until a little over 100 years ago that definite proof of disease transmission through water was established. For many years following, the major consideration was to produce adequate supplies that were hygienically safe. The public has been more exacting in its demands as time has passed, and today water engineers are expected to produce finished waters that are free of color, turbidity, taste, odor, and harmful metal ions. In addition, the public desires water which is low in hardness and total solids, noncorrosive, and non-scale-forming. To provide such water, chemists, biologists, and engineers must combine their efforts and talents. The chemist, through his knowledge of colloidal and physical chemistry, is especially helpful in solving problems related to the removal of color, turbidity, hardness, and harmful metal ions and to the control of corrosion and scaling. The biologist is often of great help in taste and odor problems that derive from aquatic growths. In a true sense, therefore, all who cooperate in the effort regardless of discipline are environmental engineers.

As populations increase, the demand for water grows accordingly and at a much more rapid rate if the population growth is accompanied by improved living standards. The combination of these two factors is placing greater and greater stress upon water engineers to find adequate supplies. In many cases inferior-quality, and often polluted, water supplies must be utilized to meet the demand. It is to be expected that this condition will continue and grow more complicated as long as population growth occurs. In many situations in water-short areas, purposeful recycling of treated wastewaters will be required in some degree to avoid serious curtailing of per-capita usage. The ingenuity of scientists and engineers will be taxed to the limit to meet this need.

1-2 WASTEWATER AND STREAM POLLUTION CONTROL

The disposal of human wastes has always constituted a serious problem. With the development of urban areas, it became necessary, from public health and aesthetic considerations, to provide drainage or sewer systems to carry such wastes away from the area. The normal repository was usually the nearest watercourse. It soon became apparent that rivers and other receiving bodies of water have a limited ability to handle waste materials without creating nuisance conditions. This led to the development of purification or treatment facilities in which chemists, biologists, and engineers have played important roles. The chemist in particular has been responsible for the development of test methods for evaluating the effectiveness of treatment processes and providing a knowledge of the biochemical and physicochemical changes involved. Great strides have been made in the art and science of waste treatment in the past few decades. These have been made possible by the fundamental knowledge of wastewater treatment established by scientists with a wide variety of training. It has been the responsibility of the engineers to synthesize this basic knowledge into practical systems of wastewater treatment that are effective and economical.

It has long been known that all natural bodies of water have the ability to oxidize organic matter without the development of nuisance conditions, provided that the organic loading is kept within the limits of the oxygen resources of the water. It is also known that certain levels of dissolved oxygen must be maintained at all times if certain forms of aquatic life are to be preserved. A great deal of research has been conducted to establish these limits, and undoubtedly a great deal more is needed. Such surveys require the combined efforts of biologists, chemists, and engineers if their full value is to be realized. In the past, streams have been classified into the following four broad categories: (1) those to be used for the transportation of wastes without regard to aquatic growths but maintained to avoid the development of nuisance conditions, (2) those in which the pollutional load will be restricted to allow fish to flourish, (3) those to be used for recreational purposes, and (4) those that are used for water supplies.

The consensus at the present time in the United States is to require the highest practical degree of treatment of all wastewaters, regardless of stream purification capability. Thus, effluent quality or effluent standards have superseded stream standards.

1-3 INDUSTRIAL WASTES

Perhaps the most challenging field in environmental engineering practice at the present time is the treatment and disposal of industrial wastes. Because of the great variety of wastes produced from established industries and the introduction of wastes from new processes, a knowledge of chemistry is essential to a solution of most of the problems. Some may be solved with a knowledge of inorganic chemistry; others may require a knowledge of organic, physical, or colloidal chemistry, biochemistry, or even radiochemistry. It is to be expected that, as further technological advances are made and industrial wastes of even greater variety appear, chemistry will serve as the basis for the development and selection of treatment methods.

1-4 ENVIRONMENTAL SANITATION

Although the problems of water supply and liquid-waste disposal are of major importance to urban populations, their solution alone does not ensure a completely satisfactory environment. Pollution of the atmosphere increases in almost direct ratio to the population density and is largely related to the products of combustion from heating plants, incinerators, and automobiles, plus gases, fumes, and smokes arising from industrial processes. The intensity of most air pollution problems is usually related to the amount of particulate matter emitted into the atmosphere and to the atmospheric conditions that exist. In general, visible particulate matter can be controlled by adequate regulations. The most serious situations develop where local conditions favor atmospheric inversions and the products of combustion and of industrial processing are contained within a confined air mass. A notable example is the situation at Los Angeles, where inversions occur frequently; they also occur, though less often, at several other metropolitan areas.

In cases where atmospheric inversions occur over metropolitan areas under cloudless skies, a haze commonly called "smog" is produced in the atmosphere. Under such conditions the atmosphere is often highly irritating to the eyes and to the respiratory tract and is far too intense to be accounted for by the materials emitted to the atmosphere from the separate sources. Research on this problem has been extensive in the Los Angeles area. Many theories have been advanced as to the cause, but the consensus at present is that photochemical action between nitrogen dioxide and unsaturated hydrocarbons from automobile exhaust gases combine to form the irritating substance. This substance condenses on particulate matter in the atmosphere to form a fog. A knowledge of chemistry has played an important role in finding the cause of this enigma.

Air pollution of quite another type was of major concern a few years ago. This resulted from radioactive materials that gained entrance to the atmosphere through nuclear explosions. The nuclides that were dispersed and settled as "fallout" varied greatly in their effect upon living plants and animals. Although limitations on atmospheric nuclear testing have greatly lessened this problem, other uses of nuclear energy have raised new fears over release of radioactive materials to the atmosphere. These fears have resulted from the development and installation of nuclear power plants. Experience has indicated these particular fears have little basis, however. The major threat to the environment results from the transportation of "nuclear ash," separation, and safe disposal of the waste radioactive materials. This constitutes a major challenge to the environmental engineering profession, for with ever-diminishing supplies of fossil fuels, nuclear power (perhaps together with renewable sources such as solar energy) must replace them or living standards will have to be lowered.

1-5 OTHER TECHNOLOGICAL DEVELOPMENTS

During the past few years, many new chemicals have been produced for agricultural purposes. Some of them are used for weed control, others for pest control. Residues of these materials are often carried to water courses during periods of heavy rainfall and have had serious effects upon the biota of streams. A great deal of research by chemists and biologists has demonstrated which of the materials have been most damaging to the environment and many products have been outlawed. Continuing studies will be needed, but hopefully new products will be kept from general use until proven equal or even less harmful than those in current use.

1-6 SUMMARY

From the discussions presented it should be apparent that the solution of many problems in environmental engineering has required the concerted efforts of scientists and engineers and that chemists, in many instances, have played an indispensable role. It is to be expected that problems arising in the future will be fully as complex as those of the past and that chemistry will continue to be an important factor. Engineers with sound chemical training should find that their knowledge is a great aid and advantage in conquering unsolved problems and that liaison with scientists working on the same or allied problems will be facilitated. The chapters following are dedicated to that purpose.

TWO

BASIC CONCEPTS FROM GENERAL CHEMISTRY

The factual information and basic concepts taught in freshman chemistry vary considerably, depending upon the institution and the interests of the students. In many schools, engineers are given a considerably different course from that given to science majors. Because of these differences and because certain fundamental information is essential for environmental engineering, a review of certain phases of general chemistry is indicated.

2-1 ELEMENTS, SYMBOLS, ATOMIC WEIGHTS, GRAM ATOMIC WEIGHTS

Remembering the names of the common elements poses no particular problem to the average student. However, the proper symbol does not always come to mind. This is mainly because many of the symbols are derived from Latin, Greek, or German names of the elements, and sometimes because of a similarity of names which makes a multiple choice of symbols possible. This similarity is well illustrated by the symbols for magnesium, Mg, and manganese, Mn, which are commonly confused.

To remember the symbols for magnesium, manganese, and those derived from Latin or other foreign names, one must rely entirely upon memory or association with the uncommon name. A list of the elements whose symbols are derived from Latin, Greek, or German names is given in Table 2-1.

Table 2-1 List of elements whose symbols are derived from Latin, Greek, or German names

Element	Name from which symbol is derived	Symbol
Antimony	Stibium, L.	Sb
Copper	Cuprum, L.	Cu
Gold	Aurium, L.	Au
Iron	Ferrum, L.	Fe
Lead	Plumbum, L.	Pb
Mercury	Hydrargyrum, Gr.	Hg
Potassium	Kalium, L.	K
Silver	Argentum, L.	Ag
Sodium	Natrium, L.	Na
Tin	Stannum, L.	Sn
Tungsten	Wolfram, G.	W

Atomic weights of the elements refer to the relative weights of the atoms as compared with some standard. In 1961 the C^{12} isotope of carbon was adopted as the atomic weight standard with a value of exactly 12. According to this standard, the atomic weight of oxygen is 15.9994, or 16 for all practical purposes.

It is not necessary to remember the atomic weights of the elements, because tables giving these values are readily available. It will save time, however, to remember the weight of the more commonly used elements such as hydrogen, oxygen, carbon, calcium, magnesium, sodium, sulfur, aluminum, chlorine, and a few others. It is usually sufficient for all practical purposes to round off the atomic weights at three significant figures; thus aluminum is called 27.0, chlorine 35.5, gold 197, iodine 127, and so on.

In general, elements do not have atomic weights that are whole numbers because they consist of a mixture of isotopes. Chlorine is a good example. Its atomic weight of 35.45 is due to the fact that it consists of two isotopes with atomic weights of 35 and 37. Cadmium contains eight isotopes whose atomic weights range from 110 to 116.

The *gram atomic weight* of an element refers to a quantity of the element in grams corresponding to the atomic weight. It has principal significance in the solution of problems involving weight relationships.

2-2 COMPOUNDS, FORMULAS, MOLECULAR WEIGHTS, GRAM MOLECULAR WEIGHTS, MOLE

Although the concept of chemical compounds is readily established, association of the proper and correct formula for each compound does not always follow. This difficulty is sometimes due to faulty use of symbols but much more often to a lack of knowledge regarding valence. The subject of valence will be discussed presently. If strict attention is paid to correct symbols and valences, errors in writing formulas will be eliminated.

Calculation of molecular weights poses no real problem except when rather complex formulas are involved. Most difficulties in this regard can be overcome by writing structural formulas and applying some effort in the form of practice. The importance of correct molecular weights as the basis for engineering calculations should be emphasized.

The term *gram molecular weight* (GMW) refers to the molecular weight in grams of any particular compound. It is also referred to as a *mole*. Its chief significance is in the preparation of *molar* or *molal* solutions. A molar solution consists of 1 gram molecular weight dissolved in enough water to make 1 liter of solution, whereas a molal solution consists of 1 gram molecular weight dissolved in 1 liter of water, the resulting solution having a volume slightly in excess of 1 liter.

2-3 AVOGADRO'S NUMBER

A significant fact is that the gram molecular weight contains the same number of molecules, whatever the compound. The number of molecules present is called Avogadro's number, and is approximately equal to 6.02×10^{23}. By definition, a mole of a substance contains an Avogadro's number of elementary entities. It is expressed as atoms per mole, molecules per mole, ions per mole, electrons per mole, or particles per mole, depending on the context.

$$6.02 \times 10^{23} \text{ O atoms} \quad = 16.0 \text{ g O}$$

$$6.02 \times 10^{23} \text{ H atoms} \quad = 1.01 \text{ g H}$$

$$6.02 \times 10^{23} \text{ H}_2\text{O molecules} = 18.0 \text{ g H}_2\text{O}$$

$$6.02 \times 10^{23} \text{ OH}^- \text{ ions} \quad = 17.0 \text{ g OH}^-$$

The enormous size of this number is incomprehensible. Some concept of its magnitude may be gained from a consideration that the life span of the average United States citizen is of the order of 2.2×10^9 seconds (s) and that a person would have to live about 3×10^{14} lives to count to Avogadro's number.

2-4 VALENCY, OXIDATION STATE, AND BONDING

A knowledge of valency and bonding theory serves as the key to correct formulas. In general, the writing of formulas with elements and radicals that have a fixed valence (or *oxidation state*) is easy, if a knowledge of electrostatics is applied. The real difficulty stems from elements that can assume several oxidation states, the variety of radicals that results, and a lack of knowledge of nomenclature, which is not always consistent.

Molecules, some ions, and radicals consist of two or more atoms bonded together in some definite manner. In general, the bonds may be *ionic* or *covalent*. An ionic bond is formed by the transfer of electrons from one atom to the other. One atom then takes on a positive charge (the *cation*) and the other a negative charge (the *anion*). The *ion pair* that results is held together loosely by electrostatic attraction. In other cases, electrons are not transferred, but are shared between atoms. In elementary molecules with identical atoms, such as Cl_2, N_2, and O_2, the electrons are shared equally to form a *covalent bond*. On the other hand, in heteronuclear molecules which consist of unlike atoms, the electrons forming the bond are shared unequally. For this case the bonding is termed *polar covalent*.

The valency or oxidation number of an atom is determined by the number of electrons that it can take on, give up, or share with other atoms. According to valency theory, most atoms consist of neutrons, protons (+), and electrons (−). The neutrons and protons are contained within the nucleus, and a number of electrons, corresponding to the number of protons (atomic number) in the nucleus, are arranged in orderly rings outside. The outer ring contains the valence electrons. If electrons are lost, the atom becomes a positively charged ion, and if electrons are gained, the atom becomes a negatively charged ion. Except for inert elements (such as argon) that already have complete rings, atoms tend to gain or lose electrons so as to assume or approach complete rings. To do this, they must team up with another atom in some manner. In the formation of ions, atoms of two elements undergo reduction and oxidation: one gains electrons and the other loses electrons. In the exchange, the metal or metallike element loses electrons to gain or approach a stable condition with no electrons in its outer ring. The nonmetal steals electrons from the metal to complete its outer ring to eight electrons, a stable configuration. This exchange is normally accomplished by the release of a great deal of energy. This simple type of reaction is well illustrated by the one between sodium and chlorine, as shown in Fig. 2-1.

The chlorine atom also serves as an example of polar covalent bonding in its various possible combinations with oxygen. The chlorine atom contains several

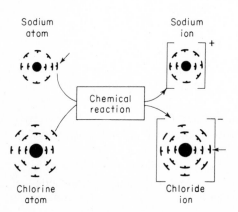

Sodium atom

Sodium ion

Chemical reaction

Chlorine atom

Chloride ion

Figure 2-1 Electron transfer during a chemical reaction, producing a sodium ion with an oxidation state of + and a chloride ion with an oxidation state of −.

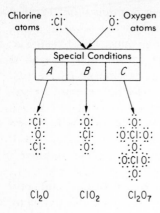

Figure 2-2 Multiple oxidation states of chlorine due to sharing of electrons. Chlorine forms similar compounds in all states of oxidation from 1+ to 7+, except for 6+.

electrons in its outer ring. Oxygen has six electrons in its outer ring and needs two more to complete the ring. These it can obtain in various ways by sharing electrons with the chlorine atom, forming various molecular species which may or may not be charged as illustrated in Fig. 2-2. The electrons contained in the outer shells are represented by dots for simplicity. With oxygen, chlorine tends to share one, three, four, five, or seven of its electrons, to form Cl_2O, ClO_2^-, ClO_2, ClO_3^-, and Cl_2O_7. The oxides from which ClO_2^- and ClO_3^- are derived have never been isolated. However, compounds of chlorine with oxidation numbers of 1+, 3+, 4+, 5+, and 7+ are well defined. Sulfur, nitrogen, and the halogens are nonmetals that are capable of exhibiting a wide range of oxidation numbers because of their ability to take on or share electrons to complete the outer shell to eight or to give up one or more electrons to reach a stable configuration. Manganese, chromium, copper, and iron are examples of metals that can obtain several oxidation states by yielding or sharing one or more electrons. Manganese is an extreme case in that it can yield or share two, three, four, six, or seven electrons.

2-5 NOMENCLATURE

There are only a very few hard-and-fast rules concerning nomenclature of inorganic compounds. One concerns binary compounds; they all have the ending -ide. For example, anhydrous HCl is hydrogen chloride. Most nomenclature problems arise from the acids containing oxygen. In general, the nomenclature is related to the oxidation state of the element that characterizes the acid. The acids having the highest oxidation state are usually called -ic, e.g., sulfuric, phosphoric, and chromic. They give rise to -ate salts. The acids that exist in the next lowest oxidation state are called -ous, e.g., sulfurous, phosphorous, and chromous. They give rise to -ite salts. If acids of a lower oxidation state exist, they are called hypo $\cdots$ ous, e.g., hypochlorous or hypophosphorous, and their salts are called hypo $\cdots$ ites.

Occasionally, as with the oxy acids of the halogens, more than three acids are known. In such cases the acid in the highest oxidation state is given the prefix per, e.g., perchloric or periodic, and their salts are called per $\cdots$ ates. The acid derived from maganese in an oxidation state of 7 ($HMnO_4$) is known as permanganic acid, and its salts are the familiar permanganates. All per $\cdots$ acids contain an element that has an oxidation state of 7, which appears to be the reason that $HMnO_4$ is given a prefix per. The only other well-defined acid of manganese is manganic, in which the manganese has an oxidation state of 6.

Acids are also named in terms of their degree of hydration: ortho, meta, and pyro. The ortho acids consist of the highest hydrated form of the acid anhydride, e.g., sulfuric (H_2SO_4), phosphoric (H_3PO_4), phosphorous (H_3PO_3), chromic (H_2CrO_4). The meta acids are derived from the ortho acids by removal of one molecule of water from each molecule of acid as follows:

$$\underset{\text{Ortho}}{H_3PO_4} \overset{\Delta}{\to} \underset{\text{Meta}}{HPO_3} + H_2O\uparrow$$

where Δ = heat
$\uparrow$ indicates that water escapes

A summary of nomenclature of the acids is given in Table 2-2. The ortho acids give rise to ortho salts and the meta acids to meta salts. The pyro acids may be

Table 2-2 Nomenclature of oxygen-bearing acids and their salts

On the basis of oxidation state

Name of acid	Formula	Name of salt
Sulfurous	H_2SO_3	Sulfite
Sulfuric	H_2SO_4	Sulfate
Hypochlorous	$HClO$	Hypochlorite
Chlorous	$HClO_2$	Chlorite
Chloric	$HClO_3$	Chlorate
Perchloric	$HClO_4$	Perchlorate

On the basis of hydration

Orthosulfuric	H_2SO_4	Orthosulfate
Orthophosphoric	H_3PO_4	Orthophosphate
Orthophosphorous	H_3PO_3	Orthophosphite
Metaphosphoric	HPO_3	Metaphosphate
Metaphosphorous	HPO_2	Metaphosphite
Pyrophosphoric	$H_4P_2O_7$	Pyrophosphate
Pyrochromic	$H_2Cr_2O_7$	Dichromate
Pyrosulfuric	$H_2S_2O_7$	Pyrosulfate

derived, theoretically, from ortho acids by removal of one molecule of water from two molecules of acid, as follows:

$$2H_3PO_4 \xrightarrow{\Delta} H_4P_2O_7 + H_2O\uparrow$$

$$2H_2SO_4 \xrightarrow{\Delta} H_2S_2O_7 + H_2O\uparrow$$

$$2H_2CrO_4 \xrightarrow{\Delta} H_2Cr_2O_7 + H_2O\uparrow$$
$$\text{Ortho} \qquad\qquad \text{Pyro}$$

Free pyro acids are not known, but well-defined salts are common. The pyro salts of chromic acid are commonly called dichromates, another instance of deviation from the rule.

2-6 CHEMICAL EQUATIONS; WEIGHT RELATIONSHIPS

A fundamental rule that must be observed at all times is that expressions of chemical reactions become equations only when they are balanced. In order to balance a chemical equation it is essential that it represent a reaction in true manner, and all formulas used must be correct. Unless these conditions are complied with, weight relationships are meaningless. Weight relationships serve as the basis for the sizing of chemical feeding equipment, necessary storage space for chemicals, structural design, and cost estimates in engineering considerations. Their importance should not need further emphasis.

Example A

$$NaOH + HCl \rightarrow NaCl + H_2O$$

	NaOH	HCl	NaCl	H₂O
	$(23+16+1)$	$(1+35.5)$	$(23+35.5)$	$(2\times1)+16$
Weight relationship	40	36.5	58.5	18

Thus 40 g of NaOH combines with 36.5 g of HCl to form 58.5 g of NaCl and 18 g of H₂O.

Example B

$$2HCl + Na_2CO_3 \rightarrow H_2O + 2NaCl + CO_2$$

	2HCl	Na₂CO₃	H₂O	2NaCl	CO₂
	$2(1+35.5)$	$(2\times23)+12+(3\times16)$	$(2\times1)+16$	$2(23+35.5)$	$12+(2\times16)$
Weight relationship	73	106	18	117	44

In this case two moles or 73 g of HCl combine with one mole or 106 g of NaCO₃ to form one mole or 18 g of H₂O, two moles or 117 g of NaCl, and one mole or 44 g of CO₂.

2-7 OXIDATION-REDUCTION EQUATIONS

Modern concepts of oxidation and reduction are based upon the idea of atomic structure and electron transfer as described in Sec. 2-4. An atom, molecule, or ion is said to undergo *oxidation* when it *loses* an electron, and to undergo *reduction* when it *gains* an electron. With reference to Fig. 2-1, when sodium reacts with chlorine to form sodium chloride, the sodium atom loses an electron and becomes oxidized to the sodium ion, Na^+. Chloride gains an electron and is reduced to the anion, Cl^-.

When oxidation-reduction reactions occur between atoms to form molecules or ions with polar covalent bonds, certain assumptions are required in order to maintain a consistent concept. A good illustration is the reaction that occurs when hydrogen burns in oxygen.

$$2H_2 + O_2 \rightarrow 2(H^+ - O^{2-} - H^+) \qquad (2\text{-}1)$$

or
$$2H:H + :\ddot{O}:\ddot{O}: \rightarrow 2:\overset{H}{\underset{}{\ddot{O}}}:H$$

H_2 and O_2 are *homonuclear covalent molecules*. We adopt the convention that the electrons are shared equally by the homonuclear cores: no atom gains or loses electrons in the formation of the molecule from its atoms; thus the *oxidation number* (or valence) is zero. Water is a *heteronuclear polar covalent molecule*. In H_2O, the electrons are shared unequally by hydrogen and oxygen, the oxygen atom tends to have a greater holding power on the electrons and is said to be more *electronegative* than hydrogen. This leads to a polar covalent bond in which the oxygen part of the molecule tends to take on a negative charge and the hydrogen a positive charge. To calculate oxidation number, we adopt the convention that the more electronegative element effectively acquires complete control of the shared electrons. This is equivalent to exaggerating the polar covalent bond into an ionic bond. In the formation of the water molecule, each hydrogen atom takes on an oxidation number or valence of $+1$ (becomes oxidized), and the oxygen atom takes on an oxidation number of -2 (becomes reduced). Hydrogen and oxygen when part of essentially all heteronuclear molecules and ions of interest in environmental engineering, take on these oxidation numbers.

Sometimes there is difficulty in naming compounds containing elements that can exhibit more than one oxidation number. A scheme now frequently used is one recommended by the International Union of Pure and Applied Chemistry (IUPAC). Here, the oxidation number of the element in the positive oxidation state is indicated by a roman numeral in parenthesis following the name of the element: thus, $FeCl_2$ is iron(II) chloride; $FeCl_3$ is iron(III) chloride; and Cl_2O_7 is chlorine(VII) oxide. Also when speaking of an element one can refer to Fe(III), which means iron with a $+3$ oxidation state, without specifically considering whether the iron is present as the ion, Fe^{3+}, or whether it occurs within a heteronuclear ion or molecule. There are times when use of this nomenclature can help greatly to avoid confusion.

With these concepts of oxidation and reduction, general definitions of oxidizing agents and reducing agents can be derived.

An *oxidizing agent* is any substance that can add electrons, e.g.,

$$O(O),\ Cl(O),\ Fe(III),\ Cr(VI),\ Mn(IV),\ Mn\ (VII),\ N(V),\ N(III),$$

$$S(O),\ S(IV),\ S(VI)$$

A *reducing agent* is any substance that can give up electrons, e.g.,

$$H(O),\ Fe(O),\ Mg(O),\ Fe(II),\ Cr(II),\ Mn(IV),\ N(III),\ Cl(-I),$$

$$S(O),\ S(-II),\ S(IV)$$

It will be noted that Mn(IV), N(III), S(O), and S(IV) appear in both series above. Any element in an intermediate state of oxidation can serve as a reducing or as an oxidizing agent under proper conditions.

It is a fundamental rule that oxidation cannot occur without reduction, and the gain of electrons by the oxidizing agent must equal the loss of electrons by the reducing agent.

Simple Oxidation-Reduction Reactions

$$H_2^\circ + Cl_2^\circ \rightarrow 2\ H^+Cl^- \tag{2-2}$$

$$4Fe^\circ + 3O_2^\circ \rightarrow 2\ Fe_2^{3+}O_3^{2-} \tag{2-3}$$

$$Mg^\circ + H_2^+SO_4^{2-} \rightarrow Mg^{++}SO_4^{2-} + H_2^\circ \tag{2-4}$$

$$2Fe^{2+} + Cl_2^\circ \rightarrow 2Fe^{3+} + 2Cl^- \tag{2-5}$$

$$2I^- + Cl_2^\circ \rightarrow I_2^\circ + 2Cl^- \tag{2-6}$$

In each of the equations above, the oxidizing agent gains the same number of electrons as are lost by the reducing agent.

Complex Oxidation-Reduction Reactions

Many oxidation-reduction reactions require the presence of a third compound, usually an acid or water, to progress. It is a rule that when the oxidizing agent is a compound containing oxygen, such as $KMnO_4$, $K_2Cr_2O_7$, and the like, one of the products is water. The balancing of complex oxidation-reduction equations is simplified if the following three steps are followed:

1. Write the skeleton equation. This may be in either the molecular or ionic form but must be a *true* representation of the reaction that occurs.
2. Balance the equation with respect to oxidation number change or electron transfer.
3. Complete the equation in the usual manner.

A few illustrations will serve to show how the scheme is applied.

Example A

STEP 1

$$KMnO_4 + FeSO_4 + H_2SO_4 \rightarrow Fe_2(SO_4)_3 + K_2SO_4 + MnSO_4 + H_2O \tag{a}$$

or $$MnO_4^- + Fe^{2+} + H^+ \rightarrow Fe^{3+} + Mn^{2+} + H_2O \qquad (b)$$

Equation (b) is the ionic form of Eq. (a), if one recognizes that the potassium and sulfate ions do not enter into the reaction. Note that neither equation is balanced at this stage of the example.

STEP 2

$$+5 \times 2 = 10e \text{ gain}$$

$$2KMn^{7+}O_4 + 10Fe^{2+}SO_4 + H_2SO_4 \rightarrow 5Fe_2^{3+}(SO_4)_3 + K_2SO_4 + 2Mn^{2+}SO_4 + H_2O$$

$$-1 \times 10 = 10e \text{ loss}$$

$$+5e \text{ gain}$$

or $$MnO_4^- + 5Fe^{2+} + H^+ \rightarrow 5Fe^{3+} + Mn^{2+} + H_2O$$

$$- 5e \text{ loss}$$

In step 2 a total of 10 electrons are involved in the molecular equation because there are two atoms of iron in each molecule of ferric sulfate. The least common multiple of 2 and 5 is 10.

STEP 3

$$2KMnO_4 + 10FeSO_4 + 8H_2SO_4 \rightarrow 5Fe_2(SO_4)_3 + K_2SO_4 + 2MnSO_4 + 8H_2O \quad (2\text{-}7)$$

or $$MnO_4^- + 5Fe^{2+} + 8H^+ \rightarrow 5Fe^{3+} + Mn^{2+} + 4H_2O \qquad (2\text{-}8)$$

Example B

STEP 1

$$K_2Cr_2O_7 + KI + H_2SO_4 \rightarrow Cr_2(SO_4)_3 + K_2SO_4 + I_2 + H_2O$$

or $$Cr_2O_7^{2-} + I^- + H^+ \rightarrow Cr^{3+} + I_2 + H_2O$$

STEP 2

$$+3 \times 2 = 6e \text{ gain}$$

$$K_2Cr_2^{6+}O_7 + 6KI^- + H_2SO_4 \rightarrow Cr_2^{3+}(SO_4)_3 + K_2SO_4 + 3I_2^0 + H_2O$$

$$-1 \times 6 = 6e \text{ loss}$$

$$+3 \times 2 = 6e \text{ gain}$$

$$Cr_2O_7^{2-} + 6I^- + H^+ \rightarrow 2Cr^{3+} + 3I_2^0 + H_2O$$

$$-1 \times 6 = 6e \text{ loss}$$

STEP 3

$$K_2Cr_2O_7 + 6KI + 7H_2SO_4 \rightarrow Cr_2(SO_4)_3 + 4K_2SO_4 + 3I_2 + 7H_2O \qquad (2\text{-}9)$$

or $$Cr_2O_7^{2-} + 6I^- + 14H^+ \rightarrow 2Cr^{3+} + 3I_2 + 7H_2O \qquad (2\text{-}10)$$

An even more complicated oxidation-reduction equation is involved when one element in a high oxidation state oxidizes the same element in a lower oxidation state, with all the particular element from both the oxidizing and reducing agent appearing in the same final oxidation state. The action of potassium bi-iodate with potassium iodide is an excellent example since it is the basis of a reaction commonly used to release iodine for the standardization of sodium thiosulfate solutions.

STEP 1

$$KH(IO_3)_2 + KI + H_2SO_4 \rightarrow I_2 + I_2 + K_2SO_4 + H_2O$$

Since free iodine is formed from both the oxidizing and reducing agents, it is necessary to repeat I_2 in the equation.

STEP 2

$$+5 \times 2 = 10e \text{ gain}$$

$$KH(I^{5+}O_3)_2 + 10KI^- + H_2SO_4 \rightarrow I_2^\circ + 5I_2^\circ + K_2SO_4 + H_2O$$

$$-1 \times 10 = 10e\,\text{loss}$$

STEP 3

$$2KH(IO_3)_2 + 20KI + 11H_2SO_4 \rightarrow 12I_2 + 11K_2SO_4 + 12H_2O \qquad (2\text{-}11)$$

In order to balance the equation in step 3, it is necessary to multiply by 2 the parts that are balanced in step 2 because of the limitations imposed by potassium. Ionically this equation results in

$$IO_3^- + 5I^- + 6H^+ \rightarrow 3I_2^\circ + 3H_2O \qquad (2\text{-}12)$$

Use of Half Reactions

Another procedure which can simplify the development of complex oxidation-reduction reactions, including those involving organic compounds, involves the use of half reactions. A series of half reactions is shown in Table 2-3. Half reactions are balanced oxidation-reduction reactions for a single element. They are not complete reactions because electrons are shown as one of the reactants. Free electrons cannot occur in solution. A complete reaction is made by adding one half reaction to the reverse of another. For example, Eq. (2-12) can be developed by adding the reverse of Reaction 13 to Reaction 14 from Table 2-3.

Reaction 14	$\frac{1}{5}IO_3^- + \frac{6}{5}H^+ + e^- = \frac{1}{10}I_2 + \frac{3}{5}H_2O$
Reverse of Reaction 13	$I^- = \frac{1}{2}I_2 + e^-$
Sum	$\frac{1}{5}IO_3^- + I^- + \frac{6}{5}H^+ = \frac{3}{5}I_2 + \frac{3}{5}H_2O$
$\times 5$ to give Eq. (2-12)	$IO_3^- + 5I^- + 6H^+ = 3I_2 + 3H_2O$

Table 2-3 Half Reactions

Reaction number	Element reduced	Half reaction	$\Delta G°$, kcal/mol	$E°$, volts	$pE°$
1	C	$\frac{1}{4}CO_2 + \frac{7}{8}H^+ + e^- = \frac{1}{8}CH_3COO^- + \frac{1}{4}H_2O$	-1.73	0.075	1.27
2	C	$\frac{1}{4}CO_2 + H^+ + e^- = \frac{1}{24}C_6H_{12}O_6 + \frac{1}{4}H_2O$	0.35	-0.015	-0.26
3	Cl	$\frac{1}{2}Cl_2 + e^- = Cl^-$	-31.39	1.359	23.01
4	Cl	$\frac{1}{2}ClO^- + H^+ + e^- = \frac{1}{2}Cl^- + \frac{1}{2}H_2O$	-39.9	1.728	29.3
5	Cl	$\frac{1}{8}ClO_4^- + H^+ + e^- = \frac{1}{8}Cl^- + \frac{1}{2}H_2O$	-31.6	1.37	23.2
6	Cr	$\frac{1}{6}Cr_2O_7^{2-} + \frac{7}{3}H^+ + e^- = \frac{1}{3}Cr^{3+} + \frac{7}{6}H_2O$	-30.7	1.33	22.5
7	Cu	$\frac{1}{2}Cu^{2+} + e^- = \frac{1}{2}Cu$	-7.78	0.337	5.71
8	Fe	$\frac{1}{2}Fe^{2+} + e^- = \frac{1}{2}Fe$	9.45	-0.409	-6.93
9	Fe	$Fe^{3+} + e^- = Fe^{2+}$	-17.78	0.770	13.04
10	Fe	$\frac{1}{3}Fe^{3+} + e^- = \frac{1}{3}Fe$	0.84	-0.036	-0.62
11	H	$H^+ + e^- = \frac{1}{2}H_2$	0.00	0.000	0.00
12	Hg	$\frac{1}{2}Hg^{2+} + e^- = \frac{1}{2}Hg$	-19.7	0.851	14.4
13	I	$\frac{1}{2}I_2 + e^- = I^-$	-12.4	0.535	9.06
14	I	$\frac{1}{5}IO_3^- + \frac{6}{5}H^+ + e^- = \frac{1}{10}I_2 + \frac{3}{5}H_2O$	-27.6	1.195	20.24
15	Mn	$\frac{1}{2}MnO_2 + 2H^+ + e^- = \frac{1}{2}Mn^{2+} + H_2O$	-27.9	1.208	20.46
16	Mn	$\frac{1}{5}MnO_4^- + \frac{8}{5}H^+ + e^- = \frac{1}{5}Mn^{2+} + \frac{4}{5}H_2O$	-34.4	1.491	25.25
17	Mn	$\frac{1}{3}MnO_4^- + \frac{4}{3}H^+ + e^- = \frac{1}{3}MnO_2 + \frac{2}{3}H_2O$	-39.2	1.695	28.70
18	N	$\frac{1}{6}NO_2^- + \frac{4}{3}H^+ + e^- = \frac{1}{6}NH_4^+ + \frac{1}{3}H_2O$	-20.75	0.898	15.21
19	N	$\frac{1}{8}NO_3^- + \frac{5}{4}H^+ + e^- = \frac{1}{8}NH_4^+ + \frac{3}{8}H_2O$	-20.33	0.880	14.91
20	N	$\frac{1}{6}NO_2^- + \frac{4}{3}H^+ + e^- = \frac{1}{12}N_2 + \frac{2}{3}H_2O$	-35.16	1.522	25.78
21	N	$\frac{1}{5}NO_3^- + \frac{6}{5}H^+ + e^- = \frac{1}{10}N_2 + \frac{3}{5}H_2O$	-28.73	1.244	21.06
22	O	$\frac{1}{4}O_2 + H^+ + e^- = \frac{1}{2}H_2O$	-28.35	1.227	20.79
23	S	$\frac{1}{6}SO_4^{2-} + \frac{4}{3}H^+ + e^- = \frac{1}{6}S + \frac{2}{3}H_2O$	-8.24	0.357	6.04
24	S	$\frac{1}{8}SO_4^{2-} + \frac{5}{4}H^+ + e^- = \frac{1}{8}H_2S + \frac{1}{2}H_2O$	-7.00	0.303	5.13
25	S	$\frac{1}{4}SO_4^{2-} + \frac{5}{4}H^+ + e^- = \frac{1}{8}S_2O_3^{2-} + \frac{5}{8}H_2O$	-7.00	0.303	5.13
26	S	$\frac{1}{2}SO_4^{2-} + H^+ + e^- = \frac{1}{2}SO_3^{2-} + \frac{1}{2}H_2O$	0.93	-0.039	-0.66
27	Zn	$\frac{1}{2}Zn^{2+} + e^- = \frac{1}{2}Zn$	17.6	-0.763	-12.9

The significance of the value for ΔG, $E°$, and $pE°$ listed in Table 2-3 will be discussed later in Sec. 5-10. A great number of oxidation-reduction reactions of interest in water chemistry can be produced through combinations of the half reactions listed. Additional half reactions can be readily developed as follows, using for illustration I_2 as the reduced species and IO^- as the oxidized species for iodine.

STEP 1 Begin the half reaction with the species containing the more oxidized form of the element on the left and the more reduced form on the right, and balance for the element.

$$2IO_3^- = I_2$$

STEP 2 Add a sufficient number of moles of water to either side of the equation to produce an oxygen balance.

$$2IO_3^- = I_2 + 6H_2O$$

STEP 3 Add sufficient H^+ to either side of the reaction to produce a hydrogen balance.

$$2IO_3^- + 12H^+ = I_2 + 6H_2O$$

STEP 4 Add electrons to the left side of the reaction to make a charge balance. This results in a balanced half reaction.

$$2IO_3^- + 12H^+ + 10e^- = I_2 + 6H_2O$$

STEP 5 Divide the equation by the number of electrons indicated by Step 4 to normalize the reaction to one electron equivalent.

$$\tfrac{1}{5}IO_3^- + \tfrac{6}{5}H^+ + e^- = \tfrac{1}{10}I_2 + \tfrac{3}{5}H_2O$$

2-8 METALS AND NONMETALS

The division between metals and nonmetals is not as distinct as we often desire it to be. In general, those elements that easily lose electrons to form positive ions are called metals. In the free state, metals usually conduct electric current readily. Elements that hold electrons firmly and tend to gain electrons to form negative ions are called nonmetals. Two tests are commonly applied in making a decision: (1) Most metals will form the cation portion of salts having oxygen-bearing anions, such as nitrates and sulfates, and nonmetals do not. (2) Metals form at least one oxide with reasonably strong basic characteristics. The latter is not of general application because of the limited solubility of some oxides and hydroxides.

2-9 THE GAS LAWS

The gas laws, particularly their influence on the solution or removal of gases from liquids, are of particular significance to the environmental engineer.

Boyle's Law

Boyle's law states: *The volume of a gas varies inversely with its pressure at constant temperature*. This law is so simple and usually so well understood that further elaboration seems unnecessary. Its principal application is in reducing observations of gas volumes from field conditions to some standard condition. This is particularly significant at high altitudes, such as at Denver and Salt Lake City.

Charles' Law

Charles' law states: *The volume of gas at constant pressure varies in direct proportion to the absolute temperature*. Interpretation of this law poses no problems, provided that the absolute-temperature scale is used. Charles' law finds its greatest use in the calculation of pressures in fixed-volume containers with variable temperature. In conjunction with Boyle's law, it serves as the basis for sizing gas holders.

Generalized Gas Law

For a given quantity of a gas, Boyle's law and Charles' law can be combined in the form

$$PV = \beta T \tag{2-13}$$

where β is a constant that is proportional to the weight of the gas, and P, V, and T are the pressure, volume, and absolute temperature of the gas, respectively. It has been shown that the constant β is a function of the number of moles of gas present, and that a more universal, idealized gas law which is quite general for any gas can be expressed as

$$PV = nRT \tag{2-14}$$

where n equals the number of moles of gas in the particular sample and R is a universal constant for all gases. The numerical value for R depends on the units chosen for the measurement of P, V, and T. A useful way to evaluate R is to remember that 1 mole of an ideal gas at 1 atm pressure occupies a volume of 22.414 liters at 273K. From this R can be evaluated to be 0.082 liter atmosphere per mole per kelvin (0.082 L-atm/mol-K).

> **Example** What tank volume is required to hold 10 000 kilograms (kg) of methane gas (CH_4) at 25°C and 2 atmospheres (atm) pressure?
>   Molecular weight of CH_4 gas is $12 + 4(1) = 16$ g.
>
>   Number of moles in 10 000 kg is $\dfrac{10\ 000\ 000}{16} = 625\ 000$ mol.
>
>   According to the general gas law,
>
> $$V = \frac{nRT}{P} = \frac{625,000(0.082)(273 + 25)}{2} = 7.64 \times 10^6 \text{ liters}$$
>
> Thus a tank with a volume of 7.64×10^6 liters or 2.7×10^5 cubic feet (ft^3) would be required.

Dalton's Law of Partial Pressures

This law has been presented in a number of ways but in essence it may be stated as follows: *In a mixture of gases, such as air, each gas exerts pressure independently of the others. The partial pressure of each gas is proportional to the amount (percent by volume) of that gas in the mixture, or in other words, it is equal to the pressure that gas would exert if it were the sole occupant of the volume available to the mixture.* The basic concept of this law, in combination with Henry's law, serves in many engineering considerations and calculations.

Henry's Law

Henry's law states: *The weight of any gas that will dissolve in a given volume of a liquid, at constant temperature, is directly proportional to the pressure that the gas exerts above the liquid.* In equation form,

$$C_{equil} = \alpha p_{gas} \tag{2-15}$$

where C_{equil} is the concentration of gas dissolved in the liquid at equilibrium, p_{gas} is the partial pressure of the gas above the liquid, and α is the Henry's law constant for the gas at the given temperature.

Henry's law is undoubtedly the most important of all the gas laws in problems involving liquids. With a firm knowledge of Dalton's and Henry's laws, one should be capable of coping with all problems involving gas transfer into and out of liquids. As an example, the Henry's law constant, α, for oxygen in water at 20°C is 43.8 mg/l-atm. Since air contains 21 percent by volume of oxygen, the partial pressure of oxygen in air according to Dalton's law would be 0.21 atm when the total air pressure is 1 atm. Therefore the equilibrium concentration of oxygen in water at 20°C and in the presence of 1 atm of air would be 43.8 × 0.21 = 9.2 mg/l.

In the environmental engineering field, most of the problems related to the transfer of gases into liquids involve addition of oxygen by aeration to maintain aerobic conditions. The removal of gases from liquids is also accomplished by aeration devices of one sort or another. Usually the processes involve gas transfer at or near atmospheric pressure from air bubbles passing through a liquid, liquid drops falling through air, or thin films of liquid flowing over surfaces exposed to the air. Although Henry's law is an equilibrium law and is not directly concerned with the kinetics of gas transfer, it serves to indicate how far a liquid-gas system is from equilibrium, which in turn is a factor in the rate of gas transfer. Thus the rate of solution of oxygen is proportional to the difference between the equilibrium concentration as given by Henry's law and the actual concentration in the liquid:

$$\frac{dC}{dt} \propto (C_{equil} - C_{actual})$$

This concept serves as the basis for engineering calculations in aerobic methods of waste treatment, such as the activated sludge process, and in the evaluation of the reaeration capacity of lakes and streams.

The removal of undesirable gases, such as carbon dioxide, hydrogen sulfide, and hydrogen cyanide, from liquids is also commonly accomplished by some form of aeration. The general principles involved are the same as in the transfer of gases into the liquid. However, in this case the normal partial pressure of the gas in air is very low, so that based on Henry's law, C_{equil} is also low and much less than C_{actual}. Thus the rate of transfer given by the preceding equation is negative, and the gas leaves rather than enters the solution. The same principles apply to the removal of any volatile substance dissolved in water as long as it exerts a significant vapor pressure at the temperature involved. A knowledge of these fundamentals can often be used to good advantage in industrial waste treatment. In one instance, the biochemical oxygen demand (BOD) of a rather warm waste was reduced nearly 50 percent through removal of ethyl alcohol by aeration. The operation also reduced the temperature of the waste to a more desirable level for subsequent treatment.

Graham's Law

Graham's law is concerned with the diffusion of gases, and it states: *The rates of diffusion of gases are inversely proportional to the square roots of their densities.* This law can be illustrated by a comparison of the rates of diffusion of hydrogen, oxygen, chlorine, and bromine, which have atomic weights of approximately 1, 16, 36, and 80, respectively. On the basis of Graham's law, oxygen diffuses about one-fourth, chlorine about one-sixth, and bromine about one-ninth as fast as hydrogen. This law finds its greatest application in the field of industrial hygiene and air pollution control.

Gay-Lussac's Law of Combining Volumes

Gay-Lussac's law is basic to an understanding of gas analysis as performed in environmental engineering practice. The law states: *The volumes of all gases that react and that are produced during the course of a reaction are related, numerically, to one another as a group of small, whole numbers.* This law may be illustrated as follows:

$$C + O_2 \rightarrow CO_2 \tag{2-16}$$
$$ \underset{\text{1 vol}}{} \underset{\text{1 vol}}{}$$

One volume of oxygen combines with carbon (a solid) to yield 1 volume of carbon dioxide, or

$$CH_4 + 2O_2 \rightarrow CO_2 + 2H_2O \tag{2-17}$$
$$\underset{\text{1 vol}}{} \underset{\text{2 vol}}{} \underset{\text{1 vol}}{} \underset{\text{0 vol}}{}$$

Two volumes of oxygen combine with 1 volume of methane to form 1 volume of carbon dioxide. If the temperature of the system is held above 100°C, 2 volumes of water vapor will result. Usually the temperature of the system is brought back to room temperature, the water vapor condenses, and the volume of water is considered zero because it is segregated from the gaseous phase and does not interfere with measurement of the volume of gaseous products.

2-10 SOLUTIONS

The concepts of unsaturated, saturated, and supersaturated solutions are usually firmly entrenched in the minds of those who have studied general science or chemistry. The terms *molar* and *molal*, as described in Sec. 2-2, are not so well understood. Molal solutions are normally used when the physical properties of solutions, such as vapor pressure, freezing point, and boiling point, are involved. Molar concentrations are generally of interest for equilibrium calculations of various kinds. *Normal* solutions are commonly used for making analytical measurements and are described in Sec. 10-4.

Vapor Pressure

The presence of a nonvolatile solute in a liquid always lowers the vapor pressure of the solution. Thus when sugar, sodium chloride, or a similar substance is dissolved in water the vapor pressure is decreased. This phenomenon is believed to be due to a physical blocking effect at the surface of the liquid where particles (ions or molecules) of the solute happen to be.

Raoult's Law

The extent of the physical blocking effect or depression of the vapor pressure is directly proportional to the concentration of the particles in solution. For solutes that do not ionize, the effect is directly proportional to the molal concentration. For solutes that do ionize, the effect is proportional to the molal concentration times the number of ions formed per molecule of solute modified by the degree of ionization. Raoult's law holds true in a strict manner only for dilute solutions.

Environmental engineers are mainly interested in the effect that solutes have upon the freezing and boiling points of water. Application of Raoult's law has shown that molal solutions of nonelectrolytes in water, such as sugar containing 6.02×10^{23} (Avogadro's number) molecules or particles, have their vapor pressures decreased to the same degree, and the boiling point is raised $0.52°C$ while the freezing point is depressed $1.86°C$. A molal solution of an electrolyte such as $NaCl$, which yields two ions, produces nearly twice as great an effect because, after solution and ionization occur, the molal solution contains nearly two times Avogadro's number of particles.

2-11 EQUILIBRIUM AND LE CHATELIER'S PRINCIPLE

Nearly all chemical reactions are reversible in some degree. When a reaction proceeds to a point where the combination of reactants to form products is just balanced by the reverse reaction of products combining to form reactants, then the reaction is said to have reached *equilibrium*. The concentration of the reactants and of the products is important in determining the final state of equilibrium. For a system in equilibrium as expressed by the classical equation

$$A + B \rightleftharpoons C + D \qquad (2\text{-}18)$$

an increase in either A or B will shift the equilibrium further to the right. Conversely, an increase of either C or D will shift the equilibrium to the left.

The shifting of an equilibrium in response to a change of concentration is an example of the well-known principle of Le Chatelier, which states: *A reaction, at equilibrium, will adjust itself in such a way as to relieve any force, or stress, that disturbs the equilibrium.*

A chemical reaction in true equilibrium can be expressed as

$$\frac{[C][D]}{[A][B]} = K \qquad (2\text{-}19)$$

where K is constant for a given temperature and is called the *equilibrium constant*, and the designation [ ] *signifies the concentration* of the reacting substances. It is easily seen that in this form, called the *equilibrium relationship*, a change in concentration of any of the four factors will change the concentrations of all the others. This expression is widely used, and a clear understanding of its implications is necessary in all phases of science and applied sciences, such as environmental engineering. For the general reaction, $wA + xB + \cdots \rightarrow yC + zD + \cdots$, in which w, x, y and z are the number of molecules of the respective substances entering into the reaction, the equilibrium relationship is:

$$\frac{[C]^y[D]^z \; \cdots}{[A]^w[B]^x \; \cdots} = K \qquad (2\text{-}20)$$

If the reactants and products of a reaction are dissolved in a solvent such as water, then concentrations in the equilibrium relationship are ordinarily expressed in moles per liter. Methods of expressing concentrations for solids, gases, and solvents are discussed in the following section.

2-12 ACTIVITY AND ACTIVITY COEFFICIENTS

Equilibria relationships are frequently applied to equilibria involving salts, acids, or bases in solution. As solutions of these materials become more concentrated, their quantitative effect on equilibria becomes progressively less than that calculated solely from the change in molar concentration. Thus the effective concentration, or *activity*, of ions is decreased below that of the actual molar concentration. This has been explained partially as resulting from forces of attraction between the positive and negative ions. These forces become less at higher dilutions, since the ions are further apart.

Therefore activities or effective concentrations, rather than molar concentrations, should be used in equilibrium relationships, for accurate results.

$$\frac{\{C\}^y\{D\}^z}{\{A\}^w\{B\}^x} = K \qquad (2\text{-}21)$$

where the braces distinguish activity from concentration.

The activity (a) of an ion or molecule can be found by multiplying its concentration (c) by an activity coefficient (γ):

$$a = \gamma c \qquad (2\text{-}22)$$

The activity coefficient is a factor that converts concentration to a value which expresses quantitatively the true equilibrium effect. Unfortunately the activity coefficient is not usually an easy number to determine with precision. This is partly due to the fact that the activity coefficient is influenced by the presence of other ions in the solution.

Although numerical calculations from equilibria relationships may be in error if actual concentrations are used in place of activities, the error is not very great for dilute aqueous solutions. Also, a high degree of precision is seldom required in equilibrium analytical computations. For this reason, activity coefficients in general will not be used for equilibrium calculations in this book. A more detailed discussion of this subject, together with a method for estimating the activity coefficient for ions, is given in Sec. 5-3.

It is up to the individual to decide whether concentrations or activities are used in the equilibrium relationship he may adopt. However, certain conventions for expressing the concentrations or activities of solvents, solutes, solids, and gases need to be understood, especially if published values for equilibrium constants are to be used. The methods of expression are as follows:

1. *Ions and molecules in dilute solution.* Concentration or activity is expressed in moles per liter.
2. *The solvent in a dilute solution.* Concentration or activity is taken to approximately equal unity. Thus, with reactions in aqueous solution involving water as a reactant or product, the water concentration is assumed to equal one. For this reason water is generally left out of an equilibrium constant expression for dilute aqueous solutions.
3. *Pure solids or liquids in equilibrium with the solution.* The concentration or activity is equal to unity. As with water, the concentration or activity of pure solids or liquids does not need to be included in the equilibrium constant expression.
4. *Mixture of liquids.* The concentration or activity is expressed in terms of *mole fraction,* which is the ratio of the number of moles of a given liquid divided by the total number of moles of all liquids in the solution.
5. *Gases.* The concentration or activity is expressed as partial pressure in atmospheres.

2-13 VARIATIONS OF THE EQUILIBRIUM RELATIONSHIP

Equations (2-20) and (2-21) are general forms of the equilibrium relationship. They are useful in helping to understand the various ways in which substances may be distributed in aqueous solution and methods for their control. *Homogeneous chemical equilibria* are characterized by all reactants and products of the reaction occurring in the same physical state or phase, such as reactions between gases or between materials dissolved in water. Examples of homogeneous equilibria in water are the ionization of weak acids and bases, and complex formation.

Heterogeneous chemical equilibria are characterized by substances occurring in two or more physical phases. Examples are equilibria for the solubility of a gas in a liquid, the solubility of solids in water, the distribution of a material between two different solvents, the equilibrium of a substance between its liquid phase and gaseous phase, or between its liquid phase and solid phase. Examples of homogeneous and heterogeneous equilibria of particular concern to environmental engineering are given below. These are considered in greater detail in Chaps. 3 and 5.

Ionization

The theory of ionization stems from a doctoral dissertation completed by Svante Arrhenius in 1887. In many respects this theory serves satisfactorily for most concepts of environmental engineering that involve electrolytic dissociation. According to the original theory of Arrhenius, all acids, bases, and salts dissociate into ions when placed in solution in water. He noted that equivalent[1] solutions of different compounds often varied greatly in conductivity. This phenomenon was attributed by Arrhenius to a difference in degree of dissociation or ionization and serves today to explain many of the observed phenomena in aqueous solution.

Ion Product of Water

One of the most important equilibria of concern in dealing with aqueous solutions is the dissociation of water into a hydrogen ion or proton and hydroxyl ion.

$$H_2O \rightleftharpoons H^+ + OH^- \tag{2-23}$$

A proton is a very small particle and as such would have an extremely large charge-to-volume ratio. As a result, it will attach itself to almost anything that does not have a large positive charge. In aqueous solution, it readily becomes attached to water molecules, so that the following is a more correct description of water dissociation than Eq. (2-23):

$$2H_2O \rightleftharpoons H_3O^+ + OH^- \tag{2-24}$$

where H_3O^+ is called the *hydronium ion.* The hydronium ion can also react with water to form other hydrated species so that even Eq. (2-24) is not a completely accurate description for the dissociation of water. For many practical purposes, the simple dissociation indicated by Eq. (2-23) can be assumed. This leads to the ion product for water, which can be written as follows:

$$\frac{[H^+][OH^-]}{[H_2O]} = K_W \tag{2-25}$$

[1] See Sec. 10-4 for definition.

However, by convention [H_2O] is taken to equal one, so that the ion product for water in its simplest form is

$$[H^+][OH^-] = K_W = 10^{-14} \qquad \text{at } 25°C \qquad (2\text{-}26)$$

In satisfying this equilibrium, the numerical values of [H^+] and [OH^-] include *all* the H^+ and OH^- ions present, regardless of whether these ions are produced by the water alone or are contributed by other constituents in the water.

Ionization of Weak Acids and Weak Bases

Arrhenius' theory of ionization can help explain the variation in strength of acids and bases. All strong acids and bases are considered to approach 100 percent ionization in dilute solutions. The weak acids and bases, however, are so poorly ionized that in most cases it is impractical to express the degree of ionization as a percentage. The equilibrium relationship, however, can be used as illustrated below:

For a typical monobasic acid (acetic acid),

$$HAc \rightleftharpoons H^+ + Ac^-$$

$$\frac{[H^+][Ac^-]}{[HAc]} = K_A = 1.75 \times 10^{-5} \qquad \text{at } 25°C \qquad (2\text{-}27)$$

where Ac^- is used to designate the acetate ion.

For a typical dibasic acid (carbonic acid),

$$H_2CO_3 \rightleftharpoons H^+ + HCO_3^-$$

$$\frac{[H^+][HCO_3^-]}{[H_2CO_3]} = K_1 = 4.45 \times 10^{-7} \qquad \text{at } 25°C \qquad (2\text{-}28)$$

$$HCO_3^- \rightleftharpoons H^+ + CO_3^{2-}$$

$$\frac{[H^+][CO_3^{2-}]}{[HCO_3^-]} = K_2 = 4.69 \times 10^{-11} \qquad \text{at } 25°C \qquad (2\text{-}29)$$

For a typical base (ammonium hydroxide),

$$NH_3 + H_2O \rightleftharpoons NH_4^+ + OH^-$$

$$\frac{[NH_4^+][OH^-]}{[NH_3]} = K_B = 1.75 \times 10^{-5} \qquad \text{at } 25°C \qquad (2\text{-}30)$$

The concentration of water was not included in Eq. (2-30) for the reason discussed in Sec. 2-12. Tables giving ionization constants of weak acids, bases, and salts may be found in the usual handbooks and many textbooks of quantitative analysis or physical chemistry.

> **Example** If 3 g of acetic acid is added to enough distilled water to make one liter of solution, what will be the acetate ion concentration?
> The molar concentration of the solution is $\frac{3}{60}$ or 0.05 M.

If x moles of the acetic acid ionize to form H^+ and Ac^- ions, then

$$[HAc] = (0.05 - x) \qquad mol/l$$

and $\qquad [H^+] = [Ac^-] = x \qquad mol/l$

$$K_A = \frac{[H^+][Ac^-]}{[HAc]} = \frac{(x)(x)}{0.05 - x} = 1.75 \times 10^{-5}$$

Solving, we obtain $x = [Ac^-] = 9.27 \times 10^{-4}$ mol/l.
Thus 0.05 M acetic acid is $9.27(10^{-4})(100)/0.05$ or 1.85 percent ionized.

Complex Ions

Complex ions are soluble species formed through combination of other simpler species in solution. For example, if Hg^{2+} and Cl^- are present in water, they will combine to form the undissociated but soluble species $HgCl_{2(aq)}$, where (aq) is used to designate that the species is in solution. Chloride can also combine with mercury in other proportions to form a variety of complexes. Equilibrium relationships can be developed for the various mercury-chloride species from the following reactions:

$$Hg^{2+} + Cl^- \rightleftharpoons HgCl^+ \tag{2-31}$$

$$HgCl^+ + Cl^- \rightleftharpoons HgCl_2 \tag{2-32}$$

$$HgCl_2 + Cl^- \rightleftharpoons HgCl_3^- \tag{2-33}$$

$$HgCl_3^- + Cl^- \rightleftharpoons HgCl_4^{2-} \tag{2-34}$$

Equilibrium relationships associated with the above reactions are as follows:

$$\frac{[HgCl^+]}{[Hg^{2+}][Cl^-]} = K_1 = 5.6 \times 10^6 \tag{2-35}$$

$$\frac{[HgCl_2]}{[HgCl^+][Cl^-]} = K_2 = 3 \times 10^6 \tag{2-36}$$

$$\frac{[HgCl_3^-]}{[HgCl_2][Cl^-]} = K_3 = 7.1 \tag{2-37}$$

$$\frac{[HgCl_4^{2-}]}{[HgCl_3^-][Cl^-]} = K_4 = 10 \tag{2-38}$$

The usual convention is to write complex reactions as indicated above, as formation rather than as dissociation reactions. The equilibrium constant is then called a *stability constant*. The numerical index on the stability constant indicates the number of chloride ions in the complex formed by the reaction.

It is sometimes convenient to consider overall reactions for the formation of complexes, and these can be developed by combination of the stepwise reactions indicated above.

$$Hg^{2+} + Cl^- \rightleftharpoons HgCl^+ \qquad \frac{[HgCl^+]}{[Hg^{2+}][Cl^-]} = \beta_1 \tag{2-39}$$

$$Hg^{2+} + 2Cl^- \rightleftharpoons HgCl_2 \qquad \frac{[HgCl_2]}{[Hg^{2+}][Cl^-]^2} = \beta_2 \qquad (2\text{-}40)$$

$$Hg^{2+} + 3Cl^- \rightleftharpoons HgCl_3^- \qquad \frac{[HgCl_3^-]}{[Hg^{2+}][Cl^-]^3} = \beta_3 \qquad (2\text{-}41)$$

$$Hg^{2+} + 4Cl^- \rightleftharpoons HgCl_4^{2-} \qquad \frac{[HgCl_4^{2-}]}{[Hg^{2+}][Cl^-]^4} = \beta_4 \qquad (2\text{-}42)$$

Here the numerical index on the equilibrium constant has the same meaning as in Eqs. (2-35) to (2-38). It can easily be shown that $\beta_1 = K_1$, $\beta_2 = K_1 K_2$, $\beta_3 = K_1 K_2 K_3$, and $\beta_4 = K_1 K_2 K_3 K_4$. Sometimes the reciprocal of the overall formation constant of the highest complex is given as the *instability constant* for that reaction. Thus,

$$HgCl_4^{2-} \rightleftharpoons Hg^{2+} + 4Cl^- \qquad \frac{[Hg^{2+}][Cl^-]^4}{[HgCl_4^{2-}]} = K_{inst} \qquad (2\text{-}43)$$

Example The chloride concentration in a typical fresh water stream is $10^{-3} M$. If the $HgCl_{2(aq)}$ concentration is $10^{-8} M$ (about the accepted limit for Hg in drinking water), what will be the concentration of Hg^{2+}, $HgCl^+$, $HgCl_3^-$, and $HgCl_4^{2-}$?

From Eq. (2-36)

$$[HgCl^+] = \frac{[HgCl_2]}{K_2[Cl^-]} = \frac{(10^{-8})}{3(10^6)(10^{-3})} = 3.3 \times 10^{-12} M$$

From Eq. (2-35)

$$[Hg^{2+}] = \frac{[HgCl^+]}{K_1[Cl^-]} = \frac{3.3(10^{-12})}{5.6(10^6)(10^{-3})} = 5.9 \times 10^{-16} M$$

From Eq. (2-37)

$$[HgCl_3^-] = K_3[HgCl_2][Cl^-] = 7.1(10^{-8})(10^{-3}) = 7.1 \times 10^{-11} M$$

From Eq. (2-38)

$$[HgCl_4^{2-}] = K_4[HgCl^-][Cl^-] = 10(7.1 \times 10^{-11})(10^{-3}) = 7.1 \times 10^{-13} M$$

These calculations indicate that most of the mercury in natural waters occurs as the neutral $HgCl_{2(aq)}$ species. Little of the mercury occurs in ionized forms.

Other soluble molecules or ions which can act as chloride does to form complexes with metals such as mercury are called *ligands*. Among the ligands are H^+, OH^-, CO_3^{2-}, NH_3, F^-, CN^-, $S_2O_3^{2-}$, and many other inorganic and organic species. NH_3 complexes with metals are common. For example, with silver

$$Ag(NH_3)_2^+ \rightleftharpoons Ag^+ + 2NH_3$$

$$\frac{[Ag^+][NH_3]^2}{[Ag(NH_3)_2^+]} = K_{instab} = 6.8 \times 10^{-8} \qquad (2\text{-}44)$$

All such ions are readily destroyed by creating conditions, physically or chemically, that will remove one of the dissociation products.

The silver-ammonia complex ion can be destroyed by adding a source of hydrogen ions. In this case destruction is caused by the formation of a more stable complex ion, NH_4^+.

The ammonium ion (NH_4^+) exists in equilibrium as follows:

$$NH_4^+ \rightleftharpoons NH_3 + H^+$$

$$\frac{[NH_3][H^+]}{[NH_4^+]} = K_{instab} = 5.7 \times 10^{-10} \qquad (2\text{-}45)$$

Addition of a strong base such as sodium hydroxide will decrease the [H^+] concentration through the formation of poorly ionized water, and the equilibrium will be shifted far to the right but not completely. The equilibrium may be completely destroyed by boiling the solution to expel ammonia. This is the basis of the separation and determination of ammonia nitrogen by the distillation technique.

Solubility Product

A fundamental concept is that all solids, no matter how insoluble, are soluble to some degree. For example, silver chloride and barium sulfate are considered to be very insoluble. However, in contact with water they do dissolve, slightly, and form the following equilibria:[2]

$$\underline{AgCl} \rightleftharpoons Ag^+ + Cl^- \qquad (2\text{-}46)$$

$$\underline{BaSO_4} \rightleftharpoons Ba^{2+} + SO_4^{2-} \qquad (2\text{-}47)$$

According to modern concepts of crystal structure, crystals of compounds consist of ions arranged in an orderly manner. Thus when crystals of a compound are placed in water, the ions at the surface migrate into the water, and will continue to do so until the salt is completely dissolved or a condition of saturation is attained. With so-called insoluble substances, the saturation value is very small and is reached quickly.

In silver chloride, barium sulfate, and other insoluble compounds, the ionic concentrations that can exist in equilibrium with the solid or crystalline material are very small.

The equilibrium that exists between crystals of a compound in the solid state and its ions in solution is amenable to consideration under the equilibrium relationship and can be treated mathematically as though the equilibrium were homogeneous in nature. For example, consider silver chloride at equilibrium, as shown in Eq. (2-46):

$$\frac{[Ag^+][Cl^-]}{[\underline{AgCl}]} = K \qquad (2\text{-}48)$$

but [$\underline{AgCl}$] represents the silver chloride that exists in the solid state. It can be

---

[2] The underscore represents solid or precipitated material.

readily demonstrated that we can regard the concentration of a solid substance as a constant in equilibrium relationships. This is related to the fact that the surface area of a solid is the only part that can be considered in equilibrium with the ions, and in a saturated solution, the rate at which these ions leave the surface is equal to the rate at which they are deposited by the solution. Thus, in the above equation, [AgCl] can be assumed equal to $[K_s]$; then

$$\frac{[Ag^+][Cl^-]}{[K_s]} = K \tag{2-49}$$

or
$$[Ag^+][Cl^-] = KK_s = K_{sp} \tag{2-50}$$

The constant K_{sp} is called the *solubility-product* constant.

For more complex substances, such as tricalcium phosphate, that ionize as follows:

$$Ca_3(PO_4)_2 \rightleftharpoons 3Ca^{2+} + 2PO_4^{3-} \tag{2-51}$$

the solubility-product expression in accordance with (Eq. 2-20) is

$$[Ca^{2+}]^3[PO_4^{3-}]^2 = K_{sp} \tag{2-52}$$

Solubility products for nearly all insoluble substances may be obtained by reference to qualitative and quantitative textbooks or chemical handbooks. Typical solubility-product constants of interest in environmental engineering are listed in Table 2-4.

Table 2-4 Typical solubility-product constants

Equilibrium equation	K_{sp} at 25°C	Significance in environmental engineering
$MgCO_3 \rightleftharpoons Mg^{2+} + CO_3^{2-}$	4×10^{-5}	Hardness removal, scaling
$Mg(OH)_2 \rightleftharpoons Mg^{2+} + 2OH^-$	9×10^{-12}	Hardness removal, scaling
$CaCO_3 \rightleftharpoons Ca^{2+} + CO_3^{2-}$	5×10^{-9}	Hardness removal, scaling
$Ca(OH)_2 \rightleftharpoons Ca^{2+} + 2OH^-$	8×10^{-6}	Hardness removal
$CaSO_4 \rightleftharpoons Ca^{2+} + SO_4^{2-}$	2×10^{-5}	Flue gas desulfurization
$Cu(OH)_2 \rightleftharpoons Cu^{2+} + 2OH^-$	2×10^{-19}	Heavy metal removal
$Zn(OH)_2 \rightleftharpoons Zn^{2+} + 2OH^-$	3×10^{-17}	Heavy metal removal
$Ni(OH)_2 \rightleftharpoons Ni^{2+} + 2OH^-$	2×10^{-16}	Heavy metal removal
$Cr(OH)_3 \rightleftharpoons Cr^{3+} + 3OH^-$	6×10^{-31}	Heavy metal removal
$Al(OH)_3 \rightleftharpoons Al^{3+} + 3OH^-$	1×10^{-32}	Coagulation
$Fe(OH)_3 \rightleftharpoons Fe^{3+} + 3OH^-$	6×10^{-38}	Coagulation, iron removal, corrosion
$Fe(OH)_2 \rightleftharpoons Fe^{2+} + 2OH^-$	5×10^{-15}	Coagulation, iron removal, corrosion
$Mn(OH)_3 \rightleftharpoons Mn^{3+} + 3OH^-$	1×10^{-36}	Manganese removal
$Mn(OH)_2 \rightleftharpoons Mn^{2+} + 2OH^-$	8×10^{-14}	Manganese removal
$Ca_3(PO_4)_2 \rightleftharpoons 3Ca^{2+} + 2PO_4^{3-}$	1×10^{-27}	Phosphate removal
$CaHPO_4 \rightleftharpoons Ca^{2+} + HPO_4^{2-}$	3×10^{-7}	Phosphate removal
$CaF_2 \rightleftharpoons Ca^{2+} + 2F^-$	3×10^{-11}	Fluoridation
$AgCl \rightleftharpoons Ag^+ + Cl^-$	3×10^{-10}	Chloride analysis
$BaSO_4 \rightleftharpoons Ba^{2+} + SO_4^{2-}$	1×10^{-10}	Sulfate analysis

A prediction of relative solubilities of compounds cannot be made by a simple comparison of solubility-product values because of the squares and cubes that enter into the calculation when more than two ions are derived from one molecule, as shown in Eqs. (2-51) and (2-52). Barium sulfate, which yields two ions, and calcium fluoride, which yields three ions, may be used to illustrate the point. The solubility of these compounds at 20°C is

$$BaSO_4 = 1.1 \times 10^{-5} M$$

$$CaF_2 = 2.05 \times 10^{-4} M$$

It will be noted that calcium fluoride is about twenty times more soluble than barium sulfate. In saturated solutions of poorly soluble substances, it is assumed that ionization of the dissolved material is complete. Therefore in a saturated solution of barium sulfate, both the $[Ba^{2+}]$ and the $[SO_4^{2-}]$ are equal to 1.1×10^{-5}, and in a saturated solution of calcium fluoride the $[Ca^{2+}]$ is equal to 2.05×10^{-4} and the $[F^-]$ is twice as great, or 4.1×10^{-4}. When these values are substituted into the solubility-product equation we obtain

$$[Ba^{2+}][SO_4^{2-}] = [1.1 \times 10^{-5}][1.1 \times 10^{-5}] = 1.2 \times 10^{-10} \qquad (2\text{-}53)$$

and
$$[Ca^{2+}][F^-]^2 = [2.05 \times 10^{-4}][4.1 \times 10^{-4}]^2 = 3.4 \times 10^{-11} \qquad (2\text{-}54)$$

From this it is obvious that the most soluble material (CaF_2) has the smallest solubility product because of the squaring of the fluoride concentration. The case of compounds that yield more than three ions is even more exaggerated.

There are two corollary statements related to the solubility-product principle, an understanding of which is basic to explaining the phenomena of precipitation and solution of precipitates. They may be expressed as follows:

1. In an unsaturated solution, the product of the molar concentration of the ions is less than the solubility-product constant, or for a species AB, $[A^+][B^-] < K_{sp}$.
2. In a supersaturated solution, the product of the molar concentration of the ions is greater than the solubility-product constant, or $[A^+][B^-] > K_{sp}$. In the former case, if undissolved AB is present, it will dissolve to the extent that $[A^+][B^-] = K_{sp}$, and a saturated solution results. In the second case, nothing will happen until such time as crystals of AB are introduced into the solution or internal forces allow formation of crystal nuclei; then precipitation will occur until the ionic concentrations are reduced equal to those of a saturated solution.

Common Ion Effect

The advantage of relating solubilities to the equilibrium relationship is that this allows mathematical treatment of the equilibrium and prediction of the effect of adding a common ion to a solution containing a slightly soluble salt. For example, consider a solution that has been saturated with barium sulfate. As indicated in Eq. (2-53), both $[Ba^{2+}]$ and $[SO_4^{2-}]$ would equal 1.1×10^{-5}. Now, if the barium ion concentration should be increased by addition from an outside source, such as $BaCl_2$, the concentration of sulfate ion must decrease and the amount of precipitated $BaSO_4$ must increase in order for K_{sp} to remain the same. To illustrate, assume that 10×10^{-5} mol/l of $BaCl_2$ is added to the above solution. This will result in the formation of an additional x moles of precipitated $BaSO_4$. The following changes in $[Ba^{2+}]$ and $[SO_4^{2-}]$ must then take place:

$$\underset{\substack{(1.1 \times 10^{-5} + \\ 10 \times 10^{-5} - x)}}{Ba^{2+}} + \underset{(1.1 \times 10^{-5} - x)}{SO_4^{2-}} \rightarrow \underset{x}{\underline{BaSO_4}}$$

According to the solubility-product principle,

$$(11.1 \times 10^{-5} - x)(1.1 \times 10^{-5} - x) = K_{sp} = 1.2 \times 10^{-10}$$

By solving for x, it is found that an additional 0.98×10^{-5} mol/l of precipitated $BaSO_4$ is formed and that the new equilibrium concentrations of barium and sulfate ions are

$$[Ba^{2+}] = (11.1 \times 10^{-5}) - (0.98 \times 10^{-5}) = 10.1 \times 10^{-5}$$

$$[SO_4^{2-}] = (1.1 \times 10^{-5}) - (0.98 \times 10^{-5}) = 0.12 \times 10^{-5}$$

These calculations indicate that the $[SO_4^{2-}]$ is reduced considerably. The above is an application of the *common ion effect,* which is used extensively in environmental engineering practice as well as in qualitative and quantitative analysis to accomplish essentially complete precipitation of desired ions.

Diverse Ion Effect

The *diverse ion effect* describes the adverse effect that unrelated ions often have upon the solubility of some relatively insoluble substances. Such ions, theoretically, play no part in the chemical equilibrium involved but often increase the solubility of desired precipitates to such an extent that quantitative results cannot be obtained. The explanation for this effect is that the forces of attraction caused by the charge on the unrelated ions decrease the effective concentration or activity of the slightly soluble ions, as discussed in Sec. 2-12. Nitrate ion has such an effect on silver chloride. The classic example, described in many physical chemistry texts, is the influence of nitrate ion on the solubility of thallous chloride. Because of the diverse ion effect, it is general practice to keep the concentration of extraneous ions as low as possible during qualitative and quantitive work.

Sometimes a common ion may serve very well up to certain concentrations, but when used at higher concentrations, appears to have a diverse ion effect. In this case, the usual explanation is that complex ion formation is taking place, as described previously. Hydrochloric acid acts in such a manner when it is used as the agent to precipitate silver ion. When hydrochloric acid is added in excess, the soluble AgCl$_2^-$ complex is formed:

$$AgCl + Cl^- \rightarrow AgCl_2^- \qquad (2\text{-}55)$$

Therefore the amount of hydrochloric acid used should be carefully controlled.

The solubility of metal ions is also increased by the presence of "chelating agents." These substances have the ability to seize or "sequester" metal ions and hold them in a clawlike grip (the word is from the Greek, *chele,* meaning claw). Like a claw, a chelating molecule forms a ring in which the metal ion is held by a pair of pincers so that it is not free to form an insoluble salt. The pincers of a chelating molecule consist of "ligand" atoms (usually nitrogen, oxygen, or sulfur), each of which donates two electrons to form a "coordinate" bond with the ion. There are many natural chelates such as hemoglobin (containing iron), vitamin B-12 (containing cobalt), and chlorophyll (containing magnesium). Many well-known substances such as aspirin, citric acid, adrenaline, and cortisone can act as chelating agents; EDTA is a chelating agent which has a remarkable affinity for calcium and is used for the determination of water hardness (Chap. 18).

2-14 WAYS OF SHIFTING CHEMICAL EQUILIBRIA

The environmental engineer deals routinely with materials that are in either homogeneous or heterogeneous equilibrium. He, like the analytical chemist, must be able to apply stresses to his system in accordance with Le Chatelier's principle to bring about desired changes. Many of the stresses that he applies are exactly the same in character as those used by the chemist. Therefore it is important to consider the ways in which equilibria can be shifted to bring about essentially complete reactions. Five methods are commonly employed.

Formation of Insoluble Substances

All precipitation reactions are examples of this method of equilibrium shift. In this case a knowledge of the solubility-product principle, solubility-product constants, and the common ion effect is brought into service. The removal of metal ions from industrial wastes, such as copper and brass wastes, by precipitation with calcium hydroxide, and the softening of hard waters by lime–soda ash treatment are excellent examples of how the engineer applies this method of shifting a chemical equilibrium to gain his objective. Equations associated with these precipitation reactions are listed in Table 2-4.

Formation of a Weakly Ionized Compound

Certain systems that are in equilibrium can be destroyed by adding a reagent that will combine with one of the ions to form a poorly ionized compound. The neutralization of acid and of caustic wastes is based upon such formation, since the reaction involved is between hydrogen ions and hydroxyl ions to form poorly ionized water:

$$Na^+ + OH^- + H^+ + Cl^- \rightarrow H_2O + Na^+ + Cl^- \qquad (2\text{-}56)$$

The environmental engineer frequently uses similar reactions to bring the pH of industrial wastes into a favorable range for subsequent biological treatment.

The analytical chemist uses this method of equilibrium shifting routinely to dissolve precipitates of the metallic hydroxides, such as ferric and aluminum hydroxide:

$$Fe(OH)_3 + 3H^+ \rightarrow Fe^{3+} + 3H_2O \qquad (2\text{-}57)$$

$$Al(OH)_3 + 3H^+ \rightarrow Al^{3+} + 3H_2O \qquad (2\text{-}58)$$

Formation of Complex Ion

The chemist uses complex-ion formation to dissolve insoluble salts and hydroxides. Silver chloride, for example, dissolves readily in ammonium hydroxide solution. This occurs because silver ion combines with molecular NH_3 contained in the ammonium hydroxide to form $[Ag(NH_3)_2]^+$; as a result the solution becomes unsaturated with respect to silver and chloride ions and solid silver chloride passes into solution in an attempt to form a saturated solution. The net effect is as follows:

$$AgCl + 2NH_3 \rightarrow [Ag(NH_3)_2]^+ + Cl^- \qquad (2\text{-}59)$$

If enough ammonium hydroxide is present, all silver chloride passes into solution. The equilibrium relationship for the above may be written

$$\frac{[Ag(NH_3)_2^+][Cl^-]}{[NH_3]^2} = K \qquad (2\text{-}60)$$

A value for the constant K may be determined from solubility-product data for silver chloride, and from the instability constant for the ammonia-silver complex. Illustrations of such computations are given in most textbooks on qualitative analysis.

Zinc and copper hydroxides dissolve in ammonium hydroxide for the same reason given above. The complex ions formed are $[Zn(NH_3)_4]^{2+}$ and $[Cu(NH_3)_4]^{2+}$. These reactions illustrate why ammonium hydroxide would not be a good reagent for precipitating copper and zinc ions from a brass-mill waste.

Industrial wastes containing sodium cyanide are particularly toxic to fish, even though the concentration of cyanide ion may be reduced to very low levels by dilution with river water. The destruction or inactivation of cyanide ion may be accomplished by complex-ion formation. Formerly, before development of more efficient methods, cyanide ion was reduced to low levels by treatment with ferrous sulfate. In the reaction, Fe^{2+} combined with CN^- to give the complex ion $Fe(CN)_6^{4-}$, which could be precipitated as Prussian blue by subsequent oxidation of excess Fe^{2+} to Fe^{3+}, in the presence of potassium ions,

$$Fe^{3+} + K^+ + [Fe(CN)_6]^{4-} \rightarrow \underline{KFe[Fe(CN)_6]} \qquad (2\text{-}61)$$

Formation of a Gaseous Product

In reactions involving the formation of a gaseous product, the reactions go to practical completion because the gas escapes from the sphere of the reaction. The analytical chemist takes advantage of this method of forcing a reaction to completion when dissolving metallic sulfides, such as ferrous sulfide, in hydrochloric acid:

$$\underline{FeS} + 2H^+ \rightarrow H_2S\uparrow + Fe^{2+} \qquad (2\text{-}62)$$

This does not necessarily mean that all metallic sulfides are soluble in hydrochloric acid, for the sulfides of copper and mercury are not. This can be explained by considering that copper and mercury sulfides are so insoluble that the sulfide ion liberated by them at equilibrium is of such small concentration that not enough un-ionized hydrogen sulfide is formed, in the presence of concentrated hydrochloric acid, to be released as a gas. Consequently the sulfides do not dissolve.

In industrial waste treatment, cyanides were formerly removed from aqueous solution by treatment with sulfuric acid:

$$2CN^- + 2H^+ + SO_4^{2-} \rightarrow 2HCN\uparrow + SO_4^{2-} \qquad (2\text{-}63)$$

The hydrogen cyanide released as a gas was diluted with large volumes of air and forced up through tall stacks to get proper dispersion and avoid serious atmospheric-pollution problems. More modern methods accomplish destruction of cyanide ion by oxidation or other techniques.

Oxidation and Reduction

A very sure method of sending reactions to completion is oxidation and reduction. In this way one or more of the ions involved in the equilibrium reaction can be destroyed, and the reaction will proceed to completion. A classic example, familiar to environmental engineers, is the destruction of cyanide ion by chlorination according to the following equation:

$$2CN^- + 5Cl_2^0 + 8OH^- \rightarrow 10Cl^- + 2CO_2 + N_2\uparrow + 4H_2O \qquad (2\text{-}64)$$

From the discussion already presented, this reaction would go to completion for two reasons.

1. Cyanide ion is oxidized to carbon dioxide and nitrogen.
2. The nitrogen escapes as a gas.

Since formation of nitrogen is dependent upon oxidation of cyanide ion, the reaction is considered complete whether the nitrogen escapes or not.

2-15 AMPHOTERIC HYDROXIDES

The oxides or hydroxides of metals are basic in character, and react with acids to form salts. The insoluble metallic hydroxides, such as ferric hydroxide, dissolve readily in acids to form salts but are insoluble in solutions of bases. Likewise the oxides of nonmetals are acidic in character, and insoluble forms are soluble in bases but not in acids. These characteristics serve as one basis of differentiating between metals and nonmetals.

The hydroxides of aluminum, zinc, chromium, and a few other elements are soluble in both acids and bases. They are known as *amphoteric* hydroxides, and advantage is often taken of this fact to accomplish separations in qualitative analysis and in chemical processing.

An insoluble metallic hydroxide, of course, exists in equilibrium with its ions. For example, ferric hydroxide dissolves to a limited extent to produce Fe^{3+} and OH^- ions:

$$Fe(OH)_3 \rightleftharpoons Fe^{3+} + 3OH^- \tag{2-65}$$

and at saturation

$$[Fe^{3+}][OH^-]^3 = K_{sp} \tag{2-66}$$

When a strong acid is added, the OH^- ions combine with the H^+ ions of the acid to form poorly ionized water, and the $[OH^-]$ decreases so that

$$[Fe^{3+}][OH^-]^3 < K_{sp} \tag{2-67}$$

Therefore ferric hydroxide dissolves in an attempt to establish conditions represented in Eq. (2-66). If enough acid is added, eventually all the ferric hydroxide will dissolve. A similar situation holds for insoluble nonmetallic oxides that are soluble in bases, except that the insoluble oxide is in equilibrium with H^+ and some acid radical.

An amphoteric hydroxide forms complexes with hydroxide. The complexes formed under both acidic and basic conditions are charged ions and soluble in water. Those formed under conditions between these extremes are neutral in charge and insoluble in water. Aluminum hydroxide will serve as an example. Stepwise reactions for Al^{3+} and OH^- are as follows:

$$Al^{3+} + OH^- \rightleftharpoons Al(OH)^{2+} \tag{2.68}$$

$$Al(OH)^{2+} + OH^- \rightleftharpoons Al(OH)_2^+ \tag{2.69}$$

$$Al(OH)_2^+ + OH^- \rightleftharpoons Al(OH)_3 \tag{2.70}$$

$$Al(OH)_3 + OH^- \rightleftharpoons Al(OH)_4^- \tag{2.71}$$

Under strongly acidic conditions, the OH^- concentration is low and the species present will be the positively charged ions, Al^{3+} and $Al(OH)^{2+}$. If a strong base is gradually added, the OH^- concentration will increase and these ions will begin to add to the aluminum complexes, reducing the charges until the neutral and insoluble species $Al(OH)_3$ is formed, which will then precipitate from solution. As base continues to be added and the OH^- concentration in solution increases further, the negatively charged and soluble $Al(OH)_4^-$ ion is formed, and the precipitate dissolves.

Actually, metal ions such as Al^{3+} and Fe^{3+} do not occur as such in solution, just as H^+ does not. Because of their strong positive charges, they readily become hydrated to form species such as $[Fe(H_2O)_6]^{3+}$ and $[Al(H_2O)_6]^{3+}$. However, the symbols Fe^{3+} and Al^{3+} will be used in this book to represent these species in order to reduce the complexity of the equations.

The amphoteric property of aluminum hydroxide is a factor limiting its use as a coagulant in water purification and industrial waste treatment. The amphoteric properties of zinc and chromium hydroxides are important considerations in treating industrial wastes containing Zn^{2+} and Cr^{3+}.

PROBLEMS

2-1 Calculate the gram molecular weight of (a) $CaCO_3$, (b) $NaNO_3$, (c) CH_4, and (d) K_2HPO_4.
  Answer: (a) 100; (b) 85; (c) 16; (d) 174

2-2 Calculate the gram molecular weight of (a) $BaSO_4$, (b) Na_2CO_3, (c) H_2SO_4, and (d) $Mg(OH)_2$.

2-3 What is the molar concentration of a solution containing 10 g/l of (a) NaOH, (b) Na_2SO_4, (c) $K_2Cr_2O_7$, and (d) KCl?
  Answer: (a) 0.25 M; (b) 0.0704 M; (c) 0.034 M; (d) 0.134 M

2-4 Calculate the weight of $KMnO_4$ contained in 2 liters of a 0.15 M solution.

2-5 Balance the following equations:
  (a) $CaCl_2 + Na_2CO_3 \rightarrow CaCO_3 + NaCl$
  (b) $Ca_3(PO_4)_2 + H_3PO_4 \rightarrow Ca(H_2PO_4)_2$
  (c) $MnO_2 + NaCl + H_2SO_4 \rightarrow MnSO_4 + H_2O + Cl_2 + Na_2SO_4$
  (d) $Ca(H_2PO_4)_2 + NaHCO_3 \rightarrow CaHPO_4 + Na_2HPO_4 + H_2O + CO_2$

2-6 Balance the following equations:
  (a) $FeS + HCl \rightarrow FeCl_2 + H_2S$
  (b) $Cl_2 + KOH \rightarrow KCl + KClO_3 + H_2O$
  (c) $FeSO_4 + K_2Cr_2O_7 + H_2SO_4 \rightarrow Fe_2(SO_4)_3 + Cr_2(SO_4)_3 + K_2SO_4 + H_2O$
  (d) $Al_2(SO_4)_3 \cdot 14H_2O + Ca(HCO_3)_2 \rightarrow Al(OH)_3 + CaSO_4 + H_2O + CO_2$

2-7 Balance the following equations:
  (a) $Fe(OH)_2 + H_2O + O_2 \rightarrow Fe(OH)_3$
  (b) $KI + HNO_2 + H_2SO_4 \rightarrow NO + I_2 + H_2O + K_2SO_4$
  (c) $H_2C_2O_4 + KMnO_4 + H_2SO_4 \rightarrow CO_2 + MnSO_4 + K_2SO_4 + H_2O$
  (d) $SO_3^{2-} + Fe^{3+} + H_2O \rightarrow SO_4^{2-} + Fe^{2+} + H^+$

2-8 Balance the following equations:
  (a) $HClO \rightarrow HClO_3 + HCl$
  (b) $NO_2^- + MnO_4^- + H^+ \rightarrow NO_3^- + Mn^{2+} + H_2O$
  (c) $Cl^- + NO_3^- + H^+ \rightarrow Cl_2 + NO + H_2O$
  (d) $I_2 + IO_3^- + H^+ + Cl^- \rightarrow ICl_2^- + H_2O$

2-9 Using half reactions, write complete balanced oxidation-reduction equations for the following:
  (a) Oxidation of I^- to I_2 and reduction of MnO_2 to Mn^{2+}
  (b) Oxidation of $S_2O_3^{2-}$ to SO_4^{2-} and reduction of Cl_2 to Cl^-
  (c) Oxidation of NH_4^+ to NO_3^- and reduction of O_2 to H_2O

(d) Oxidation of CH_3COO^- to CO_2 and reduction of $Cr_2O_7^{2-}$ to Cr^{3+}

(e) Oxidation of $C_6H_{12}O_6$ to CO_2 and reduction of NO_3^- to N_2

Answers:

(a) $2I^- + MnO_2 + 4H^+ \rightarrow I_2 + Mn^{2+} + H_2O$

(b) $S_2O_3^{2-} + 4Cl_2 + 5H_2O \rightarrow 2SO_4^{2-} + 8Cl^- + 10H^+$

(c) $NH_4^+ + 2O_2 \rightarrow NO_3^- + H_2O + 2H^+$

(d) $3CH_3COO^- + 4Cr_2O_7^{2-} + 35H^+ \rightarrow 6CO_2 + 8Cr^{3+} + 22H_2O$

(e) $5C_6H_{12}O_6 + 24NO_3^- + 24H^+ \rightarrow 30CO_2 + 12N_2 + 42H_2O$

2-10 Using half reactions, write complete balanced oxidation-reduction equations for the following:

(a) Oxidation of Mn^{2+} to MnO_2 and reduction of O_2 to H_2O

(b) Oxidation of $S_2O_3^{2-}$ to SO_4^{2-} and reduction of I_2 to I^-

(c) Oxidation of NH_4^+ to NO_2^- and reduction of O_2 to H_2O

(d) Oxidation of $C_6H_{12}O_6$ to CO_2 and reduction of $Cr_2O_7^{2-}$ to Cr^{3+}

(e) Oxidation of CH_3COO^- to CO_2 and reduction of SO_4^{2-} to H_2S

2-11 Construct half reactions for the following reductions:

(a) SO_4^{2-} to S

(b) NO_3^- to NO_2^-

(c) CH_3COO^- to $CH_3CH_2CH_2COO^-$

Answers:

(a) $\frac{1}{8}SO_4^{2-} + \frac{4}{3}H^+ + e^- = \frac{1}{6}S + \frac{2}{3}H_2O$

(b) $\frac{1}{2}NO_3^- + H^+ + e^- = \frac{1}{2}NO_2^- + \frac{1}{2}H_2O$

(c) $\frac{1}{2}CH_3COO^- + \frac{5}{4}H^+ + e^- = \frac{1}{4}CH_3CH_2CH_2COO^- + \frac{1}{2}H_2O$

2-12 Construct half reactions for the following reductions:

(a) CO_2 to CH_4

(b) S_2 to H_2S

(c) $CH_3COO^- + CO_2$ to $CH_3CH_2COO^-$

2-13 Develop the appropriate half reactions and from these construct the complete oxidation-reduction equation for oxidation of H_2S to S and reduction of Fe^{3+} to Fe^{2+}.

Answer: $H_2S + 2Fe^{3+} \rightarrow S + 2Fe^{2+} + 2H^+$

2-14 Develop the appropriate half reactions and from these construct the complete oxidation-reduction equation for oxidation of CH_3CH_2OH to CO_2 and reduction of NO_3^- to NO_2^-.

2-15 How many moles of H_2SO_4 are required to form 65 g of $CaSO_4$ from $CaCO_3$?

Answer: 0.478

2-16 How many g of iodine (I_2) are formed from the oxidation of an excess of KI by 6 g of $K_2Cr_2O_7$, under acid conditions [see Eq. (2-9)]?

2-17 Calculate the volume in cubic feet occupied by 120 lb of carbon dioxide at 1.5 atm and 40°C.

Answer: 750 cu ft

2-18 Determine the weight in grams of oxygen contained in a 10-liter volume under a pressure of 5 atm and at a temperature of 0°C.

2-19 What volume of oxygen at 25°C and 0.21 atm is required for combustion of 25 g of methane gas?

Answer: 363 liters

2-20 If 6 g of ethane gas (CH_3CH_3) is burned in oxygen, (a) how many moles of water are formed; (b) how many moles of carbon dioxide are formed; (c) what is the volume in liters of carbon dioxide formed at 1 atm pressure and 20°C?

2-21 A gas mixture at 25°C and 1 atm contains 100 mg/l of H_2S gas. What is the partial pressure exerted by this gas?

Answer: 0.072 atm

2-22 A 30-liter volume of gas at 25°C contains 12 g of methane, 1 g of nitrogen, and 15 g of carbon dioxide. Calculate (a) the moles of each gas present, (b) the partial pressure exerted by each gas, (c) the total pressure exerted by the mixture, and (d) the percentage by volume of each gas in the mixture.

2-23 Five liters of water are equilibrated with a gas mixture containing carbon dioxide at a partial pressure of 0.3 atm. If the Henry's law constant for carbon dioxide solubility is 2.0 g/l-atm, how many grams of carbon dioxide are dissolved in the water?

Answer: 3.0 g

2-24 What is the concentration of oxygen dissolved in water at 20°C in equilibrium with a gas mixture at 0.81 atm and containing 21 percent by volume of oxygen?

2-25 Calculate the percent ionization and hydrogen ion concentration at 25°C in a solution containing (*a*) 0.10 M H_2CO_3, and (*b*) 0.01 M H_2CO_3.

Answer: (*a*) 0.211 percent, 2.11×10^{-4} mol/l; (*b*) 0.67 percent, 6.67×10^{-5} mol/l

2-26 Calculate the percent ionization and hydrogen ion concentration in a solution containing 0.05 M hypochlorous acid (HOCl), which has an ionization constant at 25°C of 2.85×10^{-8}.

2-27 (*a*) Write reactions and equilibrium relationships for the first four complexes formed between cadmium (II) and chloride.

(*b*) Write the equilibrium relationship for the instability constant for $CdCl_4^{2-}$.

2-28 (*a*) Write reactions and equilibrium relationships for the first five complexes formed between copper(II) and NH_3. A sixth complex is theoretically possible, but its formation has not yet been detected.

(*b*) Write the overall reaction and equilibrium relationship for the fourth complex between copper(II) and NH_3.

2-29 Equilibrium constants for the complexes between cadmium(II) and chloride are $K_1 = 21$, $K_2 = 8$, $K_3 = 1.2$, and $K_4 = 0.35$. Calculate the molar concentration of each of the first four cadmium chloride complexes in a water sample if $Cd^{2+} = 10^{-8} M$ and $Cl^- = 10^{-3} M$. Identify the most prevalent cadmium species.

Answer: 2.1×10^{-10}, 1.7×10^{-12}, 2×10^{-15}, 7×10^{-19}, Cd^{2+}

2-30 Do Prob. 2-29, but assume that the chloride concentration is similar to that of seawater or about 0.5 M.

2-31 Calculate the concentration in (*a*) moles per liter, (*b*) milligrams per liter, and (*c*) number of ions per liter, for the sulfate ion in a saturated solution of barium sulfate to which barium chloride is added until $[Ba^{2+}] = 0.0001 M$.

Answer: (*a*) 10^{-6} mol/l; (*b*) 0.096 mg/l; (*c*) 6.02×10^{17} ions/l

2-32 Calculate the concentration in (*a*) moles per liter, (*b*) milligrams per liter, and (*c*) number of ions per liter, for the chloride ion in a saturated solution of silver chloride to which silver nitrate is added until $[Ag^+] = 0.0001 M$.

2-33 (*a*) Write the expression for the solubility-product constant of (1) AgCl, (2) CuS, (3) $MgNH_4PO_4$, (4) $Au(OH)_3$, (5) Ag_2CrO_4, and (6) $BaCO_3$. (*b*) What ions can be added to solutions containing the compounds listed in (*a*) which, in each case, will lower the concentration of the cation? Explain why.

2-34 Why does solubility often increase as still larger quantities of a common ion are added?

2-35 Tell what is meant by (*a*) common ion effect, (*b*) complex ion, (*c*) solubility-product constant, (*d*) amphoteric hydroxide, (*e*) diverse ion effect, (*f*) heterogeneous equilibrium, (*g*) homogeneous equilibrium, (*h*) driving reaction to completion, (*i*) saturated solution.

2-36 From each of the following values of water solubility, evaluate the corresponding solubility-product constant: (*a*) $Mg_3(PO_4)_2$, 6.1×10^{-5} mol/l, (*b*) FeS, 6.3×10^{-9} mol/l, (*c*) $Zn_3(PO_4)_2$, 1.6×10^{-7} mol/l, and (*d*) CuF_2, 7.4×10^{-3} mol/l.

2-37 Which is the better chemical for removing calcium ions from solution, sodium hydroxide or sodium carbonate? Why?

2-38 Which is the better chemical for removing magnesium ions from solution, sodium hydroxide or sodium carbonate? Why?

2-39 Given the reaction $NH_3 + H_2O \rightleftharpoons NH_4^+ + OH^-$, how many ways can you think of for forcing the reaction (*a*) to the right? (*b*) to the left? (*c*) Under what conditions might it go to completion in either direction?

2-40 Determine how many mg/l of magnesium ion will dissolve in water which is (a) $10^{-5} M$ in OH^-, (b) $10^{-3} M$ in OH^-.

2-41 A metal-plating waste contains 20 mg/l Cu^{2+} and it is desired to add $Ca(OH)_2$ to precipitate all but 0.5 mg/l of the copper. To what concentration in moles per liter must the hydroxide concentration be raised to accomplish this?

Answer: 1.6×10^{-7} mol/l

2-42 A waste contains 50 mg/l of Zn^{2+}. How high must the pH be raised to precipitate all but 1 mg/l of the zinc? What adverse effect might occur if the pH were raised too high?

2-43 Calculate the pH required to decrease the iron concentration in a water supply to 0.03 mg/l if (a) the iron is in the Fe^{2+} form, (b) the iron is in the Fe^{3+} form.

Answer: (a) 10.0; (b) 3.7

2-44 Calculate the pH required to decrease the manganese concentration in a water supply to 0.01 mg/l if (a) the manganese is in the Mn^{2+} form, (b) the manganese is in the Mn^{3+} form.

2-45 It is desired to fluoridate a water supply by adding sufficient sodium fluoride to increase the fluoride concentration to 1 mg/l. Will this fluoride concentration be soluble in a water containing 200 mg/l of calcium? Show your computations.

2-46 At what Ca^{2+} concentration in mg/l will precipitation of $Ca_3(PO_4)_2$ occur in a solution containing $10^{-6} M$ PO_4^{3-}?

2-47 If the CO_3^{2-} concentration in a water sample is 100 mg/l, what is the maximum solubility in mg/l of (a) Ca^{2+}? (b) Mg^{2+}?

Answer: (a) 0.12 mg/l; (b) 583 mg/l

2-48 (a) If excess AgCl is mixed in distilled water, what will be the concentration of silver in solution in milligrams per liter at equilibrium? (b) If ammonium hydroxide is added to the above solution so that the resultant NH_3 concentration is 0.01 mol/l, what will the total concentration of silver in solution then be? Assume K in Eq. (2-60) equal to 5×10^{-3}.

REFERENCES

Dean, J. A.: "Lang's Handbook of Chemistry," 11th ed., McGraw-Hill, New York, 1973.

Dickerson, R. E., H. B. Gray, and G. P. Haight, Jr.: "Chemical Principles," 2d ed., W. A. Benjamin, Menlo Park, Cal., 1974.

Eastman, R. H.: "Essentials of Modern Chemistry," Rinehart, Corte Madera, Cal., 1975.

Hildebrand, J. H., and R. E. Powell: "Principles of Chemistry," 7th ed., Macmillan, New York, 1964.

Hutchinson, E.: "Chemistry: The Elements and Their Reactions," 2d ed., Saunders, Philadelphia, 1964.

Mahan, B. H.: "University Chemistry," 3d ed., Addison-Wesley, Reading, Mass., 1975.

Pauling, L.: "College Chemistry," 3d ed., Freeman, San Francisco, 1964.

Pauling, L.: "General Chemistry," 3d ed., Freeman, San Francisco, 1970.

Sienko, M. J., and R. A. Plane: "Chemistry," 5th ed., McGraw-Hill, New York, 1976.

Sorum, C. H.: "Fundamentals of General Chemistry," 2d ed., Prentice-Hall, Englewood Cliffs, N.J., 1963.

Weast, R. C. (ed.): "Handbook of Chemistry and Physics," 56th ed., Chemical Rubber Publishing Co., Cleveland, 1975.

THREE

BASIC CONCEPTS FROM PHYSICAL CHEMISTRY

INTRODUCTION

That portion of science dealing with laws or generalizations related to chemical phenomena is called *physical chemistry*. A sound knowledge of physics is fully as important in the study of physical chemistry as a good grounding in chemistry. Some physical chemistry has always been taught in general, qualitative, and quantitative chemistry without being described as such. The trend has been to include more and more physical chemistry in such courses, and even in high-school chemistry, as those in the teaching profession have realized the need and gained stimulation of teaching "why" as well as "what" chemical reactions occur.

The subject of valency and bonding, oxidation–reduction reactions, the gas laws, Raoult's law, equilibrium relationships, the theory of ionization, the solubility-product principle, and other discussions pertaining to heterogeneous and homogeneous equilibria are all examples of topics in physical chemistry which have been developed previously. Some of them require further amplification, and a great many additional concepts need to be developed.

3-2 THERMODYNAMICS

Thermodynamics is the study of energy changes accompanying physical and chemical processes. The energy changes associated with chemical reactions are of considerable importance and will be briefly discussed to indicate the relationships of most interest in environmental engineering. First, it is necessary to review the relationship between heat and work.

Heat and Work

Heat and work are related forms of energy. Heat energy can be converted into work, and work can be converted into heat energy. Steam engines and frictional losses are examples of each conversion.

   Heat is that form of energy which passes from one body to another solely as a result of a difference in temperature. On the molecular scale it is known that the temperature of a substance is related to the average translational energy of the molecules, and that flow of heat results from transfer of this molecular energy. The basic unit of heat is the calorie, the heat required to raise the temperature of one gram of water one degree Celsius. In engineering it is common to measure heat in British Thermal Units (Btu), which is the heat required to raise one pound of water one degree Fahrenheit. One Btu is equivalent to 252 calories.

   The *specific heat* of a substance is the heat required to raise one gram of the material one degree Celsius, or

$$C = \frac{q}{M\Delta T} \qquad (3\text{-}1)$$

where C is the specific heat, q is the heat added in calories, M is the weight of the material in grams, and ΔT is the rise in temperature of the material in degrees Celsius. For water the specific heat is 1.000 calories per gram-degree Celsius only at 15°C, but varies from this value by less than 1 percent over the entire range from 0 to 100°C. Therefore the assumption of a constant specific heat over a limited range in temperature is frequently a good one. Two values for specific heat are frequently reported: C_v is the specific heat at constant volume, and C_p is the specific heat at constant pressure. For liquids and solids there is little difference between these two values. For gases, however, C_p is greater than C_v because of the extra heat energy required to expand a gas against a constant pressure.

   The *heat of fusion* is the heat required to melt a substance at its normal melting temperature. The *heat of vaporization* is the heat required to evaporate the substance at its normal boiling point. The values for water are 79.7 and 539.7 calories per gram, respectively. Environmental engineers encounter a wide variety of heating, evaporation, drying, and incineration problems involving a knowledge of specific heat, heat of fusion, and heat of vaporization. In addition, many industrial processes involve evaporation or drying in which cooling water is needed. In essence, the heat of vaporization is transferred to the coolant during condensation, and the warmed water may become a thermal pollution problem in terms of the receiving body of water. This is a matter of growing concern as water use by industry and the public increases.

   Work in chemical systems usually involves work of expansion. The system may either do work on its surroundings or have work done on itself. This depends upon whether the volume of the system is expanding or contracting. Work is normally measured in terms of force times distance, which for a closed system is equivalent to pressure times the change in volume:

$$dw = P \, dV \tag{3-2}$$

Work is measured in foot-pounds or in joules (J). Since heat and work are both forms of energy, they can be equated. Thus 1 calorie of heat is equivalent to 4.187 J. In other units, 1 Btu of heat is equivalent to 778 ft-lb.

Energy

The first law of thermodynamics states that energy can be neither created nor destroyed. In chemical systems the energy involved is most easily handled in terms of three quantities: the work that is performed, the heat that flows, and the energy stored in the system. The chemical system may range from the contents of a laboratory beaker to a full-scale activated sludge plant. According to the conservation-of-energy law, any heat or work which flows into or out of the system must result in a change in the total energy stored in the system. In equation form,

$$\Delta E = q - w \tag{3-3}$$

Here ΔE is the change in *internal* energy of the system, q is the heat flowing *into* the system, and w is the work done *by* the system. It is important to note that if heat is absorbed by the system, q has a positive value. If the system gives off heat, q has a negative value. Also, if the system does work on the surroundings, w has a positive value, but if the surroundings do work on the system, w has a negative value.

In chemical systems, work performed is usually expansion work as given by Eq. (3-2). When the system expands in volume, it does work on its surroundings, and w is positive. When it contracts in volume, work is done on the system, and w is negative. If the volume of the system remains constant, then no expansion work can be done, and w equals zero. For this case,

$$\Delta E = q_v \qquad (V \text{ constant}) \tag{3-4}$$

where q_v is the heat absorbed in a constant-volume system. Thus, in a constant-volume system, the change in internal energy is just equal to the heat absorbed. Although this is convenient for constant-volume systems, most chemical systems of interest to environmental engineers are open to the atmosphere and so operate under constant pressure, rather than constant volume. For such systems, the concept of enthalpy has been developed.

Enthalpy

The enthalpy, H, of a system is defined as follows:

$$H = E + PV \tag{3-5}$$

where E is the internal energy of the system, P is the pressure on the system, and V is the volume of the system. Consider a constant-pressure system in which some chemical change has taken place, resulting in a change in internal energy. The heat absorbed at constant pressure is q_p, and the work done by the system is given by an integration of Eq. (3-2). For constant temperature and pressure this integration gives $w = P(V_2 - V_1)$, where V_1 is the initial volume of the system and V_2 is the final volume after the change. The change in internal energy of such a system then becomes

$$E = E_2 - E_1 = q_p - w = q_p - P(V_2 - V_1)$$

Rearranging, we obtain

$$(E_2 + PV_2) - (E_1 + PV_1) = q_p$$

The terms in parentheses are just equal to the final and initial enthalpy of the system, and thus

$$H_2 - H_1 = q_p$$

or
$$\Delta H = q_p \quad (T \text{ and } P \text{ constant}) \tag{3-6}$$

Thus *the quantity of heat absorbed by a system at constant temperature and pressure is equal to the change in system enthalpy*. Chemical changes that are accompanied by the absorption of heat, making ΔH positive, are called *endothermic* reactions. Those accompanied by the evolution of heat, making ΔH negative, are called *exothermic*.

The total enthalpy (H) of a system would be difficult to measure. We are normally interested, however, only in the *change* in enthalpy and not in its absolute value. By developing a standard basis for comparison, it is possible to calculate the change in enthalpy or the heat of a given reaction from tabulated measurements from quite different reactions.

A convenient standard state for a substance may be taken as the stable state of the compound at 25°C and 1 atm pressure. For example, under these conditions, the standard state for oxygen is a gas, for mercury a liquid, and for sulfur rhombic crystals. By convention, the enthalpies of the *chemical elements* in this standard state are set equal to zero. The standard enthalpy of any *compound* is then the heat of the reaction by which it is formed from its elements; reactants and products all being in the standard state at 25°C and 1 atm.

For example,

$$\text{(1)} \quad H_2(g) + \tfrac{1}{2}O_2(g) \rightarrow H_2O(lq) \qquad \Delta H^0_{298} = -68,320 \text{ cal}$$

$$\text{(2)} \quad C(c) + O_2(g) \rightarrow CO_2(g) \qquad \Delta H^0_{298} = -94,052 \text{ cal}$$

The symbols in parentheses after each element or compound indicate the standard state of the elements or compounds. The superscript zero on the enthalpy indicates a standard heat of formation with reactants and products at 1 atm. The absolute temperature is written as a subscript. Standard enthalpies for many compounds of interest are listed in Table 3-1. Many other values can be found in chemical handbooks such as those listed at the end of the chapter.

Table 3-1 Standard enthalpies and free energies of formation at 25°C

Sub-stance	State*	ΔH^0_{298} kcal/mole	ΔG^0_{298} kcal/mole	Sub-stance	State*	ΔH^0_{298} kcal/mole	ΔG^0_{298} kcal/mole
Ca^{2+}	aq	− 129.77	− 132.18	H_2O	lq	− 68.32	− 56.69
$CaCO_3$	c	− 288.45	− 269.78	H_2O	g	− 57.80	− 54.64
CaF_2	c	− 290.3	− 277.7	HS^-	aq	− 4.22	3.01
$Ca(OH)_2$	c	− 235.80	− 214.33	H_2S	g	− 4.82	− 7.89
$CaSO_4 \cdot 2H_2O$	c	− 483.06	− 429.19	H_2S	aq	− 9.4	− 6.54
CH_4	g	− 17.89	− 12.14	H_2SO_4	lq	− 193.91	· · ·
CH_3CH_3	g	− 20.24	− 7.86	Na^+	aq	− 57.28	− 62.59
CH_3COOH	aq	− 116.74	− 95.51	NH_3	g	− 11.04	− 3.98
CH_3COO^-	aq	− 116.84	− 88.99	NH_3	aq	− 19.32	− 6.37
CH_3CH_2OH	lq	− 66.36	− 41.77	NH_4^+	aq	− 31.74	− 19.00
$C_6H_{12}O_6$	aq	· · ·	− 217.02	NO_2^-	aq	− 25.4	− 8.25
CO_2	g	− 94.05	− 94.26	NO_3^-	aq	− 49.37	− 26.41
CO_2	aq	− 98.69	− 92.31	OH^-	aq	− 54.96	− 37.60
CO_3^{2-}	aq	− 161.63	− 126.22	S^{2-}	aq	10	20
F^-	aq	− 78.66	− 66.08	SO_4^{2-}	aq	− 216.90	− 177.34
HCO_3^-	aq	− 165.18	− 140.31	Zn^{2+}	aq	− 36.43	− 35.18
H_2CO_3	aq	− 167.0	− 149.00	ZnS	c	− 48.50	− 47.40

* aq—aqueous, c—crystal, g—gas, lq—liquid

Standard enthalpy values can be used to determine the heat given off by a variety of reactions. In making such calculations it is very important to note the state in which the products and reactants exist, as this can make a significant difference in the heat of reaction. The procedure used is to write a balanced equation for the reaction. The heat of the reaction is then equal to the sum of the standard enthalpies of the products minus the sum of the standard enthalpies of the reactants. It should be noted that standard enthalpies of formation are given in terms of calories per mole, and these values must be multiplied by the number of moles entering into the reaction.

Example A Calculate both the net and gross heat of combustion of methane gas.

The *gross heat* is the heat released if the water vapor formed upon combustion is condensed to form liquid water.

$$\underset{\Delta H^0_{298}}{CH_4(g)} + \underset{}{2O_2(g)} \rightarrow \underset{}{CO_2(g)} + \underset{}{2H_2O(lq)}$$
$$\qquad -17.89 \qquad 0 \qquad -94.05 \qquad 2(-68.32)$$

ΔH^0_{298} for combustion = $(-94.05) + 2(-68.32) - (-17.89)$

$$= -212.80 \text{ kcal/mol of methane}$$

The *net heat* is the heat released if the water remains as a vapor or gas. This is the value of usual interest.

$$\underset{\Delta H^0_{298}}{CH_4(g)} + \underset{}{2O_2(g)} \rightarrow \underset{}{CO_2(g)} + \underset{}{2H_2O(g)}$$
$$\qquad -17.89 \qquad 0 \qquad -94.05 \qquad 2(-57.80)$$

ΔH^0_{298} for combustion = $(-94.05) + 2(-57.80) - (-17.89)$

$$= -191.76 \text{ kcal/mol of methane}$$

Since the values of ΔH^0_{298} for the combustions are negative, heat is given off by the combustion of methane gas.

Example B Calculate the approximate rise in solution temperature if 1 liter of 1 N H_2SO_4 is mixed with 1 liter of 1 N NaOH.

Sulfuric acid is a strong acid and sodium hydroxide is a strong base and so they are completely ionized in solution, as is the Na_2SO_4 which is formed when the above solutions are mixed. Therefore the enthalpies of the aqueous solutions are equal to the sum of the enthalpies for the individual ions.

$$\Delta H^0_{(H_2SO_4)} = 2\Delta H^0_{(H^+)} + \Delta H^0_{(SO_4^{2-})} = 2(0) + (-216.90) = -216.90$$

$$\Delta H^0_{(NaOH)} = \Delta H^0_{(Na^+)} + \Delta H^0_{(OH^-)} = (-57.28) + (-54.96) = -112.24$$

$$\Delta H^0_{(Na_2SO_4)} = 2\Delta H^0_{(Na^+)} + \Delta H^0_{(SO_4^{2-})} = 2(-57.28) + (-216.90) = -331.46$$

The neutralization reaction which occurs is

$$\tfrac{1}{2}H_2SO_4(aq) + NaOH(aq) \rightarrow H_2O(lq) + \tfrac{1}{2}Na_2SO_4(aq)$$

ΔH^0_{298} $\tfrac{1}{2}(-216.90)$ -112.24 -68.32 $\tfrac{1}{2}(-331.46)$

ΔH^0_{298} for neutralization $= (-68.32) + \tfrac{1}{2}(-331.46) - \tfrac{1}{2}(-216.90)$

$$- (-112.24) = -13.36 \, \text{kcal}$$

The final volume of the mixed solution is 2 liters, which would weigh about 2000 g. Since this solution is mainly water, the specific heat would be 1 cal/g-°C, and, from Eq. (3-1),

$$T = \frac{13\,360}{2000 \times 1} = 6.68°C$$

Thus, if the initial temperatures of the solutions were 25°C, the temperature would rise to a value somewhat over 31°C.

A knowledge of heats of reaction, as well as heats of fusion, vaporization, and specific heats is used in environmental engineering in incineration, combustion, heating of digesters, chemical handling, and thermal pollution studies.

Entropy

A large part of chemistry is concerned, in one way or another, with the state of equilibrium and the tendency of systems to move spontaneously in the direction of the equilibrium state. The concept of *entropy* was developed from the search for a thermodynamic function that would serve as a general criterion of spontaneity for physical and chemical changes. The concept of entropy is based on the second law of thermodynamics, which in essence states that all systems tend to approach a state of equilibrium. The significance of the equilibrium state is realized from the fact that work can be obtained from a system only when the system is not already at equilibrium. If a system is at equilibrium, no process tends to occur spontaneously, and no chemical or physical changes are brought about.

The chemist's interest in entropy is related to the use of this concept to indicate something about the position of equilibrium in a chemical process. Entropy is defined by the following differential equation:

$$dS = \frac{dq_{rev}}{T} \tag{3-7}$$

where S is the entropy of the system, and T is the absolute temperature. The quantity q_{rev} is the amount of heat that the system absorbs if a chemical change is brought about in an infinitely slow, reversible manner. As with enthalpy, it is the change in entropy in a system which is of usual interest, and this is evaluated as follows:

$$\Delta S = S_2 - S_1 = \int_1^2 \frac{dq_{rev}}{T} \tag{3-8}$$

On the basis of the third law of thermodynamics, the entropy of a substance at 0 K is zero. Because of this, the absolute entropy of elements and compounds at some standard state can be determined by integration of the above equation, using the initial state to represent the equilibrium state of 0 K.

The significance of entropy is that when a spontaneous change occurs in a system, it will always be found that if the total entropy change for everything involved is calculated, a positive value is obtained. Thus *all spontaneous changes in an isolated system occur with an increase of entropy*. If one wanted to determine whether a chemical or physical change from one state *a* to another state *b* could occur in a system, a calculation of entropy change would give the desired information. If ΔS for the whole system were positive, the change could occur spontaneously; if ΔS were negative, the change would tend to occur in the reverse direction, i.e., from *b* to *a*. However, if ΔS were zero, the system would be at equilibrium, and the change could not take place spontaneously in either direction.

On the molecular scale, entropy has a statistical basis. Systems tend to move from a highly ordered state to a more random state. The more highly probable or random a system becomes, the higher will be its entropy. For example, if two different ideal gases are placed together in a closed container, the gas molecules will not remain isolated, but will become randomly mixed. This spontaneous process occurs without a change in internal energy in the container, but with an increase in entropy.

Free Energy

In the original concepts of thermodynamics it was incorrectly assumed that energy given out by a reaction was a measure of the driving force. It is now seen that in an isolated system, where energy cannot be gained or lost, the entropy change is the driving force. In more general systems, such as those used in environmental engineering practice, both energy and entropy factors must be considered in order to determine what processes will occur spontaneously. For this purpose the concept of *free energy* has been developed. Free energy (G) is defined as

$$G = H - TS \tag{3-9}$$

Here H is the enthalpy expressed in calories, T is the absolute temperature in Kelvins ($°C + 273.2$), and S is the entropy expressed in calories per Kelvin.

At constant temperature and pressure, the change in free energy for a given reaction is

$$\Delta G = \Delta H - T\Delta S \qquad (T \text{ and } P = \text{constant}) \qquad (3\text{-}10)$$

From Eq. (3-5) and for $P = $ constant,

$$\Delta H = H_2 - H_1 = (E_2 + P_2 V_2) - (E_1 + P_1 V_1) = \Delta E + P\Delta V$$

Combining with Eq. (3-3) for E, we obtain $\Delta H = q - w + P\Delta V$.

From Eq. (3.8) at constant T, we see that $T\Delta S$ is equal to q_{rev}. If the system change is brought about very slowly so that energy losses are at a minimum, then q becomes q_{rev}, and w becomes w_{max}, the maximum quantity of work that could be obtained from the change.

Therefore

$$\Delta G = q_{rev} - w_{max} + P\Delta V - q_{rev}$$

and

$$-\Delta G = w_{max} - P\Delta V$$

$P\Delta V$ gives the portion of the work that must be "wasted" in expanding the system against the confining pressure. Therefore $-\Delta G$ gives the difference between the maximum work and the wasted work; in other words, it gives the *useful* work available from the system change:

$$-\Delta G = w_{useful} \qquad (T \text{ and } P = \text{constant}) \qquad (3\text{-}11)$$

In principle, any process that tends to proceed spontaneously can be made to do useful work. Since the free energy change measures the useful work that might be obtained from a constant-pressure process, it is a measure of the spontaneity of the process.

Consider a change from a to b in a constant-pressure system. If ΔG for such a change is negative, then the process can proceed; if ΔG is positive, the process can proceed, but in the reverse direction, i.e., from b to a; if ΔG is zero, the system is in equilibrium, and the process cannot proceed in either direction. This is a particularly significant relationship, and for the chemist it is one of the most important in thermodynamics.

In order for the concept of free energy to be useful, a reference point for determining free energy changes must be available. As in the case of enthalpy, a zero value is assigned to free energies of the stable form of the elements at 25°C and 1 atm pressure. In addition, the hydrogen ion at unit activity (approximately one normal solution) is assigned a standard free energy of zero.

The standard free energy of a compound (ΔG_{298}^0) is the free energy of formation of that compound from its elements, considering reactants and products all to be in the standard state at 25°C and 1 atm. A list of standard free energies of formation of various compounds is given in Table 3-1.

Free energy changes accompanying a chemical reaction can now be calculated, and the direction in which the reaction will proceed can be determined in a qualitative manner from the sign of the free energy change. It is apparent from a knowledge of equilibrium reactions that a reaction will proceed in a given direction only until the system reaches a state of equilibrium. It can be shown that free energies can also be used to determine the equilibrium state to which the reaction carries the system, as well as the reaction direction. Without going into the details, it should also be apparent that the direction of a reaction is dependent upon the concentration of reactants and products, and so this must be considered in free energy calculations. Consider the following reaction:

$$aA + bB \rightleftharpoons cC + dD$$

The free energy of this reaction, considering the concentration of the various reactants and products, is given by the following equation:

$$\Delta G = \Delta G^\circ + RT \ln \frac{\{C\}^c \{D\}^d}{\{A\}^a \{B\}^b} \qquad (3\text{-}12)$$

where ΔG = reaction free energy change in calories
ΔG^0 = standard free energy change in calories
R = universal gas constant = 1.99 cal/deg-mol
T = absolute temperature in Kelvins

The terms in the braces are the activities of the various reactants and products. The convention outlined in Sec. 2-12 for expressing activities must be followed.

The development of Eq. (3-12) is beyond the scope of this book, but is given in most books on physical chemistry. This equation is important, as it allows the prediction of reaction direction for any activity (or concentration, approximately) of products and reactants of interest. If a reaction is allowed to proceed to a state of equilibrium, it will reach a position for which no further driving force is operative. At this point the free energy change (ΔG) will be zero, and thus

$$\Delta G^0 = -RT \ln \left(\frac{\{C\}^c \{D\}^d}{\{A\}^a \{B\}^b} \right)_{equilibrium} \qquad (3\text{-}13)$$

The subscript "equilibrium" is to indicate that this equation is true only when the system is at equilibrium. It should now be noted that at equilibrium the ratio of product concentrations to reactant concentrations is equal to the equilibrium constant:

$$\frac{\{C\}^c \{D\}^d}{\{A\}^a \{B\}^b} = K$$

Thus Eq. (3-13) can be written in the following form:

$$\Delta G^0 = -RT \ln K \qquad (3\text{-}14)$$

This equation is one of the most important results of thermodynamics. It shows the relationship between the equilibrium constant of a reaction and its thermochemical properties. By use of this equation, the equilibrium constant for a reaction can be calculated from thermochemical properties of the reactants and products which may have been determined from entirely different reactions.

Example A Determine the first ionization constant for carbonic acid at 25°C from free energy considerations.

The equation and free energies for the first ionization are

$$H_2CO_3(aq) \rightleftharpoons H^+(aq) + HCO_3^-(aq)$$

$$\Delta G^0_{29_8} = \underset{-149.00}{} \qquad \underset{0}{} \qquad \underset{-140.31}{}$$

$$\Delta G^0 \text{ for reaction} = (-140.31) + (0) - (-149.00) = 8.69 \text{ kcal}$$

From Eq. (3-14), $\ln K_1 = -\Delta G^0/RT = -8690/1.99(298) = -14.65$

Therefore

$$K_1 = 4.35 \times 10^{-7}$$

So

$$\frac{[H^+][HCO_3^-]}{[H_2CO_3]} = 4.35 \times 10^{-7}$$

Example B Determine the solubility constant for carbon dioxide gas in water at 25°C from free energy considerations.

The equation and free energies of interest are

$$CO_2(g) \rightleftharpoons CO_2(aq)$$

$$\Delta G^0_{29_8} = \underset{-94.26}{} \qquad \underset{-92.31}{}$$

$$\Delta G^0 \text{ for reaction} = (-92.31) - (-94.26) = 1.95 \text{ kcal}$$

From Eq. (3-14), $\ln K_{sol} = -1950/1.99(298) = -3.29$

Therefore

$$K_{sol} = 0.0373$$

So

$$\frac{CO_2(aq) \text{ in moles/liter}}{CO_2(g) \text{ in atmospheres}} = 0.0373$$

Thus the Henry's law constant discussed in Sec. 2-9 can be determined from free energy calculations.

Example C Calculate the solubility product for calcium fluoride at 25°C from free energy considerations.

The equation and free energies for this solubility are

$$CaF_2(c) \rightleftharpoons Ca^{2+}(aq) + 2F^-(aq)$$

$$\Delta G^0_{29_8} = \underset{-277.7}{} \qquad \underset{-132.18}{\phantom{Ca^{2+}}} \qquad \underset{2(-66.08)}{}$$

$$\Delta G^0 \text{ for reaction} = (-132.18) + 2(-66.08) - (-277.7) = 13.4 \text{ kcal}$$

From Eq. (3-14), $\ln K_{sp} = -13,400/1.99(298) = -22.6$

Therefore

$$K_{sp} = 1.6 \times 10^{-10}$$

And so, on the basis of these calculations,

$$[Ca^{2+}][F^+]^2 = 1.6 \times 10^{-10}$$

Temperature Dependence of Equilibrium Constant

From a consideration of the relationships between free energy and the equilibrium constant and between free energy and enthalpy, a thermodynamic basis for predicting the change in equilibrium constant with temperature can be obtained. In differential form, this relationship is

$$\frac{d \ln K}{dT} = \frac{\Delta H^0}{RT^2} \tag{3-15}$$

This equation indicates that for exothermic reactions the equilibrium constant decreases with increasing temperature, while for endothermic reactions it increases. Over the rather limited temperature range of interest in environmental engineering, ΔH^0 is normally constant enough that the integrated form can be used:

$$\ln \frac{K_{(T_2)}}{K_{(T_1)}} = - \frac{\Delta H^0}{R} \left(\frac{T_1 - T_2}{T_1 T_2} \right) \tag{3-16}$$

Thus, if the equilibrium constant for a reaction is known for one temperature, the value for another temperature can be calculated from standard enthalpy values for the products and reactants of the reaction.

Example Calculate the ionization constant, K_1, at 10°C for carbonic acid, assuming $K_1 = 4.35 \times 10^{-7}$ at 25°C as calculated in Example A above. The equation and enthalpies of formation for this reaction are

$$H_2CO_3(aq) \rightleftharpoons H^+(aq) + HCO_3^-(aq)$$
$$\Delta H^0_{298} = \quad -167.0 \qquad\qquad 0 \qquad -165.18$$

$$\Delta H^0 \text{ for reaction} = (-165.18) - (-167.0) = 1.8 \text{ kcal}$$

Using Eq. (3-16), we obtain

$$\ln \frac{K_{T(2)}}{K_{T(1)}} = \frac{-1800}{1.99} \left[\frac{298 - 283}{(298)(283)} \right] = -0.161$$

Therefore $K_{(T_2)} = 0.85(4.35 \times 10^{-7}) = 3.7 \times 10^{-7}$

3-3 VAPOR PRESSURE OF LIQUIDS

According to the kinetic theory, liquids as well as gases are in constant agitation, and molecules are constantly flying from the surface of the liquid into the atmosphere above. In open systems most of these particles never return, and the liquid is said to be undergoing evaporation. In a closed system, however, particles return to the liquid phase in proportion to their concentration in the gaseous phase. Eventually the rate of return equals the rate of flight, and a condition of equilibrium is established. The vapor is then said to be saturated. The pressure exerted by the vapor under these conditions is known as the *vapor pressure*. The vapor pressure of all liquids increases with temperature. The vapor-pressure values of water and a few organic liquids are given in Table 3-2. It will be noted that vapor pressures do not rise in a regular manner. For rough approximations, the vapor pressure may be considered to increase about 1.5 times for each 10°C rise in temperature.

The Rankine formula,

$$\log p = \frac{A}{T} + B \log T + C \tag{3-17}$$

or one of its modifications is commonly used to calculate vapor pressures. Environmental engineers will find vapor-pressure data for many of the compounds with which they are concerned in the "International Critical Tables" or

Table 3-2 Vapor pressure of liquids, mm Hg

Temp., °C	Water	Ethyl alcohol	n-Hexane	Benzene
0	4.58	12.2	45.4	26.5
10	9.21	23.6	75.0	45.4
20	17.54	43.9	120.0	74.7
30	31.82	78.8	185.4	118.2
40	55.32	135.3	276.7	181.1
50	92.51	222.2	400.9	269.0
60	149.4	352.7	566.2	388.6
70	233.7	542.5	787.0	547.4
80	355.1	812	1062	735.6
90	525.8	1187	1407	1016
100	760.0	1693	1836	1344

standard handbooks of chemistry. Appropriate formulas and constants are usually given to allow calculation of vapor-pressure values for any temperature.

When the vapor pressure of a liquid becomes equal to the pressure of the atmosphere above it, the liquid is said to have reached its *boiling point*. Violent agitation of the liquid occurs under these conditions as a result of the transformation of liquid to gas at the source of heat, migration of the bubbles of vapor through the liquid, and their escape from the liquid surface.

Liquids with appreciable vapor pressure may be caused to boil over a wide range of temperatures by decreasing or increasing the pressure. Water boils at room temperature if the pressure above it is reduced to about 0.4 psia. On the other hand, the boiling point of water in a steam boiler operating at 200 psia is 194.4°C. Environmental engineers make particular application of this principle when they employ the wet oxidation process for combustion of organic sludges. In this process part of the organic matter is chemically oxidized in an aqueous phase by dissolved oxygen in a specially designed reactor in which the water temperature is elevated to between 250 and 300°C. To maintain a temperature in this range without boiling the water requires pressures between 577 and 1246 psia, respectively. The oxidizability of sludge solids, as well as the oxidation rate, increases markedly with temperature. However, the maximum temperature that can be used in such a reactor is set by the *critical temperature* above which water can no longer exist in a liquid phase, regardless of pressure. This temperature is 374°C. The *critical pressure* that suffices to keep water in liquid form just below this critical temperature is 3200 psia.

3-4 SURFACE TENSION

According to the kinetic theory, molecules of a liquid attract each other. At the surface, the molecules are subjected to an unbalanced force, since the molecules in the gaseous phase are so widely dispersed. As a result, the molecules at the *surface* are under *tension* and form a thin skinlike layer that adjusts itself to give a minimum surface area. This property of surface tension causes liquid droplets to assume a spherical shape; water to rise in a capillary tube; and liquids, such as water, to move through porous materials that they are capable of wetting. The movement of water through soils is an excellent example.

Surface tension may be most accurately determined by measuring the height to which a liquid will rise in a capillary tube. Most liquids, like water, wet the walls of a glass tube and the liquid adhering to the walls pulls liquid up into the tube to decrease the total surface area in relation to its surface tension. This is the basis of capillary action so important in supplying water and nutrients to plant and animal tissues. Under static conditions, such as occur in a glass tube used to measure surface tension, the opposing forces are equal. The downward force may be expressed as $\pi r^2 h \rho g$, and the upward force as $2\pi r \gamma \cos \theta$.

$$\pi r^2 h \rho g = 2\pi r \gamma \cos \theta \tag{3-18}$$

or
$$\gamma = \frac{h \rho g r}{2 \cos \theta} \tag{3-19}$$

In Eqs. (3-18) and (3-19), γ is the surface tension in dynes per centimeter (mN/m) when the height h and the radius r are expressed in centimeters. Rho (ρ) is the density of the liquid, g is the acceleration due to gravity in cm/sec^2, and θ is the angle of contact that the liquid makes with the wall of the capillary tube. For water and for many other liquids, θ is so small that $\cos \theta$ may be considered equal to 1. Then

$$\gamma = \frac{1}{2} h \rho g r \qquad \text{or} \qquad h = \frac{2\gamma}{\rho g} \frac{1}{r} \tag{3-20}$$

and the relationship between the height to which a liquid will rise in a capillary and its radius is readily apparent. The capillaries in the giant sequoia trees, which reach a height of over 300 ft, must be extremely small.

The surface tension of liquids is commonly measured by means of the *Du Nuoy tensiometer* (Fig. 3-1). This method employs a platinum ring, and the force required to pull the ring through the surface film is measured. The method is satisfactory for most measurements needed in environmental engineering practice.

Poiseuille's Law

The behavior of liquids when flowing through capillary tubes, in relation to their viscosity, was studied by Poiseuille. He summarized his findings in the equation

$$\mu = \frac{\pi P r^4}{8 V l} t \tag{3-21}$$

where V denotes the volume of liquid of viscosity μ, flowing through a capillary tube of length l and radius r, in time t, and under pressure P.

Figure 3-1 The Du Nuoy tensiometer used for surface-tension measurements. *(Central Scientific Co.)*

Environmental engineers are often confronted with problems involving the flow of liquids through capillaries. A notable example is in the filtration of sludge in which the void areas are considered as tortuous capillaries. Since water of a uniform viscosity is the liquid to be removed, and the volume or rate of movement is of interest, *Poiseuille's equation* is important and is usually written

$$V = \frac{\pi P r^4}{8\mu l} \tag{3-22}$$

where P is pressure, r is the radius of the capillaries, μ is the viscosity of the liquid, and l is the length of the capillaries. The value of t is usually dropped, as unit time is understood. From this equation, the importance of the diameter of the capillaries is immediately apparent as being the principal factor in determining the pressure differential needed to maintain a constant liquid volume flow. A knowledge of Poiseuille's law is helpful in explaining how filter aids and chemical-conditioning agents such as lime are beneficial in filtration operations and as a basic concept for planning research on filtration or related problems.

3-5 BINARY MIXTURES

Binary mixtures of miscible liquids such as water and ethanol are of interest because of the differences in vapor pressure which they exhibit and the influence which vapor pressure has upon their separation by distillation. All mixtures fall into one of three classes, and their properties are considerably different.

Class I

Class I includes all mixtures whose vapor pressure, regardless of the composition of the mixture, is always less than that of the most volatile component and always more than that of the least volatile component; consequently the boiling point of Class I mixtures is always between those of the two components. The composition of the vapor is always richer in the more volatile component than the liquid from which it distills. Such mixtures are amenable to essentially complete separation by means of fractional distillation.

A diagram showing the composition of the liquid and vapor phases, and how separation of the two components of Class I mixtures can be accomplished by fractional distillation, is given in Fig. 3-2. If a mixture of A and B having the composition represented by x is heated to its boiling point, the liquid will have a temperature corresponding to l, and the vapor produced will have a composition corresponding to v on the vapor curve. If the vapor at v is condensed, a liquid corresponding to x', much richer in B, is obtained. Redistillation of the mixture x' results in a vapor with a composition x''. Through successive condensations and evaporations, normally accomplished by a fractionating column, a distillate of essentially pure B can be obtained, and A will remain as a relatively pure residue in the still. A wide variety of compounds form Class I binary mixtures.

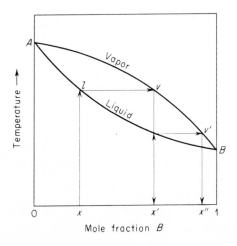

Figure 3-2 Composition of liquid and vapor phases during distillation of Class I binary mixtures.

Class II

Class II binary mixtures include those that at certain mole ratios have vapor pressures less than either of the components and, consequently, at these ratios have boiling points that are greater than that of either of the components. Upon distillation of such mixtures, one or the other of the components may be fractionated into relatively pure form until the liquid mixture reaches a composition of minimum vapor pressure or maximum boiling point. From that point on, a constant boiling mixture is obtained, the compositions of vapor and liquid are identical, and further separation is impossible by this means.

A diagram showing the composition of the liquid and vapor phases of a Class II binary mixture is given in Fig. 3-3. If a mixture corresponding to x' is distilled, the vapor formed is rich in component A and may be recovered in part in relatively pure form by fractionation. However, as A is removed from the liquid phase, the liquid phase grows richer in B until it equals the composition shown by x. At this point the composition of the vapor and liquid phases is identical, and a constant boiling mixture that cannot be fractionated results. Likewise, if a mixture with a composition represented by x'' is distilled, a distillate of B can be obtained, the liquid remaining will approach a composition equal to x, and a constant boiling mixture will result. Hydrochloric, hydrobromic, hydroiodic, hydrofluoric, nitric, and formic acid in aqueous solution are all binary mixtures of Class II. The constant boiling mixture of hydrochloric acid at 760 mm pressure contains 20.2 percent HCl and is often used as a primary standard in quantitative analysis.

Class III

Class III binary mixtures include those that at certain mole ratios have vapor pressures greater than that of either of the components, and therefore the boiling points at such mole ratios are lower than that of either component. Upon

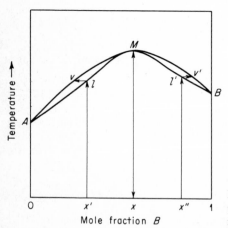

Figure 3-3 Composition of liquid and vapor phases during distillation of a Class II binary mixture.

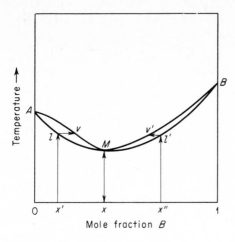

Figure 3-4 Composition of liquid and vapor phases during distillation of a Class III binary mixture.

distillation of Class III mixtures, the results are opposite to those obtained with Class II mixtures. A distillate is obtained that contains both components in a constant ratio, and the residue remaining in the flask consists of one or the other component in pure form.

A diagram showing the composition of the liquid and vapor phases of a Class III binary mixture is given in Fig. 3-4. A mixture corresponding to x' will produce a vapor with composition v. Fractionation of this vapor will produce a distillate with composition x, and the liquid phase will grow richer in component A. Distillation of a mixture corresponding to x'' will produce a vapor with a composition v'. Fractionation will yield a distillate with composition x, and the liquid will become richer in component B. Eventually either pure A or pure B will remain in the liquid phase. Ethyl alcohol and water form a binary mixture of this class. The distillate, regardless of the composition of the original mixture, will always contain 95.6 percent of alcohol at 760 mm pressure as long as both components are present in the liquid phase.

3-6 SOLUTIONS OF SOLIDS IN LIQUIDS

The amount of a solid that will dissolve in a liquid is a function of the temperature, the nature of the solvent, and the nature of the solute. The broad concepts of unsaturated, saturated, and supersaturated solutions are generally treated quite adequately in courses in general science and chemistry. However, the significance of crystal or particle size and the influence of temperature on solubility are not always adequately discussed.

Significance of Particle Size

The solubility of solids has been shown to increase as particle size diminishes. For example, coarse granular $CaSO_4$ dissolves to the extent of 2.08 g/l at 25°C, whereas finely divided $CaSO_4$ dissolves to the extent of 2.54 g/l. This phenomenon is considered to be due to the increased ratio of surface area to mass and to an increase in vapor pressure of the solid as particle size decreases. Particles of colloidal size are considered to have the greatest solubility because of their submicroscopic size.

In quantitative analysis, advantage is often taken of the fact that solubility varies with particle size. In gravimetric analysis, such as in the determination of sulfate by precipitation as $BaSO_4$ the precipitate first formed is highly colloidal in nature. However, if some care is used in the precipitation procedure, a few crystals of larger size will be formed. If the precipitated material is allowed to stand for a period of time before filtration, the colloidal-size particles will pass into solution and precipitate out on the large crystals present. The rate of transfer is a function of the number of crystals present, the differential in solubilities, and the temperature. The difference in solubility is known to increase with temperature, as well as the rate of exchange; consequently "digestion" of precipitates is normally done at temperatures near the boiling point of the solvent.

Temperature Relationships

In general, the solubility of solids in liquids increases as the temperature increases. There are a number of exceptions, however. The influence of temperature on solubility depends mainly upon the total heat effects of the solution. If the heat of solution is endothermic, the solubility increases with an increase in temperature; if the heat of solution is exothermic, the solubility decreases with

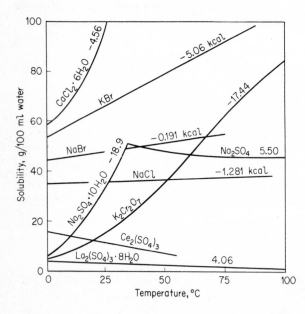

Figure 3-5 Relationship between solubility in water and heats of solution.

an increase in temperature; and if there is little thermal change, the solubility is influenced very little by change of temperature. These considerations are all in accord with Le Chatelier's principle and the thermodynamic principles discussed in Sec. 3-2.

Figure 3-5 shows solubility curves for a number of solids in water, and illustrates the relationship to heat of solution. The solubility curves for some solids, such as sodium sulfate, show abrupt changes because of a change in molecular composition and heat of solution.

3-7 OSMOSIS

Osmosis is the movement of a solvent through a membrane which is impermeable to a solute. The direction of flow is from the more dilute to the more concentrated solution. For example, if a salt solution is separated from water by means of a semipermeable membrane, as shown in Fig. 3-6, water will pass through the membrane in both directions, but it will pass more rapidly in the direction of the salt solution. As a result, a difference in hydrostatic pressure develops. The tendency for the solvent to flow can be opposed by applying pressure to the salt solution. The excess pressure that must be applied to the solution to produce equilibrium is known as the *osmotic pressure* and is denoted by π.

The net flow of solvent across a membrane results in response to a driving force which can be estimated by the difference in vapor pressure of the solvent on either side of the membrane. The transfer of solvent across the membrane from the less concentrated to the more concentrated solution will continue until the effect of hydrostatic pressure overcomes the driving force of the vapor-

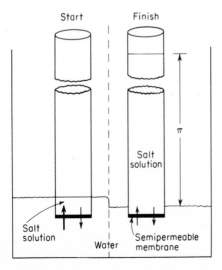

Figure 3-6 The process of osmosis and the development of osmotic pressure.

pressure differential. For an incompressible solvent, the osmotic pressure at equilibrium can be estimated from the following:

$$\pi = \frac{RT}{V_A} \ln \frac{P_A^0}{P_A} \tag{3-23}$$

Here π is expressed in atmospheres, R is 0.0882 l-atm/mol · K, T is in kelvins, and P_A^0 and P_A are the vapor pressure of solvent in the dilute and concentrated solutions, respectively. Also, V_A is the volume per mole of the solvent and is equal to 0.018 liter for water.

Raoult's law, discussed in Sec. 2-10, indicates that for dilute solutions the reduction in vapor pressure of a solvent is directly proportional to the concentration of particles in solution. From this fact, Eq. (3-23) can be rearranged to relate osmotic pressure to the molar concentration of particles, c, in the concentrated solution

$$\pi = cRT \tag{3-24}$$

This equation is valid in a strict sense only for dilute solutions in which Raoult's law holds true.

An application of osmotic pressure principles in environmental engineering is in the demineralization of salt-laden (brackish) water by the *reverse osmosis* process. As the name implies, this process is the reverse of osmosis, and water is caused to flow in a reverse manner through a semipermeable membrane from brackish to dilute fresh water. This is accomplished by exerting a pressure on the brackish water in excess of the osmotic pressure. The semipermeable membrane acts like a filter to retain the ions and particles in solution on the brackish water side, while permitting water alone to pass through the membrane. Theoretically the process will work if a pressure just in excess of the osmotic pressure is used. In practice, however, a considerably higher pressure is necessary to obtain an appreciable flow of water through the membrane. Also, as fresh water passes through the membrane, the concentration of salts in the brackish water remaining increases, creating a greater osmotic pressure differential. The theoretical minimum energy required to remove salts from water in such a process is equal to the osmotic pressure multiplied by the volume of water being demineralized.

Example. The molar concentration of the major ions in a brackish ground water supply are as follows: Na^+, 0.02; Mg^{2+}, 0.005; Ca^{2+}, 0.01; K^+, 0.001; Cl^-, 0.025; HCO_3^-, 0.001; NO_3^-, 0.002; and SO_4^{2-}, 0.012.

(*a*) What would be the approximate osmotic pressure difference across a semipermeable membrane which had brackish water on one side and mineral-free water on the other, assuming the temperature is 25°C?

The molar concentration of particles in the brackish water is

$$c = 0.02 + 0.015 + 0.01 + 0.001 + 0.025 + 0.001 + 0.002 + 0.012$$

$$= 0.075 \ M$$

From Eq. (3-24)

$$\pi = cRT = \frac{0.075 \text{ mole}}{\text{liter}} \times \frac{0.082 \text{ l-atm}}{\text{K-mole}} \times (273 + 25) \text{ K}$$

$$= 1.83 \text{ atm or } 26.9 \text{ psi}$$

(*b*) If in the above example, a yield of 75 percent fresh water were desired, what minimum pressure would be required to balance the osmotic pressure difference that will develop?

For a 75 percent yield, the salts originally present in 4 volumes of brackish water would be concentrated in one volume of brackish water left behind the membrane after three volumes of fresh water have passed through the membrane. Thus, the particle concentration in the remaining brackish water would be four times that of the original brackish water or 0.30 *M*. Then,

$$\pi = 0.30 \times 0.082 \times 298 = 7.33 \text{ atm or } 108 \text{ psi}$$

At this point the pressure required to push the fresh water through the membrane would be in excess of 108 psi.

3-8 DIALYSIS

The environmental engineer does not have much opportunity to apply the principle of osmosis in its strictest sense. However, he does make use of a related phenomenon referred to as *dialysis*. By choice of a membrane of a particular permeability, which is wetted by the solvent, it is possible to cause ions to pass through the membrane while large molecules of organic substances or colloidal particles are unable to pass. Thus a separation of solutes can be accomplished, and the term dialysis is justified.

Dialysis is used extensively to remove electrolytes from colloidal suspensions to render the latter more stable. Chemical and environmental engineers use dialysis to recover sodium hydroxide from certain industrial wastes that have become contaminated with organic substances, as shown in Fig. 3-7. In

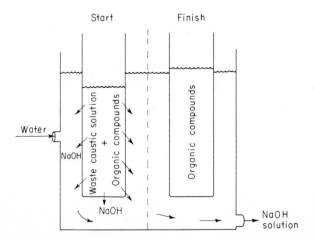

Figure 3-7 A simple dialysis cell for recovery of sodium hydroxide from an industrial waste.

the process, the waste material is placed in cells with permeable membranes, and the cells are surrounded with water. The sodium and hydroxide ions pass through the cell wall into the surrounding water. The water is evaporated to recover the sodium hydroxide, and the organic waste remaining in the cells is disposed of separately. Waste caustic solutions must be quite concentrated before recovery by dialysis can be justified economically. Mercerizing wastes of the cotton textile industry are an example.

In another application, the dialysis principle can be used for demineralization of brackish water. In this case, the brackish water is placed both inside and outside the cell. Electrodes are placed in the water outside the cell and when a current is applied, the ions within the cell are caused to flow through the semipermeable membrane and to concentrate in the water outside. The cations flow toward the cathode and the anions flow toward the anode, as discussed in Sec. 3-10. By this method, the water within the cell is demineralized. This process, termed *electrodialysis,* uses electrical energy to cause the flow of ions against a concentration gradient. In practice, a large number of thin, continuous-flow cells are used to make the process efficient for large-scale usage.

3-9 PRINCIPLES OF SOLVENT EXTRACTION

Industrial wastes often contain valuable constituents which can be recovered most effectively and economically by means of extraction with an immiscible solvent, such as petroleum ether, diethyl ether, benzene, chloroform, or some other organic solvent. Also, many methods for water analysis involve extraction of a constituent or complex from the water sample as one step in the determination. This is true of some procedures for measurement of surface-active agents and various heavy metals. Because of the importance of this operation in environmental engineering practice, a discussion of the principles involved is merited.

When an aqueous solution is intimately mixed with an immiscible solvent, the solutes contained in the water distribute themselves in relation to their solubilities in the two solvents. For low to moderate concentrations of solute, the ratio of distribution is always the same:

$$\frac{C_{\text{solvent}}}{C_{\text{water}}} = \frac{C_s}{C_w} = K \tag{3-25}$$

The equilibrium constant K, or the ratio of distribution, is known as the *distribution coefficient.* In actual practice the immiscible solvent is selected for its ability to dissolve the desired material, and the values for K are normally greater than 1.

If the volume of solvent used is equal to the volume of the sample being extracted, the mathematics involved is rather simple. For a system with a distribution coefficient of 9, 90 percent of the material would be extracted in the first step, and 90 percent of the material remaining in each successive step. After three extractions with fresh solvent, 99.9 percent of the material would be removed.

In actual practice it is seldom feasible to use a volume of solvent equal to the waste volume, and calculations become somewhat involved. The question in industrial waste treatment which usually requires answering is this: How much remains in the aqueous phase after n extractions? The expression defining the distribution coefficient may be written in terms of the amounts of the substance extracted and the volumes of the liquids involved,

$$K = \frac{C_s}{C_w} = \frac{(W_0 - W_1)/V_s}{W_1/V_w} \tag{3-26}$$

where W_0 is the weight of the substance originally present in the aqueous phase, W_1 is the weight remaining in the water after one extraction, and V_s and V_w are the volumes of solvent and water, respectively. Simplifying, we obtain

$$W_1 = W_0 \frac{V_w}{KV_s + V_w} \tag{3-27}$$

In the second step of the extraction,

$$W_2 = W_1 \frac{V_w}{KV_s + V_w} \tag{3-28}$$

or, in terms of the original sample,

$$W_2 = W_0 \frac{V_w}{KV_s + V_w} \frac{V_w}{KV_s + V_w}$$

$$= W_0 \left(\frac{V_w}{KV_s + V_w}\right)^2 \tag{3-29}$$

and after n extractions the weight of substance remaining in the water is

$$W_n = W_0 \left(\frac{V_w}{KV_s + V_w}\right)^n \tag{3-30}$$

Equation (3-30) has general application and may be used to calculate the volume of a solvent needed to reduce the concentration of a material in the aqueous phase to definite levels with a fixed number of extractions, or the number of extractions needed with a fixed volume of a solvent, provided that the distribution coefficient is known.

3-10 ELECTROCHEMISTRY

Electrochemistry is concerned with the relationships between electrical and chemical phenomena. A knowledge of electrochemistry has several applications in environmental engineering. It is germane to an understanding of corrosion as well as to a study of solutions of electrolytes and the phenomena occurring at electrodes immersed in such solutions. Many analytical procedures of interest to environmental engineers are based on electrochemical measurements. Also, automatic continuous stream monitors, which have wide application, use electrochemical methods to translate chemical characteristics into electrical impulses which can be recorded. A brief review of the more fundamental concepts of electrochemistry will be presented. More detailed information can be obtained from standard texts on physical chemistry.

Current Flow in Solution

An electrical current can flow through a solution of an electrolyte as well as through metallic conductors. However, there are some basic differences which are of importance and are summarized below:

Characteristics of current flow through a metal

1. Chemical properties of metal are not altered.
2. Current is carried by electrons.
3. Increased temperature increases resistance.

Characteristics of current flow through a solution

1. Chemical change occurs in the solution.
2. Current is carried by ions.
3. Increased temperature decreases resistance.
4. Resistance is normally greater than with metals.

A significant feature of current flow through a solution is that the current is carried by ions which move toward electrodes immersed in the solution. Also, a chemical change takes place in the solution at the electrodes, and this alters the chemical properties of the solution. These two phenomena, current flow or conductivity and chemical change at electrodes, will first be considered separately.

Conductivity

The conductivity of a solution is a measure of its ability to carry an electrical current, and varies both with the number and type of ions the solution contains. Conductivity can be measured in a *conductivity cell* connected to a Wheatstone bridge circuit as shown in Fig. 3-8. Such an arrangement allows measurement of the electrical resistance provided by the cell. The measurement consists of altering the variable resistance until no current flows through the detecting circuit containing the meter A. When this state of balance is achieved, the potential at D must be the same as that at E, and the resistance offered by the solution is determined by the relationship

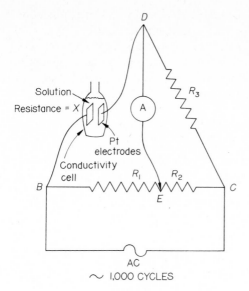

Figure 3-8 Conductivity-measuring apparatus.

$$X = R_3 \frac{R_1}{R_2} \tag{3-31}$$

Special care must be taken if this measured resistance is to be meaningful. If direct current is used, the apparent resistance changes with time, because of a *polarization* effect at the electrodes. This unwanted effect can be overcome by rapidly changing the direction of current flow. This is done by using an alternating current of several thousand cycles per second. In addition, a more reproducible state of balance is normally obtained when the platinum electrodes of the cell are coated with platinum black.

When these precautions are taken, it can be shown that the conductivity cell filled with an electrolytic solution obeys Ohm's law:

$$E = IR \tag{3-32}$$

where E is the electromotive force in volts, I is the current in amperes, and R is the resistance of the cell contents in ohms. The resistance depends upon the dimensions of the conductor:

$$R = \rho \frac{l}{A} \tag{3-33}$$

where l is the length and A the cross-sectional area of the conductor. The value ρ (ohm-cm) is called the *specific resistance* of the conductor. Our interest is normally in the *specific conductance* of a solution rather than its *specific resistance*. These quantities are reciprocally related as follows:

$$\kappa = \frac{1}{\rho} \tag{3-34}$$

where κ is the specific conductance and has units of 1/ohm-cm, which is usually referred to as mho/cm. The specific conductance can be thought of as the conductance afforded by 1 cc of a solution of electrolyte.

In practice, a conductivity cell is calibrated by determining the resistance, R_s, of a standard solution, and from this the cell constant C is determined:

$$C = \kappa_s R_s \tag{3-35}$$

Normally, 0.0100 N KCl is used as a standard solution for this calibration and has a specific conductance (κ_s) of 0.0014118 mho/cm at 25°C, or in more convenient units, 1411.8 micromhos/cm. The specific conductance of an unknown sample can be determined by measuring its resistance, R, in the cell and then using the following relationship:

$$\text{Specific conductance} = \frac{C}{R} \tag{3-36}$$

Specific conductance has a marked temperature dependence, and caution must be taken to measure the resistance of the standard and the unknown at the same temperature.

Specific conductance measurements are frequently used in water analysis to obtain a rapid estimate of the dissolved solids content of a water sample. If a flow-through cell is used and water from a river or waste stream is pumped through the cell, a continuous recording of specific conductance can be obtained. The dissolved solids content can be approximated by multiplying the specific conductance by an empirical factor varying from about 0.55 to 0.9. The proper factor to use depends upon the ionic components in the solution, as will be indicated by the introduction of a new parameter, the *equivalent conductance*, Λ, which is defined as follows:

$$\Lambda = \frac{1000}{N} \kappa \tag{3-37}$$

where N is the normality of the salt solution. For an ideal ionic solution, κ should vary directly with N, and thus Λ should remain constant with varying solution normality. However, because of deviation from ideal behavior, Λ decreases somewhat as the salt concentration increases.

Current is carried by both anions and cations of a salt, but to a different degree. The equivalent conductance of a salt is thus the sum of the equivalent ionic conductances of the cation, λ_0^+, and the anion λ_0^-:

$$\Lambda_0 = \lambda_0^+ + \lambda_0^- \tag{3-38}$$

The zero subscript is used to indicate equivalent conductance at infinite dilution, where the deviation from ideal behavior is at a minimum. Several values for equivalent ionic conductance are shown in Table 3-3. It should be noted that these values are strongly temperature-dependent. It is apparent that the equivalent ionic conductances are in general of the same order of magnitude, with the exception of the hydrogen ion and the hydroxyl ion. The latter two are more mobile than the others in aqueous solution and so can carry a larger portion of

Table 3-3 Equivalent ionic conductance at infinite dilution at 25°C in mho-cm²/equivalent*

Cation	λ_0^+	Anion	λ_0^-
H^+	349.8	OH^-	198.0
Na^+	50.1	HCO_3^-	44.5
K^+	73.5	Cl^-	76.3
NH_4^+	73.4	NO_3^-	71.4
$\frac{1}{2}Ca^{2+}$	59.5	CH_3COO^-	40.9
$\frac{1}{2}Mg^{2+}$	53.1	$\frac{1}{2}SO_4^{2-}$	79.8

* D. A. MacInnes, "The Principles of Electrochemistry," Reinhold, New York, 1939.

the current. This fact should be considered when estimating the solids concentration from conductivity measurements of solutions with either a high or a low pH. If the approximate chemical composition of a water solution is known, the equivalent ionic conductance values will allow a better choice of the appropriate factor for conversion from conductance to dissolved solids concentration.

Another important point is that only ions can carry a current. Thus the un-ionized species of weak acids or bases will not carry a current, although they are a portion of the total dissolved solids in a water sample. Also, uncharged soluble organic materials, such as ethyl alcohol and glucose, cannot carry a current and so are not measured by conductance.

Example The specific conductance (κ) of a sodium chloride solution at 25°C is 125×10^{-6} mho/cm. What is the approximate concentration of sodium chloride in mg/l?

The equivalent conductance for sodium chloride, if we use Eq. (3-38) and the values from Table 3-3, is

$$\Lambda_0 = \lambda_0^+ + \lambda_0^- = 50.1 + 76.3 = 126.4 \text{ mho-cm}^2/\text{equiv}$$

The approximate normality of the solution, if we use Eq. (3-37), is

$$N \cong 1000(\kappa)/\Lambda_0 = 1000(125 \times 10^{-6})/126.4 = 0.99 \times 10^{-3} \text{ equiv/l.}$$

The equivalent weight of NaCl is 58.5, and so

$$\text{NaCl concentration} \cong (0.99 \times 10^{-3})(58.5 \times 10^3) = 58 \text{ mg/l.}$$

Current and Chemical Change

When electrodes are introduced into a water solution in such a way as to allow a direct current to flow through the solution, a chemical change will take place at the electrodes. The nature of the chemical change depends upon the composition of the solution, the nature of the electrodes, and the magnitude of the imposed electromotive force.

Consider first the chemical changes occurring when platinum electrodes are introduced into a solution of HCl as indicated in Fig. 3-9. When a voltage of

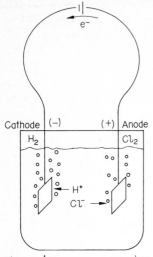

$H^+ + e^- \rightarrow \frac{1}{2}H_2$ $Cl^- \rightarrow \frac{1}{2}Cl_2 + e^-$ **Figure 3-9** Electrolysis of a hydrochloric acid solution.

about 1.3 volts is applied, it is found that H_2 is evolved at the cathode and Cl_2 at the anode. The current movement is as follows: Electrons flow through the external metallic conductor in the direction shown, as a result of the driving force of the battery. Such a flow maintains the negative charge at the cathode and the positive charge at the anode. The flow of current through the solution is maintained by movement of the cations (H^+) to the negatively charged cathode and anions (Cl^-) to the positively charged anode.

When a H^+ ion reaches the cathode, it picks up an electron and is reduced to H_2 gas, according to the half reaction

$$H^+ + e^- \rightarrow \tfrac{1}{2}H_2 \tag{3-39}$$

When the Cl^- ion reaches the anode, it gives up an electron and is oxidized to Cl_2 gas by the following half reaction:

$$Cl^- \rightarrow \tfrac{1}{2}Cl_2 + e^- \tag{3-40}$$

The electrons released by the Cl^- ions are "pumped" through the external circuit by the driving force of the battery to be picked up by the H^+ ions at the cathode. Thus the battery acts as a driving force to keep the current flowing and the reaction going. As indicated above, *reduction* takes place at the cathode and *oxidation* at the anode. The overall chemical change which takes place in the solution is as follows:

Reduction at cathode	$H^+ + e^- \rightarrow \tfrac{1}{2}H_2$
Oxidation at anode	$Cl^- \rightarrow \tfrac{1}{2}Cl_2 + e^-$
Net change	$H^+ + Cl^- \rightarrow \tfrac{1}{2}H_2 + \tfrac{1}{2}Cl_2$

The flow of electrons in the external circuit is necessary to bring about the chemical change. It is apparent that in order to bring about an equivalent of chemical change at an electrode, an Avogadro's number of electrons must flow through the external circuit. This quantity of electrons is called the faraday, F. The rate of flow of electrons gives the current, I, which is normally measured in amperes. One faraday is equivalent to an ampere of current flowing for 96,500 seconds. An ampere is also defined as a coulomb per second, so that a faraday is equivalent to 96,500 coulombs.

Example If a current is passed through a sodium chloride solution, hydrogen gas is evolved at the cathode and chlorine gas at the anode according to the following equations:

$$\text{Cathode} \qquad H_2O + e^- \rightarrow OH^- + \tfrac{1}{2}H_2$$
$$\text{Anode} \qquad \underline{Cl^- \rightarrow \tfrac{1}{2}Cl_2 + e^-}$$
$$\text{Net} \qquad H_2O + Cl^- \rightarrow OH^- + \tfrac{1}{2}H_2 + \tfrac{1}{2}Cl_2$$

How many grams of chlorine are produced if a 0.2-A current is passed through the solution for 24 h?

The amount of current flowing in this time period is

$$(0.2 \text{ coulomb/sec})(24 \times 60 \times 60 \text{ sec}) = 17,300 \text{ coulombs}$$

Thus the equivalents of chemical change taking place is 17,300/96,500 or 0.179 equivalent. Since the equivalent weight of chlorine is 35.5,

$$\text{Chlorine formed} = 0.179(35.5) = 6.35 \text{ g}$$

Types of Electrodes

As indicated previously, the nature of a chemical change occurring at an electrode is partially dependent upon the type of electrode used. A single electrode dipping into a solution is said to constitute a *half-cell;* the combination of two half-cells as indicated in Fig. 3-9 is a typical electrochemical cell. Some of the more common half-cell systems are described below.

Gas electrode A gas electrode consists of a strip of nonreactive metal, such as platinum or gold, in contact with both the solution and a gas stream. Both electrodes in Fig. 3-9 are of this type, and the reactions indicated by Eqs. (3-39) and (3-40) are typical of gas electrodes. One gas electrode of particular importance is the *hydrogen electrode,* which is used as the standard to which the potentials of other electrodes are related. It consists of a strip of sheet platinum coated with platinum black, immersed in a solution that is $1.0N$ with respect to hydrogen ions and bathed with a stream of hydrogen gas under 1 atm of pressure. The reaction occurring at this electrode is that given by Eq. (3-39). The hydrogen electrode is assigned a value of zero, and potentials related to it are designated by the prefix E_H. The hydrogen electrode is cumbersome to use, and in practice other electrodes are commonly used for reference.

In shorthand notation the hydrogen electrode can be described as follows:

$$Pt \,|\, H_2 \,(P \text{ atm}) \,|\, HCl(c \text{ mol/l})$$

The vertical lines separate the different phases of the electrode, and the pressure of the gas and composition of the solution can be included as indicated.

Metal electrode A metal electrode consists of a metal in contact with its ions in solution as indicated in Fig. 3-10. An example would be an iron wire immersed in water. The notation for this electrode would be $Fe | Fe^{2+} (c)$, and the half-cell reaction at the electrode would be $\frac{1}{2}Fe \rightleftharpoons \frac{1}{2}Fe^{2+} + e^-$. Such an electrode would be simulated by an iron pipe in contact with water, and the reactions which occur are of great interest in corrosion control.

Oxidation-reduction electrode This electrode consists of a nonreactive electrode immersed in a solution of ions in both reduced and oxidized form. An example would be a platinum wire immersed in a solution containing both ferrous and ferric chloride. Such an electrode could be designated as $Pt | FeCl_2(c)$, $FeCl_3(c)$. The reaction occurring at this electrode would be $Fe^{2+} \rightleftharpoons Fe^{3+} + e^-$. Sometimes oxidation-reduction electrodes are used as internal indicators to show the stoichiometric end point during the oxidation-reduction type of titrations, as discussed in Sec. 10-4. They are also of interest as a means of showing the conditions which exist in biological systems, although here measurements may be somewhat empirical because of the complexity of the system.

Electrode with metal contacting slightly soluble salt This type of electrode consists of a metal in contact with one of its slightly soluble salts, while the salt, in turn, is in contact with a solution containing a common anion. The example of greatest interest is the calomel electrode, which is used extensively as a reference electrode. The elements of a calomel electrode are shown in Fig. 3-11. The electrode contains mercury in contact with the slightly soluble Hg_2Cl_2, which in turn is in contact with a solution of KCl; $Hg | Hg_2Cl_2 | KCl(c)$. The reaction that occurs at this electrode is $Hg + Cl^- \rightleftharpoons \frac{1}{2}Hg_2Cl_2 + e^-$. The calomel electrode is the reference electrode used in electrometric pH determinations, in oxidation-

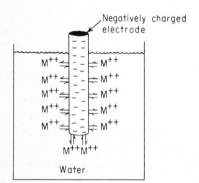

Figure 3-10 Development of single-electrode potential on a metal electrode immersed in water.

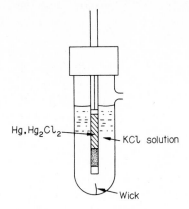

Hg, Hg$_2$Cl$_2$

KCl solution

Wick

Figure 3-11 Calomel reference electrode.

reduction measurements, and in most other electrochemical analyses for which a stable easy-to-use reference electrode is desired.

Calomel electrodes are of three types, normal, tenth normal, and saturated, depending on the concentration of KCl solution used in preparing them. The potential that each develops is a function of the concentration of the potassium chloride solution used. The potential of each with respect to the hydrogen electrode is shown in Table 3-4. From the data listed, it becomes apparent that knowledge of the concentration of KCl used in preparing a calomel reference electrode is necessary for proper interpretation of information gathered with its aid. In general, the saturated type of calomel electrode is employed, and its use is so common that results are often reported in terms of E_{cal} instead of E_H.

Table 3-4 Standard potential of calomel reference electrodes at 25°C

Concentration of KCl	E_H, volts
0.1 N	−0.334
1.0 N	−0.281
Saturated	−0.242

Other electrodes Other special-purpose electrodes have been designed for specific analytical procedures. The major ones of importance are the glass electrode for pH measurements, the membrane-covered electrode for dissolved oxygen measurements, and specific ion electrodes for measurement of the concentration of ions of interest in water quality such as sodium, calcium, ammonium, chloride, fluoride, and nitrate. These are discussed in the chapter on Instrumental Methods of Analysis.

Electrochemical Cell

When two half-cells are connected so that ions can pass between them, an electrochemical or galvanic cell is obtained. The electrochemical cell shown in Fig. 3-9 is quite simple, as only one solution is involved. However, in most electrochemical analyses the solutions associated with each half-cell are different, and they must be kept from mixing. In such a case, some type of salt bridge is used which allows passage of ions while keeping interdiffusion of the solutions to a minimum (see Fig. 3-12). As indicated in Fig. 3-11, the calomel electrode has a small pinhole opening or wick so that a small quantity of ions can diffuse into or out of another electrode solution in which it might be placed.

If the two electrodes of an electrochemical cell are connected through a metallic conductor, electrons will flow through the external circuit, and a chemical change will begin to take place in the solutions. If a voltmeter is connected across the half-cells as indicated in Fig. 3-12, it will be found that electromotive force is being generated by the cell. This emf is a measure of the driving force of the chemical reaction that is occurring in the half-cell solutions. Thus it gives a measure of the chemical potential or free energy of the reaction. From this fact, a relationship between electrical potential and chemical free energy can be found. Electrical energy is measured in terms of the *joule,* which is the energy generated by the flow of one ampere in one second against an emf of one volt. Electrical energy is given by the product, EIt and has units of the volt coulomb.

The electrical energy expended in bringing about one mole of chemical change is zEF, where z is the number of equivalents per mole, F is the faraday or coulombs per equivalent, and E is the emf of the cell in volts. By convention, if the reaction proceeds, E is positive, so that the relation between free energy and electrical energy is

$$\Delta G = -zFE \qquad (3\text{-}41)$$

Consider the following chemical reaction:

$$aA + bB \rightleftharpoons cC + dD$$

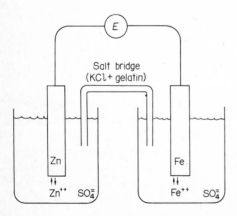

Figure 3-12 An electrochemical cell.

If we substitute the relationship for E from Eq. (3-41) into the free-energy Eq. (3-12), the following relationship between cell emf and concentration of reactants and products results:

$$E = E^0 - \frac{RT}{zF} \ln \frac{\{C\}^c\{D\}^d}{\{A\}^a\{B\}^b}$$ (3-42)

Here the value of the gas constant R, in electrical units, is 8.314 J/K·mol. This important equation indicates the relationship between the standard electrode potential of a cell and the activities of the products and reactants. When the activities of products and reactants are unity, the logarithmic term is unity, and $E = E^0$ (that is, the emf for the standard state).

The emf of a cell can be calculated from tabulated values just like free-energy and enthalpy values. Such tables list the standard potentials of various half-cells with respect to the standard hydrogen electrode, which is assigned by convention the value $E^0 = 0$. By taking the difference between the standard potentials of two half-cells, the potential of the whole cell can be determined. Standard potentials for various half-cells of interest are listed in Table 3-5.

If an electrochemical cell has reached a state of equilibrium, no current can flow, and the emf of the cell is zero. For this case a relationship between standard-cell potential and the equilibrium constant for the reaction can be obtained by using Eqs. (3-14) and (3-41):

Table 3-5 Standard electrode potentials in water at 25°C*

Half-cell reaction	E^0, volts
$O_2 + 4H^+ + 4e^- \rightarrow 2H_2O$	1.229
$Ag^+ + e^- \rightarrow Ag$	0.799
$Fe^{3+} + e^- \rightarrow Fe^{2+}$	0.771
$Ag_2CrO_4 + 2e^- \rightarrow 2Ag + CrO_4^{2-}$	0.446
$Cu^{2+} + 2e^- \rightarrow Cu$	0.337
$AgCl + e^- \rightarrow Ag + Cl^-$	0.222
$S + 2H^+ + 2e^- \rightarrow H_2S$	0.141
$2H^+ + 2e^- \rightarrow H_2$	0.000
$Pb^{2+} + 2e^- \rightarrow Pb$	−0.126
$Sn^{2+} + 2e^- \rightarrow Sn$	−0.136
$Fe^{2+} + 2e^- \rightarrow Fe$	−0.441
$S + 2e^- \rightarrow S^{2-}$	−0.48
$Zn^{2+} + 2e^- \rightarrow Zn$	−0.763
$Zn(OH)_2 + 2e^- \rightarrow Zn + 2OH^-$	−1.245
$ZnS + 2e^- \rightarrow Zn + S^{2-}$	−1.44
$Al^{3+} + 3e^- \rightarrow Al$	−1.66
$Mg^{2+} + 2e^- \rightarrow Mg$	−2.37
$Mg(OH)_2 + 2e^- \rightarrow Mg + 2OH^-$	−2.69

* W. M. Latimer, "Oxidation Potentials," 2d ed., Prentice-Hall, Englewood Cliffs, N.J., 1952.

$$E^0 = -\frac{\Delta G^0}{zF} = \frac{RT}{zF} \ln K \tag{3-43}$$

Example Determine the solubility product constant at 25°C for silver chloride, using standard electrode potentials.

The equilibrium of interest is

$$AgCl \rightleftharpoons Ag^+ + Cl^-$$

and

$$[Ag^+][Cl^-] = K_{sp}$$

A cell consisting of the following two half-cells from Table 3-5 will produce the overall reaction of interest:

		E^0
		Volts
	$Ag \rightarrow Ag^+ + e^-$	-0.799
	$AgCl + e^- \rightarrow Ag + Cl^-$	0.222
Net	$AgCl \rightarrow Ag^+ + Cl^-$	-0.577

Thus E^0 for the net reaction as written is -0.577 volt, and K_{sp} can be determined by using Eq. (3-43):

$$E^0 = \frac{RT}{zF} \ln K_{sp}$$

$$-0.577 = \frac{8.314(273 + 25)}{(1)(96,500)} \ln K_{sp}$$

$$\ln K_{sp} = -22.5 \quad \text{and} \quad K_{sp} = 1.7 \times 10^{-10}$$

Galvanic Protection

If a zinc metal electrode is connected to an iron metal electrode by means of a conducting salt bridge and an external metallic conductor as indicated in Fig. 3-12, a cell will result. The cell reaction and standard potential of the cell will be

		E^0
Zn electrode	$Zn \rightarrow Zn^{2+} + 2e^-$	0.763
Fe electrode	$Fe^{2+} + 2e^- \rightarrow Fe$	-0.441
Net	$Zn + Fe^{2+} \rightarrow Zn^{2+} + Fe$	0.322

Since the potential is positive, the reaction will proceed as written when products and reactants are near unit activity. Zinc ions will tend to pass into solution, and iron ions will tend to plate out on the iron electrode. Thus the iron is kept from passing into solution, while the zinc acts in a *sacrificial* manner. As electrochemical cells are sometimes called galvanic cells, this method of corrosion prevention is called *galvanic protection*. It is the basic principle involved in the protection of iron by galvanizing with zinc.

From this consideration, it is apparent that when two metals are in electrical contact, the metal with the greater single-electrode potential will sacrifice itself to protect the other. Since the protected electrode assumes a negative charge, it is the cathode, and this can be called *cathodic protection*. From these considerations, the engineer can explain why discontinuous coatings of tin aggravate the rusting of iron. These principles are also the basis of regulations prohibiting the joining of copper and iron pipe without the use of insulating connectors.

If a battery is placed in the external circuit connecting two half-cells, either electrode can be made to be the cathode and thus be protected. Hence electrical energy can be made to counterbalance the chemical energy of the cell, and so reverse the reaction. Such electrochemical principles are widely used for the cathodic protection of steel pipelines, tanks, and structures by means of sacrificial anodes or artificially impressed negative potentials.

3-11 CHEMICAL KINETICS

Chemical kinetics is concerned with the speed or velocity of reactions. Many reactions have rates which at a given temperature are proportional to the concentration of one, two, or more of the reactants raised to a small integral power. For example, if a reaction is considered in which A, B, and C are possible reactants, then the rate equations which express the concentration dependence of the reaction rate may take one of the following forms.

$$\text{Rate} = kC_a \qquad\qquad \text{1st order}$$
$$\text{Rate} = kC_a^2 \text{ or } kC_aC_b \qquad\qquad \text{2d order}$$
$$\text{Rate} = kC_a^3 \text{ or } kC_a^2C_b \text{ or } kC_aC_bC_c \qquad\qquad \text{3d order}$$

where C_a, C_b, and C_c represent the concentrations of reactants A, B, and C, respectively. Reactions that proceed according to such simple expressions are said to be reactions of the first, second, or third order as indicated, with the *order* of the reaction being defined as the sum of the exponents of the concentration terms in the reaction equation. Not all reactions have such simple rate equations. Some involve concentrations raised to a fractional power, while others consist of more complex algebraic expressions.

The environmental engineer deals with many reactions that proceed slowly and require rate expressions so that the reactions can be dealt with on a practical basis. This is true for BOD removal, aeration, biological growth, radioactive decay, and disinfection. In general, first-order reactions are the most common, although reactions of other orders or of a more complex nature are sometimes involved.

Zero-Order Reactions

Many biologically induced reactions, particularly those involving soluble substrates, appear to occur in a linear manner over fairly large ranges of concentrations. Thus, the rate of substrate change is independent of substrate concentration. Such reactions are referred to as being *zero order*. Of particular note to environmental engineers is the oxidation of ammonia to nitrite. All such reactions, however, become slower as the substrate concentration approaches zero, as will be discussed under Enzyme Reactions.

First-Order Reactions

The decomposition of a radioactive element is the simplest example of a true first-order reaction. In such a reaction the rate of decomposition is directly proportional to the amount of undecayed material and may be expressed mathematically as

$$- \frac{dC}{dt} = kC \tag{3-44}$$

where the minus sign indicates a loss of material with time, C is its concentration, and k is the rate constant for the reaction and has units of reciprocal time.

If the initial concentration, at time $t = 0$, is C_0, and if at some later time t the concentration has fallen to C, the integration of Eq. (3-44) gives

$$- \int_{C_0}^{C} \frac{dC}{C} = k \int_{0}^{t} dt$$

and
$$- \ln \frac{C}{C_0} = \ln \frac{C_0}{C} = kt \tag{3-45}$$

or
$$C = C_0 e^{-kt} \tag{3-46}$$

Converting to $\log_{10}$, we see that Eqs. (3-45) and (3-46) become

$$\log_{10} \frac{C_0}{C} = \frac{kt}{2.303} = k't \quad \text{and} \quad C = C_0 10^{-k_1 t} \tag{3-47}$$

From Eq, (3-47) it becomes evident that for a first-order reaction a plot of $\log_{10} (C_0/C)$ versus t will yield a straight line. This is a common way of proving whether or not a reaction is first-order. The rate constant k can be evaluated by multiplying the slope of the plotted line by 2.303.

For radioactive substances it is customary to express decomposition rates in terms of *half-life*, or the time required for the amount of substance to decrease to half its initial value. For a first-order reaction, the half-life, denoted by $t_{1/2}$, can be found from Eq. (3-45) by inserting the requirement that at $t = t_{1/2}$ the concentration of $C = \frac{1}{2}C_0$. This gives

$$t_{1/2} = \frac{\ln 2}{k} = \frac{0.693}{k} \tag{3-48}$$

There is a growing tendency to use the half-life mode of expression for a variety of phenomena in environmental engineering practice.

Environmental engineers also find application for the concepts of first-order reactions in areas that do not involve decomposition reactions. For example, the solution of oxygen in water, under a given set of conditions, is a first-order reaction. In other cases, reactions which may in fact not be true first-order reactions can be approximated as such. The rate of death of microorganisms by disinfection is frequently considered to be a first-order reaction and dependent upon the concentration of live microorganisms remaining. Also, the decomposition of organic matter by bacteria in the BOD test is normally considered to be a first-order reaction dependent only upon the concentration of organic matter remaining. In actual fact, this is a very complex reaction, and the limitations of a first-order assumption should be well understood to prevent misinterpretation of BOD data. The kinetics of the BOD test are discussed in more detail in Sec. 22-2.

Example Strontium 90 (Sr^{90}) is a radioactive nuclide of public health significance and has a half-life of 29 years. How long would a given amount of Sr^{90} need to be stored to obtain a 99.9 percent reduction in quantity? From Eq. (3-52), we have $k = 0.693/t_{1/2} = 0.693/29$ yr $= 2.39 \times 10^{-2}$ yr^{-1}. Time required is that for C to be reduced 99.9 percent to $0.001C_0$. Thus, from Eq. (3-49),

$$kt = \ln (C_0/0.001C_0) = \ln 1000$$

and $\qquad\qquad t = \ln 1000/2.39 \times 10^{-2} = 289$ years

Second–Order Reactions

A second-order reaction is one in which the rate of the reaction is proportional to the square of the concentration of one of the reactants or to the product of the concentrations of two different reactants. Thus, if the overall second-order reaction were of the form

$$A + B \rightarrow \text{products}$$

then the rate law for this situation might be

$$-\frac{dC_a}{dt} = kC_a^2 \quad \text{or} \quad = kC_aC_b \qquad (3\text{-}49)$$

where C_a and C_b are the concentrations of A and B, respectively. The decrease in B could be formulated in a similar manner:

$$-\frac{dC_b}{dt} = kC_b^2 \quad \text{or} \quad = kC_aC_b \qquad (3\text{-}50)$$

The interested student can find the integrated forms of Eqs. (3-49) and (3-50) in most textbooks on physical chemistry. However, environmental engineers have had little occasion to use second-order-reaction kinetics in practice.

Consecutive Reactions

Consecutive reactions are complex reactions of great importance in environmental engineering, and so equations describing the kinetics of such reactions are of real interest. In consecutive reactions, the products of one reaction become the reactants of a following reaction:

$$A \xrightarrow{k_1} B \xrightarrow{k_2} C \qquad (3\text{-}51)$$

Here reactant A is converted to product B at a rate determined by rate constant k_1. Product B in turn becomes the reactant for the second step and is converted to product C as determined by rate constant k_2. If the rates of each of the consecutive reactions are considered to be first-order, then the differential equations which describe the rates of decomposition and formation of the reactants and products are as follows:

$$-dC_a/dt = k_1 C_a \qquad (3\text{-}52)$$

$$dC_b/dt = k_1 C_a - k_2 C_b \qquad (3\text{-}53)$$

and

$$dC_c/dt = k_2 C_b \qquad (3\text{-}54)$$

If at $t = 0$ we have $C_a = C_a^\circ$, $C_b = C_b^\circ$, and $C_c = C_c^\circ$, then a solution for the concentration of each constituent at some time t is as follows:

$$C_a = C_a^\circ e^{-k_1 t} \qquad (3\text{-}55)$$

$$C_b = \frac{k_1 C_a^\circ}{k_2 - k_1} (e^{-k_1 t} - e^{-k_2 t}) + C_b^\circ e^{-k_2 t} \qquad (3\text{-}56)$$

$$C_c = C_a^\circ \left(1 - \frac{k_2 e^{-k_1 t} + k_1 e^{-k_2 t}}{k_2 - k_1} \right) + C_b^\circ (1 - e^{-k_2 t}) + C_c^\circ \qquad (3\text{-}57)$$

Equation (3-56) is widely used to describe the oxygen deficit in a stream caused by organic pollution. In this case C_b can be considered the oxygen deficit being created in the first step by the biological oxidation of organic matter with concentration C_a. At the same time, the oxygen deficit is being decreased by atmospheric reaeration to give the second step in the consecutive reactions.

A consecutive reaction can also be used to describe the bacterial nitrification of ammonia. Here ammonia is oxidized by *Nitrosomonas* bacteria to nitrite, which is then oxidized in the second step by *Nitrobacter* bacteria to nitrate as indicated by the following sequence:

$$NH_3 \xrightarrow[\textit{Nitrosomonas}]{O_2} NO_2^- \xrightarrow[\textit{Nitrobacter}]{O_2} NO_3^- \qquad (3\text{-}58)$$

The buildup and decay of the various forms of nitrogen in this consecutive reaction is sometimes assumed for simplicity to follow first-order kinetics. The changes in nitrogen forms which would occur with this assumption are illustrated in Fig. 3-13. The concentrations of NO_2^- and NO_3^- were set equal to zero when $t = 0$, and k_1 was assumed to equal $2k_2$. In actual fact, the kinetics for nitrification are much more complex so that one should consider the limitations of this assumption before applying these equations in practice. The changes in nitrogen forms indicated are typical of those frequently noted in wastes flowing through trickling filters or in rivers downstream from a waste discharge.

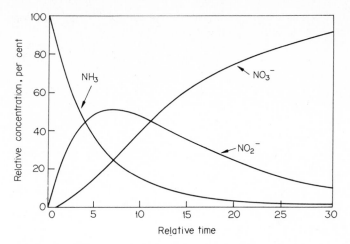

Figure 3-13 Nitrogen changes during nitrification, assuming consecutive first-order reactions.

Consecutive-type kinetics are frequently used to describe the growth and decay of microorganisms in waste treatment plants. They can also describe the consecutive steps in the decomposition of organic matter as it occurs in anaerobic waste treatment. Thus consecutive-type kinetics are widely applicable in environmental engineering practice.

Enzyme Reactions

Another complex kinetics expression to describe the rate of biological waste treatment was first used by Michaelis and Menten[1] to describe enzyme reactions. Since bacterial decomposition involves a series of enzyme-catalyzed steps, the Michaelis-Menten expression can be empirically extended to describe the kinetics of bacterial growth and waste decomposition. Figure 3-14 indicates the normally observed relationship between substrate or waste concentration, designated as S, and speed of waste utilization per unit mass of enzyme or bacteria, designated as V/E.

The Michaelis-Menten relationship for enzyme reactions in a simplified form assumes the following reaction, where E_f is free enzyme, S is substrate, and E_cS is enzyme-substrate complex:

$$E_f + S \underset{k_{-1}}{\overset{k_1}{\rightleftharpoons}} E_cS \overset{k}{\rightarrow} E_f + products \qquad (3\text{-}59)$$

E_cS is formed at rate k_1 when free enzyme and substrate combine. The complex is unstable and decomposes either back to the original free enzyme and substrate at rate k_{-1}, or into free enzyme and reaction products at rate k. The total enzyme concentration in the system, E, remains constant and is equal to

[1] Michaelis and Menton, *Biochem. Zeit.*, **49**: 333 (1913).

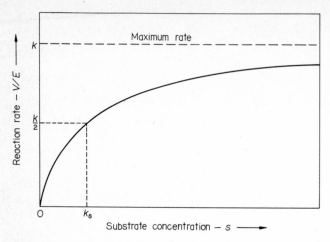

Figure 3-14 The relationship between substrate concentration and reaction rate for enzyme-type reactions.

$[E_f] + [E_cS]$. On the basis of the above relationship, the rate of formation of enzyme-substrate complex is

$$d[E_cS]/dt = k_1[E_f][S] - (k_{-1} + k)[E_cS]$$
$$= k_1S([E] - [E_cS]) - (k_{-1} + k)[E_cS] \qquad (3\text{-}60)$$

For a particular system of continuous substrate addition and enzyme concentration, the concentration of enzyme-substrate complex will reach a constant value; so $d[E_cS]/dt = 0$. Therefore

$$k_1S([E] - [E_cS]) = (k_{-1} + k)[E_cS]$$

Rearranging, we obtain

$$\frac{S([E] - [E_cS])}{[E_cS]} = \frac{(k_{-1} + k)}{k_1} = K_s \qquad (3\text{-}61)$$

or

$$[E_cS] = \frac{ES}{K_s + S} \qquad (3\text{-}62)$$

The rate of product formation is equal to the overall velocity of the reaction and is given by $V = k[E_cS]$, and thus from the relationship given by Eq. (3-62), the overall rate as a function of E and S becomes

$$V = \frac{kES}{K_s + S}$$

or

$$\frac{V}{E} = \frac{kS}{K_s + S} \qquad (3\text{-}63)$$

The significance of the constants k and K_s is indicated in Fig. 3-16. The constant k gives the maximum rate of the reaction, and K_s is equal to the substrate concentration at which the reaction rate is one-half of maximum. Two limiting cases for Eq. (3-63) are apparent:

$$V/E \cong k'S \qquad \text{when } S \ll K_s \qquad (3\text{-}64)$$

and

$$V/E \cong k \qquad \text{when } S \gg K_s \qquad (3\text{-}65)$$

Equation (3-64) indicates that when the substrate concentration is low compared to K_s, the rate of the enzyme reaction is directly proportional to S. Therefore the reaction can be described as *first-order*. However, when S is much greater than K_s, the reaction rate is a maximum and independent of the concentration S. The reaction is then said to be *zero-order*.

Both the continuous Eq. (3-67) and the discontinuous set of Eqs. (3-64) and (3-65) have been frequently used to describe biological reaction rates.[2] All are somewhat empirical when used to describe complex biological processes, but they give a sufficiently adequate description of the overall process to yield practical results.

> **Example** A study was made to evaluate the constants so that the Michaelis-Menten relationship could be used to describe waste utilization by bacteria. It was found that 1 g of bacteria could decompose the waste at a maximum rate of 20 g/day when the waste concentration was high. Also, it was found that this same quantity of bacteria would decompose waste at a rate of 10 g/day when the waste concentration surrounding the bacteria was 15 mg/l. What would be the rate of waste decomposition by 2 g of bacteria if the waste concentration were maintained at 5 mg/l?
>
> The constant k gives the maximum rate of waste utilization, and so for this case it is equal to 20 g/day-g. The constant K_s is equal to the substrate concentration at which the rate is $\frac{1}{2}$ of maximum or 10 g/day-g. Therefore, for this example, $K_s = 15$ mg/l, and from Eq. (3-63),
>
> $$\frac{V}{E} = \frac{20S}{15 + S}$$
>
> and assuming E = weight of bacteria = $2g$, and $S = 5$ mg/l,
>
> $$V = 2(20)(5)/(15 + 5) = 10 \text{ g/day}$$

Temperature Dependence of Reaction Rates

In general, the rates of most chemical and biological reactions increase with temperature. An approximate rule is that the rate of a reaction will about double for each 10°C rise in temperature. In biological reactions, this rule will hold more or less true up to a certain optimum temperature. Above this the rate decreases, probably owing to destruction of enzymes at the higher temperatures.

The change in rate constant with temperature can be expressed mathematically by the Arrhenius equation,

$$d \ln k/dT = E_a/RT^2 \qquad (3\text{-}66)$$

[2] J. Monod, *Ann. Inst. Pasteur,* 79: 390 (1950); D. Herbert, R. Elsworth, and R. C. Telling, *J. Gen. Microbiol.,* 14: 601 (1956).

where $d \ln k/dT$ represents the change in the natural log of the rate constant with temperature, R is the universal gas constant, and E_a is a constant for the reaction termed the *activation energy*. Integrating between limits gives

$$\ln \frac{k_2}{k_1} = \frac{E_a(T_2 - T_1)}{RT_2T_1} \tag{3-67}$$

where k_2 and k_1 are the rate constants at temperatures T_2 and T_1, respectively. Temperature is expressed in kelvins.

Most processes of concern to environmental engineers operate over a small temperature range near ambient temperatures. For this case the product T_2T_1 changes very little, and for practical purposes it can be considered constant. Thus E_a/RT_2T_1 can be considered equal to a constant θ, so that an approximate formula for temperature dependence of reaction rates can be used:

$$\ln k_2/k_1 = \theta(T_2 - T_1) \tag{3-68}$$

or

$$k_2 = k_1 e^{\theta(T_2 - T_1)} \tag{3-69}$$

By using the expanded form of e^x, Eq. (3-73) can also be written as

$$k_2 = k_1[1 + \theta(T_2 - T_1) + \theta^2(T_2 - T_1)^2/2 + \cdots]$$

or approximately

$$k_2 = k_1[1 + \theta(T_2 - T_1)] \tag{3-70}$$

Both Eqs. (3-69) and (3-70) are commonly used in environmental engineering to express the effect of temperature on reaction rates. Although θ is supposed to be a constant, it sometimes varies significantly even over a limited temperature range. For example, for the BOD reaction rate (see Chap. 22), θ has been indicated[3] to vary from 0.135 in the temperature range from 4° to 20°C, down to 0.056 in the temperature range from 20° to 30°C. Thus caution must be exercised in using a θ value beyond the temperature range for which it was evaluated.

3-12 CATALYSIS

Catalysts have the power to change the rate of a chemical reaction. They may be positive or negative in effect. Regardless of their actual role in the reaction, they are recoverable in their original form at the end of the reaction. It is important to remember that catalysts have no influence on the final equilibrium of a reaction. They simply alter the speed with which the equilibrium is attained by changing the energy of activation. Positive catalysts have one other property of interest to environmental engineers. They can initiate and maintain reactions at concentration levels below those at which ordinary reactions would occur.

Catalysts are used in environmental engineering for the control of air pollution. Hydrogen sulfide is catalytically oxidized to sulfur dioxide at concen-

[3] G. J. Schroepfer, M. L. Robins, and R. H. Susag, The Research Program on the Mississippi River in the Vicinity of Minneapolis and St. Paul, "Advances in Water Pollution Research," vol. I, Pergamon, London, 1964.

trations normally incapable of supporting combustion. Catalytic devices are now used to oxidize olefinic compounds in the exhaust gases of automobiles, trucks, and buses as one means of controlling smog problems. Enzymes produced by bacteria and other microorganisms are organic catalysts which permit the occurrence at room temperature of a great many reactions of importance to environmental engineers such as hydrolysis, oxidation, and reduction of both inorganic and organic pollutants.

3-13 ADSORPTION

Adsorption is the process by which ions or molecules present in one phase tend to condense and concentrate on the surface of another phase. Adsorption of contaminants present in air or water onto activated carbon is frequently used for purification of the air or water. The material being concentrated is the *adsorbate,* and the adsorbing solid is termed the *adsorbent.* There are three general types of adsorption, *physical, chemical,* and *exchange* adsorption. Physical adsorption is relatively nonspecific and is due to the operation of weak forces of attraction or van der Waals' forces between molecules. Here, the adsorbed molecule is not affixed to a particular site on the solid surface, but is free to move about over the surface. In addition, the adsorbed material may condense and form several superimposed layers on the surface of the adsorbent. Physical adsorption is generally quite reversible; i.e., with a decrease in concentration the material is desorbed to the same extent that it was originally adsorbed.

Chemical adsorption, on the other hand, is the result of much stronger forces, comparable with those leading to the formation of chemical compounds. Normally the adsorbed material forms a layer over the surface which is only one molecule thick, and the molecules are not considered free to move from one surface site to another. When the surface is covered by the monomolecular layer, the capacity of the adsorbent is essentially exhausted. Also, chemical adsorption is seldom reversible. The adsorbent must generally be heated to higher temperatures to remove the adsorbed materials.

Exchange adsorption is used to describe adsorption characterized by electrical attraction between the adsorbate and the surface. Ion exchange is included in this class. Here, ions of a substance concentrate at the surface as a result of electrostatic attraction to sites of opposite charge on the surface. In general, ions with greater charge, such as trivalent ions, are attracted more strongly toward a site of opposite charge than are molecules with lesser charge, such as monovalent ions. Also, the smaller the size of the ion (hydrated radius), the greater the attraction. Although there are significant differences among the three types of adsorption, there are instances in which it is difficult to assign a given adsorption to a single type.

Since adsorption is a surface phenomenon, the rate and extent of adsorption is a function of the surface area of the solids used. Activated carbon is used extensively for adsorptive purposes because of its tremendous surface area in

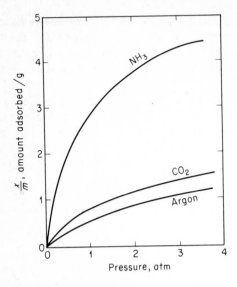

Figure 3-15 Adsorption of gases on charcoal in relation to pressure at constant temperature.

relation to mass. It is generally made from a wood product or coal by heating to temperatures between 300 and 1000°C in one of a variety of possible gaseous atmospheres such as CO_2, air, or water vapor, and then quickly quenching in air or water. The interior of the wood cells is cleaned out by this procedure, leaving a structure with remarkably small and uniform pores. Surface areas in the range of 1000 m² per gram of activated carbon results, with pore sizes in the general range of 10 to 1000 Å in diameter. At a given temperature and pressure a sample of activated carbon will adsorb a definite quantity of a gas. If the pressure is increased, it will adsorb more; if the pressure is decreased, it will adsorb less. If the quantities of adsorbed gas are plotted against pressure, curves of the sort shown in Fig. 3-15 are obtained.

Adsorption of solutes from solution follow the same general laws as gases. This is illustrated in Fig. 3-16, which shows data for the adsorption of acetic and benzoic acids. The curves are of the same nature as those shown in Fig. 3-15. From these data, it may be concluded that the quantity of substance adsorbed by a given sample of adsorbent depends upon the nature of the material and its

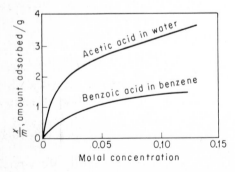

Figure 3-16 Adsorption of solutes on charcoal; temperature and pressure constant.

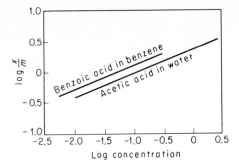

Figure 3-17 Logarithmic plot of adsorption data.

concentration. Temperature is also a factor which is not demonstrated by the data presented.

Freundlich studied the adsorption phenomenon extensively and showed that adsorption from solutions could be expressed empirically by the equation

$$y = K_F C^{1/n} \qquad (3\text{-}71)$$

where C is the concentration of solute after adsorption, y is the amount of material adsorbed per unit weight of the adsorbent, and K_F and n are constants which must be evaluated for each solute and temperature.

The Freundlich isotherm is often expressed in its logarithmic form,

$$\log y = \log K_F + \frac{1}{n} \log C \qquad (3\text{-}72)$$

Adsorption data, when plotted according to Eq. (3-72), yield straight lines, as shown in Fig. 3-17. Experimental data are often plotted in this manner as a convenient way of determining whether removal of materials from solution is accomplished by adsorption, and as a means of evaluating the constants k and n.

Other equations which are frequently used for describing adsorption isotherms and which can be derived from fundamental considerations are the Langmuir isotherm and the BET isotherm developed by Brunauer, Emmett, and Teller. The Langmuir isotherm is used to describe single-layer adsorption and can be written as follows:

$$\frac{C}{y} = \frac{a}{y_m} + \frac{C}{y_m} \qquad (3\text{-}73)$$

The value y_m represents the maximum adsorption that can take place in grams of absorbate per gram of adsorbent, when C is large relative to the constant a. When y approaches y_m, the coverage of the surface is essentially complete. If C/y is plotted against C, a straight line should be obtained, from which the constants a and y_m can be evaluated (Fig. 3-18).

The BET type of adsorption is generally more applicable than the Langmuir isotherm and corresponds to multilayer adsorption. This model assumes that a number of layers of adsorbate accumulate at the surface, and that the

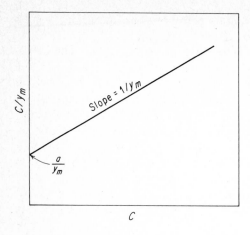

Figure 3-18 Straight-line form of the Langmuir isotherm.

Langmuir isotherm applies to each layer. It is somewhat more complex than the Langmuir isotherm and takes the form

$$\frac{y}{y_m} = \frac{bC}{(C_s - C)[1 + (b - 1)C/C_s]} \tag{3-74}$$

The value C_s represents the saturation concentrations for the adsorbate in solution. Of course, when C exceeds C_s, the solute precipitates or condenses from solution as a solid or liquid and concentrates on the surface. The BET equation can be put into the form

$$\frac{C}{y(C_s - C)} = \frac{1}{by_m} + \frac{(b - 1)}{by_m}\left(\frac{C}{C_s}\right) \tag{3-75}$$

With this equation, C_s and b can be obtained from the slope and intercept of the straight line best fitting of the plot of the left side of Eq. (3-75) versus C/C_s. The shape of the BET isotherm and its straight line form are shown in Fig. 3-19. The best isotherm equation to use in a particular instance can be determined by

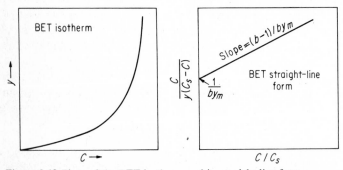

Figure 3-19 Plots of the BET isotherm and its straight-line form.

comparing the goodness of fit of the data when plotted in the form for each isotherm which should yield a straight line.

The adsorption isotherms are equilibrium equations and apply to conditions resulting after the adsorbate-containing phase has been in contact with the adsorbent for sufficient time to reach equilibrium. However, in any practical process for the removal of a contaminant from a gas or liquid, the rate at which the material is adsorbed onto the solid becomes an important consideration. Essentially three steps can be identified in the removal of a contaminant by adsorption. First, it must move from the liquid or gaseous phase through a surface film to the exterior of the adsorbent. Next, it must pass by diffusion into and through the pores of the adsorbent. Finally, it must become attached to the adsorbent. If the phase containing the adsorbent is quiescent, then diffusion through the surface film may be the slowest and rate-determining step. In this case, if the fluid is agitated, the thickness of the surface layer becomes reduced and the rate of adsorption will increase. At increased turbulence, however, a point will be reached where diffusion through the pores becomes the slowest step so that increased turbulence will not result in increased rates of adsorption. Thus, depending upon the general characteristics of the material being adsorbed and the relative rates of diffusion through the surface layer and into the pores, increased agitation of the fluid containing the material may or may not increase the rate of adsorption.

For adsorption, say, of a contaminant in water onto activated carbon, both the rate and the extent of adsorption are dependent upon the characteristics of the molecule being adsorbed and of the adsorbent.[4] The extent of adsorption is governed to some extent by the degree of solubility of the substance in water. The less soluble the material, the more likely it is to become adsorbed. With molecules containing both hydrophilic (water liking) and hydrophobic (water disliking) groups, the hydrophobic end of the molecule will tend to become attached to the surface. Next, the relative affinity of the material for the surface is a factor as already discussed in relation to the three general types of adsorption. Finally, the size of the molecule is of significance, as this affects its ability to fit within the pores of the adsorbent, and its rate of diffusion to the surface. Except for exchange adsorption, ions tend to be less readily adsorbed than neutral species. Many organics form negative ions at high pH, positive ions at low pH, and neutral species in intermediate pH ranges. Generally, adsorption is increased at pH ranges where the species is neutral in charge. In addition, pH affects the charge on the surface, altering its ability to adsorb materials. In water or wastewater samples, there are many different materials, each with different adsorption properties. Each competes in some way with the adsorption of the others. For this reason a given material may adsorb to a much less extent in a mixture of materials than if it were the only material in the solution.

---

[4] Walter J. Weber, Jr., Adsorption, chap. 5 in "Physicochemical Processes for Water Quality Control," p. 199, Wiley-Interscience, New York, 1972.

One of the most important uses of adsorption in environmental engineering has been for the removal of taste- and odor-producing organic materials from water supplies. Here activated carbon may be mixed with the water and then removed with the adsorbed materials by settling or filtration. When large quantities of organic material must be removed, more efficient usage of carbon and a higher quality of water can be obtained by passing the water through a carbon filter bed of large depth. This is a practical application of the principles expressed by the adsorption isotherms and, in some ways, can be compared to countercurrent extraction. Use of such systems to conquer severe taste and odor problems and to remove residual organic contaminants from treated wastewater effluents is becoming common practice. The development of special activated carbons which can be rejuvenated by roasting procedures has improved the economy considerably.

PROBLEMS

3-1 Determine the *net* heat of combustion of ethane gas from standard enthalpies of formation. Give the answer in kcal/mol of ethane.
   Answer: -341.26 kcal/mol

3-2 A small quantity of hydrogen gas is sometimes present in the gas from an anaerobic digester. Determine the *net* heat value in kcal/mol available from burning this gas.

3-3 A waste containing 1 percent sulfuric acid (10,000 mg/l) is neutralized by the addition of a concentrated lime slurry. If the water temperature is $15°C$ before neutralization, what is it after neutralization?
   Answer: $17.7°C$

3-4 (*a*) How many calories of heat are required to evaporate 1 liter of water at 1 atm if the initial water temperature is $20°C$?
   (*b*) If the above water sample contained 30,000 mg/l of acetic acid, would sufficient heat be liberated by combustion of the acetic acid to satisfy the heat requirements for water evaporation?

3-5 From standard free energies of formation, determine the solubility product for zinc sulfide (ZnS) at $25°C$.
   Answer: 2.5×10^{-24}

3-6 (*a*) Determine the solubility constant at $25°C$ for hydrogen sulfide gas in water from standard free energies of formation.
   (*b*) Using standard enthalpies of formation and the solubility constant from part (*a*), estimate the solubility constant for hydrogen sulfide gas at $10°C$.

3-7 From standard enthalpies and free energies of formation, determine the ionization constant for ammonia in water at $25°C$ and at $10°C$. [See Eq. (2-29).]
   Answer: 1.86×10^{-5}, 1.71×10^{-5}

3-8 Estimate the ionization constant for acetic acid at $35°C$ from thermodynamic considerations.

3-9 Estimate the first ionization constant for hydrosulfuric acid at $35°C$ from thermodynamic considerations.
   Answer: 1.33×10^{-7}

3-10 (*a*) Calculate the standard free energy of the reaction for the biological decomposition of one mole of acetate under both aerobic and anaerobic conditions:

Aerobic: $CH_3COO^-(aq) + 2O_2(g) \rightarrow HCO_3^-(aq) + H_2O(lq) + CO_2(g)$
Anaerobic: $CH_3COO^-(aq) + H_2O(lq) \rightarrow HCO_3^-(aq) + CH_4(g)$

(*b*) For a given quantity of acetate waste, which system would you expect to be capable of supporting the growth of the largest biological population? Why?

3-11 At 1 atm the boiling temperature for isopropyl alcohol is 82.5°C and for water is 100°C. A mixture containing 12.5 percent water and 87.9 percent isopropyl alcohol by weight has a boiling temperature of 80.4°C, and at this temperature the composition of the vapor is the same as the liquid. If the waste from an industry contained 10,000 mg/l of isopropyl alcohol, would it be possible to remove the alcohol from the water by fractional distillation of the waste? Why?

Answer: Yes

3-12 An industrial wastewater contains 10 percent by weight of an organic solvent and has a boiling temperature of 105°C at 1 atm. The vapor is found to be richer in organic solvent than the liquid waste. The boiling temperature of pure solvent is 80°C. Can pure solvent be obtained by fractional distillation? Pure water? Explain why.

3-13 (*a*) At 1 atm, *n*-butyl alcohol boils at 117.8°C. A binary mixture containing 2.52 mols of water per mole of this alcohol boils at 92.4°C, and with this mixture the composition of the vapor is equal to that of the liquid mixture. Sketch roughly the temperature composition diagram for the liquid-vapor equilibrium for mixtures of water and *n*-butyl alcohol at atm pressure.

(*b*) An industrial waste contains 90 percent water and 10 percent *n*-butyl alcohol by weight. What would be the composition of the distillate and the residue from fractional distillation of the waste?

Answer: (*b*) distillate 0.716 mol fraction H_2O, residue pure H_2O

3-14 The boiling temperature for butyric acid at 1 atm is 163.5°C. When wastewaters containing low concentrations of butyric acid are distilled, it is found that the distillate is richer in butyric acid than the wastewater being distilled. In what class would a butyric acid-water binary mixture be placed? Illustrate.

3-15 What approximate osmotic pressure would be created across a semipermeable membrane if water containing 0.01 M Na_2SO_4, 0.02 M $MgCl_2$, and 0.03 M $CaCl_2$ were placed on one side of the membrane and distilled water were on the other?

Answer: 4.4 atm

3-16 (*a*) 20,000 mg/l of NaCl would lower the vapor pressure of water at 100°C by about 8.4 mm Hg. Estimate the osmotic pressure across a semipermeable membrane containing a brackish water with this sodium chloride concentration on one side and distilled water on the other.

(*b*) Calculate the theoretical minimum energy requirement to remove the salt from 1000 gallons of the brackish water. Express energy required in units of liter-atmospheres, foot-pounds, and kilowatthours.

3-17 Phenol is approximately 12 times more soluble in 1 volume of isopropyl ether than it is in 1 volume of water. How many extractions are required to reduce the concentration of phenol below 100 mg/l in a waste containing 2000 mg/l of phenol if an ether-to-wastewater ratio of 0.2 by volume is used in each extraction?

Answer: 3

3-18 The solubility of picric acid at 20°C is 9.56 g per 100 g of benzene, and 1.4 g per 100 g of water. If an industrial wastewater contains 5000 mg/l of picric acid, what concentration would remain after 1 extraction with 1 lb of benzene for each 2 lb of water?

3-19 What is the approximate specific conductance at 25°C of a solution containing 100 mg/l of $CaCl_2$ and 75 mg/l of Na_2SO_4?

Answer: 384 μmhos/cm

3-20 A standard KCl solution (0.01 N), when placed in a conductivity cell at 25°C, was found to produce a resistance of 1000 ohms. A $MgCl_2$ solution was then placed in the cell, and the measured resistance at 25°C was 3000 ohms. Approximately what is the concentration of $MgCl_2$ in mg/l?

3-21 The specific conductance of a $CaCl_2$ solution is 200×10^{-6} mhos/cm. Estimate the concentration of $CaCl_2$ in mg/l.

    Answer: 81.8 mg/l

3-22 Ions can contribute to the conductivity of a solution, but un-ionized molecules cannot. On this basis what is the approximate specific conductance at 25°C of a solution containing 1000 mg/l of acetic acid, if the ionization constant for this acid is 1.75×10^{-5}?

3-23 What weight of silver will pass into solution from a silver anode by the passage of 0.02 A of current through the solution for 24 h?

    Answer: 1.93 g

3-24 When an electric current is allowed to pass through a dilute sulfuric acid solution, hydrogen gas is evolved at the cathode and oxygen gas at the anode. What volumes of gases, measured at 1 atm pressure and 0°C, will be obtained at the electrodes when 1 A of current is passed through the solution for a 1-h period?

3-25 On the basis of standard electrochemical potentials, which of the following metals could act in a sacrificial manner to protect iron from corrosion: aluminum, copper, lead, magnesium, silver, tin, and zinc?

    Answer: aluminum, magnesium, zinc

3-26 A zinc and an iron bar, connected by a copper wire, were introduced into a solution containing 2000 mg/l of Zn^{2+} ions and 5 mg/l of Fe^{2+} ions. What reaction took place?

3-27 Estimate the solubility-product constant for $Mg(OH)_2$ at 25°C from standard oxidation potentials.

    Answer: 1.5×10^{-11}

3-28 In a study of the natural die-off of coliform organisms in a stream, it was found that 36 percent of the organisms died within 10 hours and 59 percent died within 20 hours. If the rate of die-off followed first-order kinetics and were proportional to the number remaining, how long would it take to obtain a 99 percent reduction in coliform organisms?

3-29 A stream flowing with a velocity of 2 mph contains no BOD, but has an oxygen deficit of 6 mg/l. Ten miles downstream this deficit has been reduced to 4 mg/l through absorption from the atmosphere. The stream conditions are uniform throughout its length. Assuming the rate of aeration is proportional to the deficit, what would be the deficit 35 miles downstream from the original point?

    Answer: 1.45 mg/l

3-30 The radioactive nuclide P^{32} has a half-life of 14.3 days. How long would a waste containing 10 mg/l of this nuclide have to be stored in order to reduce the concentration to 0.3 mg/l?

3-31 Differentiate between chemical adsorption, exchange adsorption, and physical adsorption.

3-32 An evaluation of the ability of activated carbon to reduce the odor of a water with a threshold odor of 30 was made, using the Freundlich adsorption isotherm. By plotting the log of odor removed per unit dose of activated carbon versus residual odor, the constants K_F and n in Eq. (3-76) were found to be 0.5 and 1.0, respectively. What activated carbon dosage in mg/l would be required to reduce the threshold odor to 4 units?

REFERENCES

Barrow, G. M.: "Physical Chemistry," 3d ed., McGraw-Hill, New York, 1973.

Castellan, G. W.: "Physical Chemistry," 2d ed., Addison-Wesley, Reading, Mass., 1971.

Cockford, H. D., and S. B. Knight: "Fundamentals of Physical Chemistry," 2d ed., Wiley, New York, 1964.

Daniels, F., and R. A. Alberty: "Physical Chemistry," 4th ed., Wiley, New York, 1975.

Dean, J. A. (ed.): "Làng's Handbook of Chemistry," 11th ed., McGraw-Hill, New York, 1973.

Moore, W. J.: "Physical Chemistry," 4th ed., Prentice-Hall, Englewood Cliffs, N.J., 1972.

Weast, R. C. (ed.): "Handbook of Chemistry and Physics," 56th ed., Chemical Rubber Publishing Co., Cleveland, 1975.

FOUR

BASIC CONCEPTS FROM ORGANIC CHEMISTRY

4-1 INTRODUCTION

The fundamental information that an environmental engineer needs concerning organic chemistry differs considerably from that which the organic chemist requires. This difference is due to the fact that chemists are concerned principally with the synthesis of compounds, whereas the environmental engineer is concerned, in the main, with how the organic compounds in liquid, solid, and gaseous wastes can be destroyed. Another major difference lies in the fact that the organic chemist is usually concerned with the product of the reaction; the by-products of a reaction are of little interest to him. Since few organic reactions give better than 85 percent yields, the amount of by-products and unreacted raw materials that represent processing wastes is of considerable magnitude. In addition, many raw materials contain impurities that do not enter the desired reaction and, of course, add to the organic load in waste streams. A classic example is formaldehyde, which normally contains about 5 percent of methanol unless special precautions are taken in its manufacture.

The environmental engineer, like the biochemist, must have a fundamental knowledge of organic chemistry. It is not important for either to know a multiplicity of ways of preparing a given organic compound and the yields to be expected from each. Rather, the important consideration is how the compounds react in the realm of dilute solutions or when serving as a source of energy for living organisms. It is from this viewpoint that organic chemistry will be treated in this chapter, and considerations will be from the viewpoint of classes rather than individual compounds. Unfortunately organic chemists have presented very little information on the nature of the by-products of reactions to aid environmental engineers in solving industrial waste problems.

History

Organic chemistry deals with the compounds of carbon. The science of organic chemistry is considered to have originated in 1685 with the publication by Lémery[1] of a chemistry book which classified substances according to their origin as mineral, vegetable, or animal. Compounds derived from plants and animals became known as *organic* and those derived from nonliving sources were inorganic.

Until 1828 it was believed that organic compounds could not be formed except by living plants and animals. This was known as the *vital-force theory,* and belief in it severely limited the development of organic chemistry. Wöhler,[2] in 1828, by accident, found that application of heat to ammonium cyanate, an inorganic compound, caused it to change to urea, a compound considered organic in nature. This discovery dealt a death-blow to the vital-force theory, and by 1850 modern organic chemistry became well established. Today about a million organic compounds are known. Many of these are products of synthetic chemistry, and similar compounds are not known in nature.

Elements

All organic compounds contain carbon in combination with one or more elements. The hydrocarbons contain only carbon and hydrogen. A great many compounds contain carbon, hydrogen, and oxygen, and they are considered to be the major elements. Minor elements in naturally occurring compounds are nitrogen, phosphorus, and sulfur. Compounds produced by synthesis may contain, in addition, halogens, certain metals, and a wide variety of other elements.

Properties

Organic compounds, in general, differ greatly from inorganic compounds in seven respects:

1. Organic compounds are usually combustible
2. Organic compounds, in general, have lower melting and boiling points.
3. Organic compounds are usually less soluble in water.
4. Several organic compounds may exist for a given formula. This is known as *isomerism.*

[1] Nicholas Lémery (1645–1715), French physician and chemist.
[2] Friedrich Wöhler (1800–1882), German chemist.

5. Reactions of organic compounds are usually molecular rather than ionic. As a result, they are often quite slow.
6. The molecular weights of organic compounds may be very high, often well over 1000.
7. Most organic compounds can serve as a source of food for bacteria.

Sources

Organic compounds are derived from three sources:

1. Nature: fibers, vegetable oils, animal oils and fats, alkaloids, cellulose, starch, sugars, and so on.
2. Synthesis: A wide variety of compounds and materials prepared by manufacturing processes.
3. Fermentation: Alcohols, acetone, glycerol, antibiotics, acids, and the like are derived by the action of microorganisms upon organic matter.

The wastes produced in the processing of natural organic materials and from the synthetic organic and fermentation industries constitute a major part of the industrial waste problems that the environmental engineer is called upon to solve.

The Carbon Atom

A question commonly asked is: How is it possible to have so many compounds of carbon? There are two reasons. In the first place, carbon normally has four covalent bonds. This factor alone allows many possibilities, but the most important reason is concerned with the ability of carbon atoms to link together by covalent bonding in a wide variety of ways. They may be in a continuous open chain,

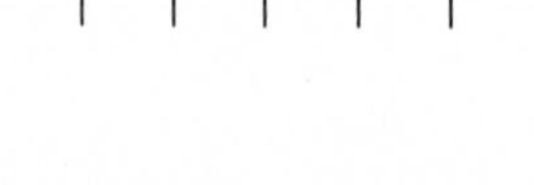

$$-\overset{|}{\underset{|}{C}}-\overset{|}{\underset{|}{C}}-\overset{|}{\underset{|}{C}}-\overset{|}{\underset{|}{C}}-\overset{|}{\underset{|}{C}}-$$

or a chain with branches,

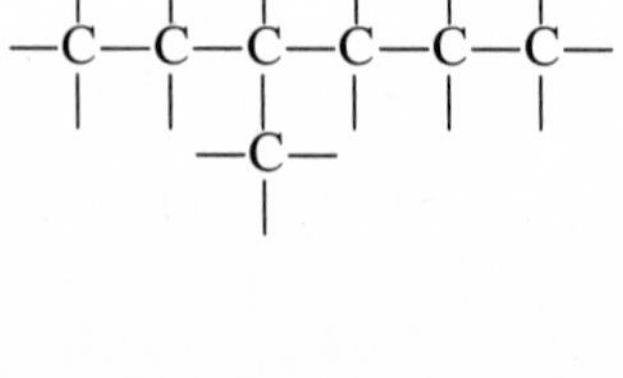

$$
\begin{array}{c}
\overset{|}{\underset{}{C}}- \\
-\overset{|}{\underset{|}{C}}-\overset{|}{\underset{|}{C}}-\overset{|}{\underset{|}{C}}-\overset{|}{\underset{|}{C}}-\overset{|}{\underset{|}{C}}-\overset{|}{\underset{|}{C}}- \\
-\overset{|}{\underset{|}{C}}-
\end{array}
$$

or in a ring,

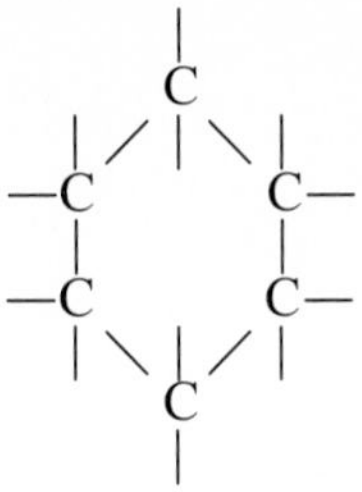

or in chains or rings containing other elements,

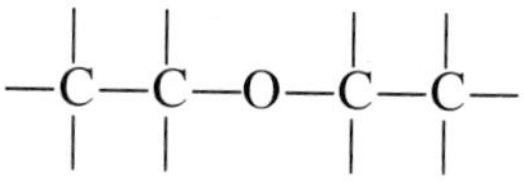

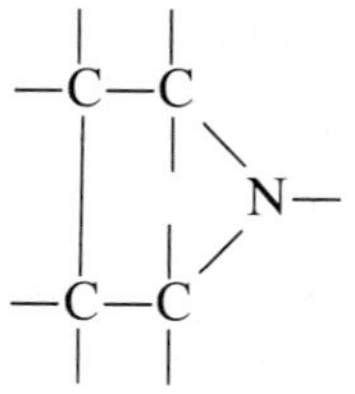

These examples will serve to show the tremendous number of possibilities that exist.

Isomerism

In inorganic chemistry, a molecular formula is specific for one compound. In organic chemistry, most molecular formulas do not represent any particular compound. For example, the molecular formula $C_3H_6O_3$ represents at least four separate compounds and therefore is of little value in imparting information other than that the compound contains carbon, hydrogen, and oxygen. Four compounds having the formula $C_3H_6O_3$ are

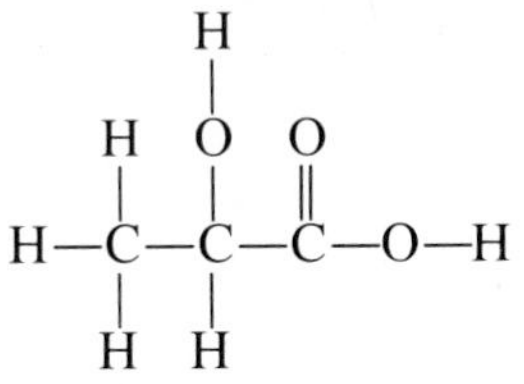

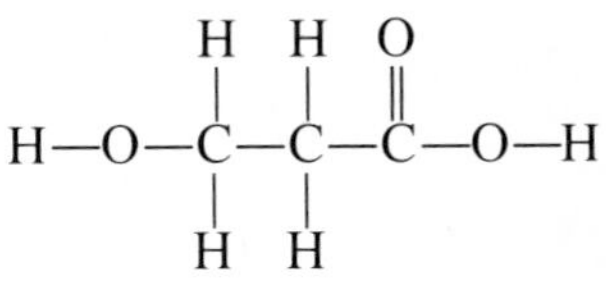

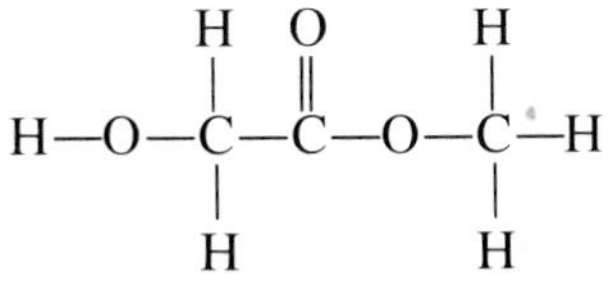

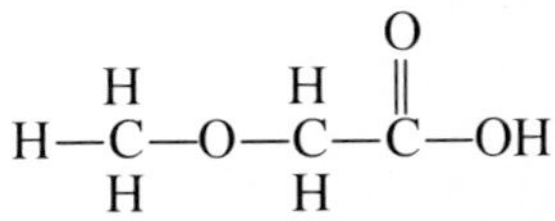

Compounds having the same molecular formula are known as *isomers*. In the case cited above, the first two isomers are hydroxy acids, the third is an ester of a hydroxy acid, and the fourth is a methoxy acid. To the organic chemist, each

of the formulas represents a chemical compound with definite physical and chemical properties. The term *structural* formulas is applied to molecular representations as given above. They are as useful to a chemist as blueprints are to an engineer.

In many cases structural formulas may be simplified as *condensed* formulas so as to use only one line. Thus the formula

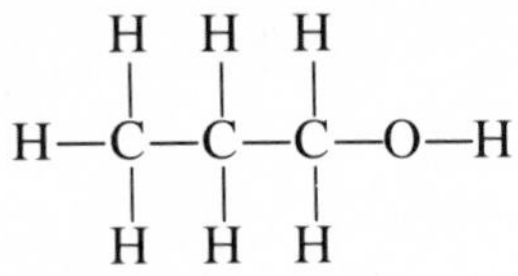

$$
\begin{array}{c}
\quad\ \text{H}\ \ \text{H}\ \ \text{H} \\
\quad\ |\ \ \ |\ \ \ | \\
\text{H}-\text{C}-\text{C}-\text{C}-\text{O}-\text{H} \\
\quad\ |\ \ \ |\ \ \ | \\
\quad\ \text{H}\ \ \text{H}\ \ \text{H}
\end{array}
$$

may be written as

$$CH_3-CH_2-CH_2OH \quad \text{or} \quad CH_3CH_2CH_2OH$$

thereby saving a great deal of space.

There are three major types of organic compounds, the *aliphatic, aromatic,* and *heterocyclic*. The *aliphatic* compounds are those in which the characteristic groups are linked to a straight or branched carbon chain. The *aromatic* compounds have these groups linked to a particular type of six-member carbon ring which contains three double bonds. Such rings have peculiar stability and chemical character, and are present in an important group of compounds. The *heterocyclic* compounds have a ring structure in which one member is an element other than carbon.

ALIPHATIC COMPOUNDS

4-2 HYDROCARBONS

The hydrocarbons are compounds of carbon and hydrogen. There are two types, saturated and unsaturated. Saturated hydrocarbons are those in which adjacent carbon atoms are joined by a single covalent bond and all other bonds are satisfied by hydrogen.

$$
\begin{array}{c}
\quad\ \text{H}\ \ \text{H}\ \ \text{H} \\
\quad\ |\ \ \ |\ \ \ | \\
\text{H}-\text{C}-\text{C}-\text{C}-\text{H} \\
\quad\ |\ \ \ |\ \ \ | \\
\quad\ \text{H}\ \ \text{H}\ \ \text{H}
\end{array}
$$

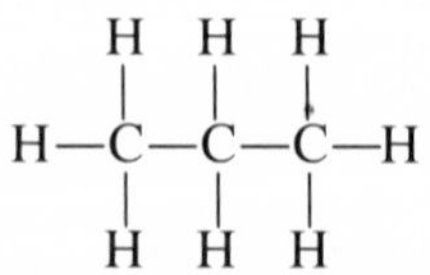

A saturated compound

Unsaturated hydrocarbons have at least two carbon atoms that are joined by more than one covalent bond and all remaining bonds are satisfied by hydrogen.

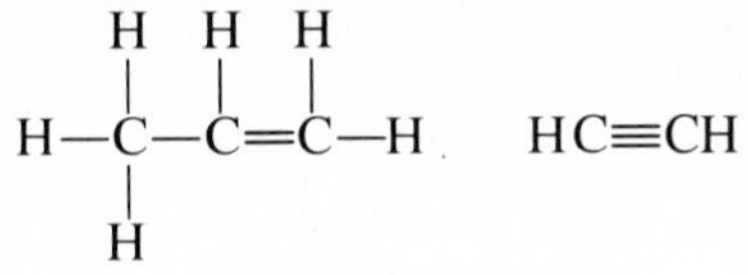

$$
\begin{array}{ccccc}
& H & H & H & \\
& | & | & | & \\
H- & C & -C & =C & -H \qquad HC{\equiv}CH \\
& | & & & \\
& H & & &
\end{array}
$$

Unsaturated compounds

Saturated Hydrocarbons

The saturated hydrocarbons form a whole series of compounds starting with one carbon atom and increasing one carbon atom, stepwise. These compounds are also known as the *paraffin* series, the *methane* series, and as *alkanes*. The principal source is petroleum. Gasoline is a mixture containing several of them; diesel fuel is another such mixture.

The hydrocarbons are known as parent compounds by organic chemists because they may be used to prepare a wide variety of organic chemicals. This knowledge serves as the basis of the great petrochemical industry within the petroleum industry. Saturated hydrocarbons are quite inert toward most chemical reagents. For this reason they were termed "paraffins" by early chemists (from the Latin *parum affinis*, meaning little affinity).

Methane (CH_4) is the simplest hydrocarbon. It is a gas of considerable importance to environmental engineers since it is a major end product of the anaerobic treatment process as applied to sewage sludge and other organic waste materials. It is a component of marsh gas and of natural gas and, in a mixture with air containing from 5 to 15 percent methane, it is highly explosive. This property allows its use as fuel for gas engines. Methane is commonly called "firedamp" by miners and makes their work particularly hazardous.

Ethane ($CH_3{-}CH_3$) is the second member of the series.

Propane ($CH_3{-}CH_2{-}CH_3$) is the third member of the series.

Butane (C_4H_{10}) is the fourth member of the series and is of interest because it occurs in two isomeric forms:

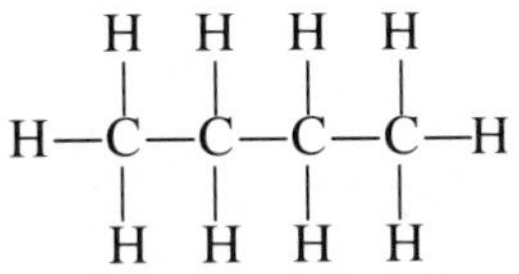

$$
\begin{array}{cccccccc}
H & H & H & H & \qquad & H & H & H \\
| & | & | & | & & | & | & | \\
H{-}C{-}C{-}C{-}C{-}H & & & & & H{-}C{-}C{-}C{-}H & & \\
| & | & | & | & & | & | & | \\
H & H & H & H & & H & HCH & H \\
& & & & & & H &
\end{array}
$$

n-Butane 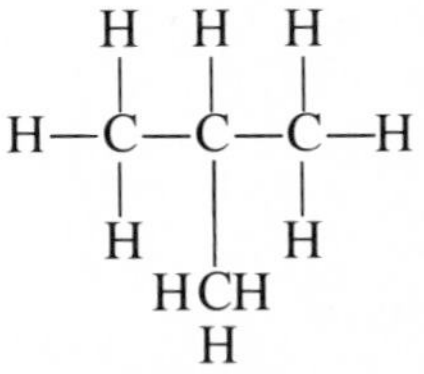   Isobutane

Pentane (C_5H_{12}) is the fifth member of the series and exists in three isomeric forms:

```
                                                           H
                                                          HCH
      H  H  H  H  H          H  H  H  H          H    |   H
      HC—C—C—C—CH            HC—C—C—CH           HC—C—CH
      H  H  H  H  H          H  |  H  H          H    |   H
                                HCH                  HCH
                                 H                    H
         n-Pentane             Isopentane          Neopentane
         bp, 36.2°C            bp, 28°C            bp, 9.5°C
```

The third isomer of pentane might also be called tetramethylmethane or dimethylpropane, as the reader will shortly recognize.

As the number of carbon atoms increases in the molecule, the number of possible isomers increases accordingly. There are five possible isomers of *hexane* (C_6H_{14}) and 75 possible isomers of *decane* ($C_{10}H_{22}$).

Physical properties Table 4-1 lists the names and physical constants of the *normal* saturated hydrocarbons of 1 to 10 carbon atoms per molecule. The term "normal" applies to the isomer that has all its carbon atoms linked in a *straight chain*. The others are referred to as *branched-chain* compounds. The branched form of butane and the simplest branched form of pentane are commonly given the prefix *iso-*.

The saturated hydrocarbons are colorless, practically odorless, and quite insoluble in water, particularly those with five or more carbon atoms. They dissolve readily in many organic solvents. At room temperature all members through C_5 are gases, those from C_6 to C_{17} are liquids, and those above C_{17} are solids.

Homologous series It will be noted from Table 4-1 that each successive member of the series differs from the previous member by CH_2. When the formulas of a series of compounds differ by a common increment, such as CH_2, the series is

Table 4-1 Physical constants of some normal paraffins*

Name	Formula	Mp, °C	Bp, °C	Sp. gr., 20°/4°	Calcd. no. of isomers
Methane	CH_4	−183	−161.5	$0.554^{0°}$	1
Ethane	C_2H_6	−172	−88.3	$0.56^{-100°}$	1
Propane	C_3H_8	−187.1	−42.2	$0.585^{-44.5°}$	1
Butane	C_4H_{10}	−135	−0.6	$0.6^{0°}$	2
Pentane	C_5H_{12}	−130	36.2	0.626	3
Hexane	C_6H_{14}	−94.3	69.0	0.660	5
Heptane	C_7H_{16}	−90.5	98.5	0.684	9
Octane	C_8H_{18}	−56.5	125.8	0.704	18
Nonane	C_9H_{20}	−53.7	150.7	0.718	35
Decane	$C_{10}H_{22}$	−30	174	0.730	75

* From E. Wertheim and H. Jeskey, "Introductory Organic Chemistry," 3d ed., McGraw-Hill, New York, 1956. Table reproduced by permission of the authors.

Table 4-2 Names of methane-series radicals (alkyl groups)

Parent compound	Radical	Formula
Methane	Methyl	$CH_3—$
Ethane	Ethyl	$C_2H_5—$
Propane	n-Propyl	$C_3H_7—$
Propane	Isopropyl	$(CH_3)_2CH—$
n-Butane	n-Butyl	$C_4H_9—$

referred to as being a *homologous series*. Such compounds can be expressed by a general formula. That for the methane series is C_nH_{2n+2}.

Radicals The inert character of the paraffin hydrocarbons has been mentioned; however, they may be made to react under the proper conditions, and a wide variety of compounds results. It becomes necessary, therefore, to establish some form of nomenclature to identify the products formed. When one hydrogen is replaced from a molecule of a methane-series hydrocarbon, the *-ane* ending is dropped and a *-yl* is added. The names of some are as shown in Table 4-2. The system serves quite well for the normal compounds but is of little value in naming derivatives of the isomers.

Nomenclature The methane series of hydrocarbons is characterized by names ending in *-ane*. The straight-chain compounds are termed *normal* compounds. The branched-chain compounds and the derivatives of both straight- and branched-chain compounds are difficult to name with any degree of specificity. The IUPAC system, as proposed by the International Union of Pure and Applied Chemistry, is commonly used. In this system the compounds are named in terms of the longest continuous chain of carbon atoms in the molecule. A few examples will illustrate the method.

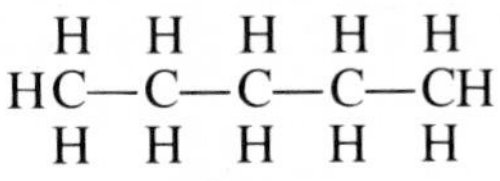

<pre>
     H H H H H
    HC—C—C—C—CH
     H H H H H
         n-Pentane
</pre>

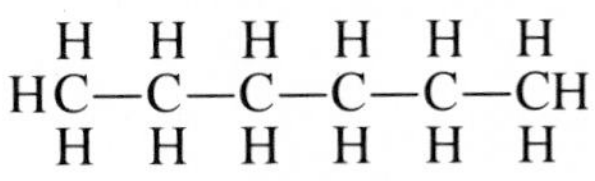

<pre>
     H H H H H H
    HC—C—C—C—C—CH
     H H H H H H
          n-Hexane
</pre>

<pre>
     H H H H H H
    HC—C—C—C—C—CH
     H H | H H H
          HCH
           H
</pre>

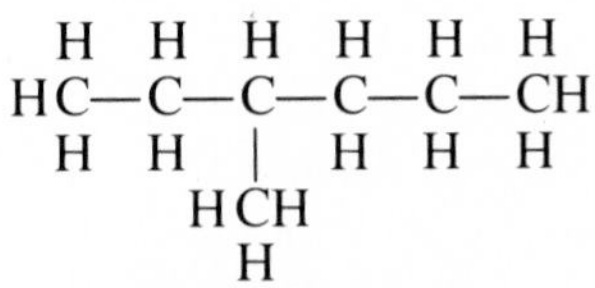

3-Methylhexane
(one of the group
of heptanes)

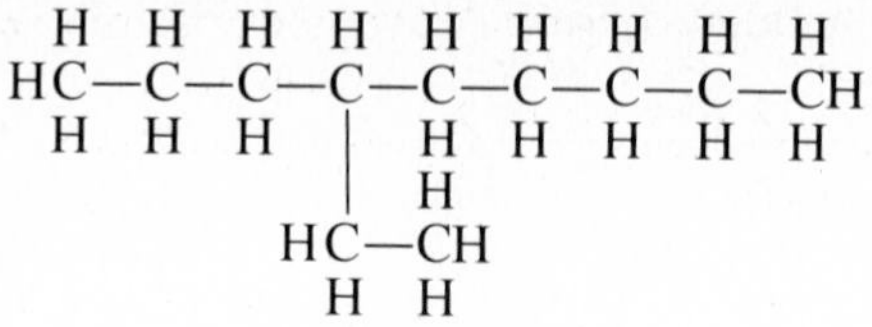

```
      H  H  H  H  H  H  H  H  H
   HC—C—C—C—C—C—C—C—CH
      H  H  H  |  H  H  H  H  H
               H
            HC—CH
            H  H
```
4-Ethylnonane
(a member of
the undecane
group)

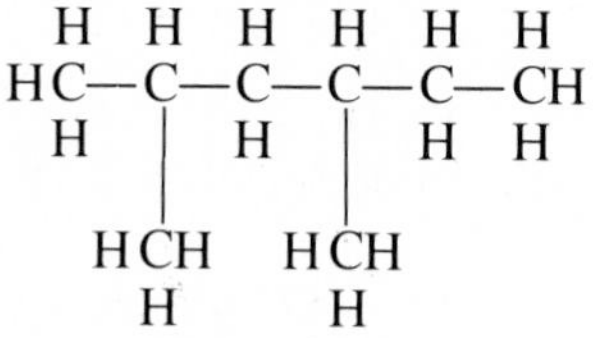

```
      H  H  H  H  H  H
   HC—C—C—C—C—CH
      H  |  H  |  H  H

      HCH   HCH
       H     H
```
2,4-Dimethyl
hexane (one of the
isomeric octanes)

It will be noted that a chain is numbered from the end nearest the attached radical. The rule is to make the numbers as small as possible. The IUPAC system is applied to other compounds as well as to hydrocarbons.

Chemical reactions Strong bases, acids, or aqueous solutions of oxidizing agents do not react with saturated hydrocarbons at room temperature. At elevated temperatures, strong oxidizing agents, such as concentrated sulfuric acid, oxidize the compounds to carbon dioxide and water. This reaction is important to environmental engineers in the determination of organic nitrogen. Other reactions of importance are as follows:

1. Oxidation with oxygen or air:

$$CH_4 + 2O_2 \xrightarrow{\Delta} CO_2 + 2H_2O \tag{4-1}$$

2. Substitution of hydrogen by halogens:

$$CH_4 + Cl_2 \rightarrow HCl + CH_3Cl \tag{4-2}$$

This reaction does not ordinarily occur in aqueous solutions and therefore is of little significance in environmental engineering.

3. Pyrolysis or cracking: High-molecular-weight hydrocarbons may be broken into smaller molecules by heat treatment. The process is used in the petroleum industry to increase the yield of light boiling fractions, suitable for sale as gasoline or for chemical synthesis. Heat treatment results in disruption of the large molecules as follows:

$$\left.\begin{array}{l}\text{High mol. wt}\\\text{paraffin compound}\end{array}\right\}\xrightarrow[\text{pressure}]{\text{heat}}\begin{array}{l}\text{paraffin compounds of lower mol. wt}\\+\text{ olefin compounds}\\+\text{ hydrogen + naphthenes}\\+\text{ carbon}\end{array}\qquad(4\text{-}3)$$

4. Biological oxidation: Hydrocarbons are oxidized by certain bacteria under aerobic conditions. The oxidation proceeds through several steps. The first step is very slow biologically and involves conversion to alcohols with attack occurring on terminal carbon atoms, i.e., omega oxidation.

$$\underset{\text{Hydrocarbon}}{2CH_3CH_2CH_3} + O_2 \xrightarrow[\text{bact.}]{} \underset{\text{Alcohol}}{2CH_3CH_2CH_2OH} \qquad(4\text{-}4)$$

The bacteria derive energy from this oxidation and, through additional oxidative steps which will be developed later, convert the hydrocarbon to carbon dioxide and water.

$$CH_3CH_2CH_3 + 5O_2 \xrightarrow[\text{bact.}]{} 3CO_2 + 4H_2O \qquad(4\text{-}5)$$

This reaction, particularly the intermediate steps, is of great interest to environmental engineers.

Unsaturated Hydrocarbons

The unsaturated hydrocarbons are usually separated into four classes.

Ethylene series The *ethylene* series corresponds to the methane series of hydrocarbons. Each member of the latter except methane can lose hydrogen to form unsaturated compounds. Because ethane is the first member capable of doing this, the series takes its name from it. The ethylene series of compounds all contain one double bond between two adjacent carbon atoms,

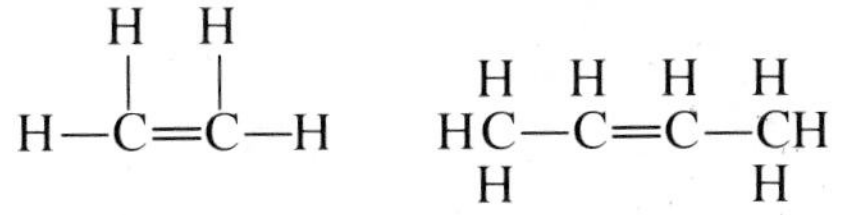

$$\begin{array}{cc}\begin{array}{c}H\ \ H\\|\ \ \ |\\H-C=C-H\end{array} & \begin{array}{c}H\ \ H\ \ H\ \ H\\HC-C=C-CH\\H\qquad\qquad H\end{array}\\[2ex]\text{Ethylene} & \text{Butylene}\\\text{or} & \text{or}\\\text{Ethene} & \text{2-Butene}\end{array}$$

and their names all end in -*ene*. The ethylene series of compounds are also called *olefins* and *alkenes*. Olefin compounds, particularly ethylene, propylene, and butylenes, are formed in great quantities during the cracking or pyrolysis of petroleum.

The names, formulas, and physical constants of a number of important alkenes are given in Table 4-3. In naming specific alkenes, the IUPAC system must be employed on all compounds with over three carbon atoms. The nomenclature becomes quite complicated with branched-chain isomers. Fortu-

Table 4-3 Physical constants of selected alkenes*

IUPAC name	Formula	Mp, °C	Bp, °C	Sp. gr., 20°/4°	Calcd. no. of isomers
Ethene	C_2H_4	−169.4	−103.9	$0.566^{-102°}$	1
Propene	$CH_2{=}CHCH_3$	−185.2	−47	$0.610^{-47°}$	1
1-Butene	$CH_2{=}CHCH_2CH_3$	−130	−5	$0.668^{0°}$	3
1-Pentene	$CH_2{=}CH(CH_2)_2CH_3$	−138	30	$0.645^{25°}$	5
1-Hexene	$CH_2{=}CH(CH_2)_3CH_3$	−98.5	64.1	0.673	13
1-Heptene	$CH_2{=}CH(CH_2)_4CH_3$	−120	95	0.699	27
1-Octene	$CH_2{=}CH(CH_2)_5CH_3$	−102.1	126	0.722	66
1-Nonene	$CH_2{=}CH(CH_2)_6CH_3$		149.9	0.730	153
1-Decene	$CH_2{=}CH(CH_2)_7CH_3$	−80	172	0.763	377

* From E. Wertheim and H. Jeskey, "Introductory Organic Chemistry," 3d ed., McGraw-Hill, New York, 1956. Table reproduced by permission of the authors.

nately, there is little reason to differentiate between normal and branched-chain compounds in this series.

Diolefins When aliphatic compounds contain two double bonds in the molecule they are called *diolefins*.

Polyenes Some organic compounds contain more than two double bonds per molecule. Such compounds are called polyenes. The red coloring matter of tomatoes, lycopene, and the yellow coloring matter of carrots are examples.

$$CH_3 \cdot \overset{\displaystyle CH_3}{\underset{|}{C}}{=}CH(CH_2)_2\overset{\displaystyle CH_3}{\underset{|}{C}}{=}CH \cdot CH{=}CH \cdot \overset{\displaystyle CH_3}{\underset{|}{C}}{=}CH \cdot CH{=}CH \cdot \overset{\displaystyle CH_3}{\underset{|}{C}}{=}CH \cdot CH$$

$$CH_3 \cdot \underset{\displaystyle CH_3}{\overset{|}{C}}{=}CH(CH_2)_2\underset{\displaystyle CH_3}{\overset{|}{C}}{=}CH \cdot CH{=}CH \cdot \underset{\displaystyle CH_3}{\overset{|}{C}}{=}CH \cdot CH{=}CH \cdot \underset{\displaystyle CH_3}{\overset{|}{C}}{=}CH \cdot CH$$

Lycopene ($C_{40}H_{56}$)

These compounds are of interest to environmental engineers because of their occurrence in industrial wastes produced in preparation of vegetables for canning. The chlorine demand of such wastes is extremely high.

Acetylene series The acetylene series of unsaturated compounds have a triple bond between adjacent carbon atoms.

$$H{-}C{\equiv}C{-}H$$

These compounds are found to some extent in industrial wastes from certain industries, particularly those from the manufacture of some types of synthetic rubber.

Chemical reactions Unsaturated hydrocarbons seldom create problems in environmental engineering. However, unsaturated linkages occur in many types of organic compounds and exhibit many properties in common, regardless of the type of compound in which they exist. For this reason, the environmental engineer should be acquainted with the chemistry of the double bond.

Unsaturated compounds undergo several reactions with relative ease.

1. Oxidation: The compounds are easily oxidized in aqueous solution by oxidizing agents such as potassium permanganate. A glycol is the normal product.
2. Reduction: Under special conditions of temperature, pressure, and catalysis, hydrogen may be caused to add at double or triple bonds. This reaction is of considerable importance commercially in the conversion of vegetable oils to more acceptable solid fats. Crisco, Spry, and many other vegetable shortenings are made by this process.
3. Addition: Halogen acids, hypochlorous acid, and halogens will add across unsaturated linkages.

$$CH_3-C=C-H + HOCl \rightarrow CH_3-\underset{H}{\overset{OH}{C}}-\underset{H}{\overset{Cl}{C}}-H \qquad (4\text{-}6)$$

The reaction with hypochlorous acid is most important to environmental engineers. Industrial wastes containing appreciable amounts of unsaturated compounds exhibit high chlorine-demand values because of such reactions.
4. Polymerization: Molecules of certain compounds having unsaturated linkages are prone to combine with each other to form polymers of higher molecular weight.

$$n\,CH_2=CH_2 \xrightarrow[\text{and pressure}]{\text{high temperature}} (C_2H_4)n \quad n = 70 \text{ to } 700 \qquad (4\text{-}7)$$
$$\text{Polyethylene}$$

Similar reactions serve as the basis for many industrial products, e.g., synthetic resins, synthetic fibers, synthetic rubber, and synthetic detergents. Industrial wastes from such industries can be expected to contain a wide variety of polymers and usually exhibit a high chlorine demand.
5. Bacterial oxidation: It is generally considered that organic compounds possessing unsaturated linkages are more prone to bacterial oxidation than corresponding saturated compounds because of the ease of oxidation at the double bonds.

4.3 ALCOHOLS

Alcohols are considered the primary oxidation product of hydrocarbons.

$$\underset{\text{Methane}}{CH_4} + \tfrac{1}{2}O_2 \rightarrow \underset{\text{Methyl alcohol}}{CH_3OH} \qquad (4\text{-}8)$$

$$CH_3—CH_2—CH_3 + \tfrac{1}{2}O_2 \rightarrow CH_3—CH_2—CH_2OH \qquad (4\text{-}9)$$

Propane   *n*-Propyl alcohol

They cannot be prepared in this manner, however, because the reaction cannot be stopped with alcohols as the end product. Therefore the reaction is of theoretical interest only, but is often used to illustrate the steps in biological degradation of hydrocarbons under aerobic conditions.

Alcohols may be considered as *hydroxy alkyl* compounds. For convenience, the alkyl group in alcohols and other organic compounds is often represented by R—, and the general formula for alcohols is R—OH. The OH group does not ionize; consequently alcohols are neutral in reaction. The chemistry of alcohols is related entirely to the OH group.

Classification

Alcohols are classified into three groups: primary, secondary, and tertiary, depending upon where the OH group is attached to the molecule. If the OH group is on a terminal (primary) carbon atom, it is a *primary* alcohol.

$$
\begin{array}{ccc}
\text{H \ H} & \text{H \ H \ H \ H} & \\
\text{HC—C—OH} & \text{HC—C—C—C—OH} & \text{R—OH} \\
\text{H \ H} & \text{H \ H \ H \ H} &
\end{array}
$$

Primary alcohols

If the OH group is attached to a carbon atom that is joined to two other carbon atoms, it is a *secondary* alcohol, and the carbon atom to which it is attached is a *secondary carbon atom.*

$$
\begin{array}{ccc}
\text{H \ H \ H} & \text{H \ H \ H \ H} & \text{R} \\
\text{HC—C—CH} & \text{HC—C—C—CH} & \text{\textbackslash} \\
\text{H \ | \ H} & \text{H \ H \ | \ H} & \text{CHOH} \\
\text{O} & \text{O} & \text{/} \\
\text{H} & \text{H} & \text{R}'
\end{array}
$$

Secondary alcohols

If the OH group is attached to a carbon atom that is joined to three other carbon atoms it is a *tertiary* alcohol, and the carbon atom to which it is attached is a *tertiary carbon atom.*

$$
\begin{array}{ccc}
 & \text{C} & \\
\text{H} & \text{|} & \\
\text{HCH} & \text{C} & \text{R} \\
\text{H \ | \ H} & \text{|} & \text{\textbackslash} \\
\text{HC—C—CH} & \text{C—C—C—C} & \text{R}'\text{—C—OH} \\
\text{H \ | \ H} & \text{|} & \text{/} \\
\text{O} & \text{O} & \text{R}'' \\
\text{H} & \text{H} &
\end{array}
$$

Tertiary alcohols

The chemistry of the primary, secondary, and tertiary alcohols differs considerably; therefore it is important to know how to differentiate among them.

Common Alcohols

The alcohols of greatest commercial importance are methyl, ethyl, isopropyl, and *n*-butyl.

Methyl alcohol (CH_3OH) Methyl alcohol is used to a considerable extent for synthesis of organic compounds. It has been used as an antifreeze for automobiles. It is prepared mainly by synthesis from natural gas and steam as follows:

$$CH_4 + H_2O \xrightarrow[\text{catalyst}]{\Delta \text{ press}} CH_3OH + H_2 \tag{4-10}$$

but may be manufactured from carbon monoxide and hydrogen.

Ethyl alcohol (CH_3CH_2OH) Ethyl alcohol is used for the synthesis of organic compounds, the production of beverages, and the manufacture of medicines. It is prepared largely by fermentation processes. Alcohol intended for beverage purposes is manufactured by fermentation of starch derived from a variety of materials, such as corn, wheat, rye, rice, and potatoes. The reactions involved are as follows:

$$\text{Starch} + \text{water} \xrightarrow[\text{of malt}]{\text{enzyme}} \text{maltose} \tag{4-11}$$

$$\underset{\text{Maltose}}{C_{12}H_{22}O_{11}} + H_2O \xrightarrow[\text{of yeast}]{\text{enzyme}} 2 \text{ glucose} \tag{4-12}$$

Fermentation of the glucose yields carbon dioxide and alcohol:

$$\underset{\text{Glucose}}{C_6H_{12}O_6} \xrightarrow{\text{fermentation}} 2CO_2 + 2C_2H_5OH \tag{4-13}$$

Industrial alcohol is produced largely from the fermentation of solutions containing sugars which are difficult to reclaim, such as molasses and, in Europe, spent sulfite liquor.

$$C_{12}H_{22}O_{11} + H_2O \xrightarrow{\text{invertase}} \underset{\text{Glucose}}{C_6H_{12}O_6} + \underset{\text{Fructose}}{C_6H_{12}O_6} \tag{4-14}$$

$$\underset{\substack{\text{Glucose} \\ \text{Fructose}}}{C_6H_{12}O_6} \xrightarrow{\text{yeast}} 2CO_2 + 2C_2H_5OH$$

The residues remaining after distillation of the desired product, ethyl alcohol, constitute some of the most potent industrial wastes with which the environmental engineer has to deal.

Isopropyl alcohol ($CH_3CHOHCH_3$) Isopropyl alcohol is widely used in organic synthesis, and considerable amounts are sold as "dry gas" to prevent separation of water in the fuel tanks of automobiles. It is prepared by hydration of propylene derived from the cracking of petroleum.

n-**Butyl alcohol (CH$_3$CH$_2$CH$_2$CH$_2$OH)** Normal butyl alcohol is used to prepare butyl acetate, an excellent solvent. It is often referred to as "synthetic banana oil" because of its odor which resembles natural banana oil, amyl acetate. Normal butyl alcohol is prepared from cornstarch by a fermentation process utilizing a particular microorganism, *Clostridium acetobutylicum.* Considerable amounts of acetone and some ethyl alcohol and hydrogen are produced during the fermentation. The liquid wastes remaining after distillation of the desired products are classed as industrial wastes, and their treatment and ultimate disposal usually fall to the lot of the environmental engineer. They are similar in character to the residues from the production of ethyl alcohol but offer less promise of by-product recovery.

Physical Properties of Alcohols

The short-chain alcohols are completely soluble in water. Those with more than 12 carbon atoms are colorless waxy solids and very poorly soluble in water. The physical constants of several alcohols are given in Table 4-4.

Nomenclature

The alcohols of commercial significance are usually called by their common names. The IUPAC system must be employed, however, to differentiate among isomers and to name the higher members, such as hexadecanol. In

Table 4-4 Physical constants of normal primary alcohols*

Name of radical	IUPAC name of alcohol	Formula	Mp, °C	Bp, °C	Sp. gr., 20°/4°	Calcd. no. of isomers
Methyl	Methanol	CH$_3$OH	$-$ 97.8	64.7	0.793	1
Ethyl	Ethanol	C$_2$H$_5$OH	$-$117.3	78.4	0.789	1
Propyl	1-Propanol	C$_3$H$_7$OH	$-$127	97.2	0.804	2
Butyl	1-Butanol	C$_4$H$_9$OH	$-$ 89.2	117.7	0.810	4
Amyl	1-Pentanol	C$_5$H$_{11}$OH	$-$ 78.5	138	0.814	8
Hexyl	1-Hexanol	C$_6$H$_{13}$OH	$-$ 51.6	157.2	0.819	17
Heptyl	1-Heptanol	C$_7$H$_{15}$OH	$-$ 34.6	176	0.822	39
Octyl	1-Octanol	C$_8$H$_{17}$OH	$-$ 16.3	195	0.825	89
Nonyl	1-Nonanol	C$_9$H$_{19}$OH	$-$ 5	213	0.827	211†
Decyl	1-Decanol	C$_{10}$H$_{21}$OH	7	231	0.829	507

* From E. Wertheim and H. Jeskey, "Introductory Organic Chemistry," 3d ed., McGraw-Hill, New York, 1956. Table reproduced by permission of the authors.

† These numbers are for all the isomers of a given carbon content.

Table 4-5 Nomenclature of alcohols*

Formula	Common name	IUPAC name
CH_3OH	Methyl alcohol	Methanol
C_2H_5OH	Ethyl alcohol	Ethanol
$CH_3CH_2CH_2OH$	n-Propyl alcohol	1-Propanol
$\begin{array}{c} CH_3 \\ \diagdown \\ CHOH \\ \diagup \\ CH_3 \end{array}$	Isopropyl alcohol	2-Propanol
$CH_3CH_2CH_2CH_2OH$	n-Butyl alcohol	1-Butanol
$\begin{array}{c} CH_3 \\ \diagdown \\ CHCH_2OH \\ \diagup \\ CH_3 \end{array}$	Isobutyl alcohol	2-Methyl-1-propanol
$\begin{array}{c} CH_3 \\ \diagdown \\ CHOH \\ \diagup \\ C_2H_5 \end{array}$	sec-Butyl alcohol	2-Butanol
$(CH_3)_3COH$	tert-Butyl alcohol	2-Methyl-2-propanol

* From E. Wertheim and H. Jeskey, "Introductory Organic Chemistry," 3d ed., McGraw-Hill, New York, 1956. Table reproduced by permission of the authors.

this terminology, the names of all alcohols end in -ol. The formulas, common names, and IUPAC names of several alcohols are given in Tables 4-4 and 4-5. In this system, the longest carbon chain containing the hydroxyl group determines the name. The location of alkyl groups and the hydroxyl group are described by number; e.g., isobutyl alcohol is described as 2-methyl-1-propanol.

Polyhydroxy Alcohols

Those alcohols having two hydroxyl groups per molecule are known as *glycols*. The principal glycol of commercial significance is *ethylene glycol,* which is prepared from ethylene. Ethylene adds hypochlorous acid to form ethylene chlorohydrin,

$$\begin{array}{c} H\ \ H H\ \ H \\ HC{=}CH + HOCl \rightarrow HC{-}CH \\ |\ \ \ | \\ Cl\ \ OH \end{array} \qquad (4\text{-}15)$$

Ethylene chlorohydrin

and treatment of the chlorohydrin with sodium bicarbonate produces ethylene glycol.

$$H_2C-CH_2 + NaHCO_3 \rightarrow NaCl + CO_2 + H_2C-CH_2 \qquad (4\text{-}16)$$

$$\underset{\text{Cl} \quad \text{OH}}{|} \qquad\qquad\qquad\qquad \underset{\underset{\text{Ethylene glycol}}{\text{OH OH}}}{|}$$

It is used extensively as a nonevaporative, radiator antifreeze compound, formerly sold exclusively under the trade name Prestone but now sold under a wide variety of names.

Glycerol or glycerin is a trihydroxy alcohol.

$$\begin{array}{ccc} H_2 & H & H_2 \\ C-C-C \\ | & | & | \\ OH & OH & OH \end{array}$$

It was formerly produced in large quantities in the soap industry through saponification of fats and oils. Presently considerable amounts are produced by synthesis. Glycerol is used in a wide variety of commercial products: foods, cosmetics, medicines, tobaccos, and so on. It is used for the manufacture of nitroglycerin, an important component of dynamite.

Chemical Reactions of Alcohols

Alcohols undergo two types of reaction that are of interest to environmental engineers.

Ester formation Alcohols react with acids, both inorganic and organic, to form esters. Inorganic hydroxy acids yield "inorganic" esters:

$$ROH + H_2SO_4 \rightleftharpoons H_2O + ROSO_3H \qquad (4\text{-}17)$$

Organic acids yield organic esters:

$$ROH + R_1CO_2H \rightleftharpoons H_2O + R_1CO_2R \qquad (4\text{-}18)$$

Organic esters are discussed in Sec. 4-6.

Oxidation Most alcohols are readily oxidized by strong oxidizing agents and by many microorganisms. The product of the oxidation depends upon the class of alcohol involved.

Primary alcohols are oxidized to aldehydes. The general equation is

$$\underset{\substack{\text{Primary} \\ \text{alcohol}}}{RCH_2OH} + \tfrac{1}{2}O_2 \rightarrow H_2O + \underset{\text{An aldehyde}}{\overset{H}{RC}\!\!=\!\!O} \qquad (4\text{-}19)$$

Care must be used in selecting the oxidizing agent, or the aldehyde may be oxidized still further to an acid.

Secondary alcohols are oxidized to ketones.

$$\underset{\substack{| \\ H}}{\overset{OH}{H_3C-C-CH_3}} + \tfrac{1}{2}O_2 \rightarrow H_2O + H_3C-\overset{O}{\overset{\|}{C}}-CH_3 \qquad (4\text{-}20)$$

Isopropyl alcohol   Acetone
a ketone

The ketones are not easily oxidized and can usually be recovered completely.

Tertiary alcohols are not oxidized by ordinary agents in aqueous solution. When attacked by very strong oxidizing agents they are converted to carbon dioxide and water.

Microorganisms oxidize primary and secondary alcohols readily under aerobic conditions. The end products are carbon dioxide and water, but aldehydes and ketones are believed to exist as intermediates. Present evidence indicates that microorganisms cannot attack tertiary alcohols, except through terminal methyl groups, as with the hydrocarbons.

4-4 ALDEHYDES AND KETONES

Aldehydes are the oxidation products of primary alcohols (ROH). *Ketones* are the oxidation products of secondary alcohols

$$(R-CHOH-R')$$

Aldehydes

Oxidation of primary alcohols to aldehydes is as follows:

$$\underset{\substack{| \\ H}}{\overset{H}{R-C-OH}} + \tfrac{1}{2}O_2 \rightarrow \overset{H}{R-C{=}O} + H_2O \qquad (4\text{-}21)$$

All aldehydes have the characteristic *carbonyl* group, $-\overset{H}{C}{=}O$. The general structural formula for an aldehyde is R—CHO, where R represents any alkyl group, CH_3^-, and $C_2H_5^-$, and so on.

Aldehydes can also be formed from unsaturated hydrocarbons by ozonation. The hydrocarbons are first converted to an ozonide by ozone,

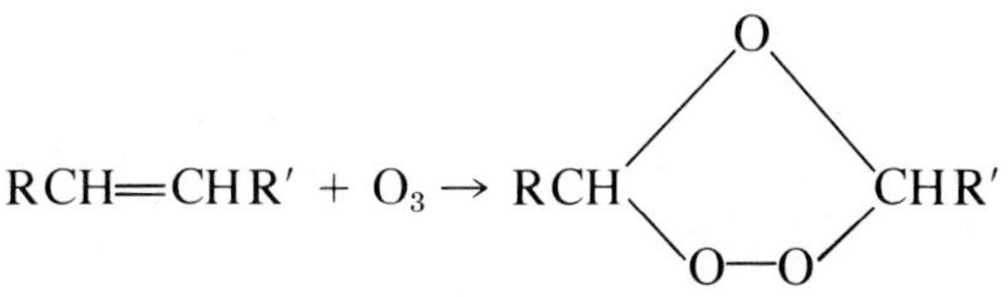

$$RCH{=}CHR' + O_3 \rightarrow RCH\overset{O}{\underset{O-O}{\diamond}}CHR' \qquad (4\text{-}22)$$

An ozonide

The ozonides react readily with water to form aldehydes,

$$RCH \begin{array}{c} O \\ \diagdown \diagup \\ \diagdown \diagup \\ O-O \end{array} CHR' + H_2O \rightarrow RCHO + R'CHO + H_2O_2 \qquad (4\text{-}23)$$

These reactions are particularly significant in air pollution, where aldehydes are formed when unsaturated hydrocarbons discharged in automobile exhausts combine with ozone catalytically produced by reactions of oxygen with sunlight in the presence of oxides of nitrogen. The aldehydes so formed cause eye irritation, one of the most serious problems associated with air pollution. The oxides of nitrogen required for catalyzing ozone formation are formed in great quantities during high-temperature combustion of fossil fuels in steam power plants and internal combustion engines. Automobile and truck engine exhaust gases are a particularly significant source because of wide distribution at ground level where contact with human, animal, and plant life is most probable.

Although a wide variety of aldehydes can be formed from primary alcohols, only a few are of commercial importance.

Formaldehyde Formaldehyde is formed by the oxidation of methyl alcohol.

$$\begin{array}{cc} H & H \\ H C-OH + \tfrac{1}{2}O_2 \rightarrow & H C{=}O + H_2O \\ H & \text{Formaldehyde} \end{array} \qquad (4\text{-}24)$$

It is used extensively in organic synthesis. It is very toxic to microorganisms, and, because of this property, it is used in embalming fluids and fluids used for the preservation of biological specimens. Industrial wastes containing formaldehyde were considered at one time to be too toxic for treatment by biological methods. Through dilution of such wastes to reduce the concentration of formaldehyde below 1500 mg/l, it was found that microorganisms could use the formaldehyde as food and oxidize it to carbon dioxide and water. This experience has led to the concept of *toxicity thresholds* in industrial waste treatment practice. It means that below certain concentrations all materials are nontoxic. The completely mixed activated sludge system employs this concept.

Acetaldehyde Acetaldehyde is formed by oxidation of ethyl alcohol.

$$\begin{array}{c} H \\ CH_3CH_2OH + \tfrac{1}{2}O_2 \rightarrow CH_3-C{=}O + H_2O \\ \text{Acetaldehyde} \end{array} \qquad (4\text{-}25)$$

It is used extensively in organic synthesis. A major industrial use involves its condensation with formaldehyde to produce pentaerythritol [$C(CH_2OH)_4$], an important intermediate for the production of a wide variety of products, including aldehyde resin paints. Development of a biological treatment process for

Table 4-6 Common aldehydes

Common name	IUPAC name	Formula
Formaldehyde	Methanal	HCHO
Acetaldehyde	Ethanal	CH_3CHO
Propionaldehyde	Propanal	C_2H_5CHO
Butyraldehyde	Butanal	C_3H_7CHO
Valeraldehyde	Pentanal	C_4H_9CHO
Caproaldehyde	Hexanal	$C_5H_{11}CHO$
Heptaldehyde	Heptanal	$C_6H_{13}CHO$
Acrolein		$CH_2{=}CHCHO$
Citral		$C_9H_{15}CHO$
Citronellal		$C_9H_{17}CHO$

the formaldehyde-bearing industrial wastes from the manufacture of pentaery-thritol led to the concept of toxicity threshold mentioned above.

A wide variety of aldehydes are of commercial interest. The names and formulas of several of them are given in Table 4-6. The IUPAC names of all aldehydes end in -al.

Ketones

Ketones are prepared by the oxidation of secondary alcohols.

$$R{-}\overset{\overset{\displaystyle H}{|}}{\underset{\underset{\displaystyle H}{|}}{C}}{-}R' + \tfrac{1}{2}O_2 \rightarrow R{-}\overset{\overset{\displaystyle O}{\|}}{C}{-}R' + H_2O \qquad (4\text{-}26)$$
$$\text{Ketone}$$

Ketones have two alkyl groups attached to the carbonyl group, $-\overset{\overset{\displaystyle O}{\|}}{C}-$, while aldehydes have one R group and a hydrogen atom. The R groups in ketones may be the same or different.

Acetone Acetone (dimethyl ketone) is the simplest ketone and is produced by oxidation of isopropyl alcohol (2-propanol).

$$CH_3{-}\overset{\overset{\displaystyle H}{\underset{\displaystyle O}{|}}}{\underset{\underset{\displaystyle H}{|}}{C}}{-}CH_3 + \tfrac{1}{2}O_2 \rightarrow CH_3{-}\overset{\overset{\displaystyle O}{\|}}{C}{-}CH_3 + H_2O \qquad (4\text{-}27)$$
$$\text{Acetone}$$

Ethyl methyl ketone Ethyl methyl ketone is prepared by the oxidation of 2-butanol.

$$CH_3-\overset{\overset{\textstyle H}{\overset{\textstyle O}{|}}}{\underset{\textstyle H}{C}}-CH_2-CH_3 + \tfrac{1}{2}O_2 \rightarrow CH_3-\overset{\overset{\textstyle O}{\|}}{C}-C_2H_5 + H_2O \qquad (4\text{-}28)$$
Ethyl methyl ketone

Ketones are used as solvents in industry and for the synthesis of a wide variety of products. The names of a few ketones are given in Table 4-7.

Chemical Properties of Aldehydes and Ketones

Aldehydes and ketones differ in ease of oxidation.

1. Aldehydes are easily oxidized to the corresponding acids.

$$R-\overset{\overset{\textstyle H}{|}}{C}{=}O + \tfrac{1}{2}O_2 \rightarrow R-\overset{\overset{\textstyle O}{\|}}{C}-OH \qquad (4\text{-}29)$$

2. Ketones are difficult to oxidize. This is because there is no hydrogen attached to the carbonyl group. As a result, further oxidation must initiate in one of the alkyl groups, the molecule is cleaved, and two or more acids are produced.

$$CH_3\overset{\overset{\textstyle O}{\|}}{C}CH_3 + 2O_2 \rightarrow CO_2 + H_2O + CH_3COOH \qquad (4\text{-}30)$$
　　Acetone　　　　　　　　　　　　　　　　　　　Acetic acid

In the case of acetone, carbon dioxide and acetic acid are formed. Theoretically, formic acid should be formed but it is so easily oxidized that it is converted under the prevailing conditions to carbon dioxide and water. Higher ketones such as diethyl ketone (3-pentanone) are oxidized as indicated by Eq. (4-31).

Table 4-7 Common ketones

Common name	IUPAC name
Acetone	Propanone
Ethyl methyl ketone	Butanone
Diethyl ketone	3-Pentanone
Methyl propyl ketone	2-Pentanone
Methyl isopropyl ketone	3-Methyl-2-butanone
n-Butyl methyl ketone	2-Hexanone
Ethyl propyl ketone	3-Hexanone
Dipropyl ketone	4-Heptanone
Dibutyl ketone	5-Nonanone

$$R-\overset{\overset{\displaystyle H}{|}}{\underset{\underset{\displaystyle H}{|}}{C}}-\overset{\overset{\displaystyle O}{\|}}{C}-R' + \tfrac{3}{2}O_2 \rightarrow R-\overset{\overset{\displaystyle O}{\|}}{C}-OH + R'-\overset{\overset{\displaystyle O}{\|}}{C}-OH \qquad (4\text{-}31)$$

Oxidation of both aldehydes and ketones is accomplished readily by many microorganisms. However, since organic acids serve as good food supply, the end products are carbon dioxide and water.

4-5 ACIDS

Acids represent the highest oxidation state that an organic compound can attain. Further oxidation results in the formation of carbon dioxide and water, which are classed as inorganic compounds, and the organic compound is considered completely destroyed.

$$CH_4 \rightarrow CH_3OH \rightarrow H_2C{=}O \rightarrow HCOOH \rightarrow H_2O + CO_2 \qquad (4\text{-}32)$$

| Hydro-carbon | Alcohol | Aldehyde | Acid | Products of complete oxidation |

All organic acids contain the $-\overset{\overset{\displaystyle O}{\|}}{C}-OH$ group. This is called the *carboxyl* group and is commonly written —COOH. Acids with one carboxyl group are known as *monocarboxylic* acids and those with more than one are *polycarboxylic* acids. The acids may be saturated or unsaturated. Some contain hydroxy groups within the molecule.

Saturated Monocarboxylic Acids

A wide variety of saturated monocarboxylic acids occur in nature as constituents of fats, oils, and waxes. Unsaturated acids are also found in these materials, and, as a result, both types are commonly known as *fatty* acids. The majority of the fatty acids derived from natural products have an even number of carbon atoms and usually have straight-chain or normal structure.

Physical properties The first nine members, C_1 to C_9, are liquids. All the others are greasy solids. Formic, acetic, and propionic acid have sharp penetrating odors; the remaining liquid acids have disgusting odors, particularly butyric and valeric. Butyric acid gives rancid butter its characteristic odor. Industrial wastes from the dairy industry must be treated with considerable care to prevent formation of butyric acid and consequent odor problems.

The names, formulas, and physical constants of the important saturated acids are given in Table 4-8. All the acids are considered weak acids from the viewpoint of ionization. Formic acid is the strongest of all.

Table 4-8 Physical constants of some normal monocarboxylic acids*

Common name	IUPAC name	Formula	Mp, °C	Bp, °C	Sp. gr., $20°/4°$	K_a at 25°
Formic	Methanoic	$HCOOH$	8.4	100.7	$1.226^{16°}$	2.14×10^{-4}
Acetic	Ethanoic	CH_3COOH	16.6	118.1	1.049	1.75×10^{-5}
Propionic	Propanoic	C_2H_5COOH	-22	141.1	0.992	1.4×10^{-5}
Butyric	Butanoic	C_3H_7COOH	-4.7	163.5	0.959	1.48×10^{-5}
Valeric	Pentanoic	C_4H_9COOH	-34.5	187	0.942	1.6×10^{-5}
Caproic	Hexanoic	$C_5H_{11}COOH$	-2	205	$0.945^{0°}$	
Enanthic	Heptanoic	$C_6H_{13}COOH$	-10	223.5	$0.913^{25°}$	
Caprylic	Octanoic	$C_7H_{15}COOH$	16	237.5	0.910	
Pelargonic	Nonanoic	$C_8H_{17}COOH$	12	254	0.906	
Capric	Decanoic	$C_9H_{19}COOH$	31.5	268–70	$0.886^{40°}$	
Palmitic	Hexadecanoic	$C_{15}H_{31}COOH$	64	339–56 (dec.)	$0.853^{62°}$	
Stearic	Octadecanoic	$C_{17}H_{35}COOH$	69.4	383	$0.847^{69°}$	

* From E. Wertheim and H. Jeskey, "Introductory Organic Chemistry," 3d ed., McGraw-Hill, New York, 1956. Table reproduced by permission of the authors.

Nomenclature The common names are usually used for most of the acids, except for those with 7, 8, 9, and 10 carbon atoms. The IUPAC names are given in Table 4-8. In naming derivatives of acids, the IUPAC system is frequently abandoned for a system using Greek letters to identify the carbon atoms. In this system the carboxyl group is the reference point, and carbon atoms are numbered from it as follows:

$$\underset{\delta}{\overset{5}{CH_3}}\ \underset{\gamma}{\overset{4}{CH_2}}\ \underset{\beta}{\overset{3}{CH_2}}\ \underset{\alpha}{\overset{2}{CH_2}}\ \overset{1}{COOH}$$

The carbon atom next to the carboxyl group is *alpha*, the next *beta*, then *gamma*, *delta*, and so on. The terminal carbon atom is also referred to as being in the *omega* position. The α-amino acids are particularly important compounds and are discussed in Sec. 4-22.

Unsaturated Monocarboxylic Acids

The principal unsaturated monocarboxylic acids are as follows:

Acrylic acid ($CH_2{=}CHCOOH$) Acrylic acid is used extensively because of its ability to polymerize, a characteristic of many compounds with unsaturated linkages. Derivatives of the acid are used to form colorless plastics such as Lucite and Plexiglas.

Oleic acid $[CH_3(CH_2)_7CH\!=\!CH(CH_2)_7COOH]$

Linoleic acid $[CH_3(CH_2)_4CH\!=\!CHCH_2CH\!=\!CH\!=\!(CH_2)_7COOH]$

Linolenic acid $[CH_3(CH_2CH\!=\!CH)_3CH_2(CH_2)_6COOH]$

    Oleic, linoleic, and linolenic acids are normal constituents of the glycerides of most fats and oils. Oleic acid is considered to be an essential acid in the diet of man and animals. Linoleic and linolenic acids as glycerides are important constituents of linseed and other drying oils. Their value for this purpose is dependent upon the multiple double bonds which they possess.

Chemical properties of acids The chemical properties of acids are determined largely by the carboxyl group. All form metallic salts that have a wide range of commercial use. In addition, the unsaturated acids have chemical properties characterized by the double bond, as described under unsaturated hydrocarbons in Sec. 4-2. The unsaturated acids may be reduced with hydrogen to give corresponding saturated acids.

    Organic acids serve as food for many microorganisms and are oxidized to carbon dioxide and water. Ease and rate of oxidation are believed to be enhanced by the presence of unsaturated linkages. The rate of biological attack on high-molecular-weight fatty acids is often limited by their solubility in water. This is a particular problem in sludge digesters where fatty materials tend to float and segregate themselves in a scum layer.

Polycarboxylic Acids

The most important of the polycarboxylic acids are those that have two carboxyl groups, one on each end of a normal chain of carbon atoms. The most important acids are listed in Table 4-9.

Table 4-9 Dicarboxylic acids

Name	Formula
Oxalic	$(COOH)_2$
Malonic	$CH_2(COOH)_2$
Succinic	$(CH_2)_2(COOH)_2$
Glutaric	$(CH_2)_3(COOH)_2$
Adipic	$(CH_2)_4(COOH)_2$
Pimelic	$(CH_2)_5(COOH)_2$
Suberic	$(CH_2)_6(COOH)_2$

Adipic acid is of some interest to environmental engineers because it is used in the manufacture of nylon fiber and may be expected to occur in the industrial wastes of that industry.

Hydroxy Acids

Hydroxy acids have OH groups attached to the molecule other than in the carboxyl group. Thus they act chemically as acids and as alcohols. A number of the hydroxy acids have special names. Some examples are

$HOCH_2COOH$ Hydroxyacetic acid, glycolic acid

$CH_3CHOHCOOH$ α-Hydroxypropionic acid, lactic acid

$HOCH_2CH_2COOH$ β-Hydroxypropionic acid, hydracrylic acid

$HOCH_2CH_2CH_2COOH$ γ-Hydroxybutyric acid

Lactic acid, α-hydroxypropionic acid, is of special interest to environmental engineers since it is formed during bacterial fermentation of milk and therefore is a normal constituent of industrial wastes from the dairy industry. Whey from cheese making contains considerable amounts of lactic acid. It is the principal acid in *sauerkraut* juice and prevents spoilage of the sauerkraut.

Lactic acid is also of interest because it is the first organic compound, to come to our attention, that possesses the property of *optical activity*. The common form of lactic acid is *levorotatory* and turns polarized light to the left. An uncommon form is *dextrorotatory* and turns polarized light to the right.

Many organic compounds are optically active. In order to be optically active, a compound must generally possess at least one *asymmetric* carbon atom. An asymmetric carbon atom is one having four dissimilar groups attached to it, as follows:

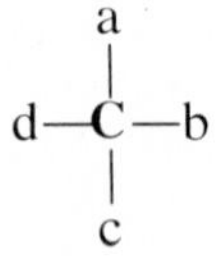

$$
\begin{array}{c}
a \\
| \\
d-C-b \\
| \\
c
\end{array}
$$

Lactic acid has one asymmetric carbon atom, and two optical isomers are possible.

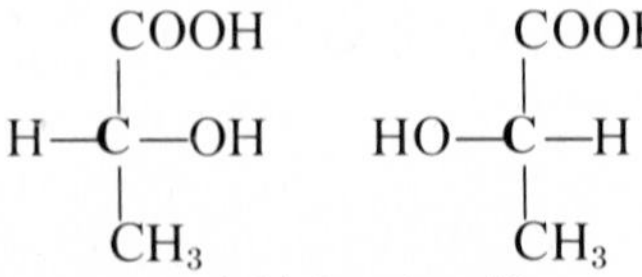

$$
\begin{array}{cc}
\text{COOH} & \text{COOH} \\
| & | \\
\text{H}-\text{C}-\text{OH} & \text{HO}-\text{C}-\text{H} \\
| & | \\
\text{CH}_3 & \text{CH}_3
\end{array}
$$

D- and L-Lactic acids

Hydroxy Polycarboxylic Acids

There are several hydroxy polycarboxylic acids.

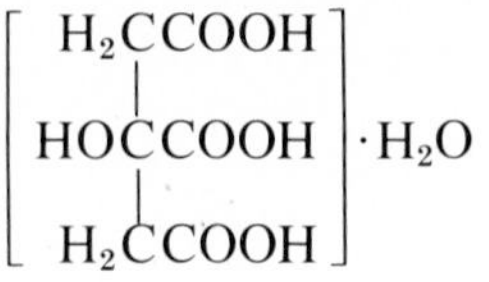

$$\left[\begin{array}{l} H_2CCOOH \\ HOCCOOH \\ H_2CCOOH \end{array}\right]\cdot H_2O$$

Citric acid

$$\begin{array}{l} COOH \\ HCOH \\ HOCH \\ COOH \end{array}$$

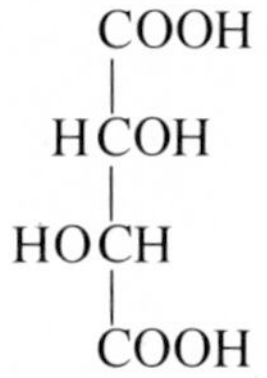

L(+)-Tartaric acid

Tartaric acid occurs in many fruits, especially grapes, and is present in canning and winery wastes. Citric acid is the major acid of all citrus fruits: oranges, lemons, limes, and grapefruit. It is, of course, a major component of the liquid wastes of the citrus industry.

4-6 ESTERS

Esters are compounds formed by the reaction of acids and alcohols. In organic chemistry, they correspond to salts in inorganic chemistry. The reaction between low-molecular-weight organic acids and alcohols is never complete. Hydrolysis occurs and a reversible reaction results. The reaction may be represented by the general equation

$$RCO-\boxed{OH + H}-OR_1 \rightleftharpoons H_2O + RCOOR_1 \tag{4-33}$$

The general formula of an ester is

$$\overset{O}{\underset{\parallel}{R-C}}-O-R'$$

A wide variety of esters are used in chemical manufacturing. Most esters have highly pleasing odors. Butyl acetate smells like banana oil (amyl acetate) and is used for solvent purposes. Many esters are used in flavoring extracts and perfumes.

Esters have been used to some extent as immiscible solvents in the separation and purification of antibiotics. Considerable quantities often reach the sewer system and become an industrial waste problem. Enzymes liberated by many microorganisms hydrolyze esters to yield the corresponding acid and alcohol.

$$\overset{O}{\underset{\parallel}{R-C}}-OR' + HOH \xrightarrow{enzyme} RCOOH + R'-OH \tag{4-34}$$

The acid and alcohol serve as bacterial food and are oxidized to carbon dioxide and water, as discussed in Secs. 4-3 and 4-5.

4-7 ETHERS

Ethers are formed by treatment of alcohols with strong dehydrating agents. In the reaction, one molecule of water is removed from two molecules of alcohol.

$$RO\,\lvert H + HO \rvert\, R' \xrightarrow[\substack{\text{dehydrating}\\ \text{agent}}]{\Delta} \underset{\text{Ether}}{R-O-R'} + H_2O \qquad (4\text{-}35)$$

The two fragments of the alcohol join to form an ether. The alkyl groups are joined through an oxygen atom; thus a carbon-to-oxygen-to-carbon bond,

$$-\overset{\displaystyle |}{\underset{\displaystyle |}{C}}-O-\overset{\displaystyle |}{\underset{\displaystyle |}{C}}-,$$ is established.

Ethers are used widely as solvents. The low-molecular-weight ethers are highly flammable. When left exposed to air, they are prone to form peroxides that are extremely explosive, particularly when recovery by distillation is practiced and the distillation is allowed to go to dryness. Diethyl ether has been used widely as an anesthetic.

Ethers are extremely resistant to biological oxidation. Fortunately, they are relatively insoluble in water and can be separated from industrial wastes by flotation or decantation procedures.

4-8 ALKYL HALIDES

The *alkyl halides* are used extensively in organic synthesis, and a few of them have important industrial uses.

Simple Alkyl Halides

The simple alkyl halides (R—X) may be prepared by treatment of an alcohol with PCl_3.

$$3ROH + PCl_3 \rightarrow P(OH)_3 + 3RCl \qquad (4\text{-}36)$$
$$3CH_3CH_2CH_2OH + PCl_3 \rightarrow P(OH)_3 + \underset{\substack{n\text{-Propyl}\\ \text{chloride}}}{3C_3H_7Cl}$$

Phosphorus bromide may also be employed:

$$3ROH + PBr_3 \rightarrow P(OH)_3 + 3RBr \qquad (4\text{-}37)$$

The alkyl halides are of great value in organic synthesis because they react with potassium cyanide to form compounds with an additional carbon atom.

$$R \cdot \overline{|I + K|} \cdot CN \rightarrow KI + R \cdot CN \qquad (4\text{-}38)$$

A nitrile
(alkyl
cyanide)

The nitrile formed can be hydrolyzed to an acid and then reduced to an alcohol, if desired. The alcohol can be converted to an alkyl halide and the process repeated. In this manner the organic chemist can increase the length of carbon-chain compounds one atom at a time.

Methyl chloride (CH_3Cl) and ethyl chloride (C_2H_5Cl) were used extensively as refrigerants in the past. Ethyl chloride is used in the manufacture of tetraethyl lead, great quantities of which are used in the production of high-octane, antiknock gasolines.

$$4C_2H_5Cl + 4NaPb \text{ (alloy)} \xrightarrow{40\text{–}60°} 4NaCl + 3Pb + (C_2H_5)_4Pb \qquad (4\text{-}39)$$

Tetraethyl
lead

During combustion the lead is converted to lead oxide and is emitted in the exhaust gases, creating a lead pollution problem. The lead also fouls catalytic devices being used to reduce the quantity of unburned fuel discharged to the atmosphere. For these reasons use of leaded gasoline is being restricted.

Polyhalogen Compounds

A wide variety of *polyhalogen* compounds are used for industrial purposes.

Ethylene bromide Ethylene bromide is formed from ethylene by addition of bromine. It has many industrial uses.

$$\begin{array}{ccc} H \ \ H & & H \ \ H \\ HC{=}CH + Br_2 \rightarrow & & HC{-}CH \\ & & Br \ \ Br \end{array} \qquad (4\text{-}40)$$

Ethylene
bromide

Chloroform ($CHCl_3$) Chloroform (trichloromethane) was one of our first anesthetics (1847) and was widely used until about 1920. It is used in industry as a solvent for oils, waxes, etc. It is nonflammable. Chloroform has been found present in microgram per liter to milligram per liter concentrations in drinking water supplies as a result of chlorination for disinfection. It apparently forms through reaction of chlorine with organics of biological origin commonly present in natural waters. Because it is a potential carcinogen, a standard to limit the concentration of chloroform and other trihalomethanes in drinking water is being proposed.

Carbon tetrachloride (CCl_4) Carbon tetrachloride (tetrachloromethane) is widely used as a fire extinguisher in small units (Pyrene) and as a solvent. It is considered a toxic compound, and its use should be restricted to well-ventilated areas. Its use as a fire extinguisher is fraught with some danger. In contact with hot iron and oxygen it is converted to phosgene ($COCl_2$), a highly toxic gas. For this reason, trained fire fighters use other types of fire-fighting equipment.

Freon (CCl_2F_2) Freon is dichlorodifluoromethane. It is the preferred refrigerant in household appliances because of its nonflammable and nontoxic properties.

4-9 SIMPLE COMPOUNDS CONTAINING NITROGEN

The simple aliphatic compounds containing nitrogen are of three types: amines, amides, and nitriles (cyanides).

Amines

The amines are alkyl derivatives of ammonia. They are of three types: primary, secondary, and tertiary.

$$
\underset{\substack{\text{Primary} \\ \text{amine}}}{R-NH_2} \qquad \underset{\substack{\text{Secondary}^3 \\ \text{amine}}}{\overset{R}{\underset{R'}{\diagdown}}NH} \qquad \underset{\substack{\text{Tertiary}^3 \\ \text{amine}}}{\overset{R}{\underset{R'}{\diagdown}}N-R''}
$$

In *primary amines,* one hydrogen atom of ammonia has been replaced by an alkyl group such as CH_3-, C_2H_5-, and so on. In *secondary amines,* two hydrogen atoms of ammonia have been replaced by alkyl groups, and in *tertiary amines,* all three hydrogens have been replaced. The amines, like ammonia, are all basic in reaction. The basicity increases from primary to tertiary.

The amines are found in certain industrial wastes, particularly those from the fish and beet-sugar industries. Very little is known about their susceptibility to biological oxidation.

Tertiary amines combine with alkyl halides to form *quaternary ammonium salts* as follows:

$$
\underset{\overset{|}{R}}{\overset{\overset{R}{|}}{R-N}} + RCl \xrightarrow{\Delta} \left[\underset{\overset{|}{R}}{\overset{\overset{R}{|}}{R-N-R}} \right]^{+} + Cl^{-} \tag{4-41}
$$

The compounds formed are actually chloride salts and ionize to form a quaternary ammonium ion and a chloride ion. The quaternary ammonium salts have bactericidal properties which can be enhanced by the proper choice of the R groups. They are therefore of interest to public health engineers, who find them

---

[3] R, R', R" represent alkyl groups. They may all be different or all alike.

useful as disinfecting agents in food- and beverage-dispensing establishments. They are also used as disinfectants in the laundering of babies' diapers to control infections of bacteria responsible for the rapid hydrolysis of urea. Solutions of the quaternary ammonium salts are sold for disinfecting purposes under a variety of trade names.

Amides

The amides may be considered as being derived from organic acids and ammonia under special conditions. The ordinary reaction between ammonia and an organic acid, of course, produces an ammonium salt.

$$RCOOH + NH_3 \rightarrow RCOO^- + NH_4^+ \tag{4-42}$$

Under special conditions, an amide results.

$$R-\overset{O}{\underset{\|}{C}}-OH + H-NH_2 \rightarrow R-\overset{O}{\underset{\|}{C}}-NH_2 + H_2O \tag{4-43}$$
$$\text{Amide}$$

Amides are of considerable significance to organic chemists in synthetic work. When they are caused to react with a halogen (Hofmann reaction) an atom of carbon is lost from the amide, and an amine with one less carbon atom is formed.

$$CH_3CONH_2 + Br_2 + 4NaOH \rightarrow 2NaBr + Na_2CO_3 + 2H_2O + CH_3NH_2 \tag{4-44}$$

This constitutes a method of reducing the length of a carbon chain by one atom.

Amides are of little importance to the environmental engineer except that the amide group $\left(-\overset{O}{\underset{\|}{C}}-NH_2\right)$ is related to the *peptide linkage* $\left(-\overset{O}{\underset{\|}{C}}-\overset{H}{\underset{}{N}}-\right)$, as discussed in Sec. 4-22.

Urea,

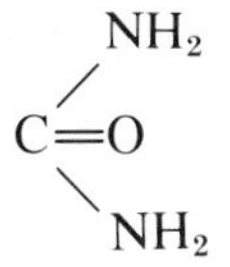

$$C=O \begin{cases} NH_2 \\ NH_2 \end{cases}$$

is an amide of considerable importance because of its many commercial uses and because it is a normal constituent of urine. It is a constituent of many agricultural fertilizers and is used in the manufacture of synthetic resins.

Although urea was originally considered an organic compound and its accidental production from ammonium cyanate by Wöhler is considered to have initiated the modern age of organic chemistry, it is in effect an inorganic compound since it cannot be used by saprophytic bacteria as a source of energy. In aqueous solutions containing soil bacteria, urea is hydrolyzed to carbon

Table 4-10 Important nitriles

Name as nitrile	Name as cyanide	Formula
Acetonitrile	Methyl cyanide	$CH_3—CN$
Propionitrile	Ethyl cyanide	$CH_3—CH_2—CN$
Acrylonitrile	Vinyl cyanide	$CH_2=CH—CN$

dioxide and ammonia. These combine to form ammonium carbonate in the presence of water.

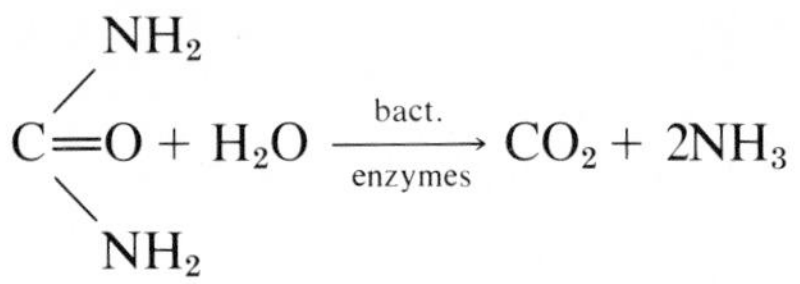

$$\underset{NH_2}{\overset{NH_2}{C}}=O + H_2O \xrightarrow[enzymes]{bact.} CO_2 + 2NH_3 \qquad (4\text{-}45)$$

and
$$CO_2 + 2NH_3 + H_2O \rightleftharpoons (NH_4)_2CO_3 \qquad (4\text{-}46)$$

The penetrating odor of latrines, privies, and some urinals is due to bacterial infections and their action on urea with subsequent release of free ammonia to the atmosphere. The use of disinfectants will control the decomposition of urea and, thereby, control odors.

Nitriles

Nitriles, or organic cyanides, are important compounds of industry. They have the general formula R—CN, and the R group may be saturated or unsaturated. The names and formulas of a few nitriles of industrial importance are given in Table 4-10.

The nitriles are used extensively in the manufacture of synthetic fibers and can be expected to be present in industrial wastes of that industry. Some are quite toxic to microorganisms.

4-10 CYCLIC ALIPHATIC COMPOUNDS

A number of *cyclic aliphatic* hydrocarbons are known. Many of these occur in petroleum and are known as naphthenes.

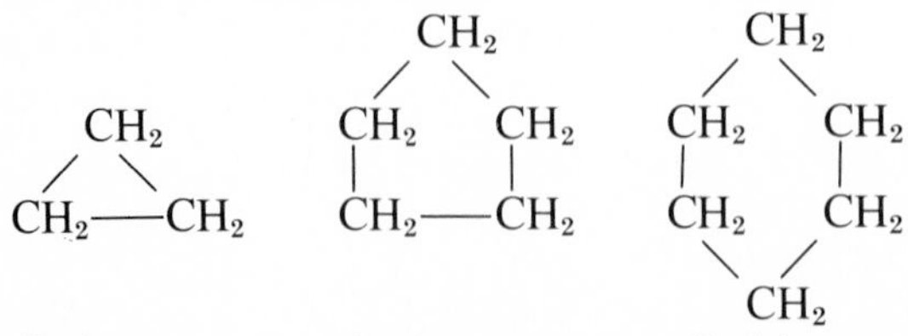

Cyclopropane   Cyclopentane   Cyclohexane

They are characterized by having two atoms of hydrogen attached to each carbon in the ring; i.e., they are saturated.

A wide variety of cyclic alcohols and ketones are known. Examples are cyclohexanol and cyclohexanone.

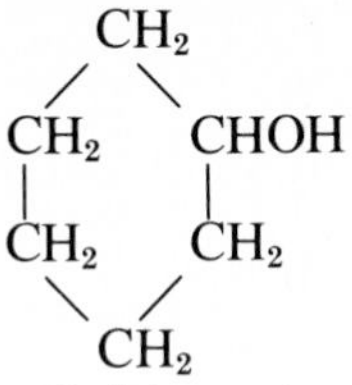

$$
\begin{array}{cc}
CH_2 & CH_2 \\
\diagup \quad \diagdown & \diagup \quad \diagdown \\
CH_2 \quad CHOH & CH_2 \quad C\!=\!O \\
| \qquad | & | \qquad | \\
CH_2 \quad CH_2 & CH_2 \quad CH_2 \\
\diagdown \quad \diagup & \diagdown \quad \diagup \\
CH_2 & CH_2 \\
\text{Cyclohexanol} & \text{Cyclohexanone}
\end{array}
$$

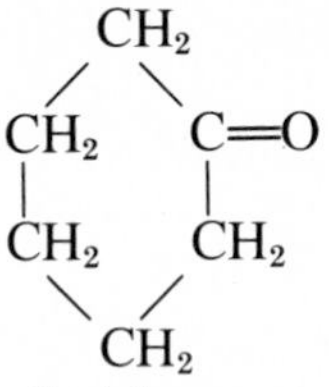

4-11 MERCAPTANS OR THIOALCOHOLS

Mercaptans or *thioalcohols* are aliphatic compounds that contain sulfur. They have a structure similar to alcohols, except that oxygen is replaced by sulfur.

$$
\begin{array}{cc}
ROH & RSH \\
\text{Alcohol} & \text{Mercaptan}
\end{array}
$$

Mercaptans are noted for their disagreeable odor and are found in certain industrial wastes, particularly those from the pulping of wood by the Kraft or sulfate process. They are considered to be quite toxic to fish. The odor of skunks is largely due to butyl mercaptan.

AROMATIC COMPOUNDS

4-12 INTRODUCTION

The *aromatic* organic compounds are all ring compounds or have cyclic groups of aromatic nature in their structure. The carbon atoms in these ring compounds have only one covalent bond, in contrast to those in aliphatic compounds with two.

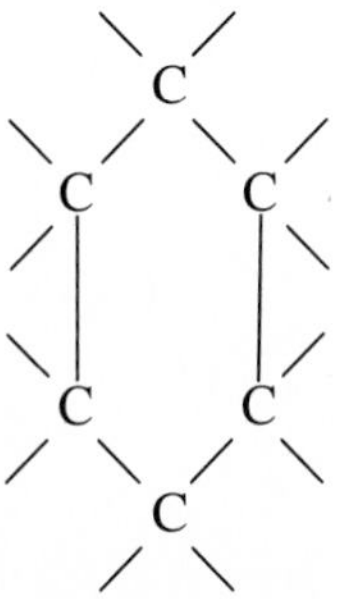

Aliphatic ring 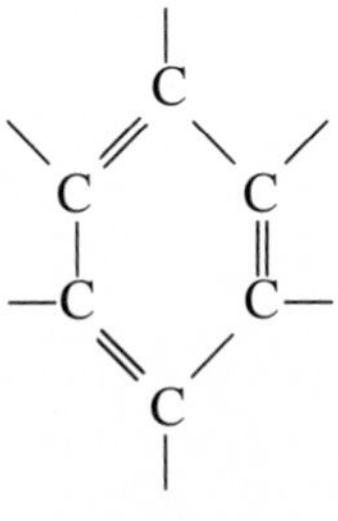   Aromatic ring

The simplest aromatic ring is made up of six carbon atoms and is known as the benzene ring. *Benzene* (C_6H_6) is known as the *parent compound* of the aromatic series. The benzene ring is usually represented by the Kekule formula.

Kekule benzene formula   Simplified formula

This formula shows double bonds between alternate carbon atoms in the ring. The double bonds, however, are not like those in the aliphatic series. For example, halogens will not add to such bonds. For purposes of simplicity, most chemists represent the benzene ring as in the simplified formula above.

Nomenclature

It is important to note that carbon atoms are not shown in the simplified benzene formula. Also, each carbon atom in a ring is like all others, and therefore, when substitution occurs on one carbon atom, the same compound is formed as though substitution had occurred on any of the other five carbon atoms. Thus, for monochlorobenzene (C_6H_5Cl) there is only one compound, no matter how the structural formula is written.

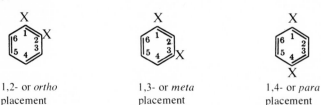

When substitution occurs on two or more carbon atoms of a benzene ring, it becomes necessary to establish some system of nomenclature. Two systems are in vogue.

X	X	X

1,2- or *ortho*   1,3- or *meta*   1,4- or *para*
placement   placement   placement

Di-substituted compounds, such as dichlorobenzene, are commonly referred to as *ortho, meta,* or *para,* depending on the point of substitution. If substitution is on adjacent carbon atoms, the term ortho is used; if on carbon atoms once removed, the term meta is used; and if on carbon atoms opposite each other, the term para is used. Tri- and other poly-substituted compounds must be named by another system. In this system, the carbon atoms of the benzene ring are numbered in a clockwise manner. Examples are

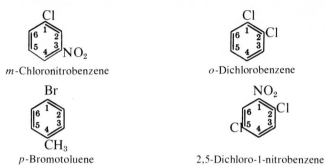

m-Chloronitrobenzene

o-Dichlorobenzene

p-Bromotoluene

2,5-Dichloro-1-nitrobenzene

When the benzene ring is attached to aliphatic compounds, the products are also called *phenyl* derivatives, the phenyl group being C_6H_5-. Thus ethyl benzene is also phenyl ethane.

4-13 HYDROCARBONS

Two series of homologous aromatic hydrocarbons are known: the benzene and the polyring series.

Benzene Series

The benzene series of homologous compounds is made up of alkyl substitution products of benzene. They are found along with benzene in coal tar and in many crude petroleums. Table 4-11 lists the benzene-series hydrocarbons of commercial importance. *Toluene,* or methylbenzene, is the simplest alkyl derivative of benzene. *Xylene* is a dimethyl derivative of benzene.

Table 4-11 Benzene-series hydrocarbons*

Name	Formula	Mp, °C	Bp, °C	Sp. gr., 20°/4°
Benzene	C_6H_6	5.51	80.09	0.879
Toluene	$C_6H_5 \cdot CH_3$	−95	110.8	0.866
o-Xylene	$C_6H_4(CH_3)_2$	−29	144	0.875
m-Xylene		−53.6	138.8	0.864
p-Xylene		13.2	138.5	0.861
Ethylbenzene	$C_6H_5 \cdot C_2H_5$	−93.9	136.15	0.867

* From E. Wertheim and H. Jeskey, "Introductory Organic Chemistry," 3d ed., McGraw-Hill, New York, 1956. Table reproduced by permission of the authors.

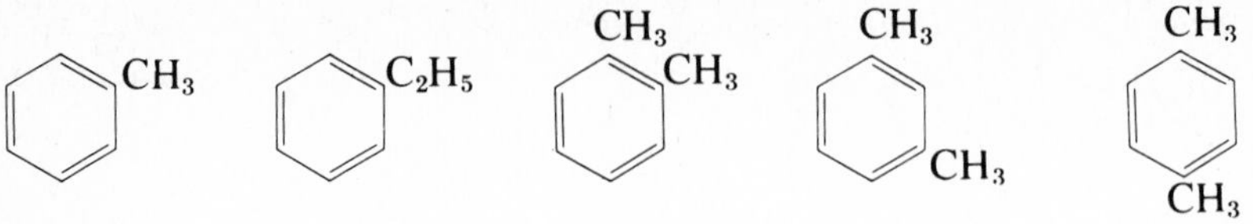

Toluene Ethylbenzene o-Xylene m-Xylene p-Xylene

It exists in three isomeric forms: ortho-xylene, meta-xylene, and para-xylene. All are isomeric with ethylbenzene in that they have the same general formula, C_8H_{10}.

The benzene-series hydrocarbons are used extensively as solvents and in chemical synthesis. Because of their relative insolubility in water, they are normally not a problem in industrial wastes.

Polyring Hydrocarbons

A wide variety of polyring aromatic hydrocarbons are known. A few examples will illustrate the possibilities.

Naphthalene ($C_{10}H_8$) *Naphthalene* is a white crystalline compound derived from coal tar and was formerly used to produce mothballs.

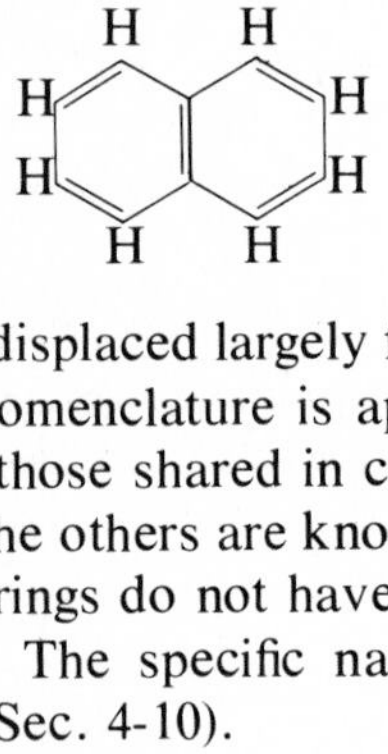

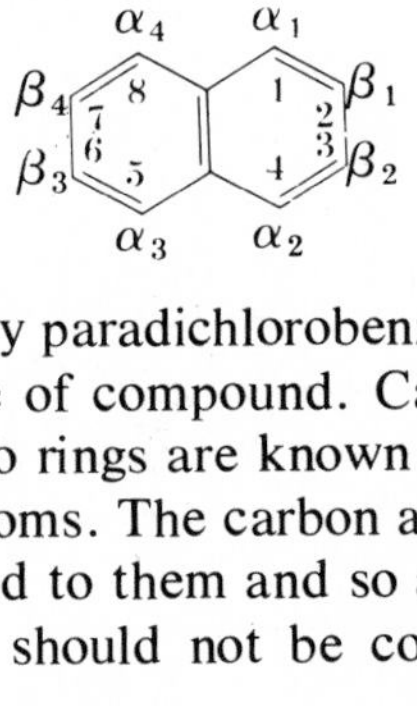

It has been displaced largely from this market by paradichlorobenzene. A new system of nomenclature is applied to this type of compound. Carbon atoms adjacent to those shared in common by the two rings are known as α-carbon atoms and the others are known as β-carbon atoms. The carbon atoms shared by the two rings do not have hydrogen attached to them and so are given no designation. The specific name, naphthalene, should not be confused with naphthene (Sec. 4-10).

Anthracene ($C_{14}H_{10}$) and phenanthrene ($C_{14}H_{10}$) *Anthracene* and *phenanthrene* are isomers.

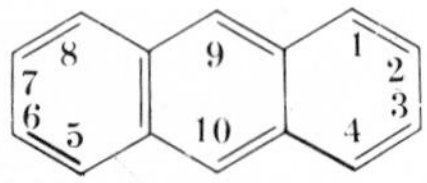

Anthracene 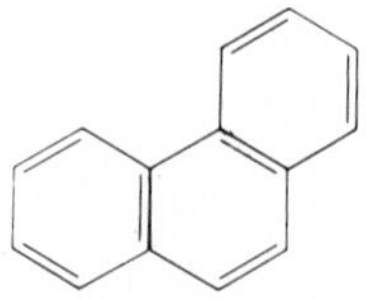   Phenanthrene

Their formulas illustrate the possible ways in which polyring aromatic hydrocarbons may occur. Many other more complex compounds, such as *chrysene* and *picene*, are known.

Chrysene   Picene

It should be remembered that hydrogen atoms occur on all carbon atoms of these compounds that are not common to two rings.

Napthalene and anthracene are widely used in the manufacture of dye-stuffs. The phenanthrene nucleus is found in important alkaloids, such as morphine, vitamin D, sex hormones, and other compounds of great biological significance.

4-14 PHENOLS

The phenols are among the most important of the aromatic compounds.

Monohydric phenols

There are several monohydric phenols of interest to environmental engineers.

Phenol (C_6H_5OH) The monohydroxy derivative of benzene is known as *phenol*.

H
O

Its formula and name indicate that it might correspond in the aromatic series to alcohols in the aliphatic series. This is not the case, however. Phenol is known to the layman as carbolic acid. It ionizes to yield H^+ to a limited extent ($K_a = 1.2 \times 10^{-10}$), and in concentrated solution is quite toxic to bacteria. It has been used widely as a germicide, and disinfectants have been rated in terms of "phenol coefficients," i.e., relative disinfecting power with respect to phenol. The system is considered archaic at the present time.

Phenol is recovered from coal tar, and considerable amounts are manufac-tured synthetically. It is used extensively in the synthesis of organic products, particularly phenolic-type resins. It occurs as a natural component in industrial wastes from the coal-gas, coal-coking, and petroleum industries as well as in a wide variety of industrial wastes from processes involving the use of phenol as a raw material.

Biological treatment of wastes containing more than 25 mg/l of phenol was considered impossible until relatively recent times. Research at the Dow Chemical Co. plant in Midland, Michigan, and elsewhere[4] has shown that phenol will serve as a bacterial food without serious toxic effects at levels as high as 500 mg/l. Studies with it and with formaldehyde have established the concept of toxicity thresholds. At levels below the threshold, bacteria use the material as food, but above the threshold they find it too toxic for use as food and reproduction of the organisms.

Cresols The next higher homologs of phenol are cresols.

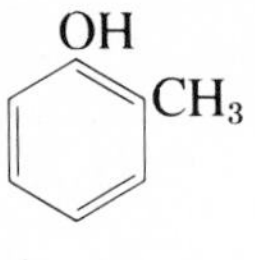

o-Cresol 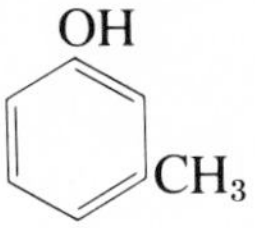   *m*-Cresol    *p*-Cresol

They are found in coal tar and have a higher germicidal action than phenol. They are less toxic to man. *Lysol* is a mixture of cresols which is sold as a household and sickroom disinfectant. Cresols are the major constituents of "creosote" which is used extensively for the preservation of wood.

Industrial wastes containing cresols are difficult to treat by biological methods. Research[5] has shown the pure cresols to be relatively nontoxic at concentrations of 250 mg/l. The toxicity of crude cresols is believed, therefore, to be due to other compounds.

Polyhydric Phenols

Three isomeric dihydric phenols are known. All have been shown[5] to be readily oxidized by properly acclimated activated sludges.

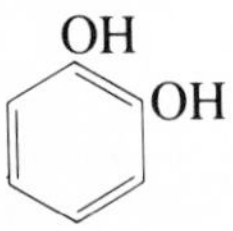

Benzene-1,2-diol,   Benzene-1,3-diol,   Benzene-1,4-diol,
Pyrocatechol,   Resorcinol   Hydroquinone
Catechol

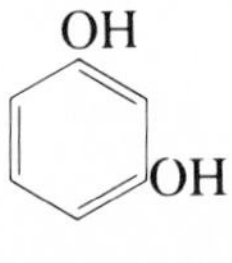

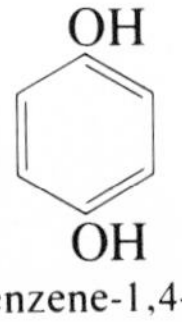

Pyrogallol *Pyrogallol,* 1,2,3-trihydroxybenzene, is known as *pyrogallic acid.*

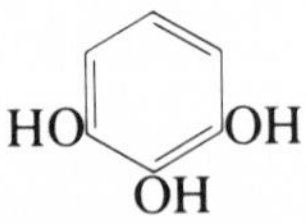

[4] R. E. McKinney, H. D. Tomlinson, and R. L. Wilcox, *Sewage and Ind. Wastes,* **28:** 547–557 (1956).

[5] *Ibid.*

It is easily oxidized and serves as a photographic developer. Environmental engineers often use alkaline solutions of it to remove oxygen from gases. It may be used in gas analysis to absorb oxygen and allow its measurement. Pyrogallol is a minor constituent of spent tan liquors. When they are discharged to streams containing iron, inky black ferric pyrogallate is formed.

4-15 ALCOHOLS, ALDEHYDES, KETONES, AND ACIDS

The aromatic alcohols, aldehydes, ketones, and acids are all formed from alkyl derivatives of benzene or one of its homologs. The active group is always in the alkyl group, and therefore the chemistry of the aromatic alcohols, aldehydes, ketones, and acids is very similar to that of the corresponding aliphatic compounds. Common names are usually employed with these compounds, but they may be more properly named as phenyl derivatives of aliphatic compounds.

Alcohols

The aromatic alcohols compose a homologous series. They are phenyl methyl, phenyl ethyl, phenyl *n*-propyl, phenyl isopropyl, and so on.

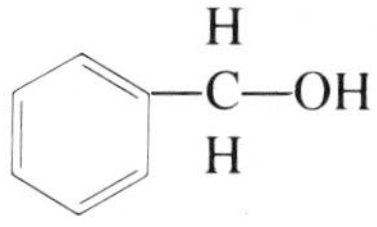

Benzyl alcohol
Phenyl methyl alcohol

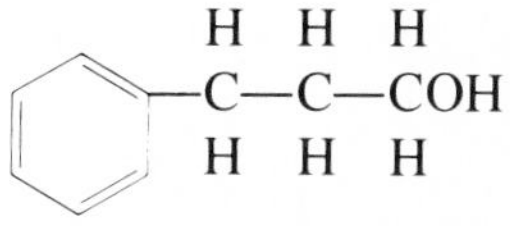

Phenyl, *n*-propyl alcohol

The aromatic alcohols are subject to chemical and biological oxidation. Oxidation of primary alcohols produces aldehydes and of secondary alcohols produces ketones.

Aldehydes

The aromatic aldehydes are important compounds in chemical synthesis.

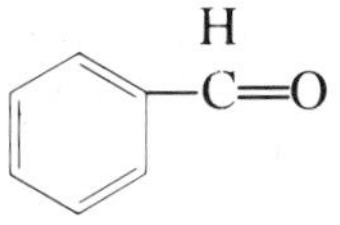

Benzaldehyde

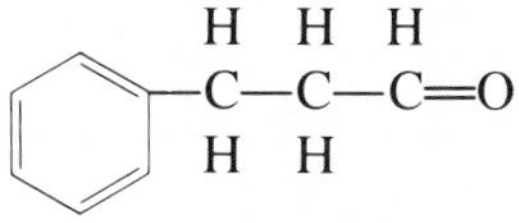

Phenyl propyl aldehyde

They are easily oxidized to the corresponding acids. Many of the more complex aldehydes have fragrant odors: coumarin, anisaldehyde, vanillin, and so on.

Ketones

The aromatic ketones are of two types: those which have one phenyl group attached to the carbonyl group and those which have two.

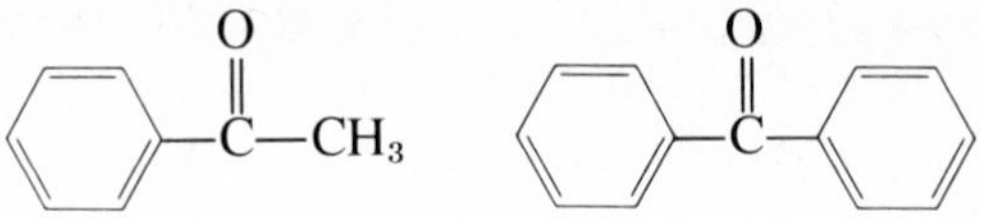

Acetophenone   Benzophenone

Chemical and biological oxidation results in disruption of the molecule, with the formation of lower-molecular-weight acids and, possibly, carbon dioxide.

Acids

A wide variety of aromatic, monocarboxylic acids are known. Oxidation of benzaldehyde produces benzoic acid. Sodium benzoate is used as a food preservative. Salicylic acid is used to prepare aspirin.

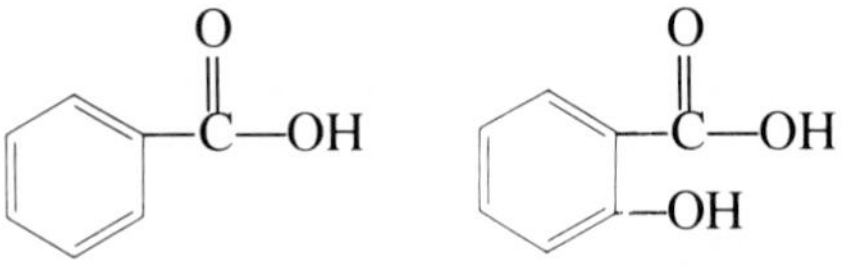

Benzoic acid   Salicylic acid

Oxidation of naphthalene produces an important dicarboxylic acid, phthalic acid.

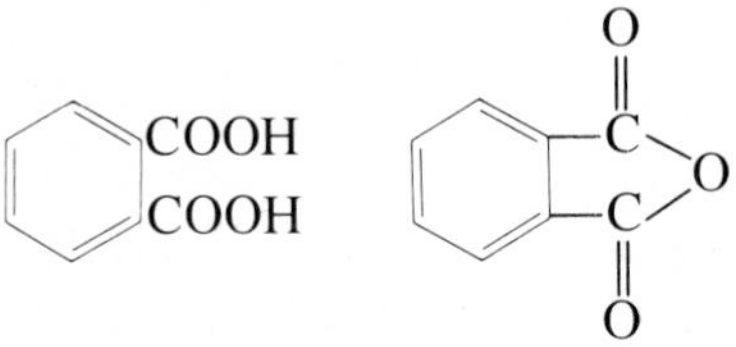

Phthalic acid   Phthalic
                        anhydride

It and its anhydride are important in the manufacture of a variety of organic compounds. Phenolphthalein,

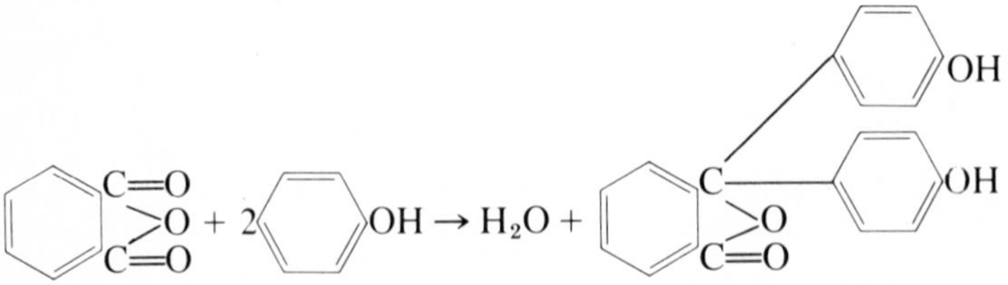

$$(4\text{-}47)$$

Phenolphthalein

used as a pH indicator in the laboratory, is an example.

The aromatic acids are subject to biological oxidation. The normal end products are carbon dioxide and water.

4-16 SIMPLE COMPOUNDS CONTAINING NITROGEN

The aromatic compounds containing nitrogen are derivatives either of ammonia or of nitric acid. The former are called *amines,* and the latter are called *nitro* compounds. In addition, many complex nitrogen compounds exist which are outside the scope of this book.

Amines

The aromatic amines are of two types: those in which the phenyl or other aromatic group is attached directly to nitrogen and those in which the nitrogen occurs in an attached alkyl group. There are three phenyl derivatives of ammonia: primary, secondary, and tertiary amines.

Aniline 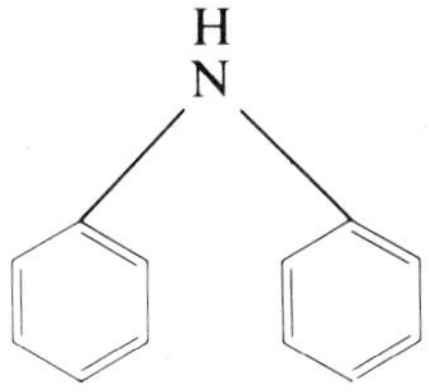   Diphenylamine

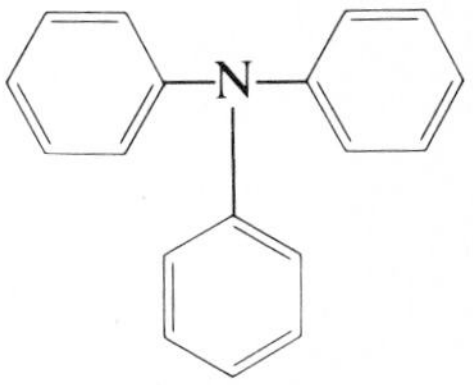

Triphenylamine

The primary form is called aniline and the secondary form is diphenylamine. They are both basic in character, react with strong acids to form salts, and are important compounds in organic synthesis. Aniline dyes are derived from aniline. Sulfanilic acid, used in the colorimetric determination of nitrites, is made from aniline.

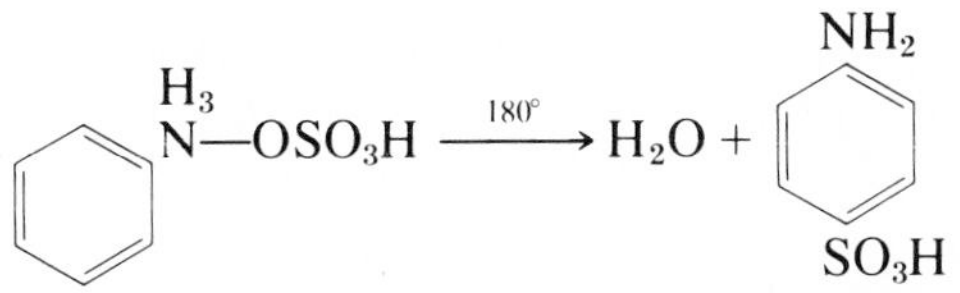

$$\text{Aniline hydrogen sulfate} \xrightarrow{180°} H_2O + \text{Sulfanilic acid} \qquad (4\text{-}48)$$

1-Naphthylamine HCl is an important compound in water analysis.

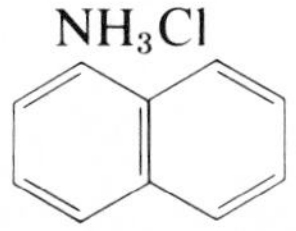

A derivative, N-(1-naphthyl)-ethylenediamine dihydrochloride is used in conjunction with sulfanilic acid in the determination of nitrite nitrogen.

Benzylamine ($C_6H_5CH_2NH_2$) is an example of an aromatic amine that has the —NH_2 group attached to the aliphatic part of the molecule. They are not important commercially.

Nitro Compounds

Nitric acid reacts with benzene and other aromatic compounds to form *nitro* compounds. The reaction is as follows:

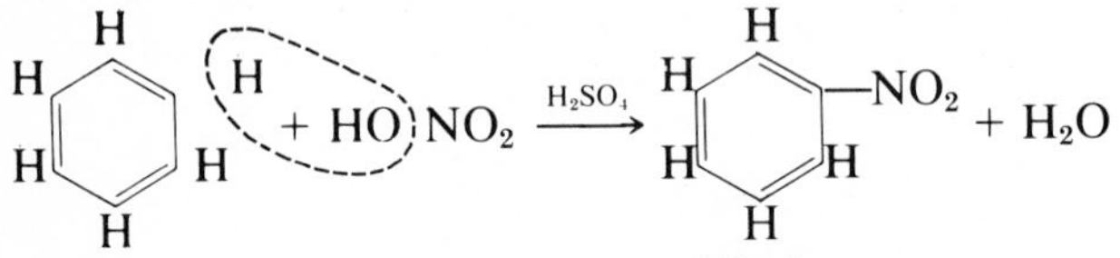

$$\text{(benzene)} + HO\,NO_2 \xrightarrow{H_2SO_4} \text{(nitrobenzene)} + H_2O \qquad (4\text{-}49)$$

Nitrobenzene

A dehydrating agent, usually sulfuric acid, must be present to remove the water that is formed. One additional nitro group may be added under proper conditions. The principal product is *m*-dinitrobenzene,

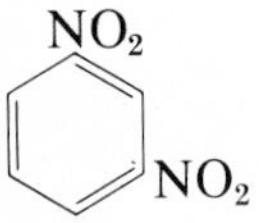

m-Dinitrobenzene 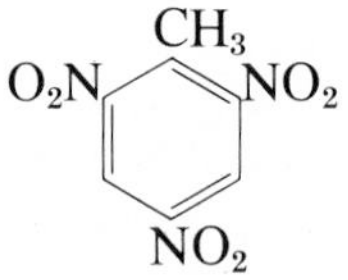   Trinitrotoluene (TNT)

as the presence of the first nitro group directs the second into the meta position. Trinitrobenzene is very difficult to prepare.

Nitration of toluene results in the formation of trinitrotoluene, or TNT. The first nitro group is directed into the ortho position by the methyl group, and additional nitro groups attach in meta position with respect to the first nitro group. TNT is widely used as an ingredient of military explosives.

HETEROCYCLIC COMPOUNDS

4-17 HETEROCYCLIC COMPOUNDS

Heterocyclic compounds have one other element in the ring in addition to carbon. A wide variety of compounds exists; some are aliphatic in character, and some are aromatic. Many are of great biological importance and several are of significance in environmental engineering.

Furaldehyde, or *furfural*, is an example of an aliphatic heterocyclic compound having a five-membered ring containing oxygen. It is produced from pentose sugars by dehydration. Commercially it is made from oat hulls and corn cobs, waste products of the cereal industry.

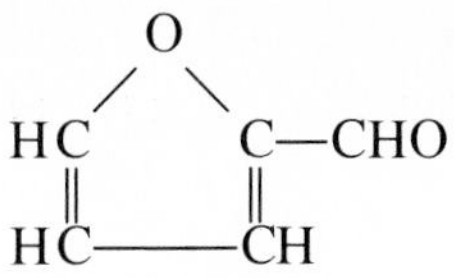

$$\underset{\text{Furaldehyde}}{\overset{\displaystyle O}{\underset{\displaystyle HC\!\!-\!\!-\!\!-\!\!CH}{HC\qquad C\!-\!CHO}}}$$

Both were formerly disposed of by burning. Manufacture of furfural yields some liquid wastes of concern to environmental engineers.

Pyrrole and *pyrrolidine* are examples of heterocyclic compounds having five-membered rings containing nitrogen.

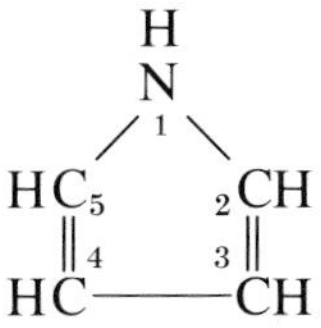

$$\underset{\text{Pyrrole}}{\overset{\displaystyle H}{\underset{\displaystyle HC\!-\!-\!-\!-\!CH}{HC_5\qquad {}_2CH}}} \qquad\qquad \underset{\text{Pyrrolidine}}{\overset{\displaystyle H}{\underset{\displaystyle H_2C\!-\!-\!-\!-\!CH_2}{H_2C\qquad CH_2}}}$$

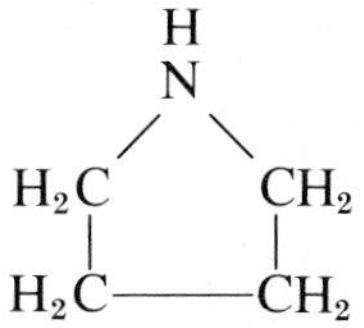

The pyrrole or pyrrolidine ring occurs in the structure of many important natural compounds, e.g., nicotine, cocaine, chlorophyll, hemoglobin.

Pyridine is an example of a six-membered aromatic heterocyclic compound with nitrogen contained in the ring.

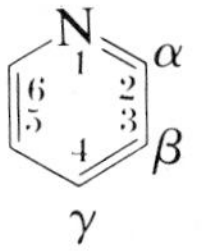

It is an especially vile-smelling liquid. It is used as a denaturant in ethyl alcohol, to make it unpalatable, and in chemical synthesis. It is weakly basic in character and forms salts with strong acids. *Nicotinic acid* is a derivative of pyridine, having a carboxyl group in the β position. It is a key component of a coenzyme, NAD, which is present in all cells and involved in oxidation of organic matter. Nicotinic acid is a vitamin required by man and many other mammals, a deficiency in the diet of man causing pellagra.

Purine and *pyrimidine* are two other ring compounds containing nitrogen and of immense biological importance.

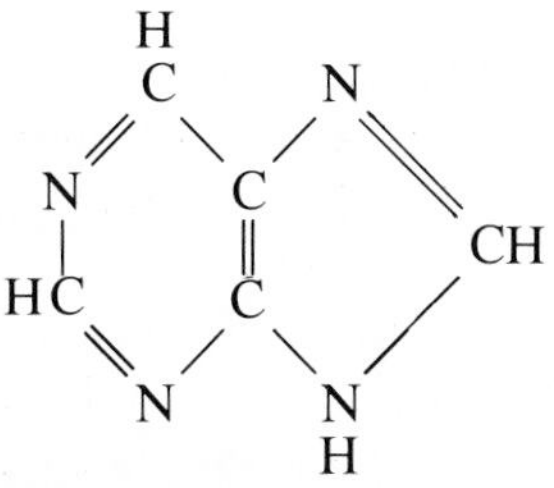

Purine

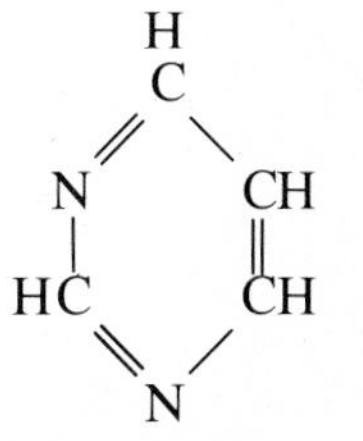

Pyrimidine

Important derivatives of purine are *adenine* and *guanine,* and of pyrimidine are *cytosine, uracil,* and *thymine.* These compounds or *bases* form major components of nucleic acids which carry the genetic information for all life. In addition they are components of key enzymes such as ATP, the primary carrier of chemical energy in all cells; and coenzyme A, which is necessary for fatty acid degradation.

Indole and *skatole* are examples of heterocyclic compounds that possess a benzene nucleus condensed with a pyrrole nucleus.

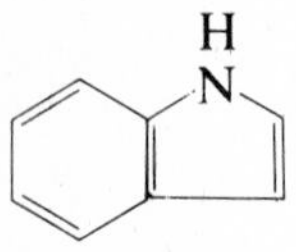

Indole 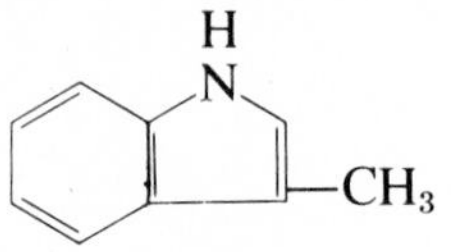   Skatole

Both possess unpleasant odors and are produced during the putrefaction of protein matter. Under controlled conditions, such as exist in well-operated sludge digesters, very little indole or skatole is formed.

4-18 DYES

The subject of *dyes* is of such magnitude and complexity that a discussion of various types will not be presented here. The environmental engineer concerned with the treatment of textile wastes, and possibly a few others, will be confronted with the need to learn more about the materials. Recourse for information should be made to standard organic chemistry texts or treatises on dyes. The sulfur dyes are noted for their toxic properties.

THE COMMON FOODS AND RELATED COMPOUNDS

4-19 GENERAL

The term *food* applies to a wide variety of organic materials that can serve as a source of energy for living organisms. In the case of bacteria, these compounds range from hydrocarbons through various oxidation products, including organic acids. In the case of higher animals and humans, the principal or common foods are restricted to *carbohydrates, fats,* and *proteins.* Other organic compounds such as ethyl alcohol, certain aldehydes, and many acids serve as food or energy sources also. The latter are sometimes referred to as exotic foods, as they are not considered part of an essential diet but are added to increase palatability or for other reasons.

4-20 CARBOHYDRATES

The term *carbohydrate* is applied to a large group of compounds of carbon, hydrogen, and oxygen in which the hydrogen and oxygen are in the same ratio as in water, i.e., two atoms of hydrogen for each atom of oxygen. The processing of carbohydrate materials occurs in the lumber, paper, and textile industries, as well as in the food industry. Wastes from these industries are major problems and tax the ingenuity of environmental engineers to find satisfactory solutions.

Carbohydrates may be grouped into three general classifications, depending upon the complexity of their structure: (1) simple sugars, or *monosaccharides;* (2) complex sugars, or *disaccharides;* (3) *polysaccharides.* In general, the *-ose* ending is used to name carbohydrates.

Simple Sugars, or Monosaccharides

The simple sugars, or *monosaccharides,* all contain a carbonyl group in the form of an aldehyde or a keto group. Those with aldehyde groups are known as *aldoses* and those with keto groups are known as *ketoses.* They are also glycols, as they possess several OH groups. Two series of simple sugars are of importance commercially: The *pentoses* are five-carbon-atom sugars and the *hexoses* are six-carbon-atom sugars.

Pentoses Pentoses have the general formula $C_5H_{10}O_5$. Two pentoses are of commercial importance, and both are *aldopentoses. Xylose* is formed by the hydrolysis of pentosans which are commonly found in waste organic materials such as oat hulls, corn cobs, and cottonseed hulls. Considerable amounts of xylose are formed in the pulping of wood through hydrolysis of hemicellulose. *Arabinose* is produced by the hydrolysis of gum arabic or wheat bran.

D(−)-Xylose[6]   D(−)-Arabinose[6]

Both xylose and arabinose are used in bacteriological work in media used to differentiate among various bacteria. Certain bacteria can ferment one but

---

[6] All sugars are optically active. The nomenclature is a bit confusing, however, and details should be obtained from a standard text on organic chemistry.

not the other, and vice versa. Mixed cultures of bacteria, such as those derived from the soil or waste, convert both sugars to carbon dioxide and water. The pentose sugars are not fermented by yeast under anaerobic conditions; therefore they cannot be used to produce ethyl alcohol. They do serve as an energy source for yeast under aerobic conditions, however, and advantage is taken of this fact in one method of treating spent sulfite liquors from the pulping of wood.

Hexoses There are four important hexose sugars with the general formula $C_6H_{12}O_6$. *Glucose, galactose,* and *mannose* are all *aldoses,* and *fructose* is a *ketose.*

Glucose Glucose is the most common of the aldohexose sugars. It is found naturally in fruit juices and in honey. It is manufactured in great quantity by the hydrolysis of corn starch. It is the principal component of corn syrup. Both corn syrup and glucose are used extensively in candy manufacture. Glucose is much less sweet than ordinary sugar and replaces it for many purposes.

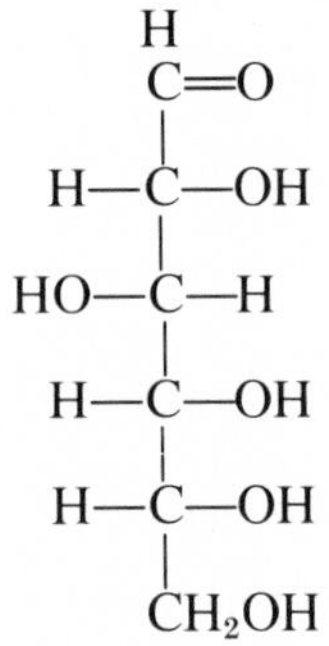

$$
\begin{array}{c}
H \\
C{=}O \\
| \\
H{-}C{-}OH \\
| \\
HO{-}C{-}H \\
| \\
H{-}C{-}OH \\
| \\
H{-}C{-}OH \\
| \\
CH_2OH
\end{array}
$$

D-Glucose

Glucose is the only hexose sugar that can be prepared in relatively pure form by hydrolysis of disaccharides or polysaccharides. All the other hexose sugars occur in combination with glucose.

Fructose Fructose is the only significant ketohexose and occurs naturally in honey. When cane or beet sugar is hydrolyzed, one molecule of fructose and one molecule of glucose are formed from each molecule of sucrose.

Galactose and mannose Galactose and mannose do not occur in free form in nature. Galactose is produced by hydrolysis of lactose, more commonly called milk sugar. Glucose is formed simultaneously. Mannose is produced by the hydrolysis of ivory nut, and glucose is formed at the same time.

Glucose and galactose are of particular interest in environmental engineering. Glucose is always one of the products and may be the sole product when di- or polysaccharides are hydrolyzed. It is therefore found in a wide variety of

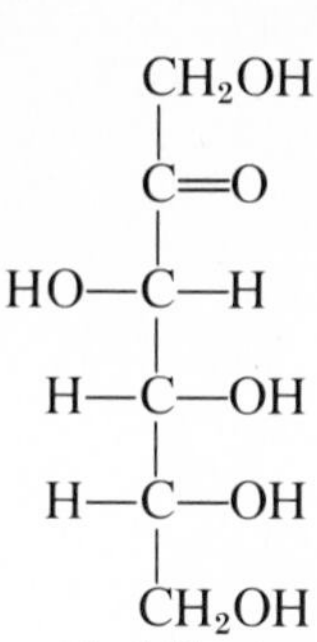

$$
\begin{array}{ccc}
\text{CH}_2\text{OH} & \text{H} & \text{H} \\
| & \text{C}{=}\text{O} & \text{C}{=}\text{O} \\
\text{C}{=}\text{O} & | & | \\
| & \text{H}{-}\text{C}{-}\text{OH} & \text{HO}{-}\text{C}{-}\text{H} \\
\text{HO}{-}\text{C}{-}\text{H} & | & | \\
| & \text{HO}{-}\text{C}{-}\text{H} & \text{HO}{-}\text{C}{-}\text{H} \\
\text{H}{-}\text{C}{-}\text{OH} & | & | \\
| & \text{HO}{-}\text{C}{-}\text{H} & \text{H}{-}\text{C}{-}\text{OH} \\
\text{H}{-}\text{C}{-}\text{OH} & | & | \\
| & \text{H}{-}\text{C}{-}\text{OH} & \text{H}{-}\text{C}{-}\text{OH} \\
\text{CH}_2\text{OH} & | & | \\
& \text{CH}_2\text{OH} & \text{CH}_2\text{OH}
\end{array}
$$

D(—)-Fructose 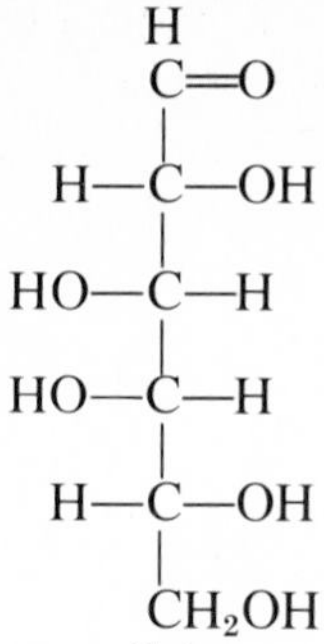   D-Galactose 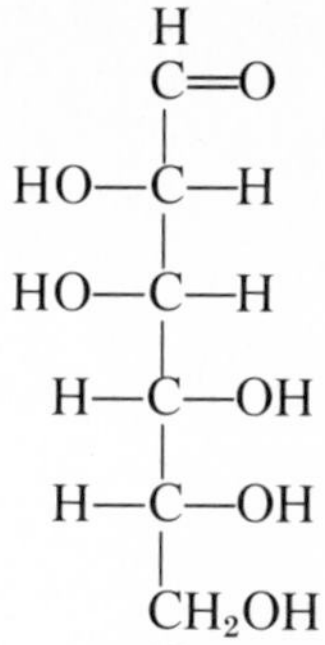   D-Mannose

industrial wastes. Galactose is formed from the hydrolysis of lactose, or milk sugar, and is found in wastes from the dairy industry. Both sugars are readily oxidized by aerobic bacteria to form acids, and the oxidation may stop at that point because of the unfavorable pH conditions produced by the acids unless precautions are taken to control the pH by means of buffers or alkaline materials. Lactic acid is an important intermediate in the oxidation of galactose. Both sugars are fermented rapidly under anaerobic conditions, with acid formation.

Complex Sugars, or Disaccharides

There are three important sugars with the general formula $C_{12}H_{22}O_{11}$: *sucrose*, *maltose*, and *lactose*. All disaccharides may be considered as consisting of two hexose sugars hooked together in one molecule. Hydrolysis results in cleavage of the molecule and formation of the hexoses.

Sucrose Sucrose is the common sugar of commerce. It is derived largely from sugarcane and sugar beets. The sap of such trees as the sugar maple contains considerable sucrose. Hydrolysis of the sucrose molecule results in the formation of one molecule of glucose and one molecule of fructose.

Glucose part / Fructose part

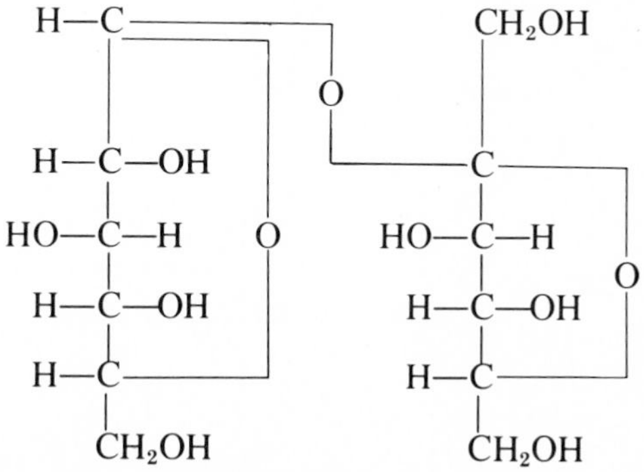

Sucrose ($C_{12}H_{22}O_{11}$)

Maltose Maltose is made by hydrolysis of starch, induced by *diastase,* an enzyme present in barley malt. The starch may be derived from a wide variety of sources, and its hydrolysis by diastase results in commercial maltose, which is used in infant foods and in malted milk.

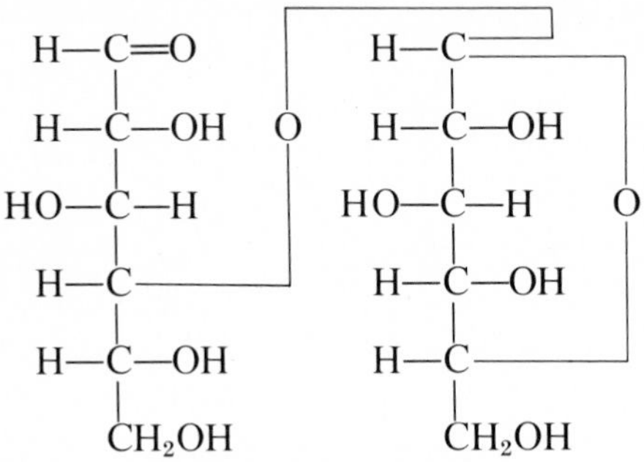

$$
\begin{array}{ll}
\text{H—C}{=}\text{O} & \text{H—C} \\
\text{H—C—OH} \quad \text{O} & \text{H—C—OH} \\
\text{HO—C—H} & \text{HO—C—H} \quad \text{O} \\
\text{H—C} & \text{H—C—OH} \\
\text{H—C—OH} & \text{H—C} \\
\text{CH}_2\text{OH} & \text{CH}_2\text{OH}
\end{array}
$$

Maltose ($C_{12}H_{22}O_{11}$)

Maltose is readily hydrolyzed to yield two molecules of glucose.

Alcohol-production by fermentation processes uses starch from a wide variety of sources. The starch is converted to maltose by the enzyme from barley malt. Enzymes from the yeast hydrolyze maltose to glucose and convert the glucose to alcohol and carbon dioxide. See Sec. 4-3.

Lactose Lactose, or milk sugar, occurs in the milk of all mammals. Upon hydrolysis, the molecule is split to yield a molecule of glucose and a molecule of galactose.

$$
\begin{array}{ll}
\text{H—C}{=}\text{O} & \text{H} \\
& \text{C} \\
\text{H—C—OH} & \text{H—C—OH} \\
\quad \text{O} & \\
\text{HO—C—H} & \text{HO—C—H} \quad \text{O} \\
\text{H—C} & \text{HO—C—H} \\
\text{H—C—OH} & \text{H—C} \\
\text{CH}_2\text{OH} & \text{CH}_2\text{OH}
\end{array}
$$

Glucose part / Galactose part

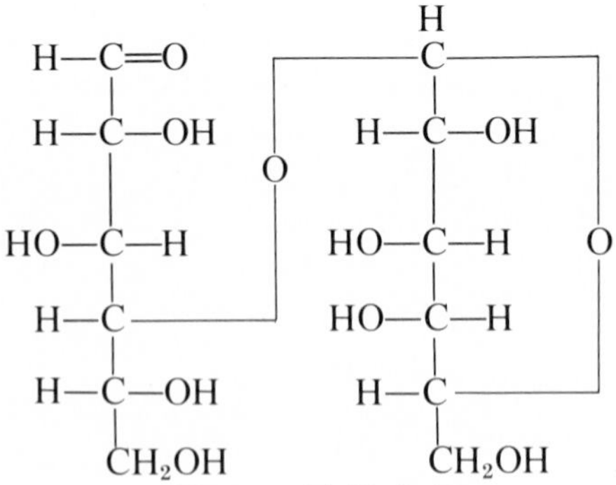

Lactose ($C_{12}H_{22}O_{11}$)

Lactose is used in infant foods and in candy making. Dried skimmed-milk solids contain about 60 percent lactose.

Polysaccharides

The polysaccharides are all condensation products of hexoses or other monosaccharides. Glucose and xylose are the most common units involved. Three polysaccharides are of interest to environmental engineers: *starch, cellulose,* and *hemicellulose.* None of them have the characteristic sweet taste of sugar because of their insolubility and complex molecular structure.

Starch Starch has the general formula $(C_6H_{10}O_5)_x$. It occurs in a wide variety of products grown for food purposes (corn, wheat, potatoes, rice, etc.). It is the cheapest foodstuff and serves mainly in human nutrition as a source of energy. Starch is used in fermentation industries to produce a wide variety of products. The structure of the starch molecule is not known definitely. Its hydrolysis yields glucose as the only monosaccharide, and its general formula may be indicated as follows:

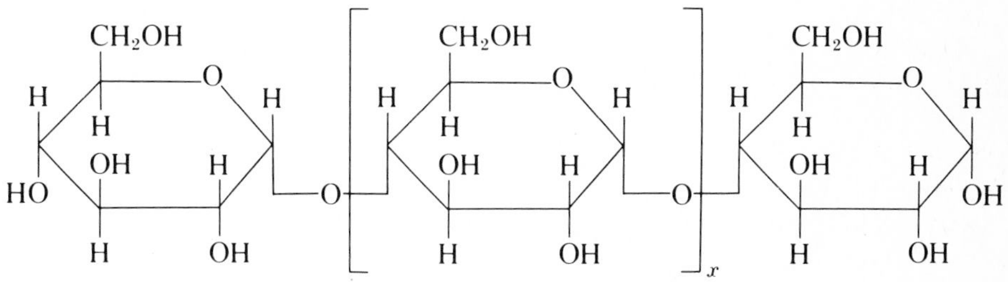

Starch (amylose, x = about 100 to 1000)

Starch consists of two major fractions. One fraction, consisting of glucose units connected in a straight chain, is termed *amylose,* and the other fraction, consisting of glucose units attached to form branched chains, is termed *amylopectin.* The amylose molecule contains 100 to 1000 glucose units. Amylose is soluble in water and absorbs up to 20 percent of its own weight in iodine to form the blue complex used as an indicator in iodimetric analysis. The amylopectin molecule, not shown, is much larger and contains about 500 to 5000 glucose units. It is not as soluble in water as amylose.

The glucose units in starch are connected by what is termed an *alpha* linkage. This linkage is readily hydrolyzed by enzymes common to all mammals as well as to microorganisms, and hence they are able to use starch as food.

The industrial wastes produced from the manufacture of starch, from the processing of carbohydrate foods, and from the industrial uses of starch are among the most difficult ones environmental engineers have had to treat.

Cellulose Cellulose forms the structural fiber of many plants. Cotton is essentially pure cellulose. High-grade cellulose can be produced from wood through the sulfite and sulfate pulping processes. Like starch, cellulose consists of glucose subunits. However, these units are connected by what is termed *beta* linkage, and mammals, including man, do not have enzymes capable of promoting the hydrolysis of this linkage. Therefore cellulose passes through the digestive tract unchanged. Certain animals, especially ruminants (cud-chewing animals) such as the cow, have bacteria in the digestive tract which can hydrolyze the beta link. This is a convenient arrangement, for the animal can digest the bacterial fermentation products and thus derive nourishment indirectly from cellulose. The beta linkage for cellulose is indicated as follows:

Portion of cellulose molecule

Industrial wastes from the paper industry usually contain considerable amounts of cellulose in suspension. This is particularly true of the wastes from the manufacture of low-grade papers involving the reuse of waste paper. Most of the wastes from other industries processing cellulose contain very little cellulose. The principal contaminants are inorganic compounds, derivatives of cellulose, and other organic compounds. Since certain bacteria can hydrolyze cellulose, biological treatment of cellulose containing wastes is possible. However, treatment by aerobic processes is slow. Since most of the cellulose will settle to produce a sludge, preliminary treatment by sedimentation is practiced, and the sludges produced are disposed of by anaerobic digestion or by physical methods such as filtration, centrifugation, and incineration.

Hemicelluloses The hemicelluloses are compounds which have characteristics somewhat like cellulose. They are composed of a mixture of hexose and pentose units, however, and upon hydrolysis yield glucose and a pentose, usually xylose. Most natural woods contain cellulose, hemicelluloses, and lignin, along with resins, pitch, and so on. In the pulping process, the lignin, hemicellulose, resins, and so on, are dissolved, leaving cellulose as the product. As a result, spent pulping liquors contain considerable amounts of glucose and xylose as well as other organic substances, principally derivatives of lignin. The lignin derivatives are very resistant to biological degradation. Glucose, xylose, and other organic substances are converted to carbon dioxide and water by yeast or bacteria under aerobic conditions. Yeast may be used to ferment the glucose to alcohol under anaerobic conditions, but the xylose, a pentose, is not fermentable to alcohol.

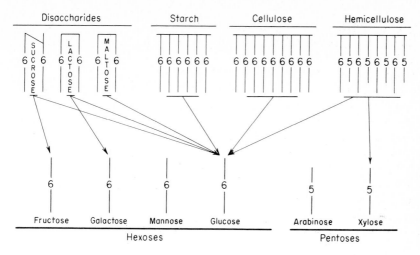

Figure 4-1 Summary of hydrolytic behavior of carbohydrates.

Summary of Hydrolytic Behavior of Carbohydrates

The hydrolytic behavior of carbohydrates is presented in a simplified graphic form in Fig. 4-1. All di- and polysaccharides yield glucose. Sucrose yields fructose, lactose yields galactose, and hemicellulose yields xylose in addition to glucose.

4-21 FATS, OILS, AND WAXES

Fats, oils, and *waxes* are all esters. Fats and oils are esters of the trihydroxy alcohol, glycerol, while waxes are esters of long-chain monohydroxy alcohols. All serve as food for humans as well as bacteria, since they can be hydrolyzed to the corresponding fatty acids and alcohols.

Fats and Oils

Fats and oils are both glycerides of fatty acids. The fatty acids are generally of 16- or 18-carbon atoms, although butyric, caproic, and caprylic acids are present to a significant extent as components of the esters of butterfat. The acids may also be unsaturated. Oleic and linoleic are important acids in cottonseed oil. Linseed oil contains large amounts of linoleic and linolenic acids. The glycerides of fatty acids that are liquid at ordinary temperatures are called *oils* and those that are solids are called *fats.* Chemically they are quite similar. The oils have a predominance of short-chain fatty acids or fatty acids with a considerable degree of unsaturation, such as linoleic or linolenic.

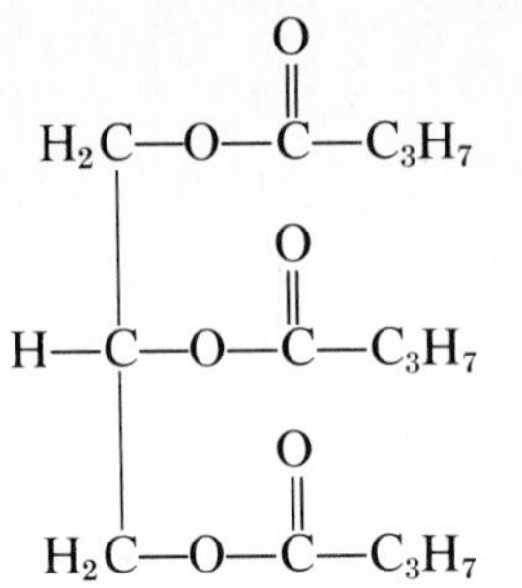

$$
\begin{array}{c}
\hspace{2cm} O \hspace{3cm} O \\
\hspace{2cm} \| \hspace{3cm} \| \\
H_2C-O-C-C_3H_7 \hspace{1cm} H_2C-O-C-C_{17}H_{35} \\
\hspace{2cm} O \hspace{3cm} O \\
\hspace{2cm} \| \hspace{3cm} \| \\
H-C-O-C-C_3H_7 \hspace{1cm} H-C-O-C-C_{17}H_{35} \\
\hspace{2cm} O \hspace{3cm} O \\
\hspace{2cm} \| \hspace{3cm} \| \\
H_2C-O-C-C_3H_7 \hspace{1cm} H_2C-O-C-C_{17}H_{35}
\end{array}
$$

Glyceryl (or glycerol)   Glyceryl (or glycerol)
tributyrate, tributyrin   tristearate, tristearin

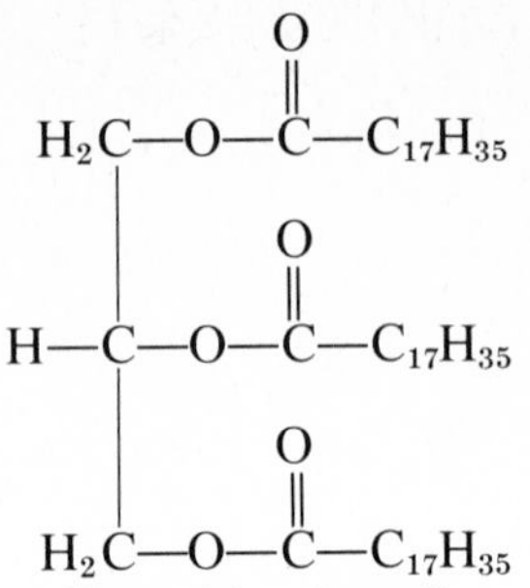

The fatty acids in a given molecule of a glyceride may all be the same, as shown above, or they may all be different.

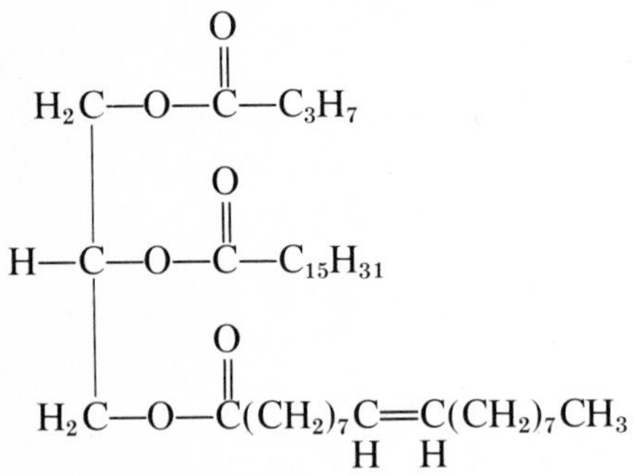

$$
\begin{array}{c}
\hspace{2cm} O \hspace{4cm} O \\
\hspace{2cm} \| \hspace{4cm} \| \\
H_2C-O-C-C_3H_7 \hspace{1cm} {}^{\alpha}H_2C-O-C-C_{17}H_{35} \\
\hspace{2cm} O \hspace{4cm} O \\
\hspace{2cm} \| \hspace{4cm} \| \\
H-C-O-C-C_{15}H_{31} \hspace{1cm} {}^{\beta}H-C-O-C-C_{15}H_{31} \\
\hspace{2cm} O \hspace{4cm} O \\
\hspace{2cm} \| \hspace{4cm} \| \\
H_2C-O-C(CH_2)_7C=C(CH_2)_7CH_3 \hspace{0.5cm} H_2C-O-C-C_{17}H_{35} \\
\hspace{4cm} H \;\; H \hspace{3cm} {}^{\alpha'}
\end{array}
$$

Glyceryl (or glycerol) oleo- β-Palmito-α,α'-distearin
butyropalmitate (found in butter
fat)

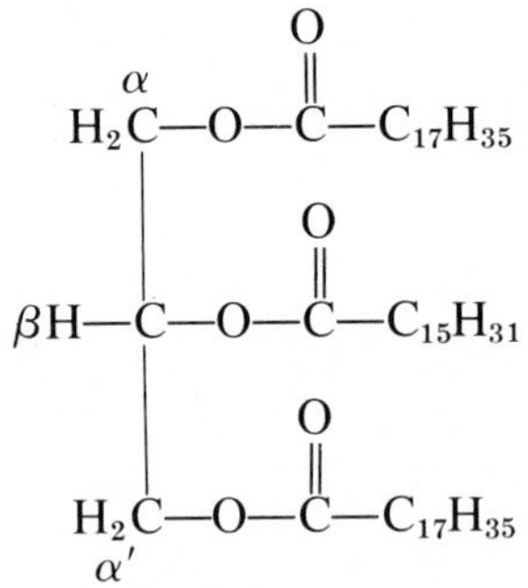

The principal acids composing the glycerides of fats and oils are shown in Table 4-12.

The relative amounts of the major fatty acids contained in various fats and oils are shown in Table 4-13.

Fats and oils undergo three types of chemical reactions that are of interest to environmental engineers.

Hydrolysis Since fats and oils are esters, they undergo hydrolysis with more or less ease. The hydrolysis may be induced by chemical means, usually by treatment with NaOH, or by bacterial enzymes that split the molecule into glycerol plus fatty acids. Hydrolysis with the aid of NaOH is called *saponification*. Hydrolysis by bacterial action produces *rancid* fats or oils and renders them unpalatable. Rancid butter and margarine are notorious.

Addition The fats and oils containing unsaturated acids add chlorine at the double bonds, as other unsaturated compounds do. This reaction is often slow because of the relative insolubility of the compounds. It may represent a significant part of the chlorine demand of some wastes.

Table 4-12 Acids of fats and oils*

Name	Formula	Mp, °C	Source
Butyric	C_3H_7COOH	− 4.7	Butter
Caproic	$C_5H_{11}COOH$	− 2	Butter, coconut oil
Caprylic	$C_7H_{15}COOH$	16	Palm oil, butter
Capric	$C_9H_{19}COOH$	31.5	Coconut oil
Lauric	$C_{11}H_{23}COOH$	44	Coconut oil, spermaceti
Myristic	$C_{13}H_{27}COOH$	58	Nutmeg, coconut oil
Palmitic	$C_{15}H_{31}COOH$	64	Palm oil, animal fats
Stearic	$C_{17}H_{35}COOH$	69.4	Animal and vegetable fats, oils
Arachidic	$C_{20}H_{40}O_2$	76.3	Peanut oil
Behenic	$C_{22}H_{44}O_2$	84	Ben oil
Oleic	$C_{18}H_{34}O_2$	14	Animal and vegetable fats, oils
Erucic	$C_{22}H_{42}O_2$	33.5	Rape oil, mustard oil
Linoleic	$C_{18}H_{32}O_2$	− 11	Cottonseed oil
Linolenic	$C_{18}H_{30}O_2$		Linseed oil
Clupanodonic	$C_{22}H_{34}O_2$	< − 78	Fish oils

* From E. Wertheim and H. Jeskey, "Introductory Organic Chemistry," 3d ed., McGraw-Hill, New York, 1956. Table reproduced by permission of the authors.

Table 4-13 Acid content of fats and oils[a]

Name	Oleic	Lino-leic	Lino-lenic	Stearic	Myristic	Palmitic	Ara-chidic
Butter[b]	27.4			11.4	22.6	22.6	
Mutton tallow	36.0	4.3		30.5	4.6	24.6	
Castor oil[c]	9	3		3			
Olive oil	84.4	4.6		2.3	trace	6.9	0.1
Palm oil	38.4	10.7		4.2	1.1	41.1	
Coconut oil[d]	5.0	1.0		3.0	18.5	7.5	
Peanut oil[e]	60.6	21.6		4.9		6.3	3.3
Corn oil[f]	43.4	39.1		3.3		7.3	0.4
Cottonseed oil	33.2	39.4		1.9	0.3	19.1	0.6
Linseed oil	5	48.5	34.1				
Soybean oil[g]	32.0	49.3	2.2	4.2		6.5	0.7
Tung oil[h]	14.9			1.3		4.1	

[a] From E. Wertheim and H. Jeskey, "Introductory Organic Chemistry," 3d ed., McGraw-Hill, New York, 1956. Table reproduced by permission of the authors.

[b] Contains caproic, 1.4 percent; caprylic, 1.8 percent; capric, 1.8 percent; butyric, 3.2 percent; lauric, 6.9 percent.

[c] Contains about 85 percent of ricinoleic acid, 12-hydroxy-9-octadecenoic acid (mp, 17°), $CH_3(CH_2)_5CHOHCH_2CH{=}CH(CH_2)_7COOH$.

[d] Contains caprylic, 9.5 percent; capric, 4.5 percent; lauric, 51 percent.

[e] Contains 2.6 percent lignoceric acid.

[f] Contains 0.2 percent lignoceric acid.

[g] Contains 0.1 percent lignoceric acid.

[h] Contains 79.7 percent eleostearic acid.

Oils that contain significant amounts of oleic and linoleic acids may be converted to fats by the process of *hydrogenation*. In this process hydrogen is caused to add at the double bonds, and saturated acids result. Thus low-priced oils such as soybean and cottonseed can be converted into a product commonly called margarine, which is acceptable as human food. Many cooking fats or shortenings such as Spry, Crisco, Fluffo, and other proprietary products are made in the same manner. The hydrogenation can be controlled to produce any degree of hardness desired in the product.

Oxidation The oils with appreciable amounts of linoleic and linolenic acids or other highly unsaturated acids, such as linseed and tung oil, are known as *drying oils*. In contact with the air, oxygen adds at the double bonds and forms a resin-like material. The drying oils are the major vehicle in all oil-base paints.

Waxes

Waxes, with the exception of paraffin wax, are esters of long-chain acids and alcohols of high molecular weight. *Beeswax* is an ester of palmitic acid and myricyl alcohol $(C_{15}H_{31}COOC_{31}H_{63})$. It also contains cerotic acid $(C_{25}H_{51}COOH)$. Spermaceti is obtained from the heads of sperm whales and is principally an ester of palmitic acid and cetyl alcohol $(C_{15}H_{31}COOC_{16}H_{33})$. Cetyl esters of lauric and myristic acids are also present to a limited extent.

4-22 PROTEINS AND AMINO ACIDS

Proteins are complex compounds of carbon, hydrogen, oxygen, and nitrogen. Phosphorus and sulfur are present in a few. They are among the most complex of the organic compounds produced in nature and are widely distributed in plants and animals. They form an essential part of all protoplasm and are a necessary part of the diet of all higher animals, in which they serve to build and repair muscle tissue. Like polysaccharides, which may be considered to be made up of glucose units, proteins are formed by the union of α-amino acids. Since more than 20 different amino acids are normally found present in proteins, the variety of proteins is considerable.

Amino Acids

The α-*amino acids* are the building blocks from which proteins are constructed. Most plants and bacteria have the ability to synthesize the amino acids from which they build proteins. Animals are unable to synthesize certain of the amino acids and must depend upon plants to supply them in the form of proteins. Such amino acids are considered to be indispensable.

The amino acids that occur in proteins all have an amino group attached to the alpha carbon atom and are therefore called α-amino acids.

$$
\begin{array}{c}
\text{NH}_2 \\
| \\
\text{R—C—COOH} \\
| \\
\text{H}
\end{array}
$$

α-Amino acid

Chemistry of amino acids The free amino acids behave like acids and also like bases because of the amino group that they contain. Thus they are *amphoteric* in character and form salts with acids or bases.

Salt formation with an acid:

$$\text{H}_2\text{NCH}_2\text{COOH} + \text{HCl} \rightarrow \text{Cl}^-\,{}^+\text{H}_3\text{NCH}_2\text{COOH} \qquad (4\text{-}50)$$

Salt formation with a base:

$$\text{H}_2\text{NCH}_2\text{COOH} + \text{NaOH} \rightarrow \text{H}_2\text{O} + \text{H}_2\text{NCH}_2\text{COO}^-\text{Na}^+ \qquad (4\text{-}51)$$

The amino acids having one amino and one carboxyl group are essentially neutral in aqueous solution. This is considered to be due to a case of self-neutralization in which the hydrogen ion of the carboxyl group migrates to the amino group and a positive-negative (dipolar) ion known as a *zwitterion* results.

$$
\begin{array}{ccc}
\text{NH}_2 & & \text{NH}_3^+ \\
| & & | \\
\text{R—C—COOH} & \rightleftharpoons & \text{R—C—COO}^- \\
| & & | \\
\text{H} & & \text{H}
\end{array}
\qquad (4\text{-}52)
$$

Zwitterion

In Sec. 4-9, it was shown that organic acids can react with ammonia to form amides. The amino and carboxyl groups of separate amino acid molecules can react in the same manner.

$$
\begin{array}{c}
\text{O} \qquad\qquad\qquad \text{R} \\
\text{H} \parallel \qquad\qquad\quad | \\
\text{R—C—C—}\boxed{\text{OH} \quad \text{H}}\text{N—C—COOH} \\
| \qquad\qquad\quad\ \ \text{H} \ \ \text{H} \\
\text{NH}_2
\end{array}
\longrightarrow
\begin{array}{c}
\text{O} \qquad \text{R} \\
\text{H} \parallel \qquad | \\
\text{R—C—C—N—C—COOH} + \text{H}_2\text{O} \\
| \qquad\quad \text{H} \ \ \text{H} \\
\text{NH}_2
\end{array}
\qquad (4\text{-}53)
$$

A dipeptide

It is this ability to form linkages between the amino and carboxyl groups that allows the large complex molecules of proteins to be formed. In the example given above, the resulting molecule contains one free amino and one free carboxyl group. Each can combine with another molecule of an amino acid. In turn, the resulting molecule will contain free amino and carboxyl groups, and the process can be repeated, presumably, *ad infinitum*. Natural forces, however, direct the synthesis to produce the type and size of protein molecules desired.

The molecule formed by the union of two molecules of amino acids is known as a *dipeptide;* if there are three units the name is *tripeptide;* if more than three units the compound is called a *polypeptide*. The particular linkage formed when amino acids join is called the *peptide link* and is formed by loss of water between an amino and a carboxyl group.

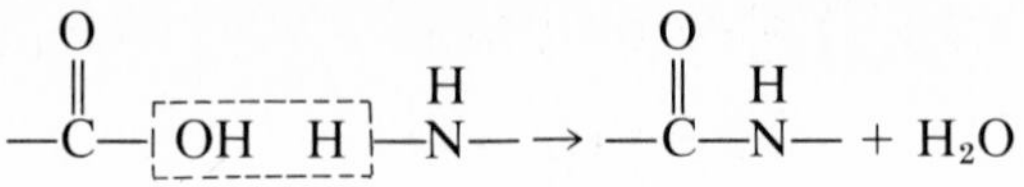

$$-\overset{\overset{\displaystyle O}{\|}}{C}-\fbox{OH \quad H}-\overset{H}{N}- \rightarrow -\overset{\overset{\displaystyle O}{\|}}{C}-\overset{H}{N}- + H_2O \qquad (4\text{-}54)$$

Peptide link

Classes of amino acids About twenty different α-amino acids can generally be isolated by hydrolysis of protein matter. The simplest have one amino group and one carboxyl group per molecule. Some have sulfur in the molecule. Some have two amino groups and one carboxyl group and consequently are basic in reaction. Some have one amino group and two carboxyl groups and are acidic in reaction. Others have aromatic or heterocyclic groups.

All the amino acids, except glycine, are optically active. In the list below those marked with an asterisk are considered indispensable in human nutrition. The generally used abbreviation for each amino acid is noted after the name.

Monoamino monocarboxy acids

$$H_2NCH_2COOH \qquad\qquad \text{Glycine (Gly)}$$

$$CH_3CHNH_2COOH \qquad\qquad \text{Alanine (Ala)}$$

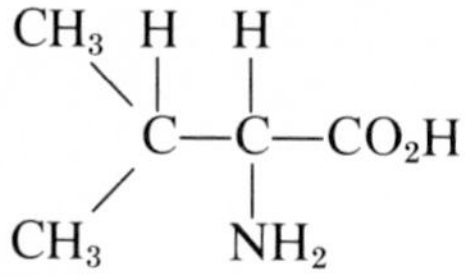

$$\text{* Valine (Val)}$$

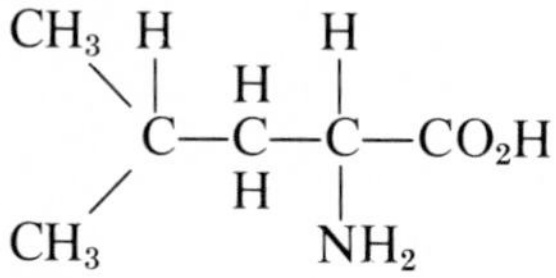

$$\text{* Leucine (Leu)}$$

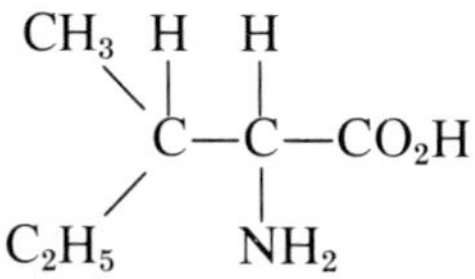

$$\text{* Isoleucine (Ile)}$$

Monocarboxy diamino acids

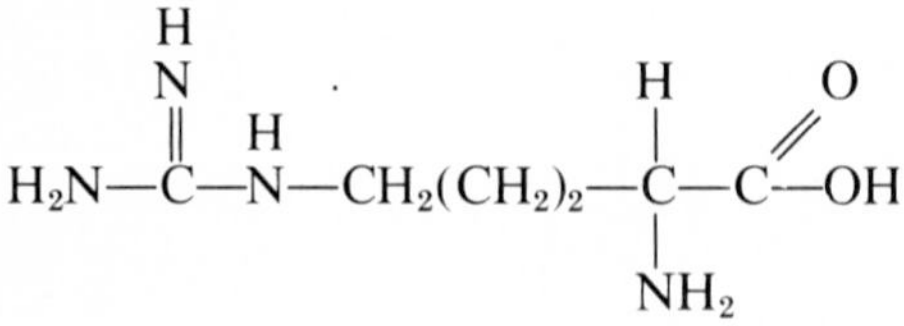

$$\text{Arginine (Arg)}$$

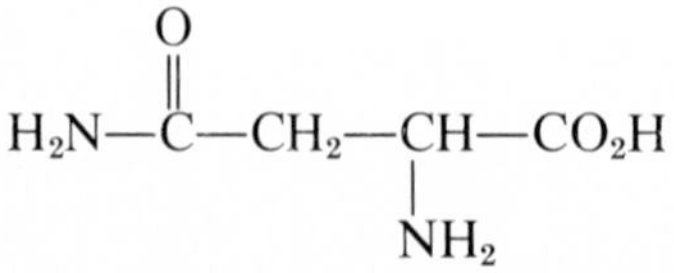

$$\text{Asparagine (Asn)}$$

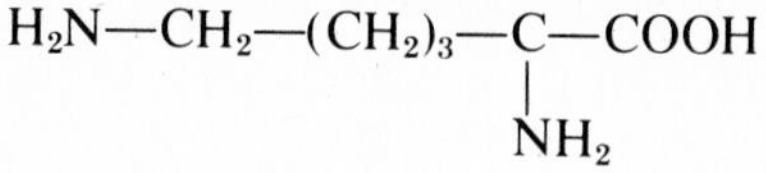

$$H_2N-CH_2-(CH_2)_3-\underset{\underset{NH_2}{|}}{C}-COOH$$

* Lysine (Lys)

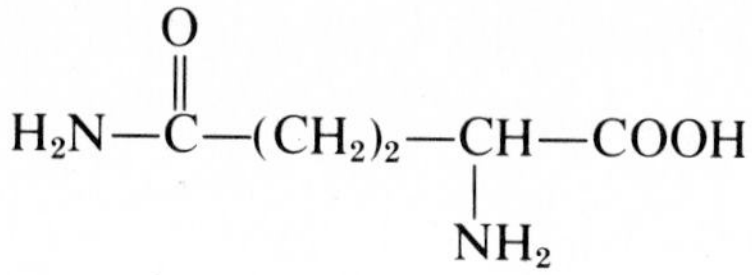

$$H_2N-\underset{\underset{}{\overset{O}{\|}}}{C}-(CH_2)_2-\underset{\underset{NH_2}{|}}{CH}-COOH$$

Glutamine (Gln)

Aromatic homocyclic acids

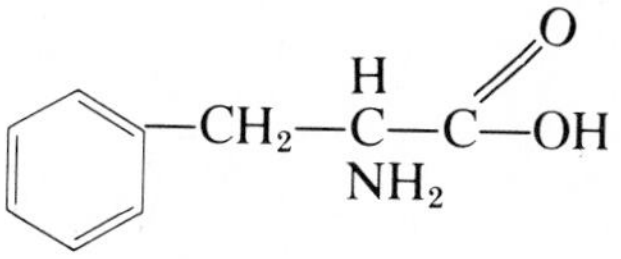

$$\langle\text{ring}\rangle-CH_2-\underset{\underset{NH_2}{|}}{\overset{H}{C}}-\overset{O}{\overset{\nearrow}{C}}-OH$$

* Phenylalanine (Phe)

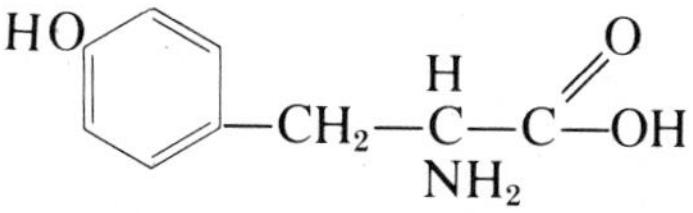

$$HO-\langle\text{ring}\rangle-CH_2-\underset{\underset{NH_2}{|}}{\overset{H}{C}}-\overset{O}{\overset{\nearrow}{C}}-OH$$

Tyrosine (Tyr)

$$HO-\langle\text{ring (I, I)}\rangle-O-\langle\text{ring (I, I)}\rangle-\overset{H_2}{C}-\underset{\underset{NH_2}{|}}{\overset{H}{C}}-CO_2H$$

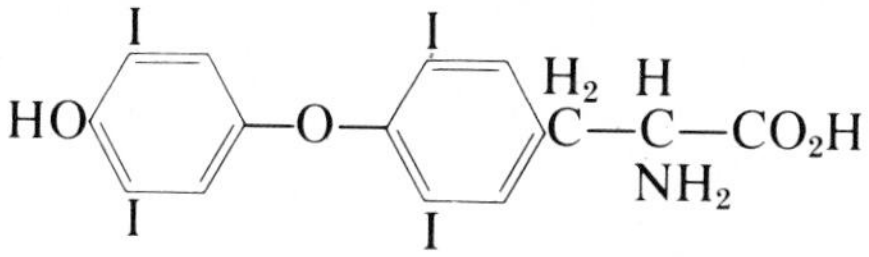

Thyroxine (thy)

Monoamino monocarboxy monohydroxy acids

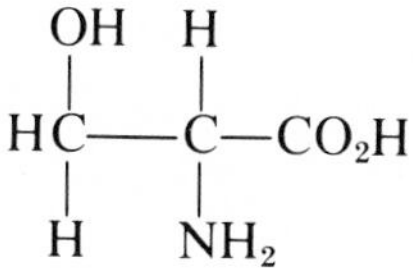

$$\underset{\underset{H}{|}}{\overset{\overset{OH}{|}}{HC}}-\underset{\underset{NH_2}{|}}{\overset{\overset{H}{|}}{C}}-CO_2H$$

Serine (Ser)

$$CH_3-\underset{\underset{\underset{H}{|}}{\overset{\overset{H}{|}}{C}}}{\overset{}{}}-\underset{\underset{NH_2}{|}}{\overset{\overset{H}{|}}{C}}-CO_2H$$

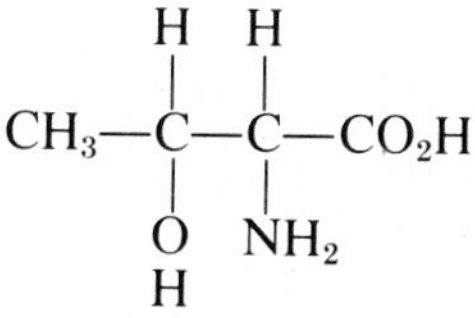

* Isoleucine (Ile)

Sulfur-containing acids

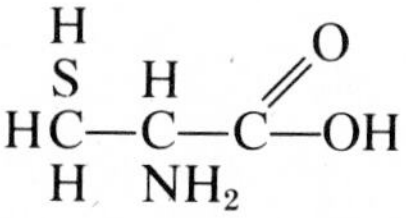

$$\underset{\underset{H}{|}}{\overset{\overset{\overset{H}{|}}{S}}{HC}}-\underset{\underset{NH_2}{|}}{\overset{\overset{H}{|}}{C}}-\overset{O}{\overset{\nearrow}{C}}-OH$$

Cysteine (Cys)

$$CH_3-S-CH_2-CH_2-\underset{\underset{NH_2}{|}}{\overset{\overset{H}{|}}{C}}-CO_2H$$

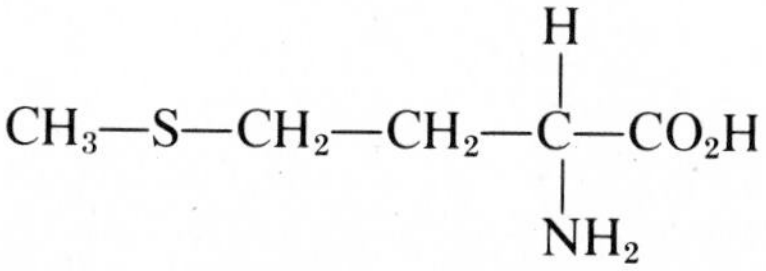

* Methionine (Met)

Dicarboxy monoamino acids

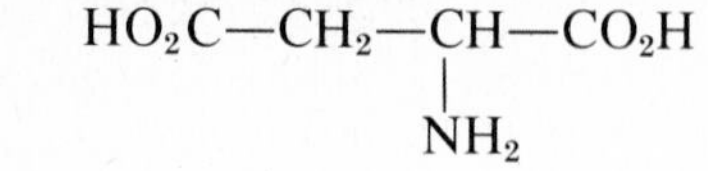

$$HO_2C-CH_2-CH-CO_2H$$
$$|$$
$$NH_2$$

Aspartic acid (Asp)

$$HO_2C-CH_2-CH_2-CH-CO_2H$$
$$|$$
$$NH_2$$

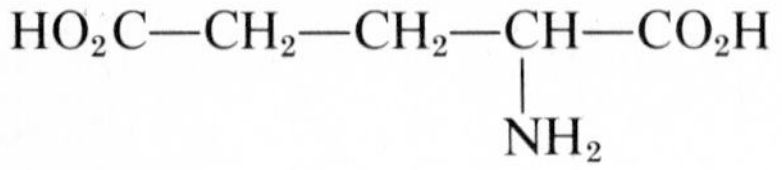

Glutamic acid (Glu)

Heterocyclic acids

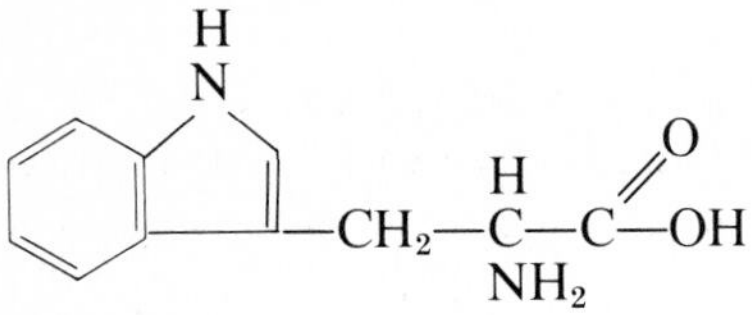

*Tryptophan (Trp)

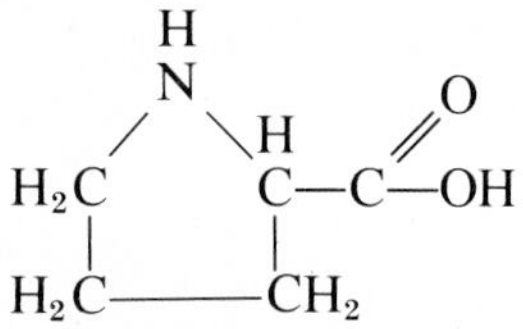

Proline (Pro)

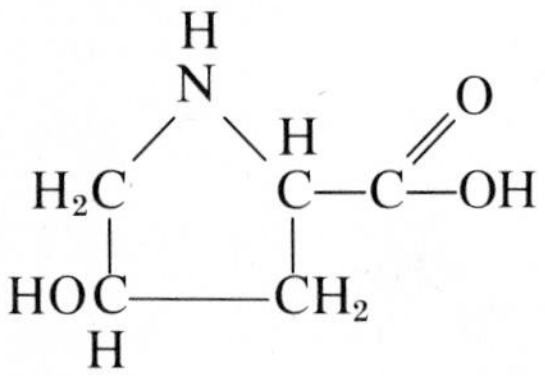

Hydroxyproline (Hypro)

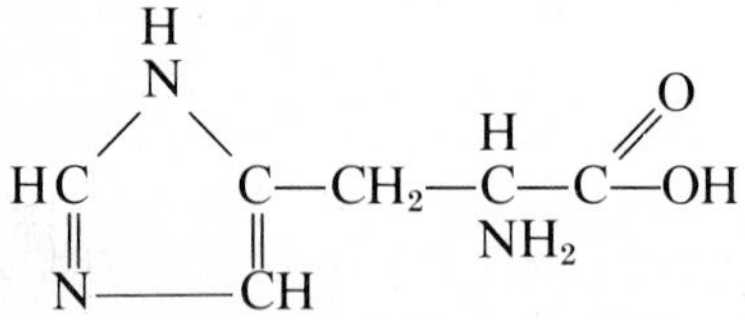

Histidine (His)

Proteins from different sources yield varying amounts of the different amino acids upon hydrolysis. The protein from a given source, however, normally yields the same amino acids and in the same ratio. All proteins yield more than one amino acid.

Proteins

Proteins constitute a very important part of the diet of humans, particularly in the form of meats, cheeses, eggs, and certain vegetables. The processing of these materials, except for eggs, results in the production of industrial wastes that are the concern of environmental engineers. Wastes produced in the meat-packing industry alone have a population equivalent in excess of 15 million.

Properties of proteins Protein molecules are very large and have complex chemical structures. Insulin, which contains 51 amino acid units per molecule, was the first protein for which the precise order of the atoms in the molecule was discovered. For this significant achievement, Frederick Sanger at the University of Cambridge received the Nobel prize in 1958. Among the several other proteins for which the order of atoms is now known are ribonuclease with 124 amino acids, tobacco mosaic virus protein with 158 amino acids, and hemoglobin with 574 amino acid units. Hemoglobin has a molecular weight of 64,500 and contains 10,000 atoms of hydrogen, carbon, nitrogen, oxygen, and sulfur, plus 4 atoms of iron. The iron atoms are more important than all the rest as they give blood its ability to combine with oxygen. Some protein molecules are thought to be 10 to 50 times larger than the above.

All proteins contain carbon, hydrogen, oxygen, and nitrogen. Regardless of their source, be it animal or vegetable, the ultimate analysis of all proteins falls within a very narrow range, as shown below:

	Percentage
Carbon	51–55
Hydrogen	6.5–7.3
Oxygen	20–24
Nitrogen	15–18
Sulfur	0.0–2.5
Phosphorus	0.0–1.0

The nitrogen content varies from 15 to 18 percent and averages about 16 percent. Since carbohydrates and fats do not contain nitrogen, advantage is taken of this fact in food analysis to calculate protein content. The value for nitrogen as determined by the Kjeldahl digestion procedure (Sec. 24-3), when multiplied by the factor 100/16 or 6.25, gives an estimate of the protein content. Environmental engineers often use this procedure to estimate the protein content of domestic and industrial wastes and of sludges from their organic nitrogen (Sec. 24-3) content. There are several other nitrogen containing organics in such wastes, however, and so such estimates should be considered only as crude approximations.

Biological treatment of protein wastes In general, satisfactory treatment of wastes containing significant amounts of proteins requires the use of biological processes. In these processes the first step in degradation of the protein is considered to be hydrolysis, induced by hydrolytic enzymes. The hydrolysis is considered to progress in steps in reverse manner to those in which proteins are synthesized.

Hydrolysis products of proteins

$$\text{Protein} \rightarrow \text{proteoses} \rightarrow \text{peptones} \rightarrow \text{polypeptides}$$
$$\downarrow$$
$$\alpha\text{-amino acids} \leftarrow \text{dipeptides} \tag{4-55}$$

The α-amino acids are then deaminized by enzymic action, and free fatty and other acids result. The free acids serve as food for the microorganisms, and they are converted to carbon dioxide and water.

TRACE ORGANICS

4-23 TRACE ORGANICS

The subject of trace organics and elements in foods has been under investigation for over half a century[7] and, since 1960 has become an item of increasing concern in the field of water supply.[8,9] Although the prime interest of environmental engineers is related to water, air, and their impurities, the effects of chemical impurities in water on human populations ordinarily can not be isolated for epidemiological or statistical study, because both food and water involve the digestive tract and vital organs of the body, particularly the kidneys and liver. In general the "food factor" is so great as to overwhelm the "water factor."

Trace organics in water supplies originate from natural sources, both plant and animal, and from the synthetic organic chemical industry. The latter are of great concern and are measured in large part, but not entirely, by the carbon adsorption method (CAM). Following adsorption on activated carbon, the organics are extracted with chloroform (CCE) or by ethyl alcohol (CAE) and weighed. Surface water supplies are most prone to carry significant amounts of trace organics, although ground waters are not necessarily immune, particularly shallow well supplies. In surface waters, natural plant exudates from growing algae and fungi carry materials of wide variety that cause taste and odors. Several are sulfur-containing compounds such as dimethyl sulfide and various mercaptans, including methyl mercaptan, isopropyl mercaptan, and butyl mercaptan. Geosmin is another such compound that has recently been isolated from the exudate of actinomycetes.[10]

---

[7] National Research Council, Committee on Food Protection, "Toxicants Occurring Naturally in Foods," National Academy of Sciences, Washington, D.C., 1973.

[8] AWWA Committee Report, Organic Contaminants in Water, *J. Am. Water Works Assoc.*, **66:** 682 (1974).

[9] AWWA T&P Council Statement, Organic Contaminants in Drinking Water, Am. Water Works Assoc. News Publ., *Willing Water:* 18 (Dec. 1974).

[10] D. Jenkins, Effects of Organic Compounds—Taste, Odor, Color, and Chelation, Organic Matter in Water Supplies; Occurrence, Significance, and Control, ed. by V. L. Snoeyink and M. F. Whelan, *Univ. Illinois Bull.*, **70:** 15 (June 4, 1973).

Geosmin

Decomposition of plant materials releases colored substances of a highly refractive nature. Except for tannins, the exact nature is not known, but they are polymeric, have high molecular weights and are phenolic in character. A symbolic description according to Christman and coworkers[11] is

They are commonly referred to as humic substances because of their similarity to humus in soil-organic matter. The excreta of humans and land animals contain soluble refractory organic matter that survives in the effluents of waste treatment plants, even those employing advanced treatment methods. It is reasonable to assume that the excreta of aquatic animals—fish, frogs, turtles, snakes, etc.—contain similar refractory materials. In human excreta, the soluble refractory organics originate largely from the unabsorbable fraction of foods and from wastes generated by kidney and liver function. Excluding aesthetic considerations, there is little evidence to indicate that water supplies containing trace organics derived from the excreta of warm or cold blooded animals have any public health significance. A possible exception is the recent finding that chlorination of humic substances in water results in the formation of chloroform and other halomethanes.

The major concern related to trace organics in water supplies, however, stems from the wide variety of synthetic chemicals that gain access to our surface waters and some ground supplies through the discharge of wastewaters from industries and municipal industrial complexes, from spills which result from accidents that occur during transportation by land and water, and from uncontrolled use or discharge of these materials to the environment. Of particular concern are insecticides and herbicides widely used in agricultural and forestry operations, polychlorinated biphenyls (PCB), and a variety of other

[11] R. F. Christman and R. A. Minear, "Organic Compounds in Aquatic Environments," ed. by S. D. Faust and J. V. Hunter, p. 119, Marcel Dekker, New York, 1971.

compounds. Until recently, much of the concern about these compounds was related to the ability of aquatic plants and animals to store and concentrate these materials in their tissues and, thereby, interfere with their reproductive processes. Of greater significance, however, was the potential hazard to humans using the animals for food. The Federal Drug Administration ban on the sale of Coho salmon taken from Lake Michigan, because of their unusual PCB content is one example.

The chlorinated organic compounds—DDT; 2,4D; 3,4,5T; PCB; chlordane, and others—have been singled out for special consideration because of their resistance to biological degradation, i.e., refractory nature. Recently, a suspected new source of chlorinated compounds has been exposed. It is related to the use of chlorine for disinfection purposes. This fear has been enhanced by reports of serious fish kills in streams below sewer outfalls where disinfection with chlorine was practiced. These results have raised serious doubt, in the minds of some people, relative to the safety of using chlorine for the disinfection of drinking water supplies, in spite of a 70-year history.

Current studies (1974) of the drinking water supplies of New Orleans and Cincinnati have isolated and identified 166 separate organic compounds or derivatives; many chloro and some bromo. Most of the organics found at New Orleans were present at concentrations of 1 μg/l (ppb) or less, a few in the 1–10-μg/l range, but chloroform was found at levels of 40–150 μg/l in six separate finished water supplies. Similar levels have been found in water supplies throughout the country and as previously indicated, evidence suggests they result from reactions between chlorine and naturally occurring organic matter.

It is generally conceded that the organic materials in surface water supplies, under normal conditions, can not be considered to be present in the toxic range capable of producing violent illness or death. Rather, some must be viewed in terms of hazardous materials because of their possible subtle, long term effects. Among the chemicals identified have been some with known effects upon the central nervous system and others with *carcinogenic* (oncogenic or tumor causing) properties. Also, the possibility exists that some with *teratogenic* (disfiguring) or *mutagenic* characteristics may be present. The importance of these materials in public water supplies is undetermined and always may remain so for the reasons previously expressed. Nevertheless, we should bend every effort to minimize their presence by source control and proper water and wastewater treatment.

DETERGENTS

4-24 DETERGENTS

The term *detergent* is applied to a wide variety of cleansing materials used to remove soil from clothes, dishes, and a host of other things. The basic ingredients of detergents are organic materials which have the property of being "surface active" in aqueous solution, and are called surface-active agents or *surfactants*. All surfactants have rather large polar molecules. One end of the molecule is particularly soluble in water and the other is readily soluble in oils. The solubility in water is due to carboxyl, sulfate, hydroxyl, or sulfonate groups. The surfactants with carboxyl, sulfate, and sulfonate groups are all used as sodium or potassium salts.

Oil-soluble part	Water-soluble part
Organic group $\left. \right\}$	$-COO^-Na^+$ $-SO_4^-Na^+$ $-SO_3^-Na^+$ $-OH$

The nature of the organic part of the molecule varies greatly with the various surfactant types.

4-25 SOAPS

Ordinary *soaps* are derived from fats and oils by *saponification* with sodium hydroxide. Saponification is a special case of hydrolysis in which an alkaline agent is present to neutralize the fatty acids as they are formed. In this way the reaction is caused to go to completion.

$$
\begin{array}{c}
H_2COOCC_{17}H_{35} \\
| \\
H\text{---}COOCC_{17}H_{35} + 3NaOH \rightarrow \\
| \\
H_2COOCC_{17}H_{35} \\
\text{Stearin}
\end{array}
\quad
\begin{array}{c}
H_2COH \\
| \\
H\text{---}COH + 3C_{17}H_{35}COONa \\
| \\
H_2COH \\
\text{Glycerol}
\end{array}
\quad (4\text{-}56)
$$

The fats and oils are split into glycerol and sodium soaps. The nature of the soap depends upon the type of fat or oil used. Beef fat and cotton seed oil are used to produce low-grade heavy-duty soaps. Coconut and other oils are used in the production of toilet soaps.

All sodium and potassium soaps are soluble in water. If the water is hard, the calcium, magnesium, and any other ions causing hardness precipitate the soap in the form of metallic soaps.

$$2C_{17}H_{35}COONa + Ca^{2+} \rightarrow (C_{17}H_{35}COO)_2Ca\downarrow + 2Na^+ \quad (4\text{-}57)$$

Soap must be added to precipitate all the ions causing hardness before it can act as a surfactant, usually indicated by the onset of frothing upon agitation.

4-25 SYNTHETIC DETERGENTS

Since 1945 a wide variety of synthetic detergents, commonly called *syndets*, have been accepted as substitutes for soap. Their major advantage is that they do not form insoluble precipitates with the ions causing hardness. As marketed, most of them contain from 20 to 30 percent of surfactant and 70 to 80 percent of *builders*. The builders are usually sodium sulfate, sodium tripolyphosphate, sodium pyrophosphate, sodium silicate, and other materials that enhance the detergent properties of the active ingredient. The synthetic surfactants are of three major types: *anionic, nonionic,* and *cationic*.

Anionic Surfactants

The anionic surfactants are all sodium salts and ionize to yield Na^+ plus a negatively charged, surface-active ion. The common ones are all sulfates and sulfonates.

Sulfates Long-chain alcohols when treated with sulfuric acid produce sulfates (inorganic esters) with surface-active properties. Dodecyl or lauryl alcohol is commonly used.

$$C_{12}H_{25}OH + H_2SO_4 \rightarrow C_{12}H_{25}\text{—}O\text{—}SO_3H + H_2O \qquad (4\text{-}58)$$

Lauryl
alcohol

The sulfated alcohol is neutralized with sodium hydroxide to produce the surfactant.

$$C_{12}H_{25}\text{—}O\text{—}SO_3H + NaOH \rightarrow C_{12}H_{25}\text{—}O\text{—}SO_3Na + H_2O \qquad (4\text{-}59)$$

Sodium lauryl sulfate

The sulfated alcohols were the first surfactants to be produced commercially. The sulfated alcohols are used in combination with other syndets to produce blends with desirable properties.

Sulfonates The principal sulfonates of importance are derived from esters, amides, and alkylbenzenes.

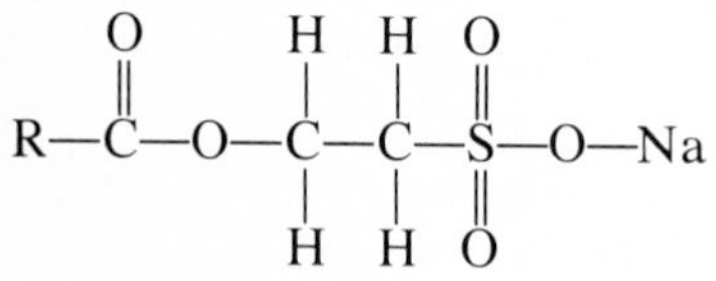

Ester

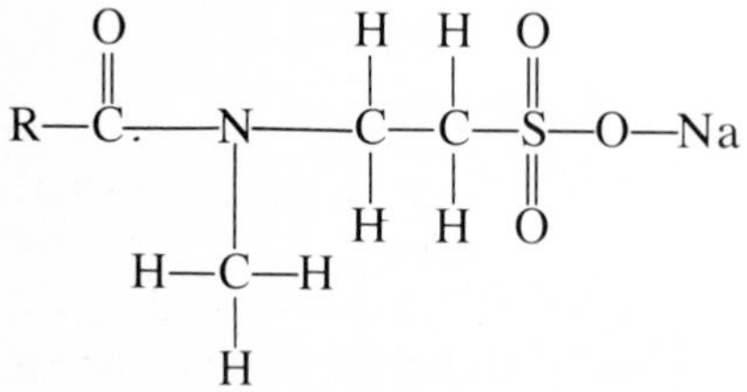

Amide

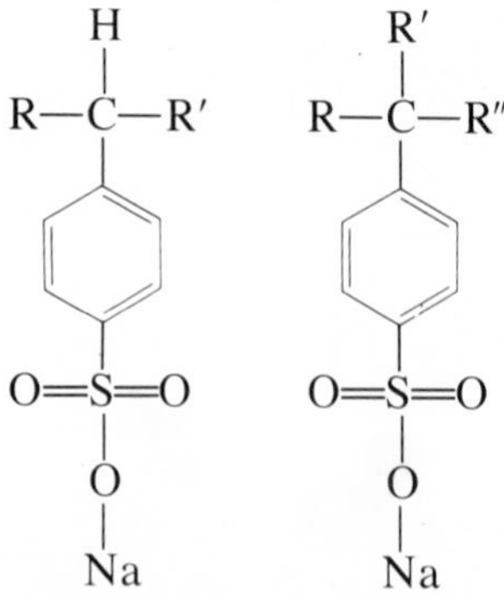

Secondary   Tertiary
Sulfonated alkylbenzenes

The esters and amides are of organic acids with 16 or 18 carbon atoms. In the past the alkylbenzene sulfonates (ABS) were derived largely from polymers of propylene, and the alkyl group, which averaged 12 carbon atoms, was highly branched. These materials are now made largely from normal (straight-chain) paraffins, and thus the alkane chain is not branched and the benzene ring is attached primarily to secondary carbon atoms. These latter materials have been labeled LAS (linear alkyl sulfonate).

Nonionic syndets

The *nonionic* detergents do not ionize and have to depend upon groups in the molecule to render them soluble. All depend upon polymers of ethylene oxide to give them this property.

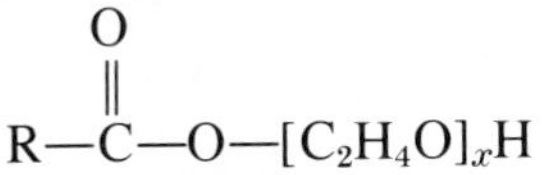

$$
\begin{array}{c}
\overset{\displaystyle O}{\underset{\displaystyle \parallel}{}} \\
R-C-O-[C_2H_4O]_xH
\end{array}
$$
Ester type

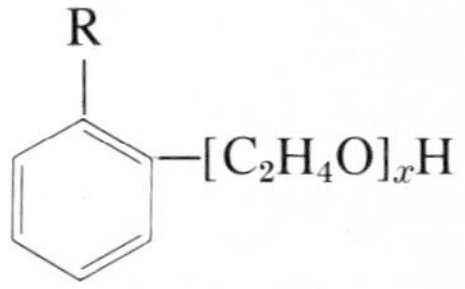

$$
\begin{array}{c}
\overset{\displaystyle O}{\underset{\displaystyle \parallel}{}} \\
R-C-N-[C_2H_4O]_xH \\
\underset{\displaystyle R}{|}
\end{array}
$$
Amide type

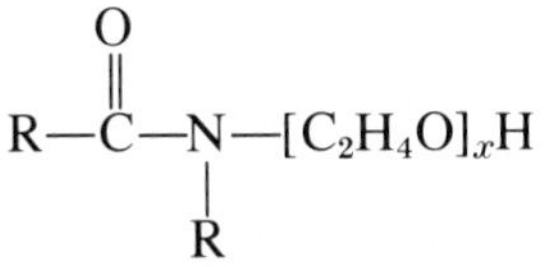

$$-[C_2H_4O]_xH$$ (R attached to benzene ring)
Aryl type

$$HO[C_2H_4O]_xH$$
Ethylene oxide polymer type

The nonionic type of syndet has been more expensive to produce than the anionic type. It is gaining in popularity, however.

Cationic Syndets

The *cationic* syndets are salts of quaternary ammonium hydroxide. In quaternary ammonium hydroxide, the hydrogens of the ammonium ion have all been replaced with alkyl groups. The surface-active properties are contained in the cation.

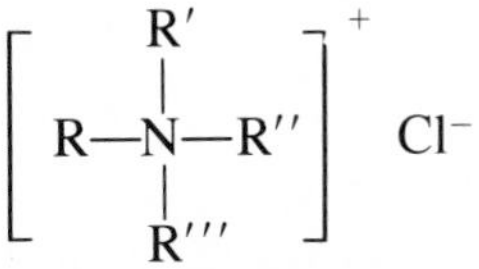

$$
\left[R-\overset{\displaystyle R'}{\underset{\displaystyle R'''}{N}}-R'' \right]^{+} Cl^{-}
$$
A cationic syndet

The cationic syndets are noted for their disinfecting (bactericidal) properties. They are used as sanitizing agents for dishwashing where hot water is unavailable or undesirable. They are also useful in the washing of babies' diapers, where sterility is important. If diapers are not sterilized by some means, bacterial infestations may occur which release enzymes that will hydrolyze urea to produce free ammonia [Eq. (4-45)]. The high pH resulting is harmful to the tender skin of babies, and the odor of free ammonia is unpleasant to all.

Biological Degradation of Detergents

Detergents vary greatly in their biochemical behavior, depending on their chemical structure.[12] Common soaps and the sulfated alcohols are readily used as bacterial food. The syndets with ester or amide linkages are readily hydrolyzed. The fatty acids produced serve as sources of bacterial food. The other hydrolysis product may or may not serve as bacterial food, depending upon its chemical structure. The syndets prepared from polymers of ethylene oxide appear susceptible to biological attack. The alkylbenzene sulfonates derived from propylene are resistant to biological attack because of branched-chain structure of the alkyl groups and because the benzene rings are attached principally to tertiary carbon atoms of the branched-chain groups. Because of this resistance, they persist after normal biological treatment and have contaminated both surface and ground-water supplies with an objectionable foaming property. For this reason, the U.S. Public Health Service established an ABS concentration limit of 0.5 mg/l in their drinking water standards and the detergent manufacturing industry changed to the production of LAS surfactants. LAS is readily degradable under aerobic conditions and its use has helped relieve the most serious problems of detergent foaming. However, unlike common soap, it is resistant to degradation under anaerobic conditions.

PESTICIDES

4-26 PESTICIDES

Pesticides are materials used to prevent, destroy, repel, or otherwise control objectionable insects, rodents, plants, weeds, or other undesirable forms of life. Common pesticides can be categorized chemically into three general groups, inorganic, natural organic, and synthetic organic. They may also be classified by their biological usefulness, viz., insecticides, algicides, fungicides, and herbicides.

The synthetic organic pesticides gained prominence during World War II, and since then their numbers have grown into the thousands, while the total annual production has increased to about a billion pounds. They are used mainly for agricultural purposes. The synthetic organic pesticides are best classified according to their chemical properties, since this more readily determines their persistence and behavior when introduced into waterways. The major types are the chlorinated pesticides, the organic phosphorus pesticides, and the carbamate pesticides.

[12] C. N. Sawyer and D. W. Ryckman, Anionic Detergents and Water Supply Problems, *J. Am. Water Works Assoc.*, **49**: 480 (1957).

4-27 CHLORINATED PESTICIDES

Chlorinated pesticides are of many types and have been widely used for a variety of purposes. DDT proved to be an extremely versatile insecticide during World War II when it was used mainly for louse and mosquito control. It is still one of the major pesticides used internationally, although use in the United States has been greatly restricted for environmental reasons. DDT is a chlorinated aromatic compound with the following structure:

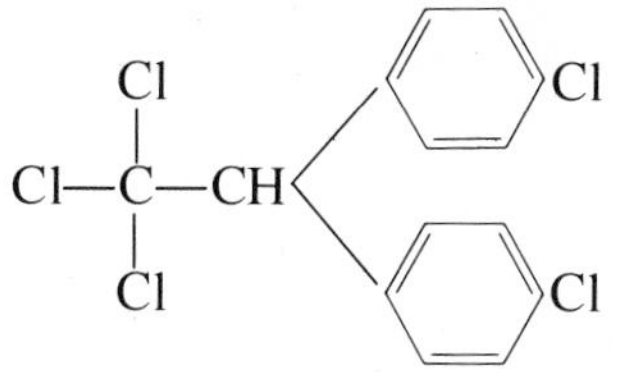

DDT

Technical grade DDT contains three isomers, the above isomer representing about 70 percent of the total. It has the long technical name 2,2-bis (*p*-chlorophenyl)-1,1,1-trichloroethane.

When benzene and chlorine react in direct sunlight, the addition product benzene hexachloride or BHC is formed. Several stereoisomers are produced, but the gamma isomer called *lindane* is by far the most effective as an insecticide.

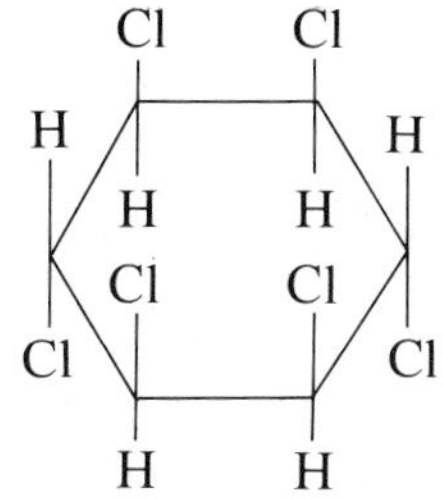

γ-Benzene hexachloride (lindane)

Two other formerly widely used chlorinated insecticides are endrin and dieldrin. They are isomers, and their structural formulas appear the same:

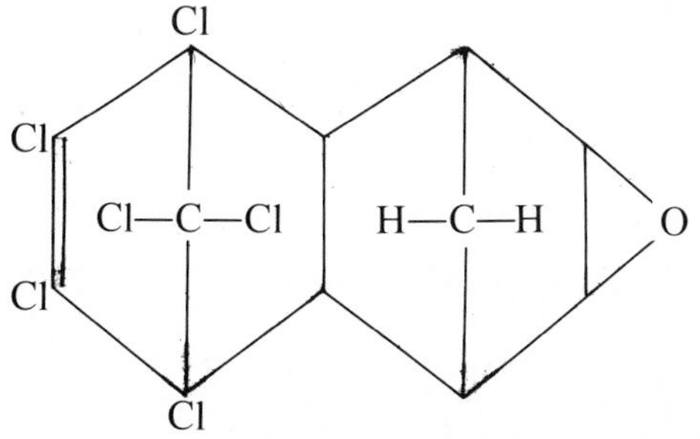

Dieldrin or endrin

However, their spatial configurations are significantly different, as are their insecticidal properties.

Chlorinated pesticides are also used as herbicides. Two of the most common are 2,4-D and 2,4,5-T.

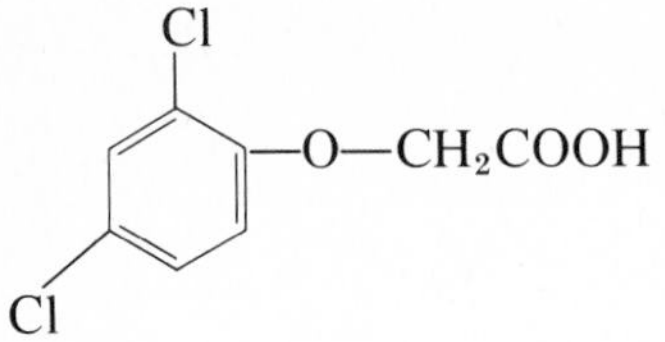

2,4-Dichlorophenoxyacetic acid
(2,4-D)

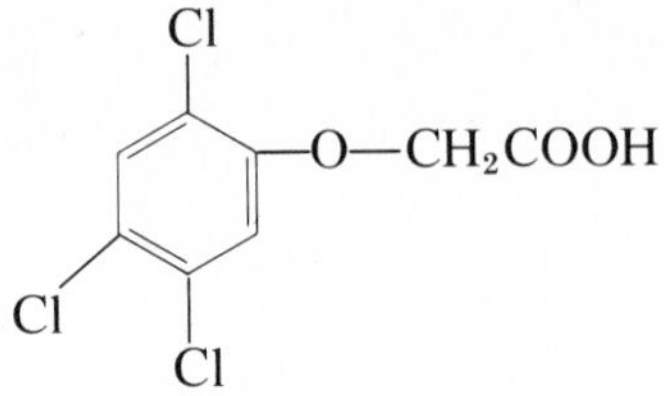

2,4,5-Trichlorophenoxyacetic acid
(2,4,5-T)

These two herbicides are effective in destroying certain broad-leaf plants while not killing grasses. They have also been used for aquatic-plant control in lakes, ponds, and reservoirs. Dioxin, an extremely toxic organic to humans, is a side-product contaminant in 2,4,5-T, which has led to restrictions in its use.

Other chlorinated pesticides of significance are aldrin, chlordane, toxaphene, heptachlor, and DDD; all have been used as insecticides or fungicides. All chlorinated pesticides have come under attack because of their persistence and high potential for creating harm to humans and the environment. For this reason aldrin and dieldrin are now banned from use in the United States and many of the other chlorinated pesticides have been greatly restricted in usage.

4-28 ORGANIC PHOSPHORUS PESTICIDES

The organic phosphorus pesticides became important as insecticides after World War II. These compounds were developed in the course of chemical warfare research in Germany and in general are quite toxic to humans as well as to pests. *Parathion* is an important pesticide which was introduced into the United States from Germany in 1946. It is an aromatic compound and contains sulfur and nitrogen as well as phosphorus in its structure, as do many of the organic phosphorus pesticides.

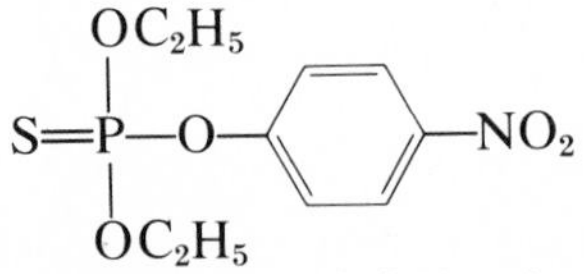

O,O-Diethyl-O-*p*-nitrophenylthiophosphate(parathion)

Parathion has been particularly effective against certain pests such as the fruit fly. However, it is also quite toxic to humans and extreme caution must be exercised in its use.

Malathion is highly toxic to a variety of insects, but unlike many organic phosphorus pesticides, it has low toxicity to mammals. This has made it particularly successful and it is widely used.

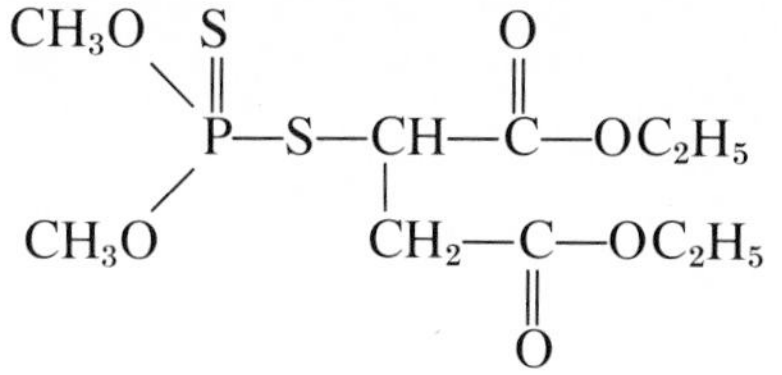

$$CH_3O \diagdown \overset{S}{\underset{\diagup}{P}} - S - \overset{}{\underset{|}{CH}} - \overset{O}{\overset{\|}{C}} - OC_2H_5$$
$$CH_3O \qquad CH_2 - \overset{}{\underset{\|}{C}} - OC_2H_5$$
$$\qquad\qquad\qquad O$$

S-(1,2-Dicarbethoxyethyl)-O,O-dimethyldithiophosphate
(malathion)

Other organic phosphorus pesticides of significance are systox, chlorthion, disyston, dicapthon, and metasystox.

4-29 CARBAMATE PESTICIDES

Carbamate pesticides are amides having the general formula RHNCOOR'. One that has received wide usage is isopropyl N-phenylcarbamate (IPC).

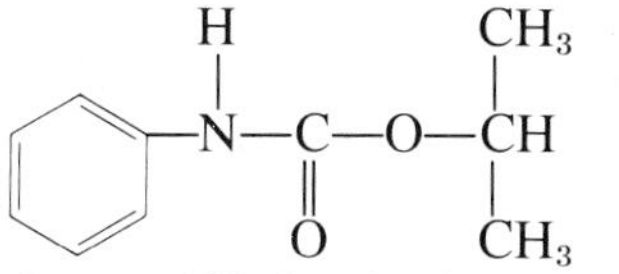

$$\text{—N—C—O—CH} \atop \text{O} \quad CH_3$$

Isopropyl N-phenylcarbamate
(IPC)

IPC is a herbicide which is effective for the control of grasses, without affecting broad-leaf crops. Other carbamates of note are captan, which is a fungicide, and ferbam and sevin, which are insecticides. Carbamates in general appear to have low toxicity to mammals.

4-30 BIOLOGICAL PROPERTIES OF PESTICIDES

Pesticides may gain access to ground and surface water supplies through direct application or through percolation and runoff from treated areas. Some pesticides are toxic to fish and other aquatic life at only a small fraction of a milligram per liter. They also tend to concentrate in aquatic plants and animals to values several thousand times that occurring in the water in which they live. Also, some pesticides are quite resistant to biological degradation and persist in soils and water for long periods of time.

In general, the chlorinated pesticides are the most resistant to biological degradation and may persist for months or years following application. Many are also highly toxic to aquatic life or to birds which feed on aquatic life. This persistance and great potential for harm has lead to restrictions on chlorinated pesticide usage. As a group, the organic phosphorus pesticides are not too toxic to fish life, and have not caused much concern in this respect, except when large accidental spills have occurred. Also, they tend to hydrolyze rather quickly at pH values above neutral, thus losing their toxic properties. Under proper conditions such as dryness, however, some have been observed to persist for many months. Their main potential for harm is to farm workers who become exposed both from contact with previously sprayed foliage as well as from direct contact during organic phosphorus pesticide application. Some of these pesticides are readily absorbed through the skin and affect the nervous system by inhibiting the enzyme cholinesterase, which is important in the transmission of nerve impulses. The carbamates are noted for their low toxicity and high susceptibility to degradation.

PROBLEMS

4-1 Why is it possible to have so many compounds of carbon?

4-2 Define homologous series, homolog, hydrocarbon, and alkyl radical.

4-3 Define isomerism and illustrate with structural formulas of compounds with molecular formula C_6H_{14}.

4-4 In general, how do organic compounds differ from inorganic compounds?

4-5 How many grams of oxygen are required to furnish just enough oxygen for the complete oxidation of 20 g of butane?

4-6 What is the difference between an aliphatic and an aromatic compound?

4-7 Define with suitable illustrations the terms primary alcohol, secondary alcohol, and tertiary alcohol, and indicate their relative ease of biodegradation.

4-8 Show the structure and name of the intermediate compounds formed in the biological oxidation of n-butane to butyric acid.

4-9 Why are oxidation reactions of organic compounds important to environmental engineers?

4-10 What might be formed from $CH_3—CH_2—CH=CH—CH_3$ in the atmosphere in the presence of sunlight, oxides of nitrogen, and water droplets?

4-11 Which of the following compounds have optical isomers?

$$CH_2OH—\underset{\underset{OH}{|}}{CH}—CHO \qquad CH_3—\underset{\underset{H}{|}}{C}=\underset{\underset{CH_3}{|}}{C}—H \qquad CH_3—CH_2—\underset{\underset{OH}{|}}{CH}—CH_3$$

$$CHO—\underset{\underset{OH}{|}}{CH}—\underset{\underset{OH}{|}}{CH}—CH_2OH \qquad CH_3—\underset{\underset{CH_3}{|}}{CH}—COOH \qquad CH_3—\underset{\underset{H}{|}}{C}=\underset{\underset{CH_3}{|}}{CH}$$

4-12 Show a general formula for fats and oils, and indicate what is the difference between the two.

4-13 What is the difference between a fat and a wax?

4-14 How many asymmetric carbon atoms are there in glucose?

4-15 Write a possible structural formula for a ketopentose and for an aldopentose.

4-16 What is the major factor contributing to the difference in biodegradability between starch and cellulose?

4-17 Show a general formula for the building blocks of proteins and illustrate how these building blocks are connected to form proteins.

4-18 If a sample of food waste contains 2.5 percent of organic nitrogen, what approximate percentage of the sample is protein?

4-19 What functional group is characteristic of each of the following: alkenes, alcohols, aldehydes, ketones, acids, amines, amides, ethers, esters, and aromatic compounds?

4-20 What monosaccharides are formed upon hydrolysis of the following: cellulose, starch, and hemicellulose?

4-21 Name the three general classes of synthetic detergents, and give an example of each.

4-22 Name the three general classes of pesticides, and describe how the pesticides in each class differ in degree of biodegradability.

4-23 Name the following:

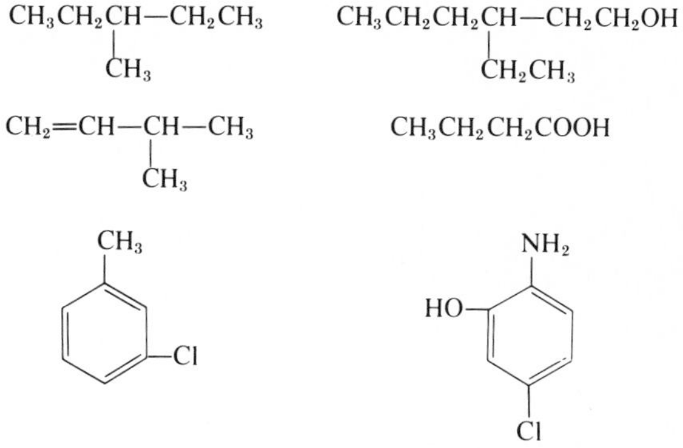

$$CH_3CH_2CH\!-\!CH_2CH_3 \atop \qquad\quad | \atop \qquad\quad CH_3 \qquad\qquad CH_3CH_2CH_2CH\!-\!CH_2CH_2OH \atop \qquad\qquad\qquad\qquad\quad | \atop \qquad\qquad\qquad\qquad\quad CH_2CH_3$$

$$CH_2\!=\!CH\!-\!CH\!-\!CH_3 \atop \qquad\qquad | \atop \qquad\qquad CH_3 \qquad\qquad CH_3CH_2CH_2COOH$$

REFERENCES

Bonner, W. A., and A. J. Castro: "Essentials of Modern Organic Chemistry," Reinhold, New York, 1965.

Cason, J.: "Principles of Modern Organic Chemistry," Prentice-Hall, Englewood Cliffs, N.J., 1966.

Roberts, J. D., R. Stewart, and M. C. Caserio: "Organic Chemistry," W. A. Benjamin, Menlo Park, Cal., 1971.

Wertheim, E., and H. Jeskey: "Introductory Organic Chemistry," 3d ed., McGraw-Hill, New York, 1956.

BASIC CONCEPTS FROM
EQUILIBRIUM CHEMISTRY

5-1 INTRODUCTION

A knowledge of equilibrium chemistry has become increasingly important for determining quantitatively the relationships between the various constituents in natural and contaminated waters, and for understanding the effect of alterations in the water on the various chemical species present. This is a matter of growing importance because of the trend toward physical-chemical methods of waste-water treatment and the hazards of heavy metals and certain organic-inorganic compounds in the environment. Thus, a good knowledge of equilibrium chemistry is helpful in understanding the effects of waste discharge on receiving waters as well as in evaluating how best to treat wastewaters in order to rid them of harmful substances. Equilibrium chemistry draws heavily upon the equilibrium relationships discussed in Chap. 2 and the thermodynamics of equilibrium relationships as discussed in Chap. 3. The reader should have a good understanding of these relationships before proceeding with this chapter.

5-2 LIMITATIONS OF EQUILIBRIUM CALCULATIONS

Before considering equilibria in water systems in detail, it is well to be aware of some of the limitations of equilibrium calculations. Most waters, especially surface waters and wastewaters, undergo dynamic changes resulting from the constant introduction of energy from the sun and materials, both inorganic and organic, from inflowing rivers and wastewater streams. Equilibrium calculations cannot be expected to describe accurately the relationships between all the various species in such a system since equilibrium may not be attained. They can, however, indicate the direction in which reactions would tend to proceed and the extent to which reactions would go if equilibrium were attained. They can indicate whether certain chemical transformations are possible.

Some reactions, such as those between soluble acids and bases occur very rapidly so that equilibrium may occur within seconds. Others, such as oxidation-reduction reactions, may occur slowly under conditions existing in natural waters so that equilibrium may not be achieved within years or sometimes centuries. On the other hand, many oxidation-reduction reactions which otherwise may tend slowly toward equilibrium may be catalyzed by bacteria which can then capture the energy released for growth. Precipitation reactions may occur readily when an excess of a required chemical constituent is added to water, but the reaction is likely to stop before true equilibrium is reached. This may be due to the initial formation of poorly crystallized or amorphous precipitates, which have greater solubilities. Reconstitution of such amorphous solids into crystalline solids occurs very slowly. Equilibrium calculations can help indicate the direction toward which the chemical nature of an aquatic system will tend to change, but cannot indicate the rate at which the change will proceed. Thus, an understanding of the kinetics of change is necessary in a realistic evaluation of the potential for change predicted from equilibrium considerations. Unfortunately, there is a lack of kinetic data for many transformations of interest in environmental engineering.

Another limitation is the lack of availability of accurate equilibrium constants for many of the reactions of interest in natural waters. While the equilibrium constant for dissociation of water is known quite accurately ($1.008 \pm 0.001 \times 10^{-14}$ at $20°C$), the best-known values for weak acids and bases generally vary in accuracy from ± 0.5 to ± 10 percent. Solubility products are sometimes known with similar accuracy, but several, such as for some sulfides, have values as reported by different investigators which differ by several orders of magnitude. Formation constants of complex ions also are of widely varying accuracy so that caution is indicated when they are used. Such poor accuracy in values generally result from inability to measure constituents at the low concentrations sometimes associated with equilibria, the uncertainty of whether true equilibria have been attained, and uncertainty over the actual reactions which occur in solution. For example, most of the earlier determinations of the solubility product for $Al(OH)_3$ (gibbsite) are inaccurate because the solid phase produced by precipitation was sometimes gibbsite, but more often a mixture of gibbsite with other hydrous and hydrated oxides.

The above uncertainties would appear to restrict the usefulness of equilibrium calculations severely. However, there are many cases where equilibrium constants are known with sufficient accuracy for most practical environmental engineering purposes and where equilibrium is readily attained. In order to approach a given problem correctly in environmental engineering, the accuracy with which predictions can be made and limitations of the predictions must be well understood.

5-3 ION ACTIVITY COEFFICIENTS

The fundamental aspects of electrolytic dissociation or ionization, were discussed in Sec. 2.13. In Sec. 2.12 it was noted that as solutions of ionized materials become more concentrated, their quantitative effect in equilibrium relationships becomes progressively less than calculated solely from the change in molar concentration. In order to overcome this deficiency, the concept of activity or effective concentration was advanced. The activity of an ion or molecule can be found by multiplying its molar concentration by an activity coefficient, γ, as indicated by Eq. (2-22).

For many practical purposes, rough calculations obtained by use of molar concentrations rather than activities are sufficient. Also, in other cases where some of the limitations discussed in Sec. 5-2 apply, the more refined use of activities may not be justified. However, in certain cases it may be practical and desirable to obtain more precise results. For these cases procedures for estimating the activities of ions, which are apt to deviate most from ideality, are desirable.

Lewis and Randall recognized that the activity coefficient for ions in an electrolyte were related to the concentration of charged particles in the solution. They introduced the concept of *ionic strength* as an empirical measure of the interactions among all the ions in a solution which caused deviation from ideal behavior. The ionic strength, μ, is a characteristic of the solution and is defined as

$$\mu = \tfrac{1}{2} \sum_i C_i Z_i^2 \tag{5-1}$$

where C_i is the molar concentration of the ith ion, Z_i is its charge, and the summation extends over all the ions in the solution. It is important to keep in mind that the ionic strength is a general property of the solution and not a property of any particular ion in the solution. Lewis and Randall found that in dilute solutions the activity coefficient of a given ion was the same in all solutions of the same ionic strength. For fresh natural waters, Langelier[1] indicated that the ionic strength could be approximated by multiplying the milligrams per liter of dissolved solids by 2.5×10^{-5}.

The concept of Lewis and Randall was extended by Debye and Hückel into a general theory which expressed the relationship between the activity coefficient for an ion and the ionic strength of the solution in which it was contained. This theory is applicable only to dilute solutions, those with ionic strength less

---

[1] W. F. Langelier, The Analytical Control of Anti-Corrosion Water Treatment, *J. Amer. Water Works Assoc.*, **28:** 1500 (1936).

than 0.1. However, excepting seawater, most water solutions of interest in environmental engineering are more dilute than this, and thus the theory is applicable. For this case the Güntelberg approximation derived from the Deybe-Hückel basic relationship is frequently used:

$$\log \gamma = -0.5Z^2 \frac{\sqrt{\mu}}{1 + \sqrt{\mu}} \tag{5-2}$$

Here Z is the charge on the ion for which the activity coefficient is being determined. This theory was developed from the concept that oppositely charged ions attract each other and cause deviations from the behavior that would be produced by an equal number of uncharged particles.

Example Calculate the activity coefficients and the activities of each ion in a solution containing 0.01 M $MgCl_2$ and 0.02 M Na_2SO_4.

   When the salts are dissolved, the following molar concentrations (C) of each ion with charge Z result:

Ion	C	Z	CZ^2
Mg^{2+}	0.01	+2	0.04
Na^+	0.04	+1	0.04
Cl^-	0.02	−1	0.02
SO_4^{2-}	0.02	−2	0.08
		$\sum C_i Z_i^2 = 0.18$	

From Eq. (5-1), we have $\mu = \frac{1}{2} \sum_i C_i Z_i^2 = 0.09$.

From Eq. (5-2),

$$\log \gamma = -0.5Z^2 \frac{\sqrt{0.09}}{1 + \sqrt{0.09}} = -0.115Z^2$$

Thus for the ions whose $Z = 1$, we have $\log \gamma = -0.115$ and $\gamma = 0.77$, and for the ions whose $Z = 2$, we have $\log \gamma = -0.46$ and $\gamma = 0.35$. Therefore the activity coefficient and activity for each ion are as follows:

Ion	Activity coefficient (γ)	Activity (γC)
Mg^{2+}	0.35	0.0035
Na^+	0.77	0.031
Cl^-	0.77	0.015
SO_4^{2-}	0.35	0.007

One can see that the effective concentration (activity) of divalent ions is significantly smaller than the analytical concentration!

For simplicity, activity corrections will not be made to equilibrium problems in this book. Activity coefficients will be assumed equal to 1.0.

5-4 SOLUTION TO EQUILIBRIUM PROBLEMS

In solving equilibrium problems it is helpful to know qualitatively what can be expected to occur. Le Chatelier's principle is helpful for this purpose. It states that a chemical system will respond to change with processes which tend to reduce the effect of the change. For example, if SO_4^{2-} is added to a solution saturated with respect to $CaSO_4$, some of the calcium ion in solution will react with added SO_4^{2-}, thus reducing the effect of the change, i.e., reducing the increase in SO_4^{2-} concentration. Le Chatelier's principle in our present context is a qualitative statement of the fact that the equilibrium constants applicable in a chemical system must be obeyed.

Conceptually, the solution to an equilibrium problem is straightforward. First, for any chemical reaction, the principal of conservation of mass must be obeyed. Conservation of mass means that the mass of each element present must remain the same before, during, and after the reaction. It also means that the quantity of a given chemical entity not destroyed by the reaction must remain constant. For example, when sodium hydroxide is added to an acetic acid solution, the acetic acid present as CH_3COOH is at least partially ionized to form the acetate ion, CH_3COO^-. Conservation of mass requires that the sum of molar concentrations of acetic acid plus acetate, $[CH_3COOH] + [CH_3COO^-]$, remains constant during the sodium hydroxide addition.

Second, electroneutrality must be maintained. This means that all positively charged species in solution must be balanced by equivalent numbers of negatively charged species. For example, in a solution of sodium chloride in water,

$$[Na^+] + [H^+] = [Cl^-] + [OH^-]$$

or in a solution of calcium hydroxide,

$$2[Ca^{2+}] + [H^+] = [OH^-]$$

The calcium concentration is multiplied by two because it is divalent, thus one mole of Ca^{2+} is equivalent to two moles of positive charge. To illustrate further, if one mole of hydrated lime, $Ca(OH)_2$, were added to one liter of water, it would dissociate to yield one mole of Ca^{2+} and two moles of OH^-. Thus, $[Ca^{2+}] = 1$ and $[OH^-] = 2$. $[H^+]$ in this case would be very small and can be ignored. Using the above equation we see that electroneutrality is maintained since $2(1) + 0 = 2$.

Third, all reactions involved must proceed toward a state of equilibrium at which all appropriate equilibrium relationships are satisfied. Steps involved in the solution of an equilibrium problem involving only the aqueous phase are as follows:

1. Carefully define the equilibrium problem being considered: e.g. What chemical reactions are taking place? What is reacting with what? What are conditions at the outset and at the end of equilibrium?
2. List all constituents of the system which are present at the outset. All systems involving water will of course include H_2O, H^+, and OH^-. The list should also include all other elements, radicals, and neutral species which are present initially.
3. For each element initially present, list all the likely forms or species containing it which are likely to be present after equilibrium is attained.
4. Identify the concentrations of each species for each element or entity under initial conditions so that appropriate mass and charge balances can be made.
5. List all appropriate equilibrium relationships between the species of concern, together with associated equilibrium constants.
6. List all mass and charge balance relationships for the system. These together with the relationships from step 5 must result in as many independent equations as there are unknown species present.
7. A simultaneous solution of the above equations will give the concentration of each species at equilibrium.

If other phases, such as gaseous or solid phases, are also involved, then equations expressing mass and charge balances between and within each phase must be written and included in the calculations.

Example A mass equal to 10^{-2} mol of acetic acid is added to sufficient water to make one liter of solution at 25°C. What is the equilibrium concentration of all species involved?

1. The problem is one of ionization of a single weak monoprotic acid in water. When acetic acid is added to water it ionizes as follows: $CH_3COOH \rightleftharpoons CH_3COO^- + H^+$
2. Elements or chemical entities originally present are CH_3COOH, H_2O, H^+, and OH^-.
3. Chemical species likely to be present after equilibrium is attained are:

   Acetic acid: CH_3COOH (HA) and CH_3COO^- (A$^-$)

   Water: H_2O, H^+, OH^-

4. Initial conditions are:

   $[HA] = 10^{-2}$ mol/l $= C$

   $[H_2O] = 1$ liter of solution (mass change will be negligible during ionization), and $[H^+] = [OH^-] = 10^{-7}$ mol/l

5. Equilibrium relationships of interest:

   $$[H^+][OH^-] = K_W = 10^{-14} \text{ at } 25°C \tag{5-3}$$

   $$\frac{[H^+][A^-]}{[HA]} = K_A = 1.8 \times 10^{-5} \text{ at } 25°C \tag{5-4}$$

6. Mass and charge balance relationships:
   The acetic acid entity is conserved during ionization such that:

   $$[HA] + [A^-] = C \tag{5-5}$$

Charge balance between all cations and anions:

$$[H^+] = [OH^-] + [A^-] \tag{5-6}$$

7. The four equations [Eqs. (5-3) through (5-6)] can be solved to determine the concentration of the four unknown species, HA, A^-, H^+, and OH^-. Solution for $[A^-]$ through a combination of Eqs. (5-3) and (5-6) yields

$$[A^-] = [H^+] - K_W/[H^+] \tag{5-7}$$

Next, combination of Eqs. (5.4) and (5.5) gives

$$[H^+][A^-] = K_A (C - [A^-]) \tag{5-8}$$

Finally, a substitution of the value for $[A^-]$ from Eq. (5-7) in Eq. (5-8) eliminates all variables but $[H^+]$. By rearranging terms, the following final equation results:

$$[H^+]^3 + K_A[H^+]^2 - (K_A C + K_W)[H^+] - K_A K_W = 0 \tag{5-9}$$

Equation (5-9) can be solved by trial and error for $[H^+]$, which can then be substituted into Eqs. (5-3) through (5-6) to determine the concentrations of the other species at equilibrium. The results are:

$$[H^+] = 4.15 \times 10^{-4}$$

$$[OH^-] = 2.42 \times 10^{-11}$$

$$[HA] = 9.59 \times 10^{-3}$$

$$[A^-] = 4.15 \times 10^{-4}$$

Trial-and-error solution of polynominal equations as in the above example can be somewhat tedious. The relatively inexpensive programmable calculators available today can make this process much more rapid. Close approximations can also often be obtained by use of simplifying assumptions, as discussed in Sec. 5-5. Graphical procedures also permit more rapid solution and have the advantage that they give a better intuitive feel for the phenomena involved. Graphical solutions are given in Secs. 5-5 to 5-10. Computer methods[2] are also available, and in fact are generally required for the solution of complex equilibria involving many species and equilibrium relationships.

5-5 ACIDS AND BASES

The concept of acids and bases was briefly introduced in Sec. 2-13. Strong acids and bases are considered to be completely ionized in dilute solution, while weak acids and bases are only partially ionized. Acids tend to increase the hydrogen-ion concentration in solution, while bases increase the hydroxide concentration. The product of the activity (or approximately, the molar concentration) of these two ions remains constant at a given temperature and equals K_W. Expression of hydrogen-ion concentration in terms of molar concentration

[2] F. Morel and J. J. Morgan, A Numerical Method for Computing Equilibria in Aqueous Chemical Systems, *Environ. Sci. and Technol.,* **6:** 58 (1972).

or activity is rather cumbersome, and so the pH method of expression was developed.

The pH of water is a highly important characteristic as it affects equilibria between most chemical species, effectiveness of coagulation, potential of water to be corrosive, suitability of water to support living organisms, and most other quality characteristics of water. The environmental engineer should thus have a good understanding of factors affecting the pH of natural water.

The pH and p(x) Concept

In 1909 Sørensen proposed to express the hydrogen-ion concentration in terms of its negative logarithm and designated such value as pH^+. His symbol has been superseded by the simple designation pH. The terms may be represented by

$$pH = -\log [H^+] \qquad \text{or} \qquad pH = \log \frac{1}{[H^+]} \qquad (5\text{-}10)$$

With water and in the absence of foreign materials, $[H^+]$ equals $[OH^-]$ as required by electroneutrality, and the product at $25°C$ equals K_W or 10^{-14}. These conditions mean that $[H^+] + [OH^-] = 10^{-7}$, and the pH equals 7, which is considered the "neutral" pH for water. The pH scale is usually represented as ranging from 0 to 14. Values of pH lower than 7 indicate the hydrogen-ion concentration is greater than the hydroxide-ion concentration, and the water is termed acidic. The opposite condition is implied when the pH exceeds 7, and the water is termed basic.

The method of expressing hydrogen-ion concentration as pH is also useful for expressing other small numbers such as the concentration of other ions or ionization constants for solutions of weak acids and bases. For this purpose, the p(x) notation is used, with p(x) defined as

$$p(x) = -\log_{10} x = \log_{10} \frac{1}{x} \qquad (5\text{-}11)$$

where the quantity x may be the concentration of a given chemical species, an equilibrium constant, or the like. Thus, just as pH is the negative logarithm of the hydrogen-ion concentration, pOH signifies the negative logarithm of the hydroxide-ion concentration, and pK_W the negative logarithm of the ionization constant for water. From the mass action equation for water,

$$[H^+][OH^-] = K_W$$

it follows that

$$-\log [H^+] - \log [OH^-] = -\log K_W$$

and that

$$pH + pOH = pK_W \qquad (5\text{-}12)$$

Since $K_W = 1 \times 10^{-14}$ at $25°C$, it follows that at this temperature $pK_W = 14$.

Table 5-1 Typical ionization constants for weak acids at 25°C

Acid	Equilibrium equation	K_A	pK_A	Significance in environmental engineering
Acetic	$CH_3COOH \rightleftharpoons H^+ + CH_3COO^-$	1.8×10^{-5}	4.7	Organic wastes
Boric	$H_3BO_3 \rightleftharpoons H^+ + H_2BO_3^-$	$5.8 \times 10^{-10}(K_1)$	9.2	Nitrogen analysis
Carbonic	$CO_2 + H_2O \rightleftharpoons H^+ + HCO_3^-$	$4.3 \times 10^{-7}(K_1)$	6.4	Many applications
	$HCO_3^- \rightleftharpoons H^+ + CO_3^-$	$4.7 \times 10^{-11}(K_2)$	10.3	
Hydrocyanic	$HCN \rightleftharpoons H^+ + CN^-$	7.2×10^{-10}	9.1	Toxicity
Hydrosulfuric	$H_2S \rightleftharpoons H^+ + HS^-$	$9.1 \times 10^{-8}(K_1)$	7.0	Odors, corrosion
	$HS^- \rightleftharpoons H^+ + S^{2-}$	$1.3 \times 10^{-13}(K_2)$	12.9	
Hypochlorous	$HOCl \rightleftharpoons H^+ + OCl^-$	2.9×10^{-8}	7.5	Disinfection
Phenol	$C_6H_5OH \rightleftharpoons H^+ + C_6H_5O^-$	1.2×10^{-10}	9.9	Tastes
Phosphoric	$H_3PO_4 \rightleftharpoons H^+ + H_2PO_4^-$	$7.5 \times 10^{-3}(K_1)$	2.1	Analytical buffer, plant nutrient
	$H_2PO_4^- \rightleftharpoons H^+ + HPO_4^{2-}$	$6.2 \times 10^{-8}(K_2)$	7.2	
	$HPO_4^{2-} \rightleftharpoons H^+ + PO_4^{3-}$	$4.8 \times 10^{-13}(K_3)$	12.3	
Propionic	$CH_3CH_2COOH \rightleftharpoons H^+ + CH_3CH_2COO^-$	1.3×10^{-5}	4.9	Organic wastes, anaerobic digestion

For weak acids and bases, pK_A is the negative logarithm of the ionization constant for weak acids and pK_B the negative logarithm of the ionization constant for weak bases. The ionization constants and pK_A and pK_B values for several weak acids and bases of interest in environmental engineering are listed in Tables 5-1 and 5-2, respectively.

Acid or Base Additions to Solution

When an acid or base is added to a solution, interactions between the different chemical species present will occur in such a way as to establish chemical equilibrium. The nature of the changes which will occur will depend upon the nature and concentration of the chemical species present. These changes can be predicted by following the procedures outlined in Sec. 5-4. However, in addition to being cumbersome to carry out, such calculations often do not give an intuitive feel for the interactions which occur. In order to help understand the

Table 5-2 Typical ionization constants for weak bases and salts of weak acids at 25°C

Substance	Equilibrium equation	K_B	pK_B	Significance in environmental engineering
Ammonia	$NH_3 + H_2O \rightleftharpoons NH_4^+ + OH^-$	1.8×10^{-5}	4.7	Disinfection, nutrient
Acetate	$CH_3COO^- + H_2O \rightleftharpoons CH_3COOH + OH^-$	5.6×10^{-10}	9.3	Organic wastes
Borate	$H_2BO_3^- + H_2O \rightleftharpoons H_3BO_3 + OH^-$	1.7×10^{-5}	4.8	Nitrogen analysis
Carbonate	$CO_3^{2-} + H_2O \rightleftharpoons HCO_3^- + OH^-$	$2.1 \times 10^{-4}(K_1)$	3.7	Many applications
	$HCO_3^- + H_2O \rightleftharpoons H_2CO_3 + OH^-$	$2.3 \times 10^{-8}(K_2)$	7.6	
Calcium hydroxide	$CaOH^+ \rightleftharpoons Ca^{2+} + OH^-$	$3.5 \times 10^{-2}(K_2)$	1.5	Softening
Magnesium hydroxide	$MgOH^+ \rightleftharpoons Mg^{2+} + OH^-$	$2.6 \times 10^{-3}(K_2)$	2.6	Softening

processes, the following discussion on acid-base titrations is presented. A *titration* is the procedure by which a measured amount of chemical or reagent is added to a solution in order to bring about a desired and measured change. Several examples will be given for illustration. In addition, simplifying assumptions to help solve the acid-base equilibria involved will be presented.

Titration of strong acids and bases When a strong base is titrated with a strong acid, the initial pH of the base is very high, usually in the range of 12 to 13. As acid is added the pH changes very little at first and then slowly declines to a pH of about 10. From then on, the pH falls very rapidly until a pH of about 4 is reached; then changes become much more gradual. A plot of such data, commonly called a titration curve, is shown in Fig. 5-1. It may be noted that the curve is essentially vertical between pH values of 10 and 4; therefore the stoichiometric end point or equivalence point lies between these values and, for all practical purposes, anywhere between them. When a strong acid is titrated with a strong base, a curve quite similar in character is produced but is, of course, a mirror image of the former, as shown also in Fig. 5-1.

The above indicates the qualitative nature of strong acid and base titrations. It is desirable to illustrate also the quantitative changes which take place during the titration of a strong acid or base. Such calculations can be readily made if the strengths of the solutions are known. These calculations will not give exact values for pH when concentrations of ions rather than activities are used, but they are sufficiently close for most practical purposes. The following example is given for illustration.

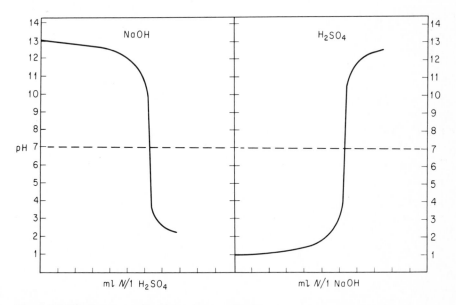

Figure 5-1 Titration curves for strong bases and acids.

Example Suppose 500 ml of 0.01 M HCl is titrated with 0.5 M NaOH. Calculate the pH values at different points in the titration to show the relative changes in pH with respect to the volume of titrating solutions used.

Since one mole of NaOH will neutralize one mole of HCl

$$NaOH + HCl \rightarrow H_2O + Na^+ + Cl^-$$

the NaOH volume required to neutralize the acid is 500 × 0.01/0.5, or 10.0 ml.

At the beginning of the titration, the $[H^+]$ concentration in the 0.01 M acid solution is 0.01. The pH is equal to $-\log[H^+]$ or 2. After 5.0 ml of base is added, 50 percent of the HCl is neutralized, and thus $[H^+] = 5 \times 10^{-3}$. The pH then would equal 2.3. By the time 9.0 ml of base is added, only 10 percent of the acid remains, and thus $[H^+] = 10^{-3}$ and the pH has increased only one unit to 3.

At the equivalence point, which is obtained after the full 10 ml of base is added, the solution contains only NaCl and water. Since NaCl is a neutral salt, the pH value of the solution is 7. If the equivalence point is passed, the solution will contain excess base and the pH will rise above 7. To illustrate, if 11.0 ml 0.5 M NaOH is added, the excess will be 1.0 ml, which represents 0.5×10^{-3} mol of OH^- ions. Thus $[OH^-] = 0.5 \times 10^{-3}/0.5 = 10^{-3}$ and the pH equals 11.

These and other values for this titration are listed in Table 5-3. By the time the pH has increased to 4, we see that 99 percent of the acid has been neutralized, and so for practical purposes the titration can be considered complete. After this point is reached, it requires only an additional 0.2 ml of NaOH to increase the pH to 10.

Titration of weak acids and bases When weak acids are titrated with strong bases, the character of the titration curve depends upon whether the acid is monobasic or polybasic, i.e., whether it yields one or more hydrogen ions. The initial pH of solutions of poorly ionized acids depends largely upon the degree of ionization, and the titration curves vary markedly, as shown in Fig. 5-2. Since the pH of natural waters is influenced mainly by weak acids and their salts, it is well to have a close familiarity with their chemistry. For this reason, it is desirable to develop the mathematical relationships which describe the influence of addition

Table 5-3 Calculated values of pH in 500 ml of 0.01 M HCl to which different amounts of 0.5 M NaOH have been added

ml 0.5 M NaOH		HCl neutralized	Moles per liter		pH
Added	Excess	(percent)	$[H^+]$	$[OH^-]$	
0.00	0.0	0	1.0×10^{-2}	1.0×10^{-12}	2.0
5.00	0.0	50	5.0×10^{-3}	2.0×10^{-12}	2.3
9.00	0.0	90	1.0×10^{-3}	1.0×10^{-11}	3.0
9.90	0.0	99	1.0×10^{-4}	1.0×10^{-10}	4.0
9.99	0.0	99.9	1.0×10^{-5}	1.0×10^{-9}	5.0
9.999	0.0	99.99	1.0×10^{-6}	1.0×10^{-8}	6.0
10.00	0.0	100.	1.0×10^{-7}	1.0×10^{-7}	7.0
10.10	0.1	100	1.0×10^{-10}	1.0×10^{-4}	10.0
11.00	1.0	100	1.0×10^{-11}	1.0×10^{-3}	11.0
20.00	10.0	100	1.0×10^{-12}	1.0×10^{-2}	12.0

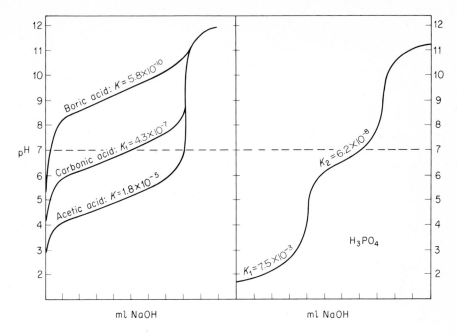

Figure 5-2 Titration curves for weak acids.

of strong acids or bases on the pH of water containing weak acids and bases. These same relationships will also indicate the factors affecting the pH of the stoichiometric end point or equivalence point of a titration.

Consider first the titration of a weak monobasic acid with a strong base. The acid ionizes as follows:

$$HA \rightleftharpoons H^+ + A^-$$ (5-13)

The equilibrium relationships which must be considered in evaluating the pH during the titration are those for the acid and for water:

$$\frac{[H^+][A^-]}{[HA]} = K_A$$ (5-14)

$$[H^+][OH^-] = K_W$$ (5-15)

When the weak acid is titrated with a strong base, the following neutralization takes place:

$$HA + B^+ + OH^- \rightarrow B^+ + A^- + H_2O$$ (5-16)

Here B^+ represents the cation associated with the strong base. For strong bases such as NaOH and KOH, the cations Na^+ and K^+ remain ionized in solution. In order to maintain electroneutrality in solution, the sum of the molar concentration times the charge of the cations must equal the sum of similar values for the anions:

$$[H^+] + [B^+] = [A^-] + [OH^-] \qquad (5\text{-}17)$$

The $[H^+]$ at any point during the titration of a monobasic weak acid with a strong base can be determined by a simultaneous solution of Eqs. (5-14), (5-15), and (5-17). If certain approximations are made, the calculations are simplified. This will be illustrated in the estimation of the pH at the beginning, the midpoint, and the stoichiometric end point or equivalence point in the titration of a solution containing C moles per liter of a weak acid, HA, with a strong base, $B^+ + OH^-$.

Beginning of titration In the initial solution containing the weak acid alone, $[B^+] = 0$. Also, the pH will be low so that $[OH^-] \ll [H^+]$. Therefore, from Eq. (5-17), we see that $[H^-]$ and $[A^-]$ are approximately equal:

$$[H^+] \cong [A^-] \qquad (5\text{-}18)$$

Substituting $[H^+]$ for $[A^-]$ in Eq. (5-14) and solving for $[H^+]$ yields

$$[H^+] \cong \sqrt{K_A[HA]} = \sqrt{K_A(C - [A^-])} \qquad (5\text{-}19)$$

For many cases at the beginning of the titration, $[A^-] \ll C$, and the following approximation is sufficient:

$$[H^+] \cong \sqrt{K_A C} \qquad (5\text{-}20)$$

From this the pH becomes

$$pH = -\log[H^+] \cong -\tfrac{1}{2}\log K_A - \tfrac{1}{2}\log C$$

or $\qquad\qquad pH \cong \tfrac{1}{2}(pK_A - \log C) \qquad (5\text{-}21)$

Example A Calculate the pH of a 0.01 M (600 mg/l) acetic acid solution. By approximate Eq. (5-21),

$$pH \cong \tfrac{1}{2}(pK_A - \log C) = \tfrac{1}{2}(4.7 + 2) = 3.35$$

Check: $[Ac^-] = [H^+] = 0.0004 \ll 0.01$; therefore approximation is good. (In the case of weak acids, the $[H^+]$ must be significantly less than the molar concentration of the acid for the approximation to be good. A comparable situation exists for bases.)

Example B Calculate the pH of a 0.01 M phosphoric acid solution. By approximate Eq. (5-21),

$$pH \cong \tfrac{1}{2}(pK_A - \log C) = \tfrac{1}{2}(2.1 + 2) = 2.05$$

Check: $[H_2PO_4^-] = [H^+] \cong 0.01 = C$; therefore approximation is poor.
Use the better approximation given by Eq. (5-19) as follows:

$$[H^+] = [H_2PO_4^-] = \sqrt{K_A(0.01 - [H_2PO_4^-])}$$

$$[H_2PO_4^-]^2 = (7.5 \times 10^{-5}) - (7.5 \times 10^{-3})[H_2PO_4^-]$$

$$[H_2PO_4^-] = \frac{-7.5 \times 10^{-3} \pm \sqrt{(7.5 \times 10^{-3})^2 + 4(7.5 \times 10^{-5})}}{2}$$

$$= 5.7 \times 10^{-3} \text{ mol/l}$$

$$[H^+] = [H_2PO_4^-] = 5.7 \times 10^{-3} \text{ mol/l}$$

$$pH = 2.25$$

The error from use of approximate Eq. (5-20) was 0.2 of a pH unit. However, the error from using approximate Eq. (5-19) is less than 0.01 pH unit.

Midpoint of titration As the titration proceeds, $[A^-]$ increases and $[HA]$ decreases, as indicated by Eq. (5-16). When the neutralization is 50 percent complete, $[B^+] \cong \frac{1}{2}C$. Also, $[B^+]$ and $[A^-]$ become quite significant in concentration, and thus $[H^+] \ll [B^+]$ and $[OH^-] \ll [A^-]$. Therefore, from Eq. (5-17),

$$[B^+] \cong [A^-] \cong \tfrac{1}{2}C \qquad (5\text{-}22)$$

Also, since

$$[HA] + [A^-] = C$$

we have

$$[HA] \cong \tfrac{1}{2}C$$

Placing the above values for $[HA]$ and $[A^-]$ in Eq. (5-14) gives

$$\frac{[H^+](\tfrac{1}{2}C)}{\tfrac{1}{2}C} \cong K_A$$

Thus

$$[H^+] \cong K_A \qquad (5\text{-}23)$$

and

$$pH \cong pK_A \qquad (5\text{-}24)$$

Equation (5-24) indicates the value of expressing the ionization constant in the pK_A form. The pK_A indicates the pH which will result when a weak acid is half neutralized. The weaker the acid, the higher the pK_A value, and thus the higher the pH at the half neutralization point.

Equivalence point of titration At the equivalence point of the titration, the equivalents of base added equal the equivalents of acid in the original solution, and thus $[B^+] = C$. Also, the pH is usually sufficiently high that $[H^+] \ll [OH^-]$. Therefore, from Eq. (5-17),

$$C \cong [A^-] + [OH^-] \qquad (5\text{-}25)$$

and

$$C - [A^-] \cong [OH^-]$$

but

$$C - [A^-] = [HA]$$

and thus

$$[HA] \cong [OH^-] = \frac{K_W}{[H^+]} \qquad (5\text{-}26)$$

Substituting the above in Eq. (5-14) results in

$$\frac{1}{[H^+]} = \left[\frac{[A^-]}{K_W K_A} \right]^{1/2} \qquad (5\text{-}27)$$

Since at the end of the titration, $[A^-] \cong C$, the following approximate solution for the equivalence point pH results:

$$pH \cong \tfrac{1}{2}(\log C + pK_A + pK_W) \tag{5-28}$$

This equation indicates that the pH at the equivalence point of the titration depends on the concentration of the acid as well as on its equilibrium constant. Frequently for practical work, a pH for the equivalence point of a titration can be chosen which will give sufficiently accurate results over a wide range of concentrations. In other cases, it may be necessary to have an idea of the concentration, so that the best pH can be chosen. It is evident from Eq. (5-28) that the weaker the acid and the higher its concentration, the higher the equivalence point pH.

In the titration of a weak base with a strong acid, the pH at various points can be determined in a fashion similar to that for a weak acid. It should be obvious, however, that such a titration is the mirror image of the acid titration and that the equations for the acid titration are applicable if pOH is substituted for pH and if pK_B is substituted for pK_A. A summary of the approximate equations for pH for both weak acid and weak base titrations is given below:

Weak acid pH during titration with strong base:

Beginning of titration	$pH \cong \tfrac{1}{2}(pK_A - \log C)$	(5-21)
Midpoint of titration	$pH \cong pK_A$	(5-24)
Equivalence point of titration	$pH \cong \tfrac{1}{2}(\log C + pK_A + pK_W)$	(5-28)

Weak base pH during titration with strong acid:

Begining of titration	$pH \cong pK_W - \tfrac{1}{2}pK_B + \tfrac{1}{2}\log C$	(5-29)
Midpoint of titration	$pH \cong pK_W - pK_B$	(5-30)
Equivalence point of titration	$pH \cong \tfrac{1}{2}(pK_W - pK_B - \log C)$	(5-31)

The inflection at the equivalence point in the titration of weak acids or bases with ionization constants less than about 10^{-7} or pK values greater than 7 is not sharp and so is difficult to detect accurately.

This is illustrated in Fig. 5-2, showing the titration curves for three different weak acids with significantly different ionization constants. Solutions of acetic acid normally have an initial pH of about 3. During titration with a strong base, the pH increases slowly to about 7, and then rapidly until pH 10 is reached. The equivalence point is usually between pH 8 and 9, on the rapidly rising portion of the curve. In the case of carbonic acid, the ionization constant of 10^{-7} is much smaller than the 10^{-5} for acetic acid, and so the initial pH is much higher. Also, for the same reason, the midpoint and equivalence point for carbonic acid are much higher. The pH for the carbonic acid titration does not break sharply until a pH of 8 is reached. All weak acids with ionization constants greater than 10^{-7}, and in the concentrations normally encountered in practice, show inflections in the curve or equivalence points at a pH of about 8.5. The second hydrogen ion of carbonic acid is so poorly dissociated ($K_2 = 4.7 \times 10^{-11}$) that its presence cannot be detected in an ordinary titration. Theoretically it begins to be released in significant amounts at pH levels above 10.

Boric acid is a good example of a very weak acid with an ionization constant much less than 10^{-7}. This acid is not completely neutralized until a pH of about 11 is reached. The titration curve for boric acid indicates that the inflection point at this pH is barely detectable, and an accurate measurement of the boric acid concentration by such a titration is not feasible.

The titration curve for phosphoric acid illustrates very well the behavior during titration of a polyprotic weak acid, yielding more than one measurable hydrogen ion (proton). Reference to Fig. 5-2 will show that the first ionization of phosphoric acid is similar to that of a strong acid and that the resulting hydrogen ions are neutralized by the time sufficient base has been added to reach a pH of about 4. The hydrogen ion resulting from the second step of ionization is neutralized by the time the pH has been raised to about 8.5. The third hydrogen ion of phosphoric acid ($K_3 = 4.8 \times 10^{-13}$), like the second of carbonic acid, is not measurable by ordinary titrations. Calculations of pH during the titration of polyprotic acids are more complex than for monoprotic acids, as the equilibrium relationships for each ionization must be considered simultaneously for an exact solution. Without going into the details, it is sufficient to indicate that the pH of the equivalence point for each ionization is approximately equal to the average of the pK value for that ionization and the pK value for the following ionization. For example, with phosphoric acid, p$K_1 = 2.1$ and p$K_2 = 7.2$, and the equivalence point pH between the first and second ionization equals about $(2.1 + 7.2)/2$ or 4.6.

Typical titration curves for weak bases are shown in Fig. 5-3. Weak bases with ionization constants greater than 10^{-7} and in the concentrations normally encountered in environmental engineering practice have equivalence points at a pH of 4 or higher. Salts of strong bases and weak acids, such as sodium carbonate and sodium acetate, are alkaline in character and behave like bases during titration (see Fig. 5-3). When considered as such, their ionization can be considered to take place as shown below for acetate:

$$CH_3COO^- + H_2O \rightleftharpoons CH_3COOH + OH^-$$

and the equilibrium relationship becomes

$$\frac{[OH^-][CH_3COOH]}{[CH_3COO^-]} = K_B$$

It follows that K_B is equal to K_W/K_A, where K_A is the ionization constant for acetic acid. It also follows that pK_B for the salt formed from a weak acid and a strong base equals p$K_W - pK_A$. With this notation, Eqs. (5-29) to (5-31) for titration of a weak base may be used to evaluate the pH during titration of salts of weak acids. Titration curves for salts of weak acids are simply acid titration curves in reverse, and the curve is a mirror image of that for the acid. This is

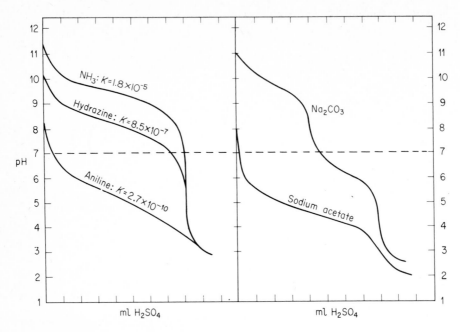

Figure 5-3 Titration curves for weak bases and for salts of weak acids.

illustrated by comparison of the titration curve for acetic acid (Fig. 5-2) with that for its salt, sodium acetate (Fig. 5-3).

The titration curve for sodium carbonate is characteristic of a base with two ionizations. The initial pH of its solution is rather high, and during titration with a strong acid, neutralization occurs in two steps, corresponding to each ionization. Addition of strong acid results in a gradual drop in pH, with a poorly defined inflection in the curve at a pH of about 8.5. This corresponds to the equivalence point for the conversion of carbonate ion to bicarbonate ion, as follows:

$$CO_3^{2-} + H^+ \rightarrow HCO_3^-$$

Further addition of acid results in a gradual lowering of the pH until a value of about 5 is reached. The curve then passes through an inflection at a pH of about 4.5. This corresponds to the equivalence point for the conversion of bicarbonate ion to carbonic acid:

$$HCO_3^- + H^+ \rightarrow H_2CO_3$$

Example A A solution containing 34 mg/l (0.001 M) H_2S is titrated with a strong NaOH solution. Calculate the pH at the beginning, the midpoint, and the equivalence point for the first ionization of the acid.

The initial pH, by approximate Eq. (5-21), is

$$pH \cong \tfrac{1}{2}(pK_A - \log C) = \tfrac{1}{2}(7.0 + 3.0) = 5.0$$

Check: $[H^+] = [HS^-] = 10^{-5} \ll 10^{-3}$; therefore approximation is good.

The midpoint pH, by approximate Eq. (5-24), is

$$pH \cong pK_A = 7.0$$

The equivalence point pH, since this is a diprotic acid, can be found as follows:

$$pH \cong \tfrac{1}{2}(pK_1 + pK_2) = \tfrac{1}{2}(7.0 + 12.9) = 9.95$$

Compare with the value given by Eq. (5-28),

$$pH \cong \tfrac{1}{2}(\log C + pK_A + pK_W) = \tfrac{1}{2}(-3 + 7.0 + 14) = 9.0$$

Since pH 9.0 would be reached before pH 9.95, pH 9.0 would be satisfactory for use as the equivalence point.

Example B Calculate the pH at the beginning, midpoint, and equivalence point for the titration of a solution containing 10^{-4} mol/l of sodium propionate with H_2SO_4.

Since sodium propionate is the salt of a weak acid, CH_3CH_2COOH, and a strong base, NaOH, it acts like a weak base,

$$pK_B = pK_W - pK_A = 14 - 4.9 = 9.1$$

The initial pH, by Eq. (5-29), is

$$pH \cong pK_W - \tfrac{1}{2}pK_B + \tfrac{1}{2}\log C = 14 - \tfrac{1}{2}(9.1) + \tfrac{1}{2}(-4) = 7.45$$

The midpoint pH, by Eq. (5-30), is

$$pH \cong pK_W - pK_B = 14 - 9.1 = 4.9$$

The equivalence point pH, using Eq. (5-31),

$$pH \cong \tfrac{1}{2}(pK_W - pK_B - \log C) = \tfrac{1}{2}(14 - 9.1 + 4.0) = 4.45$$

A check reveals that all the above approximate equations are satisfactory for use.

Logarithmic Concentration Diagrams

The approximate solutions to acid-base equilibria given in the preceding section apply only to the stoichiometric end point and midpoint of titrations. Equilibria in natural waters do not necessarily tend toward these points. Logarithmic concentration diagrams present an easier procedure for solution of many complex equilibria and for this reason are of great practical value. In this graphical procedure, pH is used as the master variable and is plotted as the abscissa. The logarithm of the concentration of each constituent of importance in a given case is plotted on the ordinate. A logarithmic concentration diagram for 0.02 M acetic acid in water is illustrated in Fig. 5-4.

Graphical solution for a monoprotic acid The relationships given by Eqs. (5-3) to (5-6) for acetic acid apply for all monoprotic acids. The line for $[H^+]$ is obtained from the relationship

$$\log [H^+] = -pH \tag{5-32}$$

That for $[OH^-]$ is obtained from the relationship

$$\log [OH^-] = pH - pK_W \tag{5-33}$$

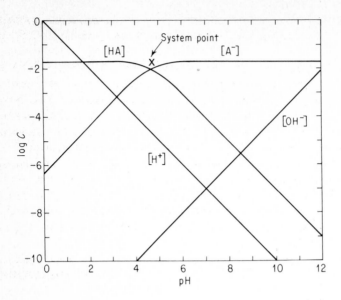

Figure 5-4 Logarithmic concentration diagram for 0.02 M acetic acid.

The concentration of acetic acid and acetate can be obtained by combining Eqs. (5-4) and (5-5):

$$[HA] = \frac{C[H^+]}{K_A + [H^+]} \tag{5-34}$$

and,

$$[A^-] = \frac{CK_A}{K_A + [H^+]} \tag{5-35}$$

The logarithms of [HA] and [A$^-$] appear as straight lines connected by a short curve. The intersection of the two straight lines is called the system point and is located where pH = pK_A. To the left of this point, [H$^+$] is greater than K_A, so that the logarithm of the concentrations from Eqs. (5-34) and (5-35) reduce to

$$\log [HA] \cong \log C \tag{5-36}$$

and

$$\log [A^-] \cong \log C - pK_A + pH \tag{5-37}$$

The first equation represents a horizontal line and the second a diagonal line with a slope of $+1$, both passing through the system point.

To the right of the system point, [H$^+$] is less than K_A, and the logarithm reduces to

$$\log [HA] \cong \log C + pK_A - pH \tag{5-38}$$

and

$$\log [A^-] \cong \log C \tag{5-39}$$

The first equation represents a line with slope -1 and the second a horizontal line passing through the system point.

Just below the system point, $[HA] = [A^-] = \frac{1}{2}C$ [from Eq. (5-5)]. The log of $\frac{1}{2}C$ equals $\log C + \log 0.5$ or $\log C - 0.3$. Thus the curves connecting the diagonal and horizontal lines intersect at 0.3 of a logarithmic unit below the system point.

Construction of a logarithmic diagram The construction of a logarithmic diagram is straightforward. Draw a horizontal line representing $\log C$. Locate the system point at $pH = pK_A$, and draw 45° lines sloping to the left and right through the system point. Locate a point 0.3 logarithm units below the system point and connect the horizontal and 45° lines with short curves passing through this point. The lines for $[H^+]$ and $[OH^-]$ are drawn as 45° lines which intersect where pH equals 7 and $\log C$ equals -7. Thus, no involved numerical calculations are required. Changing the concentration of the solution simply shifts the $[HA]$ and $[A^-]$ curves up or down. Through use of transparent overlays, solutions to acid and base equilibria can be obtained rapidly.

pH during titrations The graphical approach can be used to determine the equivalence point for acid-base titrations, and the pH at intermediate points during a titration. At all points, Eqs. (5-32) to (5-35) must be satisfied. In addition, a charge balance must be maintained:

$$[B^+] + [H^+] = [OH^-] + [A^-] \qquad (5\text{-}40)$$

Since a strong base is completely ionized, $[B^+]$ will equal the molar concentration of the base added if a monoprotic base such as NaOH is used for the titration. Initially, before NaOH is added, the condition will be

$$[H^+] = [OH^-] + [A^-] \qquad (5\text{-}41)$$

Since for acid conditions, $[OH^-]$ will be very small compared with $[H^+]$, Eq. (5-41) can be approximated as

$$[H^+] \cong [A^-] \qquad (5\text{-}42)$$

Thus, the initial pH of the solution is approximated in Fig. 5-4 by the intersection between the curves for $[H^+]$ and $[A^-]$. As base is added, $[B^+]$ soon becomes much greater than $[H^+]$ or $[OH^-]$ so that Eq. (5-40) can be approximated as

$$[B^+] \cong [A^-] \qquad (5\text{-}43)$$

and the pH corresponding to a given base addition can be found by locating the point on the graph where Eq. (5-43) is satisfied. At the midpoint in the titration, $[B^+] = \frac{1}{2}C = [A^-] = [HA]$. Thus the midpoint is located at pK. Finally, the equivalence point is reached when $[B]^+ = C = [HA] + [A^-]$, which when substituted into Eq. (5-40) gives

$$[HA] + [H^+] = [OH^-] \qquad (5\text{-}44a)$$

Under the more basic conditions, $[OH^-]$ is much greater than $[H^+]$, so that the equivalence point pH is approximated by the intersection between $[HA]$ and

[OH⁻]. From Fig. 5-4, the equivalence point occurs at a pH of approximately 8.5.

Using the above approach, the initial point, midpoint, and equivalence point for a titration are represented approximately by the intersection of curves and can readily be found. The effect of changes in the concentration of weak acid on these different points can be observed by moving the horizontal [HA] and [A⁻] curves up or down. Moving the curves up represents an increase in weak acid concentration and results in a decrease in the initial pH and an increase in the equivalence point pH. The pH of the midpoint of the titration, of course, does not change.

Logarithmic diagram for a weak base Figure 5-5 shows graphically the effect of pH on [NH₃] and [NH₄⁺] for a 0.01 M solution. In essence, the construction is similar to that for a weak acid except the system point is found from the relationship

$$pH = pK_W - pK_B \qquad (5\text{-}44b)$$

Since for ammonia, pK_B is 4.7, the pH of the system point is $14 - 4.7$ or 9.3. Initial pH and equivalence points for titration of a weak base with a strong acid are analogous to that for a weak acid and strong base. The initial pH of a 0.01 M ammonium hydroxide solution is given by the intersection of the [NH₄⁺] and [OH⁻] lines and would equal about 10.6. The equivalence point of a titration with strong acid is given by the intersection of [NH₃] and [H⁺] and would equal about 5.7.

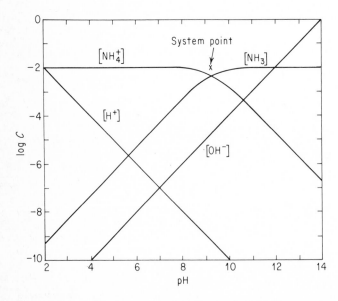

Figure 5-5 Logarithmic concentration diagram for 0.01 M ammonia solution.

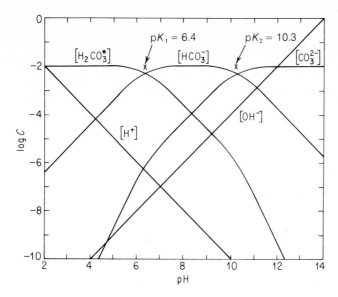

Figure 5-6 Logarithmic concentration diagram for 0.01 M carbonic acid.

Logarithmic diagram for polyprotic acids and bases Logarithmic diagrams become even more useful for solution of equilibrium problems involving polyprotic acids and bases. The carbonic acid system, which is most important for regulating the pH of most natural waters, will be used for illustration. The logarithmic diagram for a solution containing 0.01 M H_2CO_3 is illustrated in Fig. 5-6. The appropriate equations which apply are as follows:[3]

$$\frac{[H^+][HCO_3^-]}{[H_2CO_3^*]} = K_1 = 4.3 \times 10^{-7} \qquad pK_1 = 6.4$$

$$\frac{[H^+][CO_3^{2-}]}{[HCO_3^-]} = K_2 = 4.7 \times 10^{-11} \qquad pK_2 = 10.3$$

$$[H_2CO_3^*] + [HCO_3^-] + [CO_3^{2-}] = C = 10^{-2}$$

and
$$[H^+][OH^-] = K_W = 10^{-14} \qquad pK_W = 14$$

Solution of these equations for the individual carbonic acid species gives

$$[H_2CO_3^*] = \frac{C}{1 + K_1/[H^+] + K_1K_2/[H^+]^2} \qquad (5\text{-}45)$$

$$[HCO_3^-] = \frac{C}{1 + [H^+]/K_1 + K_2/[H^+]} \qquad (5\text{-}46)$$

---

[3] By convention, $[H_2CO_3^*]$ is taken to equal the sum of the actual carbonic acid concentration $[H_2CO_3]$ plus the dissolved carbon dioxide concentration $[CO_2(aq)]$.

$$[CO_3^{2-}] = \frac{C}{1 + [H^+]/K_2 + [H^+]^2/K_1 K_2} \tag{5-47}$$

It will be noted that the logarithmic diagram is constructed just as for mono-protic acids and bases, except that the slope of the line for $[CO_3^{2-}]$ changes from $+1$ to $+2$ when the pH drops below pK_1. Also, the slope of the line for $[H_2CO_3^*]$ changes from -1 to -2 when the pH becomes greater than pK_2. The reason for these slope changes is as follows. First, in reference to Eq. (5-47), when the pH drops below pK_1 ($[H^+] \gg K_1$), the last term in the denominator becomes dominant so that

$$[CO_3^{2-}] \cong \frac{C}{[H^+]^2/K_1 K_2}$$

and $\qquad \log [CO_3^{2-}] \cong \log C + \log K_1 + \log K_2 - 2 \log [H^+]$

or $\qquad \log [CO_3^{2-}] \cong \log C - pK_1 - pK_2 + 2pH$

Thus, the slope of the line for $\log [CO_3^{2-}]$ versus pH is 2. Similarly, when the pH exceeds pK_2, the last term in the denominator of Eq. (5-45) dominates and

$$[H_2CO_3^*] \cong \frac{C}{K_1 K_2/[H^+]^2}$$

The logarithmic form of this equation is

$$\log [H_2CO_3^*] = \log C + pK_1 + pK_2 - 2 \, pH$$

The slope of the curve for $\log [H_2CO_3^*]$ versus pH is thus -2.

The construction of a logarithmic diagram for a diprotic acid or base is straightforward. System points are located at intersections between a horizontal line representing $\log C$ and pH values equal to pK_1 and pK_2 (or $pK_W - pK_1$ and $pK_W - pK_2$ for a base). Diagonal lines with slopes of $+1$ and -1 are drawn from each system point to a point located just below an adjacent system point. Beyond the adjacent system point, the slope of the line changes from -1 to -2, or $+1$ to $+2$, as the case may be.

For polyprotic acids or bases with more than two exchangeable protons, the procedure is similar except that the slope of a diagonal line drawn to the left of a system point changes from $+1$ to $+2$ when it reaches below the first adjacent system point and from $+2$ to $+3$ when and if it passes below a second adjacent system point. Proof of this is left to the reader. The logarithmic concentration diagram for phosphoric acid, a triprotic acid, is illustrated in Fig. 5-7.

Figures 5-6 and 5-7 illustrate the influence of pH on the species present in solutions of two different weak acids. For example, the pH resulting from carbonic acid alone is approximated by the intersection between the lines for $[HCO_3^-]$ and $[H^+]$. If this solution is titrated with NaOH, then two equivalence points are reached. The first after an equivalent of $0.01 \, M$ of Na^+ is added is represented by the intersection of the $[H_2CO_3^*]$ and $[CO_3^{2-}]$ lines at pH 8.3. The second is reached after an additional $0.01 \, M$ Na^+ is added and is represented

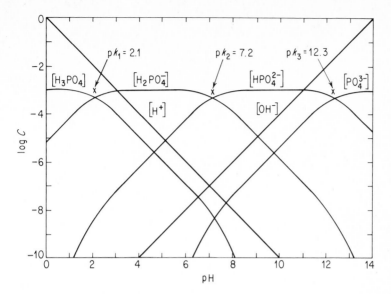

Figure 5-7 Logarithmic concentration diagram for 10^{-3} M phosphoric acid.

approximately by the intersection of the $[OH^-]$ and the $[HCO_3^-]$ lines at pH 11.1. The pH values for all intermediate points in the titration are represented by points that satisfy the charge-balance relationship

$$[Na^+] + [H^+] = [HCO_3^-] + 2[CO_3^{2-}] + [OH^-]$$

Solutions for intermediate points can be found by trial and error, using the logarithmic concentration diagram as a direct aid.

Logarithmic concentration diagram for a weak acid and a weak base A logarithmic concentration diagram for a mixture containing 0.1 M ammonium hydroxide and 0.1 M acetic acid is shown in Fig. 5-8. These constituents are commonly found in natural waters and wastewaters. They may exist in relatively high concentration in anaerobic digesters. The logarithmic concentration diagram is made by superimposing the curves for each material on a single diagram. By itself, the diagram illustrates the concentration of the various species present if the pH is varied by the addition of a strong base (NaOH) or a strong acid (HCl). The particular condition that will result from base or acid addition can be obtained, as usual, by imposing the charge-balance condition:

$$[H^+] + [NH_4^+] + [Na^+] = [OH^-] + [CH_3COO^-] + [Cl^-]$$

where $[Na^+]$ and $[Cl^-]$ represent the molar concentrations of NaOH and HCl added, respectively.

For the initial condition when acetic acid and ammonium hydroxide are added to distilled water, $[Na^+]$ and $[Cl^-]$ are zero, and $[H^+]$ and $[OH^-]$ are very small relative to $[NH_4^+]$ and $[CH_3COO^-]$.

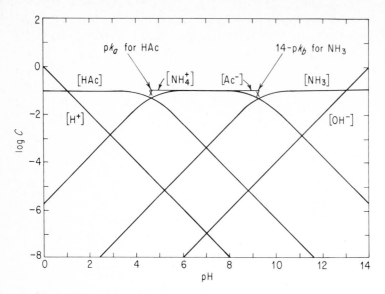

Figure 5-8 Logarithmic concentration diagram for a mixture of 0.1 M acetic acid and 0.1 M ammonia.

The resulting pH is approximated by the point where the [NH$_3$] and [HAc] curves intersect, and equals about 7.0.

More complex systems involving several weak acids and bases can be solved in a similar manner. The respective curves are superimposed on one another and then the charge-balance condition is imposed in order to obtain a solution to a particular problem. A trial-and-error solution through direct use of the logarithmic concentration diagram is relatively rapid.

5-6 BUFFERS

Buffers may be defined as *substances in solution that offer resistance to changes in pH* as acids or bases are added to or formed within the solution. There are many occasions in environmental engineering practice as well as in water chemistry when it is desirable to use buffered solutions to closely maintain a desired pH level. Buffer solutions usually contain mixtures of weak acids and their salts (*conjugate bases*) or weak bases and their salts (*conjugate acids*). The fundamental basis for an understanding of buffer action was developed in Sec. 5-5. Inspection of Figs. 5-2 and 5-3 will illustrate that at the midpoint in the titration of a weak acid or base, the slope of the titration curve is at a minimum. Thus at this point the smallest change in pH occurs for a given volume of titrant added, and hence at this point the *buffering capacity* is the greatest. At the midpoint the solution contains equal quantities of the ionized and the un-ionized acid or base. As indicated by Eqs. (5-24) and (5-30), the pH at the midpoint in the titration is given by pK_A for weak acids and by $pK_W - pK_B$ for weak bases. Thus pK values listed in Tables 5-1 and 5-2 can be used to indicate the pH at which the various acids and bases have the most effective buffering capacity.

Although weak acids and bases and their salts are most effective as buffers at a pH near the pK value, they may also be used effectively within ± 1 pH unit of the pK value by changing the relative concentration of acid to salt or base to salt. To illustrate, consider the equilibrium relationship for a weak acid,

$$K_A = \frac{[H^+][A^-]}{[HA]}$$

By rearranging, we obtain

$$\frac{1}{[H^+]} = \frac{1}{K_A} \frac{[A^-]}{[HA]}$$

Thus

$$pH = pK_A + \log \frac{[salt]}{[acid]} \tag{5-48}$$

Equation (5-48) indicates that the pH of a buffer solution made of a weak acid and its salt depends upon the *ratio* of salt concentration to acid concentration. Thus solutions of weak acids and their salts will have the same pH regardless of concentration as long as the *ratio* of concentrations remains the same. However, from the standpoint of buffering capacity, the larger the actual concentrations of salt and acid, the less they will be numerically changed by the addition of a given amount of acid or base and the more effective is the buffering action.

The pK value for the second ionization of phosphoric acid is very near 7, and hence salts of phosphoric acid are quite useful substances to buffer solutions near a neutral pH. They are commonly used in analytical tests. Of special interest in environmental engineering is the use of phosphate salts to maintain a neutral pH in the biochemical oxygen demand test (BOD). Here the monobasic potassium salt (KH_2PO_4) is used as the acid, and both the dibasic potassium salt (K_2HPO_4) and sodium salt (Na_2HPO_4) are used as the salt of the acid. These salts ionize in solution to establish the following equilibrium:

$$H_2PO_4^- \rightleftharpoons H^+ + HPO_4^{2-}$$

Although the buffering capacity of natural waters is largely due to salts of carbonic acid, the environmental engineer should remember that alkalinity determinations measure the buffering capacity of all salts of weak acids. This is particularly pertinent in the analysis of industrial wastes. The best way of evaluating the buffer capacity of an industrial waste is to perform an electrometric titration, using a standard acid or base. If observations of pH versus titrant additions are made, curves can be plotted that show its capacity and the pH range over which the buffer is especially effective.

Buffer capacity is an important characteristic of wastes that are submitted to biological treatment. The oxidation of neutral compounds—sugar, for example—results in the production of organic acids as intermediates. If the buffering capacity is not sufficient, the pH may fall to levels that inhibit the action of the bacteria. In an instance in which formaldehyde was involved, the pH was reduced to 4.5 within a matter of minutes, and the process was considered a failure until adequate buffer was applied. The initiation of anaerobic digestion to produce methane from sludge is hampered by limitations of buffer capacity. This process frequently results in the formation of considerable quantities of organic acids, and liming of digesters is often practiced to maintain favorable pH conditions. In effect, the lime combines with carbon dioxide and water to form calcium bicarbonate, a salt of carbonic acid. Hence an understanding of the buffering action of carbonic acid and its salts is of value in digester control.

In the oxidation of ammonia, nitrous and nitric acids are formed. These must be neutralized to maintain a favorable environment for nitrifying bacteria. The bicarbonates in natural waters serve this purpose. If they are not present in sufficient amounts, alkaline materials such as lime or sodium hydroxide must be added to maintain favorable pH.

5-7 BUFFER INDEX

The buffering capacity of a solution can be indicated quantitatively by the buffering index β, which is defined as the slope of a titration curve of pH versus moles of strong base added (C_B) or moles of strong acid added (C_A):

$$\beta = \frac{dC_B}{dpH} = -\frac{dC_A}{dpH} \tag{5-49}$$

The buffer index indicates the number of moles of acid or base required to produce a given change in pH. While β can be readily determined from a titration curve, it can also be calculated if the composition of a solution is known. For example, for a solution containing a concentration C of a monoprotic acid, Eqs. (5-34) and (5-35) apply. If C_B moles of base are added to the solution, then a charge balance as given by Eq. (5-40) must also hold, or

$$C_B + [H^+] = [OH^-] + [A^-] \tag{5-50}$$

Combining with Eqs. (5-34) and (5-35) gives

$$C_B = \frac{K_W}{[H^+]} - [H^+] + \frac{CK_A}{K_A + [H^+]} \tag{5-51}$$

Also,

$$\frac{dC_B}{dpH} = \frac{dC_B}{d[H^+]} \frac{d[H^+]}{dpH} \tag{5-52}$$

Since

$$pH = -\log [H^+] = -\ln [H^+]/2.303$$

then,

$$d\,[H^+]/dpH = -2.303[H^+]$$

and
$$\beta = dC_B/dpH = -2.303[H^+](dC_B/d[H^+]) \qquad (5\text{-}53)$$

Differentiating β in Eq. (5-52) with respect to $[H^+]$ and substituting into Eq. (5-53) yields

$$\beta = 2.303 \left[\frac{K_W}{[H^+]} + [H^+] + \frac{CK_A[H^+]}{(K_A + [H^+])^2} \right] \qquad (5\text{-}54)$$

Thus, if we know the pH for a given buffer solution and the molar concentration of weak acid plus its conjugate base, the buffer index can be determined by direct substitution into Eq. (5-54). When dealing with a weak base, Eq. (5-54) can also be used by substituting K_W/K_B for K_A. Solutions for β for polyprotic acids or bases are also available.[4,5]

Example A buffer solution has been prepared by adding 0.2 mol/l of acetic acid and 0.1 mol/l of acetate. The pH of the solution has been adjusted to 5.0 by addition of NaOH. How many mol/l of NaOH is required to increase the pH to 5.1?

$$C = 0.2 + 0.1 = 0.3 \text{ mol/l}$$

$$[H^+] = 10^{-5}$$

From Eq. (5-54),

$$\beta = 2.303 \left[\frac{10^{-14}}{10^{-5}} + 10^{-5} + \frac{0.3(1.8 \times 10^{-5})(10^{-5})}{(1.8 \times 10^{-5} + 10^{-5})^2} \right]$$

$$= 0.16 \text{ mol/l of base per pH unit}$$

NaOH required $= (5.1 - 5.0)(0.16) = 0.016 \text{ mol/l}$

5-8 SOLUBILITY OF SALTS

Basic concepts of salt solubility and the concept of solubility product were discussed in Sec. 2-13. Consideration of solubility relationships can aid in an understanding of the natural forces which dissolve rocks and other minerals, bringing materials into aqueous solution. Variations in the mineral characteristics of water can also be understood to some degree by considering the factors affecting solubility of minerals. A knowledge of solubility relationships can also help the environmental engineer to devise methods for treating water supplies or wastewaters to rid them of hardness causing constituents, heavy-metal contaminants, and some organic materials.

Logarithmic Concentration Diagrams for Solubility

Logarithmic concentration diagrams are useful for illustrating solubility as well as acid-base relationships. Figure 5-9 shows the relationship between carbonate concentration and the saturation concentration for various cations based upon

[4] J. N. Butler, "Ionic Equilibrium," Addison-Wesley, Reading, Mass., 1964.
[5] W. Stumm and J. J. Morgan, "Aquatic Chemistry," Wiley-Interscience, New York, 1970.

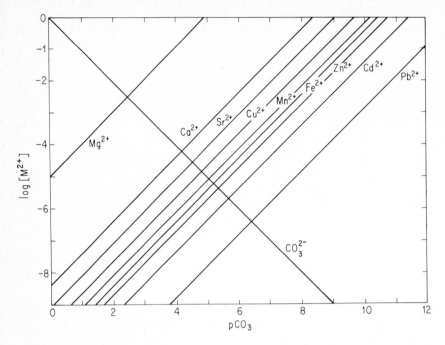

Figure 5-9 Logarithmic concentration diagram showing the solubility of various metallic carbonates at 25°C.

the solubility product constants from Table 2-4, and the solubility product relationship

$$[M^{2+}][CO_3^{2-}] = K_{sp} \qquad (5\text{-}55)$$

which yields straight lines of slope $+1$ when written and plotted from the form,

$$\log [M^{2+}] = pCO_3 + \log K_{sp} \qquad (5\text{-}56)$$

In Fig. 5-9 $[CO_3^{2-}]$ is represented as a function of pCO_3 by a line having a slope of -1. Of the cations shown, Pb^{2+} is the least soluble and Mg^{2+} is the most soluble. Relatively low carbonate concentrations can be quite effective in removal of most divalent cations from solution. Calcium hardness removal by carbonate precipitation is a commonly used water treatment process. Heavy metals are also removed quite effectively in this way by some advanced wastewater treatment processes.

The intersection of the CO_3^{2-} line with a cation solubility product line indicates the condition for the solubility of that particular salt in pure water,

$$[M^{2+}] = [CO_3^{2-}] = \text{solubility} \qquad (5\text{-}57)$$

If the cation and anion do not have the same charge, then the solubility is displaced slightly from the point of intersection since the cation molar concentration would not equal the anion molar concentration. The solubility of the salts represented in Fig. 5-9 is in the same order as that of the solubility products, which is not generally the case as discussed in Sec. 2-13. This occurs because the cations shown are all divalent. If monovalent or trivalent cations were shown, the slopes of their solubility product lines would have been $+\frac{1}{2}$ and $+\frac{3}{2}$, respectively. Use of logarithmic diagrams as in Fig. 5.9 may be no simpler than solubility product calculations for simple solubilities of the type shown. However, when complexes are present, such diagrams can give a complete and compact picture of all the equilibria involved. This is illustrated in a later section.

Example Determine the solubility of Ca^{2+} in a water sample containing 10^{-3} molar carbonate at 25°C.

$$pCO_3 = -\log 10^{-3} = 3$$

From Fig. 5-9, when pCO_3 equals 3, $\log [Ca^{2+}]$ equals -5.3; therefore,

$$[Ca^{2+}] = 10^{-5.3} = 5 \times 10^{-6}$$

Complex Solubility Relationships

Solubility relationships are generally much more complex than indicated in Sec. 2-13 or in the preceding discussion. Generally, in natural waters or wastewaters, several other factors must be considered in order to make a realistic solubility-product calculation. First, as discussed in Sec. 5-3, the ionic strength of the solution affects ion activity and must be considered if more exact calculations are desired.

Perhaps more important, other equilibria besides the solubility product affect the concentration of the ions present. Reactions of the cation or anion with water to form hydroxide complexes or protonated anion species are common. In addition, the cations or anions may form complexes with other materials in solution, thus reducing their effective concentration. Finally, other ions may form salts with less solubility than the one under consideration. If this were ignored, then solubility predictions could be considerably in error.

First, consider the effect of solution pH on the solubility of cations. Assuming that no other materials exist in solution to react with the cations, the solubility product between any cation and OH^- would be

$$[M^{z+}][OH^-]^z = K_{sp} \tag{5-58}$$

and
$$\log [M^{z+}] = \log K_{sp} - z \log [OH^-] \tag{5-59}$$

However, $\log [OH^-]$ is a function of pH,

$$pH = pK_W - pOH = pK_W + \log [OH^-] \tag{5-60}$$

so that

$$\log [M^z] = \log K_{sp} - z(pH - pK_W) \tag{5-61}$$

The above relationship between pH and $\log [M^z]$ is illustrated in Fig. 5-10 for

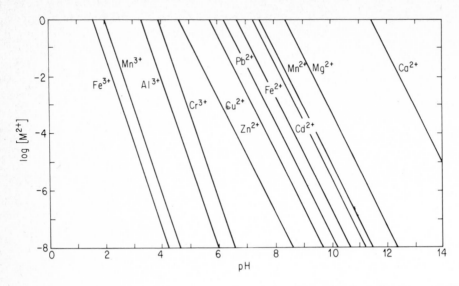

Figure 5-10 Logarithmic concentration diagram showing the solubility of various metallic hydroxides.

several cations. The slope of the lines is equal to $-z$ as indicated by Eq. (5-61). Use is made of the effect of pH on cation solubility in water and wastewater treatment. Magnesium hardness can be reduced by raising the pH to 11 or greater and precipitating $Mg(OH)_2$. Copper, zinc, and chromium (Cr III) are often removed from metal containing wastewaters by precipitation at pH of 7 or above.

The pH of water not only affects solubility of metal hydroxides as illustrated in Fig. 5-10, but also affects other equilibria in water, which in turn can affect solubility. For example, the solubility of salts of weak acids will be influenced by solution pH. The solubility of calcium carbonate will be used for illustration. It dissolves in water to give its ions,

$$CaCO_3(c) \rightleftharpoons Ca^{2+} + CO_3^{2-} \tag{5-62}$$

the equilibrium being controlled by the solubility product,

$$[Ca^{2+}][CO_3^{2-}] = K_{sp} = 5 \times 10^{-9} \tag{5-63}$$

Carbonate is an anion of the weak diprotic acid, H_2CO_3, which ionizes in water as follows:

$$H_2CO_3^* \rightleftharpoons H^+ + HCO_3^- \tag{5-64}$$

$$HCO_3^- \rightleftharpoons H^+ + CO_3^{2-} \tag{5-65}$$

Ionization products for the above are

$$\frac{[H^+][HCO_3^-]}{[H_2CO_3^*]} = K_1 = 4.3 \times 10^{-7} \tag{5-66}$$

$$\frac{[H^+][CO_3^{2-}]}{[HCO_3^-]} = K_2 = 4.7 \times 10^{-11} \tag{5-67}$$

Consider a solution formed by the addition of carbon dioxide to distilled water. Next, the pH is adjusted by addition of NaOH, while maintaining the total molar concentration of carbon and species containing it at its initial value, where

$$CO_2 + H_2O \rightarrow H_2CO_3 \tag{5-68}$$

and
$$[H_2CO_3^*] = [CO_2] + [H_2CO_3] \tag{5-69}$$

Thus, the total concentration of inorganic carbon in solution (C) can be written as

$$C = [H_2CO_3^*] + [HCO_3^-] + [CO_3^{2-}] \tag{5-70}$$

Now, if at a given pH, $CaCl_2$ is added to the solution, at what $[Ca^{2+}]$ will the solution just become saturated with respect to $CaCO_3$? How will this saturation value change as a function of pH and C? For this set of problems, Eqs. (5-63), (5-66), (5-67), and (5-70) apply. In addition,

$$[H^+][OH^-] = K_W = 10^{-14} \tag{5-71}$$

must be satisfied. In order to solve this problem in a general sense, a charge balance must be maintained,

$$2[Ca^{2+}] + [Na^+] + [H^+] = [HCO_3^-] + 2[CO_3^{2-}] + [OH^-] + [Cl^-] \tag{5-72}$$

However, for the particular problems at hand, the pH is given and so Eq. (5-72) is not required. From Eq. (5-47) $[CO_3^{2-}]$ can be found as a function of C and $[H^+]$ (and hence pH). Thus, $[Ca^{2+}]$ can be determined by substituting the value for $[CO_3^{2-}]$ into Eq. (5-63).

The relationship of pH, C, and the saturation value for $[Ca^{2+}]$ as determined by the above procedure is illustrated graphically in Fig. 5-11.

Example If $[Ca^{2+}]$ equals 10^{-4} and C equals 10^{-2}, at what pH will the solution become saturated with respect to $CaCO_3$?

From Fig. 5-11, for log $[Ca^{2+}] = -4$ and $C = 10^{-2}$, the pH of saturation equals 8.0. If the pH is raised above this value, the water will be supersaturated such that $CaCO_3$ may begin to precipitate.

Example What is the maximum solubility of calcium in water with a pH of 10 and $C = 10^{-2}$?
From Fig. 5-11, log $[Ca^{2+}] = -5.8$, so that $[Ca^{2+}] = 10^{-5.8} = 1.6 \times 10^{-6}$.

Example A water has an initial $[Ca^{2+}]$ of 4×10^{-3} and C of 10^{-2}. It is desired to reduce $[Ca^{2+}]$ to 10^{-4} by precipitation of $CaCO_3$. Based upon equilibrium relationships, what would be the final concentration of C, and how high should the solution pH be maintained in order to achieve the desired calcium removal?

The quantity of Ca^{2+} precipitated is $4 \times 10^{-3} - 10^{-4}$ or 3.9×10^{-3} mol/l of solution. Since for each mole of calcium precipitated, one mole of CO_3^{2-} is removed from solution,

$$C_{(final)} = 10^{-2} - 3.9 \times 10^{-3} = 6.1 \times 10^{-3}$$

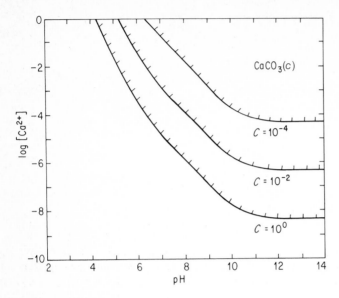

Figure 5-11 Logarithmic concentration diagram showing the relationship between pH, C (mol/l of inorganic carbon), and the equilibrium concentration of Ca^{2+} with respect to $CaCO_3(c)$.

From Fig. 5-11 when $[Ca^{2+}]$ equals 10^{-4} and C equals 6.1×10^{-3}, the pH equals about 9.0. Thus, the pH would need to be raised to at least 9.0 to obtain the desired removal. Practically, a higher pH would be used in order to ensure that sufficient removal was obtained.

Complex formation also affects the solubility of salts. For example, zinc is commonly removed from acid-plating wastewaters by adding base to increase the pH to form the insoluble $Zn(OH)_2(c)$. However, if excess base is added, zinc will form soluble complexes with OH^- and will return to or remain in solution. The appropriate equations governing this behavior of zinc are as follows:

Solubility product

$$Zn(OH)_2(c) \rightleftharpoons Zn^{2+} + 2OH^- \tag{5-73}$$

Complex formation

$$Zn^{2+} + OH^- \rightleftharpoons ZnOH^+ \tag{5-74}$$

$$ZnOH^+ + OH^- \rightleftharpoons Zn(OH)_2 \tag{5-75}$$

$$Zn(OH)_2 + OH^- \rightleftharpoons Zn(OH)_3^- \tag{5-76}$$

$$Zn(OH)_3^- + OH^- \rightleftharpoons Zn(OH)_4^{2-} \tag{5-77}$$

Water ionization

$$H_2O \rightleftharpoons H^+ + OH^- \tag{5-78}$$

Equilibrium relationships for the above equations are:

$$[Zn^{2+}][OH^-]^2 = K_{sp} = 8 \times 10^{-18} \qquad (5\text{-}79)$$

$$[ZnOH^+] / [Zn^{2+}][OH^-] = K_1 = 1.4 \times 10^4 \qquad (5\text{-}80)$$

$$[Zn(OH)_2] / [ZnOH^+][OH^-] = K_2 = 1 \times 10^6 \qquad (5\text{-}81)$$

$$[Zn(OH)_3^-] / [Zn(OH)_2][OH^-] = K_3 = 1.3 \times 10^4 \qquad (5\text{-}82)$$

$$[Zn(OH)_4^{2-}] / [Zn(OH)_3^-][OH^-] = K_4 = 1.8 \times 10 \qquad (5\text{-}83)$$

$$[H^+][OH^-] = K_W = 1 \times 10^{-14} \qquad (5\text{-}84)$$

In addition, a mass balance for zinc must be maintained:

$$C = [Zn^{2+}] + [ZnOH^+] + [Zn(OH)_2] + [Zn(OH)_3^-] + [Zn(OH)_4^{2-}]$$
$$(5\text{-}85)$$

Figure 5-12 is a logarithmic concentration diagram showing the relationships between pH and the various zinc species for a solution saturated with respect to $Zn(OH)_2(c)$. If for a given pH, the concentration of zinc in solution were found sufficiently high to be in the $Zn(OH)_2(c)$ area of the diagram, then the solution would be supersaturated and $Zn(OH)_2$ precipitation would be likely to occur.

Figure 5-12 was constructed as follows. For a given pH, $[OH^-]$ was calculated from Eq. (5-84) and used in Eq. (5-79) to determine $[Zn^{2+}]$. The latter value was in turn used in Eq. (5-80) to calculate $[ZnOH^+]$, and so on until the concentration of each complex was known. These values were then inserted in Eq. (5-85) to determine the total soluble concentration of zinc present. The total values so calculated were plotted as a function of pH and are represented by the bold curve in Fig. 5-12.

It is possible for zinc to form polynuclear complexes containing two or more zinc atoms, and these would add to the total concentration of zinc in

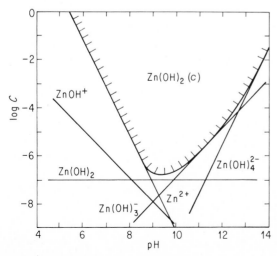

Figure 5-12 Solubility of $Zn(OH)_2$ as a function of solution pH.

solution. However, for this example the polynuclear species were not considered, and in fact, would not have affected the equilibrium concentration significantly.

Figure 5-12 illustrates that at low pH, Zn^{2+} is the predominant soluble species present, followed by $ZnOH^+$. At very high pH, $Zn(OH)_4^{2-}$ is the predominant species followed by $Zn(OH)_3^-$. Quite obviously, if complex formation were ignored and zinc solubilities were calculated solely by use of Eq. (5-79), erroneous conclusions could be drawn.

> **Example** At what pH does zinc have its minimum solubility in water?
>
> From Fig. 5-12, minimum solubility occurs at pH 9.4. At this pH the total equilibrium zinc concentration equals $10^{-6.8}$ or 1.6×10^{-7} mol/l. The predominant soluble species present is $Zn(OH)_2$ at 1.1×10^{-7} mol/l, followed by $Zn(OH)_3^-$ at 3.6×10^{-8} mol/l and Zn^{2+} at 1.3×10^{-8} mol/l.

> **Example** A plating waste has a zinc concentration of 10^{-3} mol/l. If lime is added to the solution, above what pH will zinc just begin to precipitate? What minimum pH should be used to decrease the zinc concentration below $10^{-5} M$?
>
> From Fig. 5-12, the pH for a maximum equilibrium concentration of zinc of 10^{-3} (log $C = -3$) is 6.9. At pH above this value, zinc should begin to precipitate from solution.
>
> In order to reduce the zinc concentration to 10^{-5} (log $C = -5$), the pH from Fig. 5-12 must equal 7.9. In practice, a higher pH (but not above 9.4) should be maintained to insure adequate zinc removal.

Equilibrium constants for other heavy metal hydroxide complexes are listed in Table 5-3.

Table 5-3 Stepwise formation constants for mononuclear hydroxide complexes

Metal ion	Logarithm of constants*			
	K_1	K_2	K_3	K_4
Ag^+	2.3	1.9		
Ba^{2+}	0.64			
Ca^{2+}	1.51			
Cd^{2+}	6.08	2.62	−0.32	0.04
Cu^{2+}	6.0	7.18	1.24	0.14
Fe^{3+}	11.5	9.3		
Hg^{2+}	10.3	11.4		
Mg^{2+}	2.60			
Mn^{2+}	3.4			
Ni^{2+}	4.0			
Pb^{2+}	7.82	3.06	3.06	
Zn^{2+}	4.15	6.00	4.11	1.26

* These values were taken from L. G. Sillén and A. E. Martell, "Stability Constants." *spec. publs.* 17, 1964, and 25, 1971, Chemical Society, London. With permission.

5-9 COMPLEX FORMATION

Background information and general nomenclature for complexes and complex formation are given in Sec. 2-13. The effect of complex formation on solubility of salts is discussed in Sec. 5.8. Most metals form complexes with a variety of ligands, resulting in a number of negatively, neutral, or positively charged species. Because of differences such as charge, size and shape, and mobility, the various complexes of a metal behave differently chemically. For a given metal, some complexes are more toxic to organisms, some are removed more readily by chemical flocculation, activated carbon adsorption, and ion exchange. A knowledge of complex formation can aid in the understanding of metal behavior and fate in natural water systems, and can aid the engineer in designing treatment systems for metal removal.

Mononuclear Complexes

A mononuclear complex consists of a central metal ion to which is bound a number of neutral or anionic ligands. The number of ligands attached to the central ion is called the *coordination number*. Metal ions in solution are never in the uncomplexed state, they always are bound with solvent molecules. Thus, Cu^{2+} exists in solution as the hydrated $Cu(H_2O)_6^{2+}$, and Al^{3+} as $Al(H_2O)_6^{3+}$. Formation of other complexes in aqueous solution can be thought to result from the replacement of bound water by other ligands. Thus, formation of the hydroxide complex of copper can be thought to occur as follows:

$$Cu(H_2O)_6^{2+} + H_2O \rightleftharpoons Cu(H_2O)_5(OH)^+ + H_3O^+$$

For ease in writing, water is generally dropped from the complex to give the generally used and abbreviated equation.

$$Cu^{2+} + H_2O \rightleftharpoons CuOH^+ + H^+ \tag{5-86}$$

Generally in this book, abbreviated equations will be used.

Table 5-4 lists stepwise formation constants for a number of mononuclear metal ions and ligands. When such constants are available, calculations of complex ion concentration under a given set of conditions is relatively straight forward. Consider the complexes of copper with ammonia,

$$Cu^{2+} + NH_3 \rightleftharpoons CuNH_3^{2+} \tag{5-87}$$

$$CuNH_3^{2+} + NH_3 \rightleftharpoons Cu(NH_3)_2^{2+} \tag{5-88}$$

$$Cu(NH_3)_2^{2+} + NH_3 \rightleftharpoons Cu(NH_3)_3^{2+} \tag{5-89}$$

$$Cu(NH_3)_3^{2+} + NH_3 \rightleftharpoons Cu(NH_3)_4^{2+} \tag{5-90}$$

The stepwise equilibrium relationships for the above are

$$\frac{[CuNH_3^{2+}]}{[Cu^{2+}][NH_3]} = K_1 = 9.8 \times 10^3 \tag{5-91}$$

Table 5-4 Stepwise formation constants for various ligand and metal complexes

Ligand	Metal ion	Logarithm of constants*			
		K_1	K_2	K_3	K_4
Cl^-	Ag^+	3.45	2.22	0.33	0.04
	Cd^{2+}	2.00	0.70	0.0	
	Fe^{2+}	0.36	0.04		
	Fe^{3+}	1.48	0.65	-1.0	
	Hg^{2+}	6.72	6.51	1.00	0.97
	Pb^{2+}	1.60	0.18	-0.1	-0.3
F^-	Al^{3+}	6.16	5.05	3.91	2.71
	Cd^{2+}	0.3	0.2	0.7	
	Fe^{3+}	5.25	4.00	3.00	
	Mg^{2+}	1.82			
NH_3	Ag^+	3.32	3.92		
	Cd^{2+}	2.51	1.96	1.30	0.79
	Cu^{2+}	3.99	3.34	2.73	1.97
	Hg^{2+}	8.8	8.7	1.0	0.78
	Ni^{2+}	3.0	2.18	1.64	1.16
	Zn^{2+}	2.18	2.25	2.31	1.96
SO_4^{2-}	Ag^+	1.3			
	Al^{3+}	3.73			
	Cd^{2+}	2.17	1.37		
	Cu^{2+}	2.3			
	Fe^{2+}	2.3			
	Fe^{3+}	4.04	1.30		
	Ca^{2+}	2.3			
	Pb^{2+}	3.7			
	Zn^{2+}	2.28			
HS^-	Ag^+	13.6	4.1		
	Cd^{2+}	7.55	7.06	1.88	2.36

* These values were taken from L. G. Sillén and A. E. Martell, "Stability Constants," *Spec. Publs.* 17, 1964, and 25, 1971, Chemical Society, London. With permission.

$$\frac{[Cu(NH_3)_2^{2+}]}{[CuNH_3^{2+}][NH_3]} = K_2 = 2.2 \times 10^3 \qquad (5\text{-}92)$$

$$\frac{[Cu(NH_3)_3^{2+}]}{[Cu(NH_3)_2^{2+}][NH_3]} = K_3 = 5.4 \times 10^2 \qquad (5\text{-}93)$$

$$\frac{[Cu(NH_3)_4^{2+}]}{[Cu(NH_3)_3^{2+}][NH_3]} = K_4 = 9.3 \times 10 \qquad (5\text{-}94)$$

The significance of the various ammonia complexes of copper in aqueous systems can perhaps best be illustrated with a logarithmic concentration diagram. Here, the NH_3 ligand is taken as the master variable and pNH_3 ($-\log NH_3$) is plotted as the abscissa as illustrated in Fig. 5-13. The calculations are considerably simplified if the concentration of the uncomplexed Cu^{2+} is arbitrarily held constant at some value, say $10^{-7} M$. Once the concentration is fixed, all the above equations become straight lines on a logarithmic concentration diagram. The logarithm of Eq. (5-91) is

$$\log [CuNH_3^{2+}] = \log [Cu^{2+}] + \log K_1 + \log NH_3 \qquad (5\text{-}95)$$

or since $\log K_1 = 3.99$, $\log [NH_3] = -pNH_3$, and $\log [Cu^{2+}] = -7$,

$$\log [CuNH_3^{2+}] = -3.01 - pNH_3 \qquad (5\text{-}96)$$

This line has a slope of -1 in Fig. 5-13 and intersects the $pNH_3 = 0$ ordinate at -3.01. The negative slopes for each of the succeeding complexes increase, as indicated by the respective logarithmic equations:

$$\log [Cu(NH_3)_2^{2+}] = 0.33 - 2pNH_3 \qquad (5\text{-}97)$$

$$\log [Cu(NH_3)_3^{2+}] = 3.06 - 3pNH_3 \qquad (5\text{-}98)$$

$$\log [Cu(NH_3)_4^{2+}] = 5.03 - 4pNH_3 \qquad (5\text{-}99)$$

From Fig. 5-13 it can readily be seen which complex dominates at a given ammonia concentration. If $[Cu^{2+}]$ were chosen to be some other value such as 10^{-9}, the relative concentrations of the various species would not change. The curves would all simply shift together two logarithm units downward.

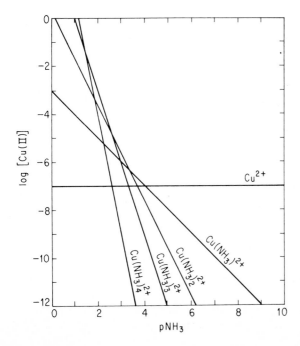

Figure 5-13 Logarithmic concentration diagram illustrating effect of ammonia concentration on relative concentration of various ammonia complexes of copper.

Ammonia is a base which ionizes in water to form NH_4^+. The distribution between NH_3 and NH_4^+ as a function of pH is illustrated in Fig. 5-5. At pH 7, only about 0.5 percent of the ammonia is in the NH_3 form, and this percentage increases to about 5 percent at pH of 8. Under these conditions in natural waters, pNH_3 would generally range from about 5 to 7, and from Fig. 5-13, Cu^{2+} would be the predominate species. In municipal wastewaters, pNH_3 would vary from about 3.5 to 5.5. Over this range, any one of the species, Cu^{2+}, $Cu(NH_3)^{2+}$, or $Cu(NH_3)_2^{2+}$ may dominate. During anaerobic digestion of sewage sludge, ammonia is released in relatively high concentration such that pNH_3 may range from 2.5 to 4. Under these conditions ammonia complexes of copper would generally predominate over Cu^{2+}. When pNH_3 equals 2.5, only about 0.1 percent of the soluble copper in the digester would be in the Cu^{2+} form. Consideration of the effect of complexes on the solubility of heavy metals in anaerobic systems and on potential toxicity to microorganisms would appear appropriate.

Another method of graphically presenting complex formation calculations is through a *distribution diagram* which shows the relative concentration of each species as a function of ligand concentration. A distribution diagram for ammonia complexes of copper is shown in Fig. 5-14. This diagram can be constructed by first assuming an arbitrary value for $[Cu^{2+}]$ such as 1, and then through use of equations similar to Eqs. (5-96) through (5-99), the concentration of each species at a given pNH_3 can be determined. The total molar concentration C of Cu(II) is then

$$C = [Cu^{2+}] + [CuNH_3^{2+}] + [Cu(NH_3)_2^{2+}] + [Cu(NH_3)_3^{2+}] + [Cu(NH_4)^{2+}] \quad (5\text{-}100)$$

The fraction of copper present as each species at a given pNH_3 is then determined as follows:

$$\alpha_0 = \frac{[Cu^{2+}]}{C} \quad (5\text{-}101)$$

$$\alpha_1 = \frac{[CuNH_3^{2+}]}{C} \quad (5\text{-}102)$$

$$\alpha_2 = \frac{[Cu(NH_3)_2^{2+}]}{C} \quad (5\text{-}103)$$

$$\alpha_3 = \frac{[Cu(NH_3)_3^{2+}]}{C} \quad (5\text{-}104)$$

$$\alpha_4 = \frac{[Cu(NH_3)_4^{2+}]}{C} \quad (5\text{-}105)$$

From such calculations at different values of pNH_3 a distribution diagram like Fig. 5-14 can be constructed. Perhaps the relative importance of the different species can be discerned more rapidly from distribution diagrams than from logarithmic concentration diagrams. As will be seen later, distribution diagrams allow pictorial representation of quite complex equilibria.

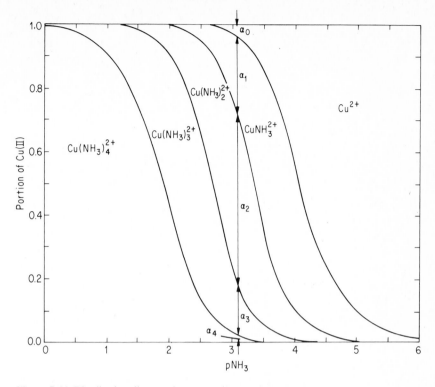

Figure 5-14 Distribution diagram for ammonia complexes of copper.

Mixed Ligand Complexes

Natural water systems and wastewaters frequently contain several ligands such as Cl^-, NH_3, S^{2-}, and OH^-, and each will compete for complex formation with the metal ions present. The solution to this kind of a problem is not complex mathematically, but requires that all stepwise equilibria for each ligand and the metal be considered. As before, a concentration for the uncomplexed ion is first assumed, and then the concentration of each complex is determined for given values of each ligand. The results can perhaps best be displayed graphically with a *predominance area diagram* as illustrated in Fig. 5-15 for complexes between mercury and the ligands Cl^- and OH^-. The abscissa is used to represent OH^- where $pH = 14 + \log [OH^-]$. The other ligand, chloride, is represented as the ordinate in the form of pCl. From calculations similar to those discussed previously, the complex which predominates under a given set of ligand concentrations can be determined. The lines drawn in the figure represent the locations where a change in predominance from one complex to another occurs.

Figure 5-15 illustrates the distribution of soluble mercury species in water when the total molar concentration of mercury is less than 10^{-1}. At higher

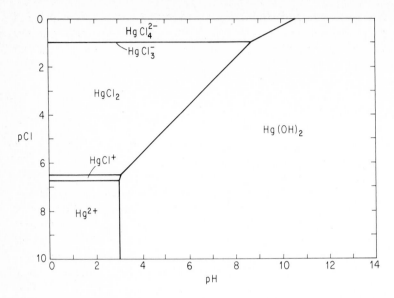

Figure 5-15 Predominance area diagram illustrating effect of pH and chloride concentration on chloride and hydroxide complexes of mercury.

mercury concentrations, precipitation of HgO might occur at high pH values. For fresh waters, pCl would normally lie between 2 and 4. With pH greater than 7, the predominant species is the neutral $Hg(OH)_2$, while below 7 it is $HgCl_2$. In seawater pCl lies between 0 and 1, and $HgCl_4^{2-}$ would be the predominant species. At low pH, mercury may occur primarily as a positively charged, a neutral, or a negatively charged complex, depending upon chloride concentration.

When more than two ligands are involved, computations for solution are similar to the above, but involve more equations. Graphical presentation, however, becomes more difficult. Three-dimensional graphs can be used for three ligand systems. However, it is rare for a given water or wastewater that more than two ligands will compete for predominance. The problem is to determine which two ligands this may be.

Polynuclear complexes of metals may also form in solution. An important example is the dimerization of Fe(III),

$$2Fe(H_2O)_5OH^{2+} \rightleftharpoons \left[(H_2O)_4Fe \begin{matrix} H \\ O \\ \diagup \diagdown \\ \diagdown \diagup \\ O \\ H \end{matrix} Fe(H_2O)_4 \right]^{4+} + 2H_2O \qquad (5\text{-}106)$$

which can be simplified as

$$2FeOH^{2+} \rightleftharpoons Fe_2(OH)_2^{4+} \qquad (5\text{-}107)$$

Polynuclear species of iron and aluminum may dominate in natural water and are believed to be important in coagulation of water and wastewater. Further discussion of mixed ligand and polynuclear species of importance in natural waters can be found elsewhere[6,7].

5-10 OXIDATION-REDUCTION REACTIONS

Oxidation-reduction reactions are among the most important with which the environmental engineer deals. Many reactions of interest in wastewater treatment such as organic oxidation and methane fermentation, nitrification, and denitrification, are of this type and are mediated by bacteria. Oxidation-reduction reactions are important in the solubilization and precipitation of iron and manganese. Oxidants such as chlorine and ozone are added to water and wastewater to bring about desired inorganic and organic transformations as well as to disinfect. The fate of materials introduced into the environment frequently depends upon the redox environment to which they become subjected. Also, many analytical tests used by the environmental engineer depend upon oxidation-reduction reactions.

As with acid-base, solubility, or complex formation, oxidation-reduction reactions tend toward a state of equilibrium. An understanding of oxidation-reduction equilibria can help to indicate whether a particular reaction is possible under given environmental conditions. It will not tell whether the reaction will in fact occur, but nevertheless such an understanding is helpful in evaluating how conditions might best be changed to encourage desirable transformations or to prevent undesirable ones. Oxidation-reduction reactions can be quite complex. Graphical approaches can be helpful to reduce this complexity and to illustrate the significant factors involved for a particular case.

Equilibrium Relationships

In Sec. 2-7 methods for balancing oxidation-reduction reactions were presented, and use of half-reactions was emphasized. A brief introduction to equilibrium in oxidation-reduction reactions was given in Sec. 3-10. These concepts are expanded in this section.

The oxidation-reduction state of an aqueous environment at equilibrium can be stated in terms of its redox potential. In the chemistry literature this is generally expressed in volts, E, or as the negative logarithm of the electron activity, pE. These terms are also directly related to the free energy for the system. A half reaction for the reduction of NO_3^- to NH_4^+ will be used for illustration,

[6] J. N. Butler, "Ionic Equilibrium," Addison-Wesley, Reading, Mass., 1964.
[7] W. Stumm and J. J. Morgan, "Aquatic Chemistry," Wiley-Interscience, New York, 1970.

$$\frac{1}{8}NO_3^- \quad + \frac{5}{4}H^+ + e^- \rightleftharpoons \frac{1}{8}NH_4^+ \quad + \frac{3}{8}H_2O \qquad (5\text{-}108)$$

$$\Delta G^0 \qquad \frac{1}{8}(-26.41) \quad 0 \quad 0 \quad \frac{1}{8}(-19.00) \quad \frac{3}{8}(-56.69)$$

The standard free energies of formation for each of the reactants and products are obtained from Table 3-1. By convention, ΔG° for H^+ and e^- are zero. The standard free energy for the reaction as written is thus:

$$\Delta G^\circ = \frac{1}{8}(-19.00) + \frac{3}{8}(-56.69) - \frac{1}{8}(-26.41) = -20.33 \text{ kcal}$$

The standard electrode potential for the reaction can be calculated from ΔG° using Eq. (3-43),

$$E^\circ = -\frac{\Delta G^\circ}{zF} = -\frac{-20,330(4.186)}{1(96,500)} = 0.88 \text{ volt}$$

where 4.186 J/cal is a conversion factor necessary to make the units consistent. As indicated by Eqs. (3-14) and (3-43), the equilibrium condition for the oxidation-reduction reaction can be determined from either ΔG° or E°:

$$\ln K = 2.3 \log K = -\frac{\Delta G^\circ}{RT} = \frac{zFE^\circ}{RT}$$

Thus for Eq. (5-108) the following relationship between concentrations must hold at equilibrium:

$$\log \frac{[NH_4^+]^{1/8}}{[NO_3^-]^{1/8}[H^+]^{5/4}[e^-]} = -\frac{\Delta G^\circ}{2.3RT} = \frac{zFE^\circ}{2.3RT} \qquad (5\text{-}109)$$

This equation can be converted into the following form since the electron activity is defined as $pE = -\log[e^-]$:

$$pE = pE^\circ - \log \frac{[NH_4^+]^{1/8}}{[NO_3^-]^{1/8}[H^+]^{5/4}} \qquad (5\text{-}110)$$

where

$$pE^\circ = -\frac{\Delta G^\circ}{2.3RT} = \frac{zFE^\circ}{2.3RT} \qquad (5\text{-}111)$$

Table 2-3 lists values of pE° and ΔG° for various half reactions of interest in water quality.

The redox potential of a system can be indicated by pE. When pE is large, the electron activity is low and the system tends to be an oxidizing one: i.e., half reactions tend to be driven to the left. When pE is low, the system is reducing and reactions tend to be driven to the right.

For a given system to be at equilibrium, the electron activity or pE for all possible oxidation-reduction reactions for constituents present must be the same.

Logarithmic Concentration Diagrams

A logarithmic concentration diagram can be used to illustrate the relative equilibrium concentrations for NH_4^+ and NO_3^- as a function of pE or E, as shown in

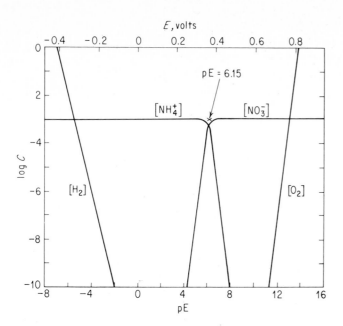

Figure 5-16 Logarithmic concentration diagram showing relationship between [NH$_4^+$] and [NO$_3$] as a function of pE for pH = 7 and [NH$_4^+$] +]NO$_3^-$] = 10^{-3} mol/l.

Fig. 5-16. This figure was constructed for 25°C and using Eq. (5-110). The value for pE° was obtained from Table 2-3. For pH = 7, [H$^+$] = 10^{-7}; thus

$$pE = 14.9 - \tfrac{1}{8} \log [NH_4^+] + \tfrac{1}{8} \log [NO_3^-] + \tfrac{5}{4} \log (10^{-7})$$

which reduces to

$$pE = 6.15 - \tfrac{1}{8} \log [NH_4^+] + \tfrac{1}{8} \log [NO_3^-] \qquad (5\text{-}112)$$

It was also assumed that $C = [NH_4^+] + [NO_3^-] = 10^{-3}$.

Figure 5-16 is constructed similar to an acid-base logarithmic concentration diagram. A horizontal line representing log C is drawn. A system point is located at pE = 6.15. Two lines are drawn through the system point, one to the left with a slope of $\tfrac{1}{8}$ representing NO$_3^-$, and one to the right with a slope of $-\tfrac{1}{8}$ representing NH$_4^+$. Since when pE = 6.15, [NH$_4^+$] = [NO$_3^-$] = $C/2$, the two diagonal lines must intersect at log $\tfrac{1}{2}$ or -0.3 logarithmic units below the system point. Curved lines through a point 0.3 unit below the system point and connecting the straight lines completes the figure.

Two additional lines are shown in the figure and represent respective values for pE at which water is reduced and oxidized:

Reduction

$$H^+ + e^- \rightleftharpoons \tfrac{1}{2} H_2$$

$$pE = 0.0 - \log \frac{[H_2]^{1/2}}{[H^+]} \qquad (5\text{-}113)$$

At pH = 7,

$$pE = -7 - \tfrac{1}{2} \log [H_2] \qquad (5\text{-}114)$$

Oxidation

$$\tfrac{1}{4}O_2 + H^+ + e^- \rightleftharpoons \tfrac{1}{2}H_2O$$

$$pE = 20.79 - \log \frac{1}{[O_2]^{1/4}[H^+]} \qquad (5\text{-}115)$$

At pH = 7,

$$pE = 13.79 - \tfrac{1}{4} \log [O_2] \qquad (5\text{-}116)$$

If pE for a given aqueous system were to the left of the $[H_2]$ line, the system would be highly reducing such that water would tend to decompose with the evolution of H_2. With pE to the right of the $[O_2]$ line, the system would be highly oxidizing with the result that water would tend to decompose with O_2 evolution. Natural aquatic systems are characterized by pE values between these two extremes.

pE–pH Diagrams

Figure 5-16 represents the conditions for a neutral pH. As indicated by Eq. (5-110), the pE in the $NH_4^+ - NO_3^-$ system is also a function of $[H^+]$ and hence pH. This is true for any oxidation-reduction reaction in which H^+ or OH^- is involved. A predominance area or $pE - pH$ diagram is frequently used to illustrate the relationship of pH, pE, and the most stable species in a given oxidation-reduction system. A $pE - pH$ diagram for the nitrogen system is illustrated in Fig. 5-17. In addition to Eq. (5-110), relationships between NO_2^- and NH_4^+ and between NO_2^- and NO_3^- were obtained from Table 2-3 and rearranged into the following logarithmic form:

$$pE = 14.13 - pH - \tfrac{1}{2} \log \frac{[NO_2^-]}{[NO_3^-]} \qquad (5\text{-}117)$$

$$pE = 14.90 - \tfrac{5}{4}pH - \tfrac{1}{8} \log \frac{[NH_4^+]}{[NO_3^-]} \qquad (5\text{-}118)$$

$$pE = 15.21 - \tfrac{4}{3}pH - \tfrac{1}{6} \log \frac{[NH_4^+]}{[NO_2^-]} \qquad (5\text{-}119)$$

In addition, Eq. (2-30) was rearranged into the following logarithmic form in order to establish the concentration of NH_3 as it is related to pH and $[NH_4^+]$:

$$pH = 9.3 - \log \frac{[NH_4^+]}{[NH_3]} \qquad (5\text{-}120)$$

The lines in Fig. 5-17 represent points of transition from predominance by one nitrogen form to predominance by another. The lines represent the location where the concentrations of the two most predominant nitrogen forms are

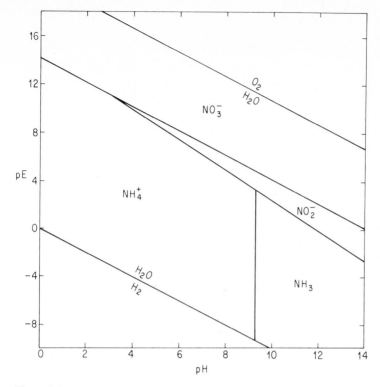

Figure 5-17 pE-pH diagram illustrating predominant nitrogen forms at equilibrium in an aqueous system.

equal. Thus, the equation of the line between NO_2^- and NO_3^- is given by the result from setting $[NO_2^-]$ equal to $[NO_3^-]$ in Eq. (5-117),

$$pE = 14.13 - pH \qquad (5-121)$$

The more reduced form, NO_2^-, is predominant in the area below the line while the oxidized form is predominant in the area above the line. The other lines in Fig. 5-17 are obtained from Eqs. (5-118) through (5-120) in a similar manner. The lines representing the boundaries where water is reduced or oxidized are obtained from Eqs. (5-113) and (5-115) by setting H_2 and O_2 equal to 1. The stable range for water occurs between these two lines.

From Fig. 5-17 at high pE, the system tends to be oxidizing and the most oxidized form, NO_3^-, is dominant. If ammonia were introduced into such a system, it would not be stable and would tend to be oxidized to nitrate. The rate of this conversion at temperatures of natural waters, however, would be extremely slow if it were not for the nitrifying bacteria which catalyze the oxidation and obtain some of the energy released for growth. If ammonia were introduced into a system with low pE, it would be stable and could not be oxidized chemically or biologically. Thus, definition of a given system in terms of pE and pH can aid in understanding which reactions may occur when materials are added to the system.

Figure 5-18 is a pE-pH diagram illustrating zones of importance in natural water systems. It was constructed using half reactions from Table 2-3 for carbon dioxide reduction to methane, for sulfate reduction to hydrogen sulfide, and for nitrogen transformations including that between N_2 and nitrate. These oxidation-reduction reactions are commonly brought about through bacterial action at normal water temperatures. The aerobic zone represents conditions in aerated waters where the partial pressure of oxygen [O_2] exceeds about 10^{-3} atm. This is the case for most natural streams, rivers, lake surfaces, and the ocean. Also represented are aerobic treatment processes such as activated sludge.

The anaerobic zone in the lower portion of Fig. 5-18 represents conditions where sulfates are reduced to hydrogen sulfide and organic materials are reduced to methane gas. Such conditions are common in organic laden sediments, the lower zones in polluted lakes, sanitary landfills, and anaerobic digesters.

A transition zone lies between the above two extremes. Reduction of nitrates to nitrogen gas through bacterial denitrification can occur in the lower

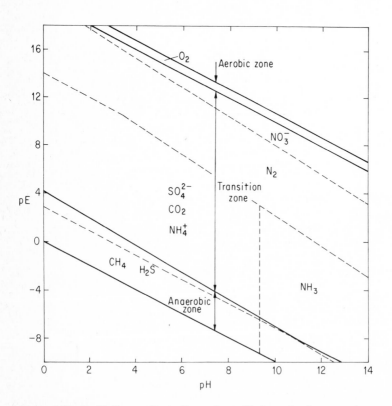

Figure 5-18 pE-pH diagram illustrating major oxidation-reduction zones in aqueous systems.

portion of this zone, a condition common at natural water-sediment interfaces, the interior of bacterial flocs in otherwise aerobic treatment processes, saturated zones in soils, and in some biological treatment processes specifically designed for this purpose. In the upper portion of the transition zone, nitrates are stable and oxidation of ammonia to nitrate by bacterial nitrification is possible.

Figure 5-18 illustrates that pE values associated with the different redox zones are dependent upon pH. At neutral pH, aerobic conditions are associated with a pE of about 13, and anaerobic by pE below -3. For each unit decrease in pH, the respective values for pE increase about one unit.

An idea of the transformations likely to occur in natural water systems when other materials such as iron, manganese, mercury, and chlorine are introduced, can be obtained by comparing their pE-pH diagrams with the zones illustrated in Fig. 5-18. Pictorial representation significantly reduces the apparent complexity of such systems and can give the water chemist and the environmental engineer a better intuitive feel for the systems with which he works.

PROBLEMS

For the following, assume that the temperature is 25°C.

5-1 Calculate the activity coefficient and activity of each ion in a solution containing 300 mg/l $NaNO_3$ and 150 mg/l $CaSO_4$.

Answer: Na^+ and NO_3^-, 0.91 and 3.21×10^{-3}; Ca^{2+} and SO_4^{2-}, 0.69 and 0.76×10^{-3}

5-2 Calculate the activity coefficient and activity of each ion in a solution containing 75 mg/l Na^+, 25 mg/l Ca^{2+}, 10 mg/l Mg^{2+}, 125 mg/l Cl^-, 50 mg/l HCO_3^-, and 48 mg/l SO_4^{2-}.

5-3 A solution is prepared by diluting 10^{-3} mol of propionic acid to one liter with distilled water. Calculate the equilibrium concentration for each chemical species in the water.

Answer: $[H^+] = 1.08 \times 10^{-4}$, $[OH^-] = 9.29 \times 10^{-11}$, $[HPr] = 8.92 \times 10^{-4}$, $[Pr^-] = 1.08 \times 10^{-4}$

5-4 A solution is prepared by diluting 10^{-2} mol of ammonia to one liter with distilled water. Calculate the equilibrium concentration for each chemical species in the water.

5-5 Calculate the pH of a solution containing 25 mg/l of the following:
(a) hydrochloric acid; (b) sodium hydroxide.
Answer: (a) 3.2; (b) 10.8

5-6 Calculate the pH of a solution containing 50 mg/l of the following:
(a) sulfuric acid; (b) potassium hydroxide.

5-7 Calculate the pH of solutions containing 100 mg/l of each of the following weak acids or weak bases: (a) acetic acid; (b) hypochlorous acid; (c) ammonia; (d) hydrocyanic acid.
Answer: (a) 3.8; (b) 5.1; (c) 10.5; (d) 5.8

5-8 Calculate the pH of a solution containing 50 mg/l of each of the following weak acids or salts of weak acids: (a) carbonic acid; (b) sodium acetate; (c) sodium hypochlorite; (d) phosphoric acid.

5-9 Calculate the pH of the equivalence point in the titration of solutions containing the following concentrations of acetic acid with sodium hydroxide: (a) 100 mg/l; (b) 1000 mg/l; (c) 10,000 mg/l.
Answer: (a) 8.0; (b) 8.5; (c) 9.0

5-10 Calculate the pH of the equivalence point in the titration of each of the following concentrations of sodium bicarbonate with sulfuric acid: (a) 10 mg/l; (b) 100 mg/l; (c) 1000 mg/l

5-11 Calculate the equivalence point pH for both ionizations in the titration of 150 mg/l of sodium carbonate with sulfuric acid.

    Answer: 8.4, 4.6

5-12 Calculate the equivalence point pH for both ionizations in the titration of 150 mg/l of sodium sulfide (Na_2S) with sulfuric acid.

5-13 A 1-liter solution contains 100 mg of HCl. Calculate the following:

    (*a*) Initial pH of the solution

    (*b*) pH after addition of 1 ml of 1 *N* NaOH

    (*c*) pH after addition of 2 ml of 1 *N* NaOH

    (*d*) pH after addition of 3 ml of 1 *N* NaOH

    (*e*) ml of 1 *N* NaOH required to reach the equivalence point

    Answer: (*a*) 2.6; (*b*) 2.8; (*c*) 3.1; (*d*) 10.4; (*e*) 2.74 ml

5-14 A 500-ml solution contains 100 mg of NaOH. Calculate the following:

    (*a*) Initial pH of the solution

    (*b*) pH after addition of 2 ml of 1 *N* H_2SO_4

    (*c*) pH after addition of 4 ml of 1 *N* H_2SO_4

    (*d*) ml of 1 *N* H_2SO_4 required to reach the equivalence point

5-15 Calculate the ratio of ammonia nitrogen in the NH_3 form to that in the NH_4^+ form in a solution with a pH of 7.4.

    Answer: 0.0125

5-16 Calculate the ratio of hypochlorous acid to hypochlorite ion in solutions with the following pH values:

    (*a*) 6.0

    (*b*) 7.0

    (*c*) 8.0

5-17 Calculate the pH of water samples containing the following mixtures:

    (*a*) 10 mg/l carbonic acid (H_2CO_3) and 100 mg/l bicarbonate ion (HCO_3^-)

    (*b*) 5 mg/l hydrosulfuric acid (H_2S) and 10 mg/l hydrosulfide ion (HS^-)

5-18 How many ml of 1 *N* NaOH must be added to a 500-ml solution containing 500 mg of acetic acid to increase the pH to 5?

5-19 (*a*) Draw a logarithmic concentration diagram for a 10^{-4} *M* solution of hypochlorous acid (25°C).

    (*b*) From the above diagram determine (1) the equilibrium pH for 10^{-4} *M* hypochlorous acid, (2) the pH after 0.5×10^{-4} mol of NaOH has been added to a liter of solution, and (3) the pH after 10^{-4} mol/l of NaOH has been added to the solution.

    Answer: (1) 5.75 (2) 7.5 (3) 8.7

5-20 (*a*) Draw a logarithmic concentration diagram for a 10^{-4} *M* solution of propionic acid.

    (*b*) From the above diagram determine (1) the equilibrium pH for 10^{-4} *M* propionic acid, (2) the pH after 0.5×10^{-4} mol of NaOH has been added to a liter of solution, and (3) the pH after 10^{-4} mol/l of NaOH has been added to the solution.

5-21 (*a*) Draw a logarithmic concentration diagram for a 10^{-3} *M* solution of hydrosulfuric acid.

    (*b*) From the diagram, determine the pH for solutions that contain the following (1) 10^{-3} *M* H_2S, (2) 10^{-3} *M* Na_2S, (3) 0.5×10^{-3} *M* HS^- and 0.5×10^{-3} *M* S^{2-}, and (4) 0.5×10^{-3} *M* H_2S and 0.5×10^{-3} *M* HS^-.

    Answer: (1) 5.0 (2) 10.9 (3) 12.9 (4) 7.0

5-22 (*a*) Draw a logarithmic concentration diagram for a 10^{-2} *M* solution of hydrosulfuric acid.

    (*b*) From the diagram, determine the pH for solutions which contain the following (1) 10^{-2} *M* H_2S, (2) 10^{-2} *M* Na_2S, (3) 0.5×10^{-2} *M* HS^- and 0.5×10^{-2} *M* S^{2-}, and (4) 0.5×10^{-2} *M* H_2S and 0.5×10^{-2} *M* HS^-.

5-23 Calculate the pH of a buffer solution prepared by mixing 2.4 g of acetic acid (CH_3COOH) and

0.73 g of sodium acetate (CH_3COONa) in 1 liter of water (equivalent to buffer solution for amperometric chlorine determination).

Answer: 4.0

5-24 Calculate the pH of a buffer solution prepared by mixing 8.5 g KH_2PO_4 with 43.5 g K_2HPO_4 in 1 liter of water (equivalent to buffer solution for BOD dilution water).

5-25 Calculate the pH (*a*) before, and (*b*) after adding 20 mg/l HCl to a buffer solution containing 100 mg/l KH_2PO_4 and 200 mg/l K_2HPO_4, assuming $pK_2' = 7.0$.

Answer: (*a*) 7.2;   (*b*) 6.7

5-26 Calculate the pH of a buffer solution prepared with 500 mg/l acetic acid (CH_3COOH) and 250 mg/l sodium acetate (CH_3COONa) under the following conditions:

(*a*) Initially

(*b*) After 20 mg/l of HCl is added to the solution

(*c*) After 20 mg/l of NaOH is added to the solution

5-27 Calculate the pH of a 200-ml buffer solution containing 20 mg/l carbonic acid and 50 mg/l bicarbonate ion, under the following conditions:

(*a*) Initially

(*b*) After 3 ml of 0.02 *N* H_2SO_4 is added

(*c*) After 3 ml of 0.02 *N* NaOH is added

Answer: (*a*) 6.8;   (*b*) 6.3;   (*c*) 8.1

5-28 Find the pH and buffer index for the following:

(*a*) 0.1 *M* acetic acid plus 0.1 *M* sodium acetate

(*b*) 0.19 *M* acetic acid plus 0.01 *M* sodium acetate

(*c*) 0.02 *M* acetic acid plus 0.18 *M* sodium acetate

5-29 Find the pH and buffer index for the following:

(*a*) 0.005 *M* HOCl plus 0.005 *M* NaOCl

(*b*) 0.001 *M* HOCl plus 0.009 *M* NaOCl

(*c*) 0.008 *M* HOCl plus 0.002 *M* NaOCl

Answer: (*a*) 7.5, 5.8×10^{-3} (*b*) 8.45, 2.2×10^{-3} (*c*) 6.9, 1.26×10^{-3}

5-30 Draw a curve of buffer index as a function of pH for 0.1 *M* acetic acid.

5-31 Draw a curve of buffer index as a function of pH for 0.1 *M* ammonium chloride.

5-32 Using a logarithmic concentration diagram, determine the pH of a solution containing 10^{-2} *M* H_2CO_3 and 2×10^{-2} *M* Na_2CO_3.

5-33 Using a logarithmic concentration diagram, determine the pH of a solution containing 10^{-2} *M* acetic acid and 2×10^{-2} *M* sodium acetate.

Answer: 5.0

5-34 Using a logarithmic concentration diagram, determine the pH of a solution containing 10^{-3} *M* acetic acid and 10^{-3} *M* hydrochloric acid.

5-35 Using a logarithmic concentration diagram, determine the pH of a solution containing 10^{-3} *M* Na_2CO_3 and 10^{-3} *M* NaOH.

Answer: 12

5-36 Using a logarithmic concentration diagram, find the pH of a solution containing 10^{-2} *M* acetic acid and 10^{-1} *M* ammonia.

5-37 Using a logarithmic concentration diagram, find the pH of a solution containing 10^{-1} *M* acetic acid and 2×10^{-1} *M* sodium bicarbonate.

Answer: 6.4

5-38 Using a logarithmic concentration diagram, find the pH of a solution formed from mixing equal volumes of 10^{-3} *M* acetic acid and 10^{-2} *M* ammonium bicarbonate.

5-39 Using a logarithmic concentration diagram, find the pH of a solution formed from mixing equal volumes of 10^{-2} *M* propionic acid, 10^{-2} *M* acetic acid, and 10^{-1} *M* sodium bicarbonate.

Answer: 7.0

5-40 From a logarithmic concentration diagram, estimate the minimum pH to which a water or wastewater need be raised to effect the precipitation as a metallic hydroxide of all but 10^{-4} mol/l of each of the following (a) Cr^{3+}, (b) Cu^{2+}, (c) Zn^{2+}, (d) Mg^{2+}, and (e) Ca^{2+}.

5-41 From a logarithmic concentration diagram, estimate the minimum pH to which a water or wastewater need be raised to effect the precipitation as a metallic hydroxide of all but 10^{-5} mol/l of each of the following (a) Fe^{3+}, (b) Fe^{2+}, (c) Mn^{2+}, (d) Al^{3+}, (e) Mg^{2+}, and (f) Cu^{2+}.

    Answer: (a) 3.2 (b) 9.2 (c) 10.0 (d) 5.0 (e) 10.8 (f) 7.2

5-42 From a logarithmic concentration diagram, estimate the minimum concentration in mol/l of CO_3^{2-} required to precipitate as a metallic carbonate all but 10^{-4} mol/l of (a) Mg^{2+}, (b) Ca^{2+}, (c) Sr^{2+}, (d) Zn^{2+}, and (e) Pb^{2+}.

5-43 From a logarithmic concentration diagram, estimate the minimum concentration in mol/l of CO_3^{2-} required to precipitate as a metallic carbonate all but 10^{-5} mol/l of (a) Ca^{2+}, (b) Cu^{2+}, (c) Fe^{2+}, (d) Cd^{2+}, and (e) Pb^{2+}.

    Answer: (a) 4×10^{-4} (b) 2×10^{-5} (c) 5×10^{-6} (d) 5×10^{-7} (e) 1.6×10^{-8}

5-44 From a logarithmic concentration diagram, determine the minimum pH at which each of the following concentrations of Ca^{2+} would be at equilibrium with $CaCO_3$ precipitate if the total concentration of inorganic carbon in solution equals 10^{-2} mol/l: (a) 10^{-3} mol/l, (b) 10^{-4} mol/l, (c) 10^{-5} mol/l.

5-45 Do Prob. 5-44, but assume that the total concentration of inorganic carbon in solution equals 10^{-4} mol/l.

    Answer: (a) 9.1 (b) 10.4 (c) Not saturated with $CaCO_3$

5-46 Construct a logarithmic concentration diagram showing the relationship between pH and the equilibrium concentration of Ca^{2+} with respect to $CaCO_3(c)$, assuming that the total concentration of inorganic carbon in solution, C, equals 10^{-1} mol/l.

5-47 A water has an initial Ca^{2+} concentration of 2×10^{-3} mol/l, and the total concentration of inorganic carbon in solution equals 2×10^{-2} mol/l. It is desired to reduce $[Ca^{2+}]$ to 2×10^{-4} mol/l by precipitation of $CaCO_3$. What minimum pH would be required to effect this removal, and what would be the final molar concentration of inorganic dissolved carbon?

5-48 Draw a diagram which shows the solubility of $Cd(OH)_2$ as a function of solution pH, and which also shows the concentration of other cadmium hydroxide complexes in a saturated solution. At what pH does $Cd(OH)_2$ have minimum solubility? K_{sp} for $Cd(OH)_2 = 2 \times 10^{-14}$.

5-49 Construct a diagram as described in Prob. 5-48, but for $Pb(OH)_2$. At what pH does $Pb(OH)_2$ have a minimum solubility? K_{sp} for $Pb(OH)_2 = 2.5 \times 10^{-16}$.

    Answer: 11.0

5-50 From a distribution diagram for $Cu-NH_3$ complexes, what is the predominant Cu(II) species when the ammonia concentration is (a) 0.1 mg/l, (b) 1 mg/l, (c) 10 mg/l, and (d) 100 mg/l.

5-51 Using a logarithmic concentration diagram, determine the concentration of various ammonia complexes of copper if $[Cu^{2+}] = 10^{-5}$ mol/l and $[NH_3] = 10^{-4}$ mol/l.

    Answer: $Cu(NH_3)^{2+}$, 10^{-5}; $Cu(NH_3)_2^{2+}$, 2.5×10^{-6}; $Cu(NH_3)_3^{2+}$, 10^{-7}; $Cu(NH_3)_4^{2+}$, 10^{-12}.

5-52 Using a logarithmic concentration diagram, determine the concentration of various ammonia complexes of copper if $[Cu^{2+}] = 10^{-4}$ mol/l and $[NH_3] = 10^{-3}$ mol/l.

5-53 Draw a logarithmic concentration diagram illustrating the effect of chloride concentration on the relative concentrations of various chloride complexes of mercury, assuming $[Hg^{2+}] = 10^{-7}$ mol/l. Which complex predominates when the chloride concentration equals (a) 0.1 mg/l, (b) 1 mg/l, (c) 10 mg/l, (d) 100 mg/l?

    Answer: $HgCl_2$ under all conditions

5-54 Draw a logarithmic concentration diagram illustrating the effect of fluoride concentration on the relative concentrations of various fluoride complexes of aluminum, assuming $[Al^{3+}] = 10^{-4}$ mol/l. Which complex predominates when the fluoride concentration equals (a) 0.1 mg/l, (b) 1 mg/l, (c) 10 mg/l?

5-55 Draw a distribution diagram for the chloride complexes of mercury.

5-56 Draw a distribution diagram for the fluoride complexes of aluminum.

5-57 From a predominance area diagram, determine which species of Hg(II) is likely to predominate (among hydroxide and chloride complexes only) when the chloride concentration is 35 mg/l and pH is (a) 6, (b) 7, (c) 8, and (d) 9.

    Answer: (a) $HgCl_2$ (b) $Hg(OH)_2$ (c) $Hg(OH)_2$ (d) $Hg(OH)_2$

5-58 Draw a predominance area diagram illustrating the effect of pH and fluoride concentration on fluoride and hydroxide complexes of Fe(III). Assume that the Fe(III) concentration is sufficiently low so that precipitation of iron does not occur.

5-59 Draw a logarithmic concentration diagram showing the relationship between $[CO_2]$ and $[CH_4]$ as a function of pE for pH = 7 and $[CO_2] + [CH_4] = 1$ atm.

5-60 Draw a logarithmic concentration diagram showing the relationship between $[SO_4^{2-}]$ and $[H_2S]$ as a function of pE for pH = 7 and $[SO_4^{2-}] + [H_2S] = 10^{-3}$ mol/l.

5-61 Draw a pE-pH diagram illustrating predominant iron forms (Fe^{3+}, Fe^{2+}, Fe) in an aqueous system.

5-62 Draw a pE-pH diagram illustrating predominant manganese forms (Mn^{2+}, MnO_2, MnO_4^-) in an aqueous system.

5-63 From a pE-pH diagram, estimate the pE range for aerobic conditions in an aqueous system at pH equal to (a) 4, (b) 7, and (c) 10.

    Answer: (a) 16.0 to 16.8   (b) 13.0 to 13.8   (c) 10.0 to 10.8

5-64 From a pE-pH diagram, estimate the pE range which is typical for sulfide and methane production in an aqueous system at pH equal to (a) 4, (b) 7, and (c) 10.

REFERENCES

Blackburn, T. R.: "Equilibrium, A Chemistry of Solutions," Holt, Rinehart and Winston, New York, 1969.

Butler, J. N.: "Ionic Equilibrium, A Mathematical Approach," Addison Wesley, Reading, Mass., 1964.

Garrels, R. M., and C. L. Christ: "Solutions, Minerals, and Equilibria," Harper & Row, New York, 1965.

Sillén, L. G.: Graphic Presentation of Equilibrium Data, chap. 8, "Treatise on Analytical Chemistry," vol. 1, I. M. Kolthoff, P. J. Elving, and E. B. Sandell (eds.), The Interscience Encyclopedia, New York, 1959.

Sillén, L. G., and A. E. Martell: "Stability Constants of Metal-Ion Complexes," *spec. publ.* 17, 1964, and *spec. publ.* 25, 1971, Chemical Society, London.

Stumm, W.: "Equilibrium Concepts in Natural Water Systems," American Chemical Society, Washington, D.C., 1967.

Stumm, W., and J. J. Morgan: "Aquatic Chemistry," Wiley-Interscience, New York, 1970.

SIX

BASIC CONCEPTS FROM COLLOID CHEMISTRY

6-1 INTRODUCTION

Colloid chemistry is concerned with dispersions. These may exist in solids, liquids, or gases. Dispersions in solids are of little consequence to environmental engineers; consequently discussion will be directed toward those that occur in liquids or in gases. Eight classes of colloidal dispersions are known, as shown in Table 6-1. Classes 4 through 8 of this table are commonly encountered in environmental engineering practice, and classes 5 through 8 are readily recognized by laymen by their common names.

Table 6-1 Classes of colloidal dispersions

Class	Dispersed phase	Dispersion medium	Common name
1	Solid	Solid	
2	Liquid	Solid	
3	Gas	Solid	
4	Solid	Liquid	
5	Liquid	Liquid	Emulsions
6	Gas	Liquid	Foams
7	Solid	Gas	Smokes
8	Liquid	Gas	Fogs

Size

Colloidal dispersions consist of discrete particles that are separated by the dispersion medium. The particles may be aggregates of atoms, molecules, or mixed materials that are considered larger than individual atoms or molecules but are small enough to possess properties greatly different from coarse dispersions. Colloidal particles normally range in size from about 1 to 100 nanometers (nm) and are not visible even with the aid of the ordinary high-powered microscope. Their relation to other dispersions is as follows:

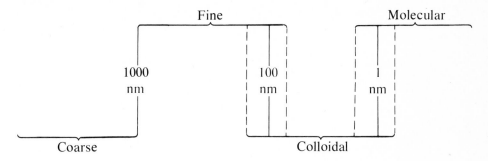

Colloidal dispersions may be considered as ultrafine dispersions and occupy a size range between fine and molecular. The boundaries between fine, colloidal, and molecular dispersions are by no means hard-and-fast values.

Methods of Formation

Any material that is reasonably insoluble in the dispersion medium can be caused to form a colloidal dispersion. Colloidal-sized particles can be produced by grinding coarse materials. Devices designed for such purposes are called *colloid mills*. Colloidal particles are formed in considerable amounts in hard-rock drilling and blasting operations. Colloidal-sized particles may be formed from ions that react to form insoluble compounds. Under the proper conditions, aggregates of molecules result that do not grow into crystals of a size large enough to settle or be filtered out. This often happens in gravimetric analysis, as discussed under crystal growth in Sec. 3-6.

Certain organic substances and compounds that are considered soluble in water do not form true solutions; instead they form colloidal dispersions. Soap, starch, gelatin, agar-agar, gum arabic, and albumin are examples. Bentonite, a volcanic clay, is an example of an inorganic material that acts likewise. In these cases the dispersion medium, water, has the ability to disintegrate the material sufficiently to carry it into colloidal suspension, but it does not necessarily have the ability to complete the dispersion into molecular particles. In certain cases, such as with gelatin, gum arabic, and albumin, the individual molecules may be so large as to fall into the colloidal range, even though dispersion might be complete.

General Properties

Because colloidal particles are so small, their surface area in relation to mass is very great. Some concept of this relation can be obtained by consideration of how the surface area of a cube 1 cm in side length increases when it is reduced to colloidal-sized cubes. If colloidal-sized cubes of 10 nm are formed, the surface area is increased from 6 cm² to 600 m², or about ⅟₇ acre. It is difficult to conceive of such a small mass of material having such a tremendous surface area. As a result of this large area, surface phenomena predominate, and control the behavior of colloidal suspensions, so much so that colloidal chemistry is often considered synonymous with surface chemistry. The mass of colloidal particles is so small that gravitational effects are unimportant.

Electrical Properties All colloidal particles are electrically charged. The charge varies considerably in its magnitude with the nature of the colloidal material and may be positive or negative, as shown in Fig. 6-1. Many colloidal dispersions are dependent upon the electrical charge for their stability. Like charges repel, and as a result, similarly charged colloidal particles cannot come close enough together to agglomerate into larger particles.

The *electrokinetic* properties of colloids are of great importance to environmental engineers, as destruction of many forms depends upon a knowledge of them. More detailed discussion is given in Sec. 6-2.

When colloidal particles are placed in an electrical field, the particles migrate toward the pole of opposite charge. This phenomenon is known as *electrophoresis* and is used extensively to determine the nature of the charge on the colloidal particle and other properties.

Brownian movement Colloidal particles are bombarded by molecules of the dispersion medium, and because of their small mass, the colloids move about under the impetus of the bombardment in a helter-skelter manner. This movement may be observed in some colloidal suspensions with the aid of a high-powered microscope but is seen best with the aid of the ultramicroscope. It was originally thought to be a characteristic of living matter, but a botanist, Robert Brown, showed in 1827 that nonliving material exhibited this same phenome-

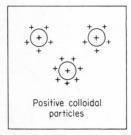

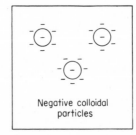

Positive colloidal particles

Negative colloidal particles

Figure 6-1 Positive and negative colloidal particles.

non. The term *Brownian movement* has been used to describe this action, whether the particles are living or inanimate.

Tyndall effect Because colloidal particles have dimensions greater than the average wavelength of white light, they interfere with the passage of light. Light which strikes them may be reflected. As a result, a beam of light passing through a colloidal suspension is visible to an observer who is at or near right angles to the beam of light. This phenomenon is called the *Tyndall effect* in honor of the English physicist who studied it extensively. This test is often used to prove the presence of a colloid, as true solutions and coarse suspensions do not produce the phenomenon. Oftentimes rays of sunlight piercing between clouds are seen when the atmosphere is charged with colloidal dust particles. Students often see the Tyndall effect illustrated in classrooms when chalk or other dust is present in colloidal form in the air. The Tyndall effect is used as one basis of determining turbidity in water.

Adsorption Colloids have tremendous surface area and, of course, great adsorptive powers. Adsorption is normally preferential in nature, some ions being chosen and others excluded. This selective action yields charged particles and is the fundamental basis of the stability of many colloidal dispersions.

Effect on freezing and boiling point Colloidal dispersions affect the freezing and boiling points of liquids as do dispersions of other particles; see Raoult's law (Sec. 2-10). Their effect, however, is not measurable with ordinary instruments, because the actual number of particles is so very few as compared with Avogadro's number. This is so because colloidal suspensions normally fall into the realm of very dilute solutions, and in addition, each particle is made up of perhaps hundreds or even thousands of molecules.

Dialysis Colloids, because of their large particle size, do not pass through ordinary semipermeable membranes. Thus a separation of crystalloids and colloids can be accomplished by dialysis, as discussed in Sec. 3-8.

Nomenclature

The nomenclature applied to colloidal systems varies considerably with the type; consequently there are few terms generally applicable.

6-2 COLLOIDAL DISPERSIONS IN LIQUIDS

Colloidal dispersions of solids, liquids, and gases in liquids are commonly encountered in environmental engineering practice. The nomenclature and behavior of each type differ somewhat; consequently discussions of each will be given.

Solids in Liquids

Colloidal dispersions of solids in liquids are generally of two types, those that bind strongly with the liquid and those that do not. Dispersions binding strongly with the liquid are generally more stable and difficult to separate from the liquid than those that do not. Colloids that bind strongly with water are termed *hydrophilic* (water-loving), and those that do not are termed *hydrophobic* (water-hating). Colloidal dispersions of solids in liquids are often referred to as *sols* or *suspensoids*.

Hydrophobic Colloids Hydrophobic colloids are all electrically charged. The primary or surface charge may be developed in several ways and may be positive or negative. The sign and magnitude of the primary charge is a function of the character of the colloid, and the pH and general ionic characteristics of the water. A lower water pH tends to make colloids more positive or less negative.

In the realm of liquids, the environmental engineer is concerned with the removal of colloidal solids from water and wastewater. These colloids are not always well-defined hydrophobic sols, but they do lend themselves to separation when treatment designed to remove hydrophobic sols is used. The natural coloring matter of surface waters and the colloidally suspended matter of

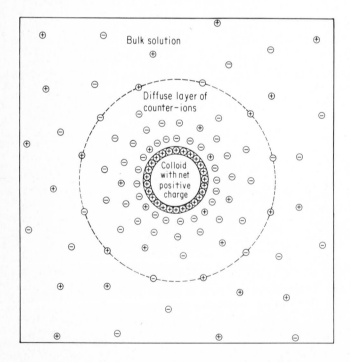

Figure 6-2 A positively charged colloidal particle with diffuse layer of ions of opposite charge.

domestic wastewaters are examples of quasi hydrophobic-hydrophilic, negatively charged colloids encountered in environmental engineering practice.

Electrokinetic properties The stability of hydrophobic colloids depends upon the electrical charge that they possess. This primary charge may result from charged groups within the particle surface or may be gained by adsorption of a layer of ions from the surrounding medium, as illustrated in Fig. 6-2.

A sol considered as a whole cannot have a net charge, so that the charge that a given particle may possess must be counterbalanced by ions of opposite charge in the solution phase. This need for electroneutrality results in an electric double layer at the interface between the solid and water consisting of (1) the charged particle, and (2) an equivalent excess of *"counter-ions"* of appropriate charge which accumulates in the water near the surface of the particle. Although the counter-ions are attracted electrostatically and are thus concentrated in the interfacial region, they are rather loosely held and may diffuse away in response to thermal agitation, to be replaced by other ions. These competing forces (electrical attraction and diffusion) spread the charge over a diffuse layer such that the concentration of counter-ions is greatest at the interface and decreases gradually with distance from the interface. When the water contains a high concentration of ions, the diffuse layer would obviously be compacted. It would thus occupy a smaller volume and would extend less far into solution.

Because of the primary charge on the particle, an electric potential exists between the surface of the particle and the bulk of the solution. The charge is a maximum at the particle surface and decreases with distance from the surface. When two similar colloidal particles with similar primary charge approach each other, their diffuse layers begin to interact. As they come closer, the similar primary charges they possess result in repulsive forces. The closer the particles approach, the stronger the repulsive forces.

The above repulsive forces which keep particles from aggregating are counteracted to some degree by an attractive force termed *van der Waals' force*. All colloidal particles possess this attractive force regardless of charge and composition. The magnitude of this force is a function of the composition and density of the colloid, but is independent of the composition of the aqueous phase. The van der Waals' force decreases rapidly with increasing distance between the particles.

Figure 6-3 illustrates the effect of separating distance between two particles on the net force that exists between them. As two similar particles approach each other, the repulsive electrostatic forces increase to keep them apart. However, if they can be brought sufficiently close together to get past this energy barrier, the attractive van der Waals' force will predominate, and the particles will remain together. If it is desired to coagulate colloidal particles, then they must be given sufficient kinetic energy to overcome the energy barrier that exists, or else the energy barrier must be lowered by some means.

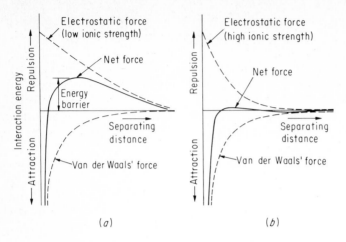

Figure 6-3 The effect of liquid ionic strength and separating distance between colloids on the forces of interaction between them.

The surface potential or primary charge on a colloid cannot be measured directly. However, a value can be computed from measurements of particle movement within an electrical field (electrophoretic mobility) which approximates, but is generally lower than, the true surface potential. This value is termed the *zeta potential*, ζ, and is defined by the equation

$$\zeta = \frac{4\pi\delta q}{D} \tag{6-1}$$

where q is the charge on the particle, δ, is the thickness of the zone of influence of the charge on the particle, and D is the dielectric constant of the liquid. Measurements of zeta potential can give an indication of the effectiveness of added electrolytes in lowering the energy barrier between colloids, and thus can serve to guide the selection of optimum conditions for coagulation.

Destruction Colloidal particles are too small to be removed by gravitational settling alone. However, if the colloids are destabilized or destroyed by causing them to aggregate or coagulate into larger particles, they can be effectively removed. There are four basic mechanisms by which colloids can be coagulated: (1) double-layer compression, (2) charge neutralization, (3) entrapment in a precipitate, and (4) interparticle bridging.

If a high concentration of an electrolyte is added to a sol, the concentration of ions within the diffuse layer will increase and hence the thickness of this layer will decrease. The addition, of counter-ions with higher charge, such as Ca^{2+} instead of Na^+, will have a similar effect. This will result in a greater decrease in charge with distance from the particle interface, resulting in a decrease or perhaps an elimination of the potential barrier, as illustrated in Fig. 6-3. With a reduced or eliminated energy barrier, particles can come together and aggregate.

The charge on a colloid can sometimes be neutralized by addition of molecules of opposite charge which have the ability to adsorb onto the colloid. For example, positively charged organic molecules such as dodecylammonium ion, $C_{12}H_{25}NH_3^+$, tend to be hydrophobic and readily adsorb to negatively charged turbidity particles. The opposite charges of the organic and the turbidity particle cancel each other out, and coagulation results. An overdose of dodecylammonium ion, however, can result in charge reversal and the formation of a stable but positively charged particle.

When sufficient metal salts of Al(III) and Fe(III) are added to solution, they may combine with OH^- to rapidly form hydroxide precipitates. Colloidal particles may provide condensation sites where the precipitates may form and hence the turbidity becomes entrapped in the precipitate and settles with it. The settling precipitates can also entrap colloids which it passes, bringing them down.

Finally, long-chained charged synthetic and natural polymers (*polyelectrolytes*) can act to destabilize colloids by forming a bridge between one colloid and another. One charge site on the long polymer can attach or adsorb to a site on one colloid, while the remainder of the polymer molecule extends out into solution. If the extended portion becomes attached to another colloid, then the two colloids are effectively tied together. Commonly, a negatively charged polymer is most effective in bridging between negative colloids. This is believed to result from interaction between the polymer and specific sites on the colloid, some of which may be positive even though the overall charge of the colloid is negative.

There are essentially four different ways in which the above mechanisms may be applied in the destruction of colloids. Only two are applied to a significant extent in environmental engineering practice. The four methods are (1) boiling, (2) freezing, (3) addition of electrolytes, and (4) mutual precipitation by addition of a colloid of opposite charge.

Boiling of a hydrophobic colloidal suspension often results in coagulation of the colloidal particles. This action is not usually attributed to a reduction in the surface potential but rather to modification in the degree of hydration of the particles, or to increased kinetic velocities, permitting the particles to overcome the energy barrier which separates them (Fig. 6-3). Chemists often boil materials to accomplish coagulation of colloids, but boiling is generally too expensive for application in environmental engineering practice.

Freezing is another method by which colloids may be coagulated. During the freezing process, crystals of relatively pure water form. Thus the colloidal and crystalloidal materials are forced into a more and more concentrated condition. Two additive effects cause coagulation to occur. As the colloidal suspension becomes more concentrated, opportunity for close contact increases. At the same time, the concentration of electrolytes increases, resulting in a decrease in the diffuse layer thickness. The net result is coagulation of the colloid.

Freezing has been proposed as a practical means of destroying the colloidal character of sludges in preparation for dewatering. However, except perhaps in cold climates it is usually less costly to condition sludges for filtration by the use of chemicals.

The common method of destroying hydrophobic colloids is by the addition of electrolytes. One way in which electroytes act is by double-layer compression. A sufficient concentration of monovalent ions, such as NaCl, can bring about coagulation in this way. However, it has been noted that salts having divalent ions of charge opposite to that of the colloidal particle exert far greater coagulation powers. Salts having trivalent ions of opposite charge are even more effective. The significance of the relation between ionic charge and precipitating power was first pointed out by Schulze and verified by Hardy. Their findings are usually called the *Schulze-Hardy rule,* which states: *The precipitation of a colloid is effected by that ion of an added electrolyte which has a charge opposite in sign to that of the colloidal particle, and the effect of such ion increases markedly with the number of charges it carries.*

Table 6-2 lists a number of electrolytes and gives their relative coagulating powers for positive and negative colloids. From the data given in Table 6-2 it becomes obvious why aluminum and iron salts are so widely used as coagulants in environmental engineering practice. The greater value of the sulfates as compared with the chlorides is not readily apparent but will be discussed shortly. Multivalent ions of opposite charge are considered able to penetrate the diffuse layer of colloidal particles and in this way neutralize, in part, the charge on the colloid.

The trivalent salts of iron and aluminum used in coagulation of water to remove colloidal color and turbidity are considered to function as coagulants by other mechanisms as well. These salts, when added to water, do ionize to yield trivalent metallic ions, the amount and life of which are a function of the pH of

Table 6-2 Relative coagulating power of several electrolytes

	Relative power* of coagulation	
Electrolyte	Positive colloids	Negative colloids
$NaCl$	1	1
Na_2SO_4	30	1
Na_3PO_4	1000	1
$BaCl_2$	1	30
$MgSO_4$	30	30
$AlCl_3$	1	1000
$Al_2(SO_4)_3$	30	>1000
$FeCl_3$	1	1000
$Fe_2(SO_4)_3$	30	>1000

* Values given are approximate and are for solutions of equivalent ionic strength.

the water. Some of the trivalent ions undoubtedly reach the target and neutralize the charge on some of the colloidal particles. The majority of the trivalent ions, however, unite with the available hydroxyl ions to give polynuclear colloidal metallic hydroxides which generally carry a positive charge and may adsorb to the colloid, causing charge neutralization. The amount of the positively charged metallic hydroxide is normally much in excess of the amount needed to react with any negatively charged color or turbidity particles that may have escaped neutralization by trivalent metallic ions. The excess colloidal metallic hydroxides must be coagulated in some manner. This is where the negative ion of the metallic salt used may be of importance. If sulfates are used instead of chlorides, the divalent sulfate ions act to compress the diffuse layer of the metal colloid and thereby help to complete the coagulation of the colloidal system. The enmeshment of the original colloid in the metal salt while coagulating and settling serves to bring about additional colloid removal. Addition of polyelectrolytes alone or in conjunction with metallic electrolytes can enhance the overall coagulation process by mechanisms already discussed.

Mutual precipitation occurs when colloids of opposite charge are mixed. If they are added in essentially equivalent amounts, in terms of electrostatic charge, coagulation occurs and is quite complete. This method is not used per se in environmental engineering practice because of the large volumes of water that would be needed to carry the second colloid and because of the relatively long time required for flocculation of the colloidal dispersions. Positively charged colloids formed when trivalent salts of iron and aluminum are added to water, however, may to some extent act in this manner to remove negatively charged colloids.

Hydrophilic colloids A wide variety of hydrophilic colloidal materials are known. Most of them are products of plant or animal life and therefore of considerable concern to environmental engineers. Soap, soluble starch, soluble proteins, protein degradation products, blood serum, agar-agar, gum arabic, pectins, and synthetic detergents are examples. These materials occur in domestic wastewater and in many industrial wastes. Soap, however, is usually precipitated by calcium and magnesium ions and does not often occur as a colloidal suspension except in laundry wastes.

The hydrophilic colloids are readily dispersed in water, and their stability depends upon their love for the solvent rather than upon the slight charge (usually negative) that they possess. This property makes it difficult to remove them from aqueous suspension. Certain of them, such as the proteins and protein degradation products, form heavy metal salts that are insoluble; thus their removal is effected by aluminum and ferric salts. Proteins, proteoses, peptones, polypeptides, and amino acids have a minimum solubility at their isoelectric point. The isoelectric point varies from about pH 4.0 to 6.5 for the majority; consequently it is usually uneconomical to attempt their removal by pH adjustment in wastes that are well buffered with bicarbonates.

Most hydrophilic colloids serve in a protective capacity for hydrophobic colloids. When acting in this capacity they are called *protective colloids*. It is believed that they envelop the hydrophobic colloid in a manner to shield it from the action of electrolytes. Coagulation of such systems requires rather drastic treatment with massive doses of coagulant salts, often ten to twenty times the amount used in conventional water treatment. Information concerning the action of coagulants on many hydrophilic colloids is lacking. Further research is needed.

Liquid-in-Liquid Systems

Colloidal systems involving the dispersion of one liquid in another are known as *emulsions*. Obviously the two liquids must be immiscible in each other. The emulsions of interest to environmental engineers usually are composed of oil and water. The oil may be dispersed in the water (the usual case), or water may be dispersed in oil. Most emulsions depend upon a third component, an *emulsifying agent,* for their stability.

Soap and synthetic detergents are excellent emulsifying agents, as would be suspected from their common use for laundering, dishwashing, and other cleaning purposes. Many natural materials such as proteins, protein degradation products, egg yolk, lanolin, saponin, and gum arabic act as emulsifying agents. Egg yolk serves as the emulsifying agent in salad dressings, as in mayonnaise, for example. Kaolin, fuller's earth, colloidal clay, and lampblack have been found to act as good emulsifiers of mineral oil in water.

Water-in-oil emulsions Water-in-oil emulsions are quite common in the petroleum industry. They can be readily broken by heating, and such treatment is practical because of the value of the oil that can be recovered.

Oil-in-water emulsions Oil-in-water emulsions are usually milky white in appearance, and their destruction depends upon treatment to inactivate the emulsifying agent. This inactivation may be accomplished in a variety of ways, but economics usually dictates the use of chemical coagulating agents such as aluminum, ferric, and calcium salts.

Gas-in-Liquid Systems

Dispersions of gas bubbles in liquids are considered to be colloidal in character regardless of the bubble size; therefore foams fall in this category. They are of concern in the treatment of certain industrial wastes such as those from the wood-pulping, textile, and meat-packing industries. Foams are sometimes a problem in anaerobic treatment, especially under unbalanced conditions.

Foams are normally stabilized by hydrophilic colloidal materials that are highly surface-active and tend to concentrate at air-water interfaces. Foams have been aptly described as a collection of interfaces separated by air bubbles.

Destruction is usually accomplished in either of two ways. Water sprays are used to break the foam by dilution and mechanical action, or antifoaming materials may be added. The antifoaming material, to be effective, must lower the surface tension more than the hydrophilic colloid. It will then displace the colloid and cause the foam to collapse.

6-3 COLLOIDAL DISPERSIONS IN AIR

The environmental engineer is becoming more and more involved in air pollution problems. The control of "smog" and of smoke are two important phases of air pollution that have colloidal aspects.

Fog and Smog

Fog consists of a colloidal dispersion of a liquid in air. The environmental engineer has no special interest in ordinary fog. However, in areas where atmospheric inversions are common photochemical reactions may occur in the polluted atmosphere near the ground to produce an artificial fog. These fogs are commonly referred to as *smog*. At the present time, smog is believed to be formed by the reaction between olefinic hydrocarbons and nitrogen dioxide in the presence of sunlight. Ozone is formed in the reaction, and a nitro derivative of the olefin is believed to be formed. The latter is suspected of condensing water from the atmosphere to produce the colloidal dispersion called smog. As with fog, the amounts involved are so tremendous that destruction is out of the question. Control appears to rest on limitation of the olefin-type compounds allowed to reach the atmosphere. Automobile exhaust gases are considered the major source.

Smoke

Dispersions of solid matter in air are called *smokes*. They may originate from a wide variety of industrial operations, the principal one being the power industry. The colloidal particles in smoke are charged, and their removal can be accomplished by passage through electrostatic precipitators.

PROBLEMS

6-1 Define the following: sol, hydrophilic, emulsion, protective colloid, smog, smoke, Brownian movement, and Tyndall effect.

6-2 What gives stability to hydrophobic colloids in water?

6-3 What factors affect the primary charge on a colloid?

6-4 By what mechanism does addition of an electrolyte such as NaCl bring about destabilization of colloids in water?

6-5 By what different mechanisms can Fe(III) and Al(III) bring about colloid destabilization?

6-6 Explain the possible mechanism by which the addition of small quantities of some materials results in destabilization of sols, but addition of larger quantities of the same material does not.

6-7 List the mechanisms which may be involved in destabilization of sols by (1) boiling, (2) freezing, and (3) addition of electrolytes.

REFERENCES

Adamson, A. W.: "Physical Chemistry of Surfaces," 2d ed., Interscience, New York, 1967.
Gregg, S. J.: "The Surface Chemistry of Solids," Chapman, London, 1961.
Mysels, K. J.: "Introduction to Colloid Chemistry," Interscience, New York, 1959.
Stumm, W., and J. J. Morgan: "Aquatic Chemistry," Wiley-Interscience, New York, 1970.

SEVEN

BASIC CONCEPTS FROM BIOCHEMISTRY

7-1 INTRODUCTION

Many environmental engineers can be classed as biological engineers because they spend considerable time and effort in designing and operating treatment facilities that utilize living organisms to bring about the destruction or transformation of waste organic and inorganic materials. Therefore an important facet of their training is biochemistry.

Biochemistry deals with chemical changes that are brought about by living organisms. The reactions may be *extracellular* or *intracellular*. Hydrolytic reactions (splitting by water) are generally extracellular in character and necessarily so, because such reactions are often required to reduce the complexity of organic compounds to a point at which they can dialyze through the cell wall. The energy requirement for hydrolytic reactions is considered nil. Oxidative reactions occur intracellularly and produce energy in accordance with the free energy of the particular reaction involved.

Biochemical reactions occur at temperatures with a normal range from about 0 to 60°C. Organisms that thrive at 0 to 10, 10 to 40, and above 40° are classed as *psychrophilic, mesophilic,* and *thermophilic,* respectively. The majority of the chemical reactions that these organisms bring about occur at far lower temperatures than would be needed in their absence. For this reason, catalysts that lower markedly the activation energy of the reactions are required. The catalysts are supplied by the living organisms as part of their life processes and are known as *enzymes.* They serve to initiate the reactions and also to control their speed in a manner that serves the best interests of the particular organism. This mechanism sometimes leads to an interference between groups of organisms when operating in mixed culture.

7-2 ENZYMES

Enzymes have been defined as temperature-sensitive catalysts of organic nature, elaborated by living cells and capable of action outside or inside the cell. Certain of the enzymes are secreted by the cell and are known as *extracellular* enzymes. Others are associated with the protoplasm of the cell and perform their function within the cell, and so they are called *intracellular* enzymes.

Enzymes are proteinaceous in character. Some are simple proteins,

Table 7-1 Classification of enzymes

Enzyme	Substrate	Products
Hydrolytic:		
1. Carbohydrases:		
a. Glycosidases (sugar splitters):		
Sucrase	Sucrose	Glucose + fructose
Maltase	Maltose	Glucose
Lactase	Lactose	Glucose + galactose
b. Amylases (starch splitters):		
Diastase ⎱		
Ptyalin ⎰	Starch	Maltose
c. Cellulase	Cellulose	Cellobiose
2. Esterases:		
a. Lipases:		
Lipase	Glycerides	Glycerol + fatty acids
b. Phosphatases	Phosphoric esters	H_3PO_4 + alcohols
3. Proteases:		
a. Proteinases:		
Pepsin ⎱		
Trypsin ⎰	Proteins	Polypeptides
b. Peptidases	Polypeptides	Amino acids
4. Amidases:		
a. Urease	Urea	$NH_3 + CO_2$
5. Deaminases	Amino acids	NH_3 + organic acids
Desmolytic or respiratory:*		
1. Oxido-reductases (oxidation-reduction reactions)		
a. Dehydrogenases		
b. Oxidases		
2. Transferases (transfer of functional groups)		
a. Transaminases		
b. ATP		
3. Lyases (addition to double bonds)		
4. Isomerases (isomerization reactions)		
5. Ligases (formation of bonds by ATP)		

* Action of desmolytic enzymes is beyond the scope of this book. Consult any standard text on biochemistry.

whereas others are of a complex conjugated type. They are highly specific for the reactions that they catalyze. Enzymes are grouped into two major classes, depending on the nature of the reaction that they control. Those that catalyze hydrolytic reactions are known as *hydrolases,* and those that catalyze the rupture of linkages that are not hydrolyzable are known as *desmolases* or *respiratory enzymes.* The enzymes involved in oxidation–reduction reactions are of the latter type. In general, hydrolases are extracellular and desmolases intracellular. The *-ase* ending is used to designate enzymes. A classification of enzymes according to their function is given in Table 7-1.

A great number of enzyme and bacterial preparations with a wide variety of trade names (Sea Chem, Bionetic, Enzymatic, Septic Aid, and so on) have been marketed as agents capable of solving problems related to the operation of septic tanks, sludge digestion units, and other treatment facilities. Research has indicated that domestic wastewater and sludges contain bacteria capable of producing the necessary enzyme systems in adequate amounts and that no beneficial effects can be demonstrated by the addition of enzymes from outside sources. In the case of malfunctioning units, reports from field studies (usually conducted without controls) have often been favorable. Laboratory studies, however, have failed to show any beneficial effects when the materials were used in amounts recommended in practice.[1]

7-3 COFACTORS

Some enzymes depend for their activity only upon their structure as proteins. Others require in addition nonprotein structures, or *cofactors.* The cofactor may be a metal ion, or it may be a complex organic molecule called a *coenzyme.* Sometimes both are required. Metal ions which sometimes serve as cofactors are zinc, magnesium, manganese, iron, copper, potassium, and sodium. The organic structures of coenzymes are generally heat-stable whereas most enzyme proteins are not. Cofactors may be bound tightly or loosely to the enzyme protein.

Approximately 12 coenzymes are known. Some of the most important are as follows:

NAD. Nicotinamide adenine dinucleotide, an enzyme that transfers hydrogen atoms or electrons and participates in the oxidation of organic materials.
NADP. Nicotinamide adenine dinucleotide phosphate, which is similar in structure and function to NAD, but has three phosphorus atoms in the structure instead of two.
Coenzyme A. A derivative of pantothenic acid, it functions in fatty-acid metabolism (oxidation) and synthesis.

[1] E. A. Pearson, et al., Biocatalytic Additives in Sludge Digestion, *Sewage and Ind. Wastes,* **29:** 1066 (1957).

Flavoproteins. Flavin mononucleotide and flavin adenine dinucleotide are known collectively as flavoproteins. They are important in the transport of hydrogen or electrons from metabolites to oxygen.

7-4 TEMPERATURE RELATIONSHIPS

Biochemical reactions, in general, follow the van't Hoff rule of a doubling of reaction rate for a 10°C increase in temperature, over a restricted temperature range. Studies with activated sludge have shown the reaction rate to be more than doubled for a 10°C rise in temperature, as shown in Fig. 7-1.[2]

In biochemistry, temperature relationships are often referred to as Q_{10} values, which are the ratio of the reaction rate at a particular temperature to the rate at 10°C lower. Table 7-2 shows Q_{10} values for aerobic processes involving activated sludge[2] and conventional anaerobic digestion.[3]

The data in Table 7-2 illustrate that the influence of temperature varies considerably, depending upon the range of temperature and also the nature of the reaction involved. A great deal of study has been given to the effect of temperature on the rate of enzyme-induced reactions. Since all biological reactions are dependent upon enzymes, the data are of considerable interest to environmental engineering. Table 7-3 shows the effect of temperature on a number of enzymes, and Fig. 7-2 shows the activity of amylase at several temperatures when hydrolyzing starch.

The data in Fig. 7-2 illustrate the point that increasing temperature has a favorable effect upon biochemical reactions, within limits. As the temperature

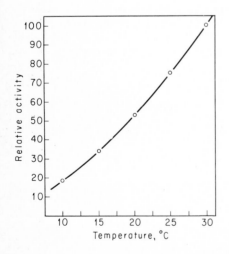

Figure 7-1 Effect of temperature on activity of activated sludge as measured by oxygen requirements per unit time.

[2] C. N. Sawyer and G. A. Rohlich, *Sewage Works J.,* **11:** 946 (1939).
[3] G. M. Fair and E. W. Moore, *Sewage Works J.,* **6:** 3 (1934).

Table 7-2 Effect of temperature on aerobic and anaerobic biochemical processes

Biological system	Temperature, °C	Q_{10}
Activated sludge	10–20	2.85
	15–25	2.22
	20–30	1.89
Anaerobic sludge	10–20	1.67
	15–25	1.73
	20–30	1.67
	25–35	1.48
	30–40	1.0

is increased, eventually a point is reached at which the enzyme becomes less active. This change is considered to be due to *denaturation* of the enzyme. In systems containing living organisms, the adverse effects of high temperature may be explained by considering that the enzymes are denaturized or that the ability of the organisms to produce enzymes has been destroyed. The net effect is the same in either case.

The importance of temperature as a factor in determining the rate of biological reactions has been recognized by environmental engineers for some time in connection with anaerobic sludge digestion. At the present time, design standards for trickling filters recognize temperature as a factor, since accepted design loadings in the United States vary with the latitude. The significance of temperature in activated sludge treatment has not been given direct recognition. This effect is a complicated matter that does not lend itself to simple mathematical treatment. It involves three major factors all of which are temperature-

Table 7-3 Effect of temperature on the activity of certain enzymes

Enzyme	Temperature, °C	Q_{10}
Amylase	10–20	1.34
	15–25	1.59
	20–30	1.44
	25–35	1.27
	30–40	1.17
Pepsin	0–10	2.60
	10–20	2.00
	20–30	1.80
	30–40	1.60
Steapsin	0–10	1.50
	10–20	1.34
	20–30	1.26

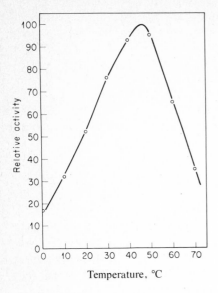

Figure 7-2 Effect of temperature on the action of malt amylase when hydrolyzing starch to glucose.

dependent: (1) required detention time, (2) oxygen requirements per unit volume per unit time, and (3) variation in solubility of oxygen with temperature.

7-5 pH

Hydrogen-ion concentration is one of the most important factors that influence the speed of biochemical reactions, and since such reactions are induced and controlled by enzymes, it is necessary to have a knowledge of how pH affects enzyme activity. The range of pH through which a particular enzyme can act effectively is usually quite narrow. Some enzymes act best at low pH levels;

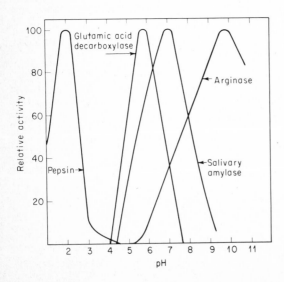

Figure 7-3 Effect of pH on enzyme activity, illustrating the rather narrow range of optimum action.

others require high pH; but the majority are most effective in neutral solutions. The optimum pH and the effective range of pH for a few enzymes are given in Fig. 7-3.

Most biological processes employed in environmental engineering practice involve the use of soil organisms operating in mixed culture. The enzyme systems of these organisms are adapted to operating in essentially neutral solutions; therefore it is important that pH be controlled over a rather narrow range of about 6 to 9. Control of pH is best accomplished by means of a buffer system, as discussed in Sec. 5-6.

7-6 MAJOR AND TRACE ELEMENTS

Bacteria are among the simplest forms of living matter, and like all living matter, they must reproduce in order to survive. Many bacteria can thrive upon very simple substrates, such as sugar, provided that certain essential elements are present in the form of ions or inorganic salts. Bacterial cell tissue has an empirical formula which approximates $C_5H_7NO_2$, corresponding to about 12 percent nitrogen.

It is known that bacteria are capable of synthesizing protein from a wide variety of carbonaceous materials. If the species is to be preserved, sufficient nitrogen must be present to produce daughter cells exactly like the parent cells. This growth would require, on an average, enough nitrogen to produce cells with about 12 percent nitrogen; therefore nitrogen is considered a major nutrient element in bacterial nutrition. Phosphorus and sulfur are other elements essential to the formation of some conjugated proteins and are considered major nutrient elements.

Certain other elements are needed in trace amounts for cell metabolism. Many are known to be important in enzyme function or in other physiological capacity. Calcium, cobalt, copper, iron, magnesium, manganese, potassium, selenium, and zinc are probably essential for most bacteria. Other elements are necessary for certain bacteria. Molybdenum, for example, is required by nitrogen-fixing bacteria.

Environmental engineers may safely assume that domestic wastewater will provide all the major and trace elements needed in its stabilization by bacteria. This may or may not be the case with industrial wastes. Many industrial wastes are deficient in the major nutrient elements nitrogen and phosphorus. Some may be deficient in sulfur and in trace elements, depending upon the nature of the carriage water.

7-7 BIOCHEMISTRY OF CARBOHYDRATES

The primary function of carbohydrate matter in higher animals is to serve as a source of energy. With microscopic organisms, however, the differentiation of foods for particular purposes is not a rigid matter. Bacteria, for example, utilize carbohydrate matter for the synthesis of fats and proteins as well as for energy. In addition, of course, carbohydrate is also used in building cell tissue and may be stored as polysaccharide inside or outside the cell wall.

The mechanisms by which bacteria and other microorganisms transform carbohydrates are believed to be essentially the same as those occurring in plants and animals. The first stage in carbohydrate metabolism involves hydrolysis (see Fig. 4-1). This degradation must progress to at least the disaccharide stage before transfer through the cell wall can occur. Once within the cell wall, the simple sugars are used for energy or synthesis.

The pathway by which bacteria metabolize simple sugars for energy depends upon whether the conditions are aerobic or anaerobic. In either case, the initial conversion is to pyruvic acid as follows for a simple hexose:

$$\underset{\text{Hexose}}{C_6H_{12}O_6} \xrightarrow[\text{enzymes}]{\text{bacteria}} \underset{\text{Pyruvic acid}}{2\,CH_3\overset{\displaystyle O}{\overset{\|}{C}}COOH + 4H} \qquad (7\text{-}1)$$

Under aerobic conditions, the pyruvic acid and hydrogen are oxidized to carbon dioxide and water for energy:

$$2\,CH_3\overset{\displaystyle O}{\overset{\|}{C}}COOH + 4H + 6O_2 \xrightarrow[\text{enzymes}]{\text{bacteria}} 6CO_2 + 6H_2O \qquad (7\text{-}2)$$

Under anaerobic conditions, however, this oxidation is not possible, and so the bacteria discharge various fermentation end products back into solution. The nature of the end products depends upon the bacterial species involved, as well as the environmental conditions imposed. A few possibilities are indicated below:

$$2\,CH_3\overset{\displaystyle O}{\overset{\|}{C}}COOH + 4H \underset{\underset{\text{enzymes}}{\text{bacteria}}}{\overset{\overset{\text{bacteria}}{\text{enzymes}}}{\rightleftharpoons}}
\begin{cases}
2\,CH_3\overset{\displaystyle OH}{\overset{|}{C}}HCOOH \\[2ex]
CH_3CH_2COOH + CH_3COOH + HCOOH \qquad (7\text{-}3) \\[2ex]
2\,CH_3CH_2OH + 2CO_2
\end{cases}$$

Other acids, alcohols, and ketones may also be formed under anaerobic conditions.

The large quantities of organic acids formed from anaerobic carbohydrate metabolism can overtax the buffering capacity of a waste, resulting in a low pH and a cessation of biological activity. This is a distinct possibility even with normally aerobic systems, which may suddenly be overloaded with a carbohydrate waste, resulting in temporary buildup of acid intermediates and a resulting drop in pH.

7-8 BIOCHEMISTRY OF PROTEINS

Proteins are essential in the diets of higher animals and are used to build and repair muscle tissue. Amounts in excess of these requirements may be consumed for energy or converted to carbohydrates and fats. Saprophytic[4] bacteria are much less demanding in their protein requirements. Most of them are capable of synthesizing protein from inorganic nitrogen and non-protein-containing organics, such as carbohydrates, fats, and alcohols. This ability is most fortunate, for many industrial wastes, including some from the food industry, have very low protein content. It would add greatly to the cost of biological treatment if proteinaceous matter had to be added.

The first step in biological utilization of proteins involves their hydrolysis, which progresses in steps as shown in Sec. 4-22. It is fairly well established that hydrolysis must yield α-amino acids before passage through the cell wall is possible. Within the cell, *deamination* of the amino acids occurs. The nature of the deamination reactions varies under aerobic and anaerobic conditions.

Deamination under Aerobic Conditions

Bacteria deaminize amino acids under aerobic conditions to produce saturated acids with one less carbon atom,

$$R-\underset{\underset{H}{|}}{\overset{\overset{NH_2}{|}}{C}}-COOH + O_2 \xrightarrow[\text{enzymes}]{\text{bacteria}} R-COOH + CO_2 + NH_3 \qquad (7\text{-}4)$$

or hydroxy acids with the same number of carbon atoms,

$$R-\underset{\underset{H}{|}}{\overset{\overset{NH_2}{|}}{C}}-COOH + H_2O \xrightarrow[\text{enzymes}]{\text{bacteria}} R-\underset{\underset{H}{|}}{\overset{\overset{OH}{|}}{C}}-COOH + NH_3 \qquad (7\text{-}5)$$

Deamination under Anaerobic Conditions

Bacterial deamination under anaerobic conditions may proceed with or without reduction to form the corresponding saturated or unsaturated acids.

$$R-\underset{\underset{H}{|}}{\overset{\overset{NH_2}{|}}{C}}-COOH + H_2 \xrightarrow[\text{enzymes}]{\text{bacteria}} R-CH_2COOH + NH_3 \qquad (7\text{-}6)$$

$$R-CH_2-\underset{\underset{H}{|}}{\overset{\overset{NH_2}{|}}{C}}-COOH \xrightarrow[\text{enzymes}]{\text{bacteria}} R-CH{=}CH-COOH + NH_3 \qquad (7\text{-}7)$$

[4] From the Greek words *sapros* meaning death or decay and *phytos* (plant).

The acids formed under aerobic or anaerobic conditions submit to further oxidation, as discussed in Secs. 7-9 and 7-10.

7-9 BIOCHEMISTRY OF FATS AND OILS

The degradation or assimilation of fatty materials is often restricted because of their relative insolubility. This difficulty is overcome in animals by emulsifying agents (bile salts) that are contained in bile secreted by the liver. It is highly improbable that bacteria are capable of secreting similar emulsifying agents. In any event, it is known that segregation of fatty materials is a serious problem in anaerobic sludge digestion units. Because of their low specific gravity, the fatty materials tend to float and complicate scum conditions. In the scum layer, these fatty materials may be rather remote from the bacteria that are capable of utilizing them.

The biological degradation of fatty materials in its initial phases is known to progress along similar lines under aerobic and anaerobic conditions. The first step is hydrolysis, with the production of glycerol and fatty acids, as discussed in Sec. 4-21. The free fatty acids derived from the hydrolysis of fatty materials, the deaminization of amino acids, and carbohydrate fermentation undergo further breakdown by oxidation. Oxidation is believed to occur at the beta carbon atom in accordance with *Knoop's theory*, sometimes called the *beta-oxidation theory*. According to this theory, oxidation proceeds in a series of steps. Coenzyme A is known to be active in these transformations. Oxidation is accomplished by enzymatic hydrogen removal and water addition:

$$R{-}\underset{\underset{H}{|}}{\overset{\overset{H}{|}}{C}}{-}\underset{\underset{H}{|}}{\overset{\overset{H}{|}}{C}}{-}COOH - 2H + H_2O \xrightarrow[\text{enzymes}]{\text{bacteria}} R{-}\underset{\underset{H}{|}}{\overset{\overset{H}{|}}{C}}{-}\underset{\underset{H}{|}}{\overset{\overset{O}{|}}{C}}{-}COOH \qquad (7\text{-}8)$$

$$R{-}\underset{\underset{H}{|}}{\overset{\overset{O}{|}\,\overset{H}{|}}{C}}{-}\overset{\overset{H}{|}}{\underset{\underset{H}{|}}{C}}{-}COOH - 2H \xrightarrow[\text{enzymes}]{\text{bacteria}} R{-}\overset{\overset{O}{\|}}{C}{-}\underset{\underset{H}{|}}{\overset{\overset{H}{|}}{C}}{-}COOH \qquad (7\text{-}9)$$

$$R{-}\overset{\overset{O}{\|}}{C}{-}\underset{\underset{H}{|}}{\overset{\overset{H}{|}}{C}}{-}COOH + H_2O \xrightarrow[\text{enzymes}]{\text{bacteria}} R{-}COOH + CH_3COOH \qquad (7\text{-}10)$$

In the final step, rupture of the molecule occurs, with formation of one molecule of acetic acid, and the original molecule of acid appears as a new acid with two less carbon atoms. Thus, by successive oxidations at the beta carbon atom, long-chain fatty acids are "whittled" into fragments consisting of acetic acid. During this oxidation four hydrogen atoms are removed for each acetic acid unit produced. Under aerobic conditions, the hydrogen is oxidized to water,

$$4H + O_2 \rightarrow 2H_2O \qquad (7\text{-}11)$$

Under anaerobic conditions, however, it is not possible for the bacteria to rid themselves of the hydrogen in this fashion, and another scheme must be used. Methods by which this may be accomplished are indicated in Sec. 7-10.

7-10 GENERAL BIOCHEMICAL PATHWAYS

Sections 7-7 to 7-9 have indicated pathways used by microorganisms for the oxidation of carbohydrates, proteins, and fats in order to obtain energy for their life processes. Oxidation involves the transfer of electrons from a reduced substance termed the *electron donor* to an oxidizing material termed the *electron acceptor*. Generally we think of the electron donor as being the "food" for the organism. Organic matter is generally used as food by bacteria and fungi as well as for animals. However, with some bacteria, reduced inorganic materials such as ammonia, sulfide, molecular hydrogen, and ferrous iron may also serve as electron donors and thus as energy sources. Bacteria that oxidize organic matter for energy are termed *heterotrophic* and those that oxidize inorganic matter are termed *autotrophic*.

Energy is transferred from the electron donor to the organism for synthesis and maintenance by a complex series of enzymatic reactions. A key compound in this energy transfer is the nucleotide, adenosine diphosphate (ADP) which has the following structure:

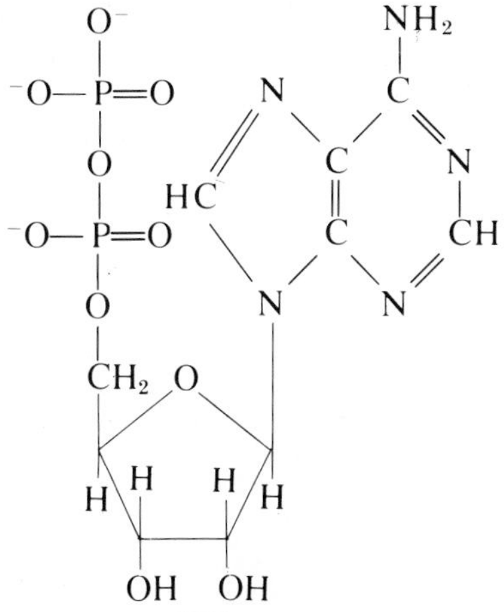

ADP

ADP uses the energy released from oxidation to form a bond with phosphate to form another nucleotide, adenosine triphosphate (ATP):

$$ADP + phosphate + energy \rightarrow ATP + H_2O \qquad (7\text{-}12)$$

The ATP so formed can travel through the cell and give up its energy for cell synthesis or maintenance by the reverse of Eq. (7-12). Transfer of energy to cell synthesis is depicted in Fig. 7-4.

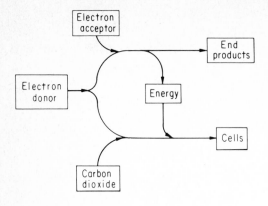

Figure 7-4 Schematic diagram of biological oxidation of an electron donor for energy and transfer of the energy for cell synthesis. Either an organic electron donor or carbon dioxide may provide the cellular need for carbon.

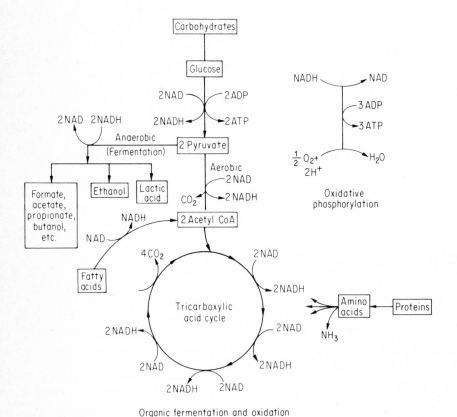

Figure 7-5 Generalized pathways for aerobic or anaerobic fermentation and oxidation of organics and transfer of energy released through NAD to form ATP.

A greatly simplified and general scheme for oxidation and fermentation of organic materials and transfer of energy to ATP is illustrated in Fig. 7-5. Carbohydrates, proteins, and fats enter the overall biochemical scheme at different locations. Not all bacteria have the capability to oxidize all organic compounds, but most are able at least to carry out some portion of the overall conversions illustrated. Organic oxidation can be thought of generally as removal of electrons (or hydrogen atoms) from organic molecules with the aid of the coenzyme NAD, which in turn is reduced and converted to NADH. In oxidative phosphorylation the electrons are transferred from NADH through a series of coenzymes and ultimately to the final electron acceptor, which is oxygen in an aerobic system. During these transfers, a portion of the energy available is transferred to ADP as previously indicated. Another portion is lost as heat because of inefficiencies in energy transfer. Some ATP may be formed directly from organic transformations such as during the conversion of glucose to pyruvate.

In aerobic systems, oxygen is the terminal electron acceptor and is reduced while organic or inorganic electron donors are being oxidized. In the absence of oxygen, other materials such as nitrates, sulfates and carbon dioxide may become electron acceptors. The use of sulfates and carbon dioxide requires strictly anaerobic conditions. Nitrate can be used, however, by facultative organisms living under intermediate conditions referred to as anoxic, which are characterized by end products of CO_2, water, and nitrogen gas. All are inoffensive as opposed to products of methane, carbon dioxide, and H_2S formed under strictly anaerobic conditions. Balanced reactions for these oxidations and reductions can be obtained by combining half reactions from Table 7-4. The reverse of Reaction 11 for acetate may be combined with the electron acceptor reactions to yield the following:

Reaction		$\Delta G°$ kcal	
Aerobic			
$\frac{1}{8}CH_3COO^- + \frac{1}{4}O_2$	$= \frac{1}{8}CO_2 + \frac{1}{8}HCO_3^- + \frac{1}{8}H_2O$	-25.28	(7-13)
Anoxic			
$\frac{1}{8}CH_3COO^- + \frac{1}{5}NO_3^- + \frac{1}{5}H^+$	$= \frac{1}{8}CO_2 + \frac{1}{8}HCO_3^- + \frac{1}{10}N_2 + \frac{9}{40}H_2O$	-23.74	(7-14)
Anaerobic			
$\frac{1}{8}CH_3COO^- + \frac{1}{8}SO_4^{2-} + \frac{3}{16}H^+ + \frac{1}{8}CO_2 + \frac{1}{8}HCO_3^- + \frac{1}{16}H_2S + \frac{1}{16}HS^- + \frac{1}{8}H_2O$		-1.52	(7-15)
$\frac{1}{8}CH_3COO^- + \frac{1}{8}H_2O$	$= \frac{1}{8}CH_4 + \frac{1}{8}HCO_3^-$	-0.85	(7-16)

The standard free energies for each reaction indicate that more energy is available for biological growth from aerobic and anoxic than from anaerobic reactions. The last two reactions of sulfate reduction and methane fermentation release the least energy of all. These reactions can be carried out only by highly specialized groups of strictly anaerobic bacteria and cannot proceed in the presence of oxygen or nitrates.

Table 7-4 Useful half reactions for bacterial systems

Reaction number	Half reaction	$\Delta G°$ (W)* kcal per electron equivalent
	Reactions for bacterial cell synthesis (R_c)	
	Ammonia as nitrogen source:	
1.	$\frac{1}{5}CO_2 + \frac{1}{20}HCO_3^- + \frac{1}{20}NH_4^+ + H^+ + e^- = \frac{1}{20}C_5H_7O_2N + \frac{9}{20}H_2O$	
	Nitrate as nitrogen source:	
2.	$\frac{1}{28}NO_3^- + \frac{5}{28}CO_2 + \frac{29}{28}H^+ + e^- = \frac{1}{28}C_5H_7O_2N + \frac{11}{28}H_2O$	
	Reactions for electron acceptors (R_a)	
	Oxygen:	
3.	$\frac{1}{4}O_2 + H^+ + e^- = \frac{1}{2}H_2O$	-18.675
	Nitrate:	
4.	$\frac{1}{5}NO_3^- + \frac{6}{5}H^+ + e^- = \frac{1}{10}N_2 + \frac{3}{5}H_2O$	-17.128
	Sulfate:	
5.	$\frac{1}{8}SO_4^{2-} + \frac{19}{16}H^+ + e^- = \frac{1}{16}H_2S + \frac{1}{16}HS^- + \frac{1}{2}H_2O$	5.085
	Carbon dioxide (methane fermentation):	
6.	$\frac{1}{8}CO_2 + H^+ + e^- = \frac{1}{8}CH_4 + \frac{1}{4}H_2O$	5.763
	Reactions for electron donors (R_d)	
	Organic donors (heterotrophic reactions)	
	Domestic wastewater:	
7.	$\frac{9}{50}CO_2 + \frac{1}{50}NH_4^+ + \frac{1}{50}HCO_3^- + H^+ + e^- = \frac{1}{50}C_{10}H_{19}O_3N + \frac{9}{25}H_2O$	7.6
	Protein (amino acids, proteins, nitrogenous organics):	
8.	$\frac{8}{33}CO_2 + \frac{2}{33}NH_4^+ + \frac{31}{33}H^+ + e^- = \frac{1}{66}C_{16}H_{24}O_5N_4 + \frac{27}{66}H_2O$	7.7
	Carbohydrates (cellulose, starch, sugars):	
9.	$\frac{1}{4}CO_2 + H^+ + e^- = \frac{1}{4}CH_2O + \frac{1}{4}H_2O$	10.0
	Grease (fats and oils):	
10.	$\frac{4}{23}CO_2 + H^+ + e^- = \frac{1}{46}C_8H_{16}O + \frac{15}{46}H_2O$	6.6
	Acetate:	
11.	$\frac{1}{8}CO_2 + \frac{1}{8}HCO_3^- + H^+ + e^- = \frac{1}{8}CH_3COO^- + \frac{3}{8}H_2O$	6.609
	Propionate:	
12.	$\frac{1}{7}CO_2 + \frac{1}{14}HCO_3^- + H^+ + e^- = \frac{1}{14}CH_3CH_2COO^- + \frac{5}{14}H_2O$	6.664
	Benzoate:	
13.	$\frac{1}{5}CO_2 + \frac{1}{30}HCO_3^- + H^+ + e^- = \frac{1}{30}C_6H_5COO^- + \frac{13}{20}H_2O$	6.892
	Ethanol:	
14.	$\frac{1}{6}CO_2 + H^+ + e^- = \frac{1}{12}CH_3CH_2OH + \frac{1}{4}H_2O$	7.592
	Lactate:	
15.	$\frac{1}{6}CO_2 + \frac{1}{12}HCO_3^- + H^+ + e^- = \frac{1}{12}CH_3CHOHCOO^- + \frac{1}{3}H_2O$	7.873
	Pyruvate:	
16.	$\frac{1}{5}CO_2 + \frac{1}{10}HCO_3^- + H^+ + e^- = \frac{1}{10}CH_3COCOO^- + \frac{2}{5}H_2O$	8.545
	Methanol:	
17.	$\frac{1}{6}CO_2 + H^+ + e^- = \frac{1}{6}CH_3OH + \frac{1}{6}H_2O$	8.965
	Inorganic donors (autotrophic reactions):	
18.	$Fe^{3+} + e^- = Fe^{2+}$	-17.780
19.	$\frac{1}{2}NO_3^- + H^+ + e^- = \frac{1}{2}NO_2^- + \frac{1}{2}H_2O$	-9.43
20.	$\frac{1}{8}NO_3^- + \frac{5}{4}H^+ + e^- = \frac{1}{8}NH_4^+ + \frac{3}{8}H_2O$	-8.245
21.	$\frac{1}{6}NO_2^- + \frac{4}{3}H^+ + e^- = \frac{1}{6}NH_4^+ + \frac{1}{3}H_2O$	-7.852
22.	$\frac{1}{6}SO_4^{2-} + \frac{4}{3}H^+ + e^- = \frac{1}{6}S + \frac{2}{3}H_2O$	4.657
23.	$\frac{1}{8}SO_4^{2-} + \frac{19}{16}H^+ + e^- = \frac{1}{16}H_2S + \frac{1}{16}HS^- + \frac{1}{2}H_2O$	5.085
24.	$\frac{1}{4}SO_4^{2-} + \frac{5}{4}H^+ + e^- = \frac{1}{8}S_2O_3^{2-} + \frac{5}{8}H_2O$	5.091
25.	$H^+ + e^- = \frac{1}{2}H_2$	9.670
26.	$\frac{1}{2}SO_4^{2-} + H^+ + e^- = \frac{1}{2}SO_3^{2-} + \frac{1}{2}H_2O$	10.595

* Reactants and products at unit activity except $[H^+] = 10^{-7}$.

Methane fermentation is important to the environmental engineer since this is the major process by which wastes are stabilized in anaerobic digestion and is an important part of the stabilization occurring in septic tanks and lagoons. The methane gas released can be used as a fuel. Increased interest in this process has been shown in recent years because it offers an alternative to the dwindling supplies of natural gas. About 70 percent of the methane resulting from the complete methane fermentation of complex wastes results from fermentation of acetic acid,[5] according to Eq. (7-16). Acetic acid is formed as an intermediate in the anaerobic fermentation of carbohydrates, proteins, and fats.

In the anaerobic fermentation of carbohydrates an external electron acceptor is not required; the organic matter itself serves as the electron acceptor. As depicted in Fig. 7-5, electrons removed from glucose by NAD are transferred to pyruvate to form the variety of possible end products, as discussed previously. A knowledge of the biochemical pathways for organic degradation can aid in understanding what transformations are possible under different environmental conditions.

7-11 ENERGETICS AND BACTERIAL GROWTH

Microorganisms oxidize inorganic and organic materials in order to obtain energy for growth and maintenance. Heterotrophic organisms use a portion of organic material metabolized for energy, which in turn is used to convert another portion of the organic matter into cells. Autotrophic organisms on the other hand oxidize inorganic materials for energy, and in turn use the energy released to reduce carbon dioxide to form cellular organics. Electrons are needed to reduce the carbon dioxide and are obtained by oxidizing another portion of the inorganic electron donor. Thus, whether heterotrophic or autotrophic growth is considered, a portion of the electron donor is used for energy and a portion is used for synthesis. These two uses for the electron donor are shown schematically in Fig. 7-4.

In the design of biological systems for wastewater treatment, the environmental engineer needs to know all the transformations that take place. He needs to know what portion of the electron donor will be converted for energy so that he can determine the quantity of oxygen, nitrates, or sulfates that will be required, or the quantity of methane that will be produced. He needs to know the portion that will be converted to microbial cells since he will need to be sure sufficient nutrients such as nitrogen and phosphorus are available. He will also need to provide sludge handling facilities to treat and dispose of the biological materials that the process produces. In other words, he needs to make a mass balance for the system.

[5] J. S. Jeris and P. L. McCarty, The Biochemistry of Methane Fermentation Using C^{14} Tracers, *J. Water Pollut. Contr. Fed.*, **37**: 178 (1965).

A balanced chemical reaction for the overall biological conversion can aid in making such a mass balance. The reaction should include terms for synthesis and for energy. For the synthesis portion an empirical formula for bacterial cells, $C_5H_7O_2N$, is commonly used.[6] A balanced overall reaction can be written through use of the half reactions listed in Table 7-4, and combining them in accordance with the following:[6]

$$R = f_s R_c + f_e R_e - R_d \qquad (7\text{-}17)$$

R_c represents the half reaction for synthesis of bacterial cells and would be either Reaction 1 or 2 from Table 7-4, depending on whether ammonia or nitrate was available, respectively, for satisfying the cell nitrogen requirements. R_e represents the half reaction for the electron acceptor, and R_d for the electron donor. The values f_s and f_e represent the portion of the electron donor used for synthesis and energy, respectively, and the sum of the two must equal 1.0. The plus sign indicates that the half reactions as written in Table 7-4 are added together, and the minus sign means that the half reaction is first switched so that the right-hand side of the equation in the table becomes the left-hand side, and then the half reaction is added.

Typical maximum values for f_s [$f_{s(max)}$] are given in Table 7-5, and are appropriate for young rapidly growing bacterial cells. In general, reactions between electron donors and acceptors that yield more energy result in higher values for $f_{s(max)}$. For old or slowly growing cultures, values for f_s would be less than $f_{s(max)}$ because a larger portion of the energy is used for cell maintenance than for synthesis. For very old cells, values for f_s may be as little as 20 percent of $f_{s(max)}$. The value for f_s also depends upon characteristics of the bacterial species, efficiencies of energy transfer, and environmental conditions.

Example (*a*) Write a balanced equation for the methane fermentation of glucose, assuming that f_s equals 0.28. (*b*) Based on this equation, what quantity of methane is produced from fermentation of 1000 g of glucose, what quantity of bacteria are produced in this conversion, and what quantity of ammonia nitrogen is needed to satisfy the growth requirements of the bacteria?

(*a*) $f_e = 1 - f_s = 0.72$

Assuming that ammonia is available for cell synthesis, R_c = Reaction 1, R_e = Reaction 6, and R_d = Reaction 9 from Table 7.4:

$f_s R_c$: $0.056CO_2 + 0.014HCO_3^- + 0.014NH_4^+ + 0.28H^+ + 0.28e^- = 0.014C_5H_7O_2N + 0.126H_2O$

$f_e R_e$: $0.09CO_2 + 0.72H^+ + 0.72e^- \qquad\qquad\qquad\quad = 0.09CH_4 + 0.18H_2O$

$-R_d$: $0.25CH_2O + 0.25\ H_2O \qquad\qquad\qquad\qquad\quad = 0.25CO_2 + H^+ + e^-$

R: $0.25CH_2O + 0.014HCO_3^- + 0.014NH_4^+ =$

$$0.104CO_2 + 0.056H_2O + 0.09CH_4 + 0.014C_5H_7O_2N$$

(*b*) Reaction R indicates that the metabolism of 0.25 mol or 7.5 g of CH_2O requires 0.014 mol or 0.196 g of ammonia nitrogen and results in the production of 0.09 mol or 2.02 liters (STP) of methane and 0.014 mol or 1.58 g of cells. Therefore, the fermentation of 1000 g of glucose will result in the following:

[6] P. L. McCarty, Stoichiometry of Biological Reactions, *Progr. in Water Technol.*, **7**: 157, Pergamon, London (1975).

Table 7-5 Typical values for $f_{s(max)}$ for bacterial reactions

Electron donor	Electron acceptor	$f_{s(max)}$
Heterotrophic reactions		
Carbohydrate	O_2	0.72
Carbohydrate	NO_3^-	0.60
Carbohydrate	SO_4^{2-}	0.30
Carbohydrate	CO_2	0.28
Protein	O_2	0.64
Protein	CO_2	0.08
Fatty acid	O_2	0.59
Fatty acid	SO_4^{2-}	0.06
Fatty acid	CO_2	0.05
Methanol	NO_3^-	0.36
Methanol	CO_2	0.15
Autotrophic reactions		
S	O_2	0.21
$S_2O_3^{2-}$	O_2	0.21
$S_2O_3^{2-}$	NO_3^-	0.20
NH_4^+	O_2	0.10
H_2	O_2	0.24
H_2	CO_2	0.04
Fe^{2+}	O_2	0.07

Ammonia-N used $= 0.196(1000)/7.5 = 26.1$ g
Methane produced $= 2.02(1000)/7.5 = 269$ liters (STP)
Cells produced $\quad = 1.58(1000)/7.5 = 211$ g

Example Write a balanced equation for the autotrophic oxidation of NH_4^+ to NO_2^-, using the value of $f_{s(max)}$ from Table 7-5. From Table 7-5, $f_s = 0.10$, and $f_e = 1.0 - 0.1 = 0.9$. Since ammonia is available for cell synthesis, use $R_c =$ Reaction 1, $R_e =$ Reaction 3, and $R_d =$ Reaction 21 from Table 7-4:

f_sR_c: $\quad 0.02CO_2 + 0.005HCO_3^- + 0.005NH_4^+ + 0.1H^+ + 0.1e^- = 0.005C_5H_7O_2N + 0.045H_2O$
f_eR_e: $\quad 0.225O_2 + 0.9H^+ + 0.9e^- \qquad\qquad\qquad\qquad = 0.45\ H_2O$
$-R_d$: $\quad 0.167NH_4^+ + 0.333H_2O \qquad\qquad\qquad\qquad = 0.167NO_2^- + 1.333H^+ + e^-$

R: $\quad 0.172NH_4^+ + 0.225O_2 + 0.005HCO_3^- + 0.02CO_2 =$
$\qquad\qquad\qquad\qquad 0.167NO_2^- + 0.005C_5H_7O_2N + 0.333H^+ + 0.162H_2O$

A similar method of writing balanced equations may be used for bacterial fermentations. Here, organics serve both as electron donor and acceptor, the donor portion is oxidized generally to carbon dioxide, and the acceptor portion is converted to a more reduced organic material. An example is the fermentation of glucose to ethanol:

$$C_6H_{12}O_6 = 2CH_3CH_2OH + 2CO_2 \qquad\qquad (7\text{-}18)$$

Example Write a balanced equation for the bacterial fermentation of glucose to ethanol, assuming that ammonia is present as a nutrient and that f_s equals 0.15:

Here, $\qquad f_e = 0.85$, $R_c =$ Reaction 1, $R_d =$ Reaction 9, and $R_e =$ Reaction 14 (ethanol is the end product and conceptually can be thought to be formed through CO_2 reduction, even though this is not the case).

f_sR_c: $\quad 0.03CO_2 + 0.0075HCO_3^- + 0.0075NH_4^+ + 0.15H^+ + 0.15e^- = 0.0075C_5H_7O_2N + 0.675H_2O$

f_eR_e: $\quad 0.142CO_2 + 0.85H^+ + 0.85e^- \qquad\qquad\qquad\quad = 0.0708CH_3CH_2OH + 0.2125H_2O$

$-R_d$: $\quad 0.25CH_2O + 0.25H_2O \qquad\qquad\qquad\qquad\qquad\quad = 0.25CO_2 + H^+ + e^-$

R: $\quad 0.25CH_2O + 0.0075HCO_3^- + 0.0075NH_4^+ =$
$$0.0708CH_3CH_2OH + 0.0075C_5H_7O_2N + 0.078CO_2 + 0.03H_2O$$

7-12 BIOCHEMISTRY OF MAN

Since environmental engineers are concerned with the disposal of human wastes, it is important that they be familiar with the major changes that organic matter, taken as food, undergoes in its passage through the body.

Carbohydrates

Much of the carbohydrate consumed by man is utilized by the body. The remainder, consisting of undigestible matter, is eliminated in the feces. Most of the rejected carbohydrate matter is cellulose and other higher polysaccharides for which the human body does not provide enzymes to accomplish its hydrolysis, or for which the detention time in the intestine is too short to complete hydrolysis. The short detention time is aggravated by improper chewing of food and by diarrhetic conditions.

The carbohydrate matter that is assimilated into the blood stream is used for energy, stored as glycogen (animal starch) in muscle tissue and the liver, or converted to fat and stored as fatty tissue. The carbohydrates that are oxidized to produce energy are converted to carbon dioxide and water. The carbon dioxide is carried away, by the blood, from the cells where it is formed. The blood is buffered to such an extent that it can carry considerable amounts of CO_2 and release it to the air in the lungs, in accordance with the principles of Henry's law.

The human body contains a remarkable mechanism for controlling the amount of sugar (glucose) in the blood stream. If excessive amounts accumulate, the excess is released into the urine. This is a part of the kidney function. Persons with diabetes suffer from improper metabolism of sugar. As a result, blood sugar exceeds the amount acceptable to the kidney (renal threshold), and the excess is separated in the kidneys and escapes in the urine. The urine of diabetics shows the presence of glucose consistently. If the carbohydrate intake of a diabetic exceeds the capacity of his kidneys to excrete sugar, blood-sugar levels build up to a point at which he may pass into a coma.

Fats

Crude fatty materials contain certain substances that are not hydrolyzed in the human alimentary system. These materials and some of the undigested fats are passed in the feces. Fats are hydrolyzed to a considerable extent by lipase in the stomach. Further hydrolysis occurs in the intestine, where the reaction is facilitated through the emulsifying properties of the bile salts. The fatty acids that enter the blood stream are oxidized to produce energy or are stored in fatty tissue for future use. The end products of oxidation are principally carbon dioxide and water, but some ketones, principally acetone, are formed. The carbon dioxide is expelled by the lungs. The ketones are excreted in the urine. Ketones are found in unusual amounts in the urine of diabetics and people suffering from faulty fat metabolism.

Proteins

Hydrolysis of proteins is started in the stomach and continues in the intestine. Amino acids, when released by hydrolysis, are absorbed into the blood stream. Fractions that are not completely hydrolyzed are excreted in the feces. The amino acids are used mainly for the building and repair of muscle tissue, and in these capacities they become fixed in body tissues.

The end products of protein metabolism that require excretion as waste products result principally from two processes: the "wearing" of muscle tissue and oxidation of amino acids to obtain energy. Deamination of amino acids precedes their use as energy sources. The ammonia is released principally as urea but small amounts of NH_4^+ are normally present. Excretion is by way of the urine. The major function of the kidneys is to separate waste nitrogen compounds from the blood. That protein metabolism involves a variety of complicated processes may be deduced from the considerable number of nitrogenous compounds present in urine. Creatine, creatinine, uric acid, hippuric acid, and traces of purine bases are normally present, in addition to urea and ammonium ion.

Vitamins

Vitamins are very potent organic substances that occur in minute quantities in natural foodstuffs. They must be supplied in the diet of animals if they are not synthesized naturally within the animal from essential dietary or metabolic precursors. Some function as precursors of enzymes; with others, the function is not well understood. In general, they exert a hormone-like or enzymic action in the control of specific chemical reactions in the animal body, and the absence or lack of a sufficient supply of certain ones leads to the development of vitamin-deficiency diseases. e.g., beriberi, rickets, pellagra, and scurvy.

A wide variety of vitamins are known. They are generally classified into two groups, the *fat-soluble* and the *water-soluble,* as shown in Table 7-6.

The role of vitamins in biological processes employed by environmental engineers has not been explored. Several of the vitamins are recovered from industrial wastes, particularly those from the fermentation industry, and their economic value has been an important factor in helping to solve the waste-disposal problem in the distilling industry. Activated sludge has been found to

Table 7-6 Classification and function of vitamins

Vitamin	Good sources	Function
Fat-soluble:		
A	Butter, liver oils	Eye health
D	Liver oils, egg	Ca metabolism, i.e., antirachitic
E	Cottonseed oil, cereals	Prevents sterility
K	Green plants, egg yolk	Clotting of blood
Water-soluble:		
B_1 Thiamine	Pork, whole wheat, peanuts	Antiberiberi
B_2 Riboflavin	Eggs, liver, cereals, milk	General health
Nicotinic acid	Meat, whole wheat, yeast	Antipellagra
B_6 Pyridoxine	Egg yolk, liver, yeast	Skin tone
Biotin	Egg yolk, liver, yeast	Skin tone
Pantothenic acid	Egg yolk, liver, milk	Skin tone, growth
Folic acid group	Green leafy vegetables	Antianemia
Inositol	Fruits, vegetables	Hair, growth
B_{12} Cobalamine	Liver, activated sludge	Antianemia
C Ascorbic acid	Citrus fruits, apples	Antiscurvy

be a rich source of vitamin B_{12}. The environmental engineer should be informed on the subject of vitamins and their economic importance.

PROBLEMS

7-1 What role do enzymes play in living organisms?

7-2 What terms are used to describe enzymes with respect to (*a*) where their action occurs, and (*b*) the nature of the reaction which they control?

7-3 How does the environmental engineer make use of temperature and pH relationships of biochemical reactions in the design and operation of biological waste treatment facilities?

7-4 Disposal of the large quantities of microorganisms produced during waste treatment is one of the most significant problems in environmental engineering. Why might this particular problem be minimized by anaerobic rather than aerobic waste treatment?

7-5 (*a*) How many acetic acid molecules are produced during the complete beta oxidation of a stearic acid molecule?

(*b*) How many hydrogen atoms are removed from stearic acid during the above beta oxidation?

7-6 Use of methanol has been proposed to rid a wastewater of nitrates by biological denitrification to N_2. (*a*) Write a balanced overall equation for nitrate removal with methanol, using $f_{s,\,max}$ and assuming nitrate serves as the nutrient source for bacteria. (*b*) If the nitrate (NO_3^-) concentration in the wastewater equals 100 mg/l, what concentration of methanol must be added for complete nitrate removal? (*c*) What percentage of the nitrate-nitrogen is used for cell synthesis?

7-7 Acetic acid is a common fatty acid in wastewaters. (*a*) Write a balanced overall equation for aerobic oxidation of acetic acid, assuming ammonia is available as a nitrogen source and that $f_{s,\,max}$ applies. (*b*) How many grams of oxygen are required to oxidize 100 grams of acetic acid? (*c*) How many grams of bacteria are produced per gram of acetic acid metabolized?

Answers: (*a*) $0.125CH_3COO^- + 0.0295NH_4^+ + 0.1025O_2 = 0.0295C_5H_7O_2N + 0.007CO_2 + 0.0955H_2O + 0.0955\ HCO_3^-$ (*b*) 44.5   (*c*) 0.45

7-8 The biological oxidation of Fe^{2+} in mine drainage waters can result in red discoloration in streams from subsequent Fe^{3+} precipitation. (*a*) Write a balanced overall equation for biological oxidation of Fe^{2+}, assuming $f_{s,max}$ and the presence of ammonia. (*b*) How many grams of oxygen are required per gram of Fe^{2+} oxidized?

7-9 Denitrification of wastewaters can be obtained through autotrophic biological reactions. (*a*) Write a balanced equation for denitrification of NO_3^- to N_2 using $S_2O_3^{2-}$ (which is oxidized to SO_4^{2-}), using $f_{s,max}$ and assuming the bacteria use ammonia as a nitrogen source for cell synthesis. (*b*) How many grams of $Na_2S_2O_3$ would need to be added to the wastewater per gram of NO_3^- reduced?

Answer: (*a*) $0.125S_2O_3^{2-} + 0.04CO_2 + 0.01NH_4^+ + 0.01HCO_3^- + 0.16NO_3^- + 0.055H_2O = 0.25SO_4^{2-} + 0.01C_5H_7O_2N + 0.08N_2 + 0.09H^+$ (*b*) 1.99

REFERENCES

Conn, E. E., and P. K. Stumpf: "Outlines of Biochemistry," 3d ed., Wiley, New York, 1972.

Florkin, M., and H. S. Mason: "Comparative Biochemistry, A Comprehensive Treatise," Academic, New York, vol. I, 1960; vol. II, 1960; vol. III, 1962; vol. IV, 1962; vol. V, 1963; vol. VI, 1963; and vol. VII, 1964.

Fruton, J. S., and S. Simmonds: "General Biochemistry," 2d ed., Wiley, New York, 1958.

Lehninger, A. L.: "Biochemistry," Worth Publishers, New York, 1970.

White, A., P. Handler, and E. Smith: "Principles of Biochemistry," 5th ed., McGraw-Hill, New York, 1973.

EIGHT

BASIC CONCEPTS FROM
NUCLEAR CHEMISTRY

8-1 INTRODUCTION

The science of *nuclear chemistry* deals with transformations in the nucleus of atoms. Atoms are the smallest particles of chemical elements. Some atoms are naturally stable, while others are not. When the instability of an atom leads to its transformation into a more stable form, the phenomenon of radioactivity results. Radioactivity is the ejection of particles or radiation from the nucleus. The environmental engineer has an interest in nuclear chemistry as the radioactivity emanating from changes in unstable elements can result in hazards to health. Radioactive elements also may be used as tracers for measuring the flow of water or for investigating the fate of materials in the environment.

Interest in radioactivity may be considered to date from 1895, when Roentgen discovered a new form of radiation from cathode-ray tubes. The radiations caused certain salts to become luminescent and also affected photographic plates. They are called *roentgen rays* or *X rays*. With a few modifications, the cathode-ray tube became the modern roentgen or X-ray tube which is used so extensively in medical and industrial applications. Gamma rays released by radioactive materials and X rays are both electromagnetic waves, the gamma rays usually having somewhat shorter wavelengths.

Roentgen's discovery of a new radiation that affected photographic plates stimulated a great deal of testing of materials for similar characteristics. Becquerel and his father had been interested in phosphorescence for some time prior to 1891. They had noted that potassium uranyl sulfate [$K_2UO_2(SO_4)_2 \cdot 2H_2O$]

exhibited pronounced phosphorescence when excited by ultraviolet light. It was natural that Becquerel would want to test his uranyl salts for emanation of X rays. He found them to do so, and subsequent observations on a wide variety of salts and materials containing uranium showed them to produce X rays in proportion to their content of uranium.

In 1898 Pierre and Marie Curie concluded that the X rays from uranium were an atomic phenomenon characteristic of the element, and they introduced the name *radioactivity*. The Curies pursued their studies of radioactive materials with much vigor. They found that compounds of thorium emitted radiation similar to that of uranium. They also noted that certain ores of uranium were more radioactive than uranium itself. This led to a search for other materials in the residues remaining after uranium extraction. Two new radioactive elements were isolated, polonium and radium. Radium is several thousand times more radioactive than uranium.

8-2 ATOMIC STRUCTURE

Modern concepts of atomic structure are largely the result of knowledge gained from the behavior of radioactive materials. It is difficult, therefore, to discuss one without considering the other. Prior to the discovery of radioactivity, atoms were considered to be indivisible. With the discovery that radioactive elements emitted positively and negatively charged particles, the foundation was laid for new concepts.

Nuclear Theory

By 1900 it was realized that atoms are not indivisible. However, it was not until 1911 that Rutherford proposed the nuclear concept of the atom. This theory held that atoms were composed of a small positively charged nucleus, containing most of the mass of the atom, with a cloud of negatively charged electrons surrounding it.

Electron Orbits

Bohr was the first to propose that the electrons about the nucleus of an atom are arranged in a methodical manner and revolve in orbits about the nucleus. Although his theory, issued in 1913, has undergone some refinements, it remains the basis of our modern-day knowledge. The present tendency is to think of the electrons as being arranged in shells about the nucleus. A major contribution was made to the Bohr theory by Sommerfeld, who has shown that the electrons within a given shell occur in several energy levels. Other contributions, particularly with respect to chemical properties, were made by Langmuir (octet theory), Mosely, G. N. Lewis, and W. Kossel.

Table 8-1 Arrangement of electrons for some common elements

Symbol	Atomic number	Number of electrons in shells						
		K	L	M	N	O	P	Q
H	1	1						
He*	2	2						
N	7	2	5					
Ne*	10	2	8					
Na	11	2	8	1				
Cl	17	2	8	7				
Ar*	18	2	8	8				
Ca	20	2	8	8	2			
Zn	30	2	8	18	2			
Br	35	2	8	18	7			
Kr*	36	2	8	18	8			
Ag	47	2	8	18	18	1		
Xe*	54	2	8	18	18	8		
Ba	56	2	8	18	18	8	2	
Hg	80	2	8	18	32	18	2	
Pb	82	2	8	18	32	18	4	
Rn*	86	2	8	18	32	18	8	
Ra	88	2	8	18	32	18	8	2

* Inert gases.

The simplest atoms, hydrogen and helium, have one shell of electrons, and the most complex have seven. The shells or rings are designated as K, L, M, N, O, P, and Q in the order of their increasing remoteness from the nucleus. The maximum number of electrons that a given shell can have is shown by the formula $2n^2$ where n is the shell number, equaling 1 for the K shell and 7 for the Q shell. The arrangement of electrons for a number of elements is given in Table 8-1. Over the course of ensuing years, the positively charged nucleus was considered to consist of protons and electrons. This theory held that the protons were always in excess of the electrons in the nucleus, and this excess was equal to the planetary electrons; thus the net charge on the atom was zero.

Neutron-Proton Concept of Nuclear Structure

In 1930, Bothe and Becker discovered a very penetrating secondary radiation when light elements, such as beryllium and lithium, were subjected to bombardment by alpha particles from polonium. The new radiation was first thought to be X rays of very short wavelength. In 1932 Chadwick showed this secondary radiation to be made up of neutral particles having a mass comparable to that of the proton. The new particles were given the name *neutrons,* and since

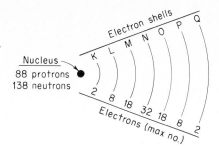

Figure 8-1 Structure of radium atom.

their source was obviously the nucleus of the bombarded atoms, a new concept of nuclear structure evolved. From this came the present "practical" model of the nucleus. Since 1932, the nucleus has been found to consist of many strange particles, but these have little significance to a useful understanding of nuclear chemistry.

According to the practical model, the nucleus of all atoms, except the simple hydrogen atom, consists of neutrons plus protons, both of which are called *nucleons*. The number of protons Z corresponds to the atomic number and is equal to the number of electrons about the nucleus of a neutral atom. The number of neutrons N is equal to the atomic weight A expressed as the nearest whole number less the number of protons: $N = A - Z$. The structure of the atom may be represented as shown in Fig. 8-1. The nucleus has a diameter of the order of 10^{-12} to 10^{-13} cm, and the atom a diameter of about 10^{-8} cm. The density of nuclear matter is tremendous. It is estimated that 1 cm^3 would weigh 10^{10} kg.

Nomenclature of Isotopes

All neutral isotopes of the same element have the same number of electrons and, of course, the same number of protons. Since the masses of the isotopes vary, the number of neutrons must vary. In order to differentiate between isotopes, a new system of symbol writing had to be developed. The IUPAC nomenclature which is growing in use in the United States lists the atomic number as a subscript just before the symbol and the atomic weight, or mass number, as a superscript also before the symbol. Earlier writings frequently placed the mass number as a superscript after the symbol. In the IUPAC system, $^{204}_{82}$Pb, $^{206}_{82}$Pb, $^{207}_{82}$Pb, and $^{208}_{82}$Pb represent four isotopic forms of lead, each of which has 82 protons and 82 electrons. Since the atomic number of a given element is always the same, it is frequently eliminated when discussing isotopes. However, in radioactive changes involving transmutation of one element into another, such as in the conversion of ^{238}U to ^{206}Pb, the change is best shown by $^{238}_{92}$U $\rightarrow$ $^{206}_{82}$Pb. Anyone familiar with nuclear chemistry knows that such a change cannot occur in one step and that several intermediate steps are involved (see Fig. 8-2).

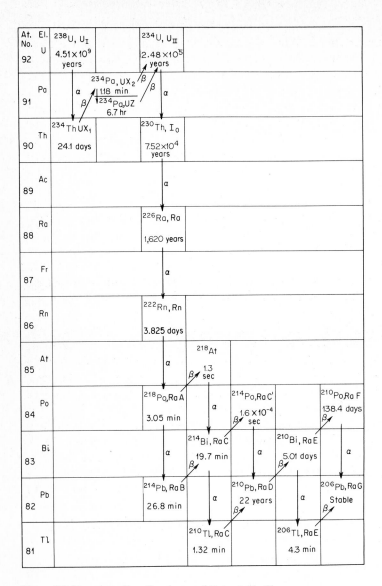

Figure 8-2 Steps in radioactive decay of U to stable Pb.

8-3 STABLE AND RADIOACTIVE NUCLIDES

The chemical identity of an atom is given by the number of protons in the nucleus Z. The stability of a nucleus can be empirically related to some degree with the ratio of the number of neutrons to the number of protons it contains: N/Z. For stable light nuclei N/Z is approximately 1, while for stable heavy nuclei the ratio approaches 2. When the nucleus is at variance with this pattern, instability results. The shakedown of an unstable nucleus to a more stable form results in the natural process of nuclear decay and the ejection of particles or radiation from the nucleus. Nuclear transformations or reactions can also be induced by bombarding the nucleus with particles or energy. Radioactive atoms may exist for times as short as 10^{-20} sec or as long as 10^{18} yr. Those that decay too slowly to be measured are termed stable. All nuclides with atomic numbers greater than 83 or atomic weights greater than 200 are unstable. Table 8-2 lists isotopes for a few elements and illustrates some compositions that are stable and some that are radioactive.

Most of the presently existing atoms on earth are stable, but some are radioactive. Many of the radioactive isotopes present decay very slowly. Others are formed as short-lived intermediates following the decay of longer-lived isotopes. Bombardment of the earth with radiation from outer space also converts some stable atoms to radioactive ones. A good example is the production of ^{14}C from ^{14}N by cosmic rays. All living forms are kept uniformly radioactive with ^{14}C since atmospheric carbon which contains this produced material is the primary source of carbon for life.

The largest terrestrial sources of radiation are ^{40}K, ^{238}U, ^{235}U, and ^{232}Th. The total energy emission from ^{40}K in the earth's crust is estimated to be 4×10^{12} W. The heat generated by the decay of these elements has resulted in a much slower cooling of the earth than would have otherwise been the case. Decay of $^{40}_{19}K$ leads to the formation of the stable isotopes $^{40}_{20}Ca$ and $^{40}_{18}Ar$, each of which is formed by a different mechanism of decay.

The heavy-metal radioactive elements fall into three series: uranium, thorium, and actinium. The uranium series has ^{238}U as its parent substance and, after 14 successive transformations have occurred, the end product is ^{206}Pb. Thorium (^{232}Th) is the parent substance of the thorium series. After 10 transformations, it remains as ^{208}Pb. The parent element of the actinium series

Table 8-2 Isotopes of a few elements ilustrating stable and radioactive forms

Element	Isotope					
Hydrogen		$^{1}_{1}H$	$^{2}_{1}H$	$^{3}_{1}H$		
		Stable	Stable	Radioactive		
Carbon		$^{11}_{6}C$	$^{12}_{6}C$	$^{13}_{6}C$	$^{14}_{6}C$	
		Radioactive	Stable	Stable	Radioactive	
Nitrogen	$^{12}_{7}N$	$^{13}_{7}N$	$^{14}_{7}N$	$^{15}_{7}N$	$^{16}_{7}N$	
	Radioactive	Stable	Stable	Stable	Radioactive	
Oxygen	$^{14}_{8}O$	$^{15}_{8}O$	$^{16}_{8}O$	$^{17}_{8}O$	$^{18}_{8}O$	$^{19}_{8}O$
	Radioactive	Radioactive	Stable	Stable	Stable	Radioactive
Phosphorous	$^{29}_{15}P$	$^{30}_{15}P$	$^{31}_{15}P$	$^{32}_{15}P$	$^{33}_{15}P$	
	Radioactive	Radioactive	Stable	Radioactive	Radioactive	
Chlorine	$^{34}_{17}Cl$	$^{35}_{17}Cl$	$^{36}_{17}Cl$	$^{37}_{17}Cl$	$^{38}_{17}Cl$	
	Radioactive	Stable	Radioactive	Stable	Radioactive	

is ^{235}U and, after 11 transformations, it remains as ^{207}Pb. This series takes its name from the fact that ^{231}Pa (protactinium) and ^{227}Ac (actinium) are long-lived isotopes formed as steps in the transformation. The uranium series is sometimes called the radium series for the same reason. The steps in the radioactive decay of ^{238}U are shown in Fig. 8-2. ^{232}Th and ^{235}U decompose through similar steps.

Nature of Radiations

Early workers with radioactive materials were cognizant of the presence of only one form of radiation, and its properties were similar to those of X rays. Later investigations established the presence of three kinds of radiations designated as alpha, beta, and gamma. Separation and identification were accomplished by directing the radiation through a magnetic field, as shown in Fig. 8-3. Some of the radiation was bent slightly toward the negative pole. This phenomenon indicated that it had a positive charge and, probably, considerable mass. This is called alpha radiation. Other radiation was bent radically toward the positive pole, showing it to be negatively charged and, probably, of small mass. This was called beta radiation. A third radiation group was unaffected by the magnetic field. It does not have a charge and is called gamma radiation.

Alpha radiation Alpha radiation is not true electromagnetic radiation as are light and X rays. It consists of particles of matter. Alpha particles are actually doubly charged ions of helium with a mass of 4 (^{4}He). Although they are propelled from the nucleus of atoms at velocities ranging from 1.4 to 2 × 10^9 cm/s (about 10 percent of the speed of light), they do not travel much more than 10 cm in air at room temperature. They are stopped by an ordinary sheet of paper. The alpha particles emitted by a particular element are all released at the same velocity. The velocity varies, however, from element to element. The alpha particles have extremely high ionizing action within their range.

Beta radiation Beta radiation consists of negatively charged particles moving at speeds ranging from 30 to 99 percent of the speed of light. *Beta particles* are

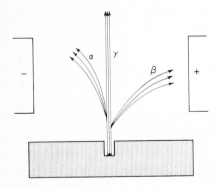

Figure 8-3 Effect of a magnetic field upon alpha, beta, and gamma rays.

actually electrons, and the velocity of flight of individual electrons varies considerably for a given element, as well as for different elements. The penetrating power of beta particles varies with their speed. They normally travel several hundred feet in air. Shielding with aluminum sheeting a few millimeters thick will stop the particles. Because of the low mass the ionizing power of beta radiation is much weaker than that of the alpha radiation.

Gamma radiation Gamma radiation is true electromagnetic radiation which travels with the speed of light. It is similar to X rays but has shorter wavelength and therefore greater penetrating power, which increases as the wavelength decreases. Proper shielding from gamma radiation requires several centimeters of lead or several feet of concrete. The unit of gamma radiation is the *photon*.

Energies of Radiations

It is important to know the energy of the various radiations produced by radioactive materials. Since alpha and beta particles have mass and gamma radiations do not, it is possible to establish a single system of expressing energies through use of Einstein's energy-mass equivalence formula

$$E = Mc^2 \tag{8-1}$$

With c as the velocity of light (2.998×10^{10} cm/s) and M the mass of the particle in grams, E has units of g·cm/s or ergs. However, a more convenient energy unit for nuclear particles is the *electron volt* (*ev*), which is the energy necessary to raise one electron through a potential difference of one volt. The conversion factor is 1 ev = 1.602×10^{-12} erg. The energies of alpha and beta particles and of gamma photons range from several thousands up to several millions of electron volts. For this reason, energies are usually expressed as million electron volts (Mev).

Equation (8-1) indicates that potential energy is stored as mass in atoms and that in a given system a change in mass is accompanied by an equivalent change in energy:

$$\Delta E = \Delta Mc^2 \tag{8-2}$$

The spontaneous radioactive decay of unstable nuclides is accompanied by the release of energy as required by the laws of thermodynamics. The energy results from a loss in mass of the nuclide during its transformation. From this requirement it follows that spontaneous radioactive decay may take place by some mode if the masses of the products are lighter than the mass of the original nuclide. The excess mass is released as radiation during the disintegration. Thus, for the following disintegration of nuclide A into nuclide B plus radiation b,

$$A \rightarrow B + b \tag{8-3}$$

the change in mass

$$\Delta M = M_A - (M_B + M_b) \tag{8-4}$$

must be positive if the disintegration is to occur spontaneously. This does not mean the disintegration will occur, only that it can occur. If ΔM is negative, the radioactive decay cannot occur spontaneously: i.e., the nuclide is stable.

Atomic Changes Resulting from Release of Radiation

The change that atoms undergo when releasing alpha particles is considerably different from the change when beta particles are released. These changes are illustrated in Fig. 8-2 and were formulated into so-called *displacement laws* by Fajans, Rutherford, and Soddy as follows:

Alpha-particle release *When an element emits an alpha particle, the product has the properties of an element two places to the left of the parent in the periodic table.* In other words, emission of an alpha particle decreases the mass number by four units and the nuclear charge, or atomic number, by two units.

Beta-particle release *When an element emits a beta particle, the product has the properties of an element one place to the right of the parent in the periodic table.* In this change the mass remains the same and the atomic number increases one unit.

Gamma radiations Gamma radiation may accompany the release of alpha or beta particles and is a result of energy released by nuclear transformations or shifts of orbital electrons.

Units of Radioactivity

The unit of radioactivity is the *curie*. Formerly, it was considered to be the number of disintegrations occurring per second in one gram of pure radium. Since the constants for radium are subject to revision from time to time, the International Radium Standard Commission has recommended the use of a fixed value, 3.7×10^{10} disintegrations per second, as the *standard curie* (Ci).

The curie is used mainly to define quantities of radioactive materials. A curie of an alpha emitter is that quantity which releases 3.7×10^{10} alpha particles per second. A curie of a beta emitter is that quantity of material which releases 3.7×10^{10} beta particles per second, and a curie of a gamma emitter is that quantity of material which releases 3.7×10^{10} photons per second. The curie represents such a large number of disintegrations per second that the millicurie (mCi), microcurie (μCi), nanocurie (nCi), and picocurie (pCi), corresponding to 10^{-3}, 10^{-6}, 10^{-9}, and 10^{-12} curie, respectively, are more commonly used.

The *roentgen* (r) is a unit of gamma or X-ray radiation intensity. It is of value in the study of the biological effects of radiation that result from ionization induced within cells by the radiations. The roentgen is defined as the amount of gamma or X radiation that will produce in one cubic centimeter of dry air, at 0°C and 760 mm pressure, one electrostatic unit (esu) of electricity. This is equivalent to 1.61×10^{12} ion pairs per gram of air and corresponds to the absorption of 83.8 ergs of energy.

The roentgen is a unit of the total quantity of ionization produced by gamma or X rays, and dosage rates for these radiations are expressed in terms of roentgens per unit time.

With the advent of atomic energy involving exposure to neutrons, protons, and alpha and beta particles which also have effects on living tissue, it has become necessary to have other means of expressing ionization produced in cells. Three methods of expression have been used.

The *roentgen-equivalent-physical* (*rep*) is defined as that quantity of radiation (other than X rays or γ radiation) which produces in one gram of human tissue ionization equivalent to the quantity produced in air by one roentgen of γ radiation or X rays (equivalent to 83.8 ergs of energy). The *rep* has been replaced largely by the term *rad,* which has wider application.

The *roentgen-absorption-dose* (*rad*) is a unit of radiation corresponding to an energy absorption of 100 ergs per gram of any medium. It can be applied to any type and energy of radiation that leads to the production of ionization. Studies on the radiation of biological materials have shown that the roentgen is approximately equivalent to 100 ergs/g of tissue and can be equivalent to 90 to 150 ergs/g of tissue, depending on the energy of the X and γ radiation and type of tissue. The *rad,* therefore, is more closely related to the roentgen than is the *rep,* in terms of radiation effects on living tissues, and is the term preferred by biologists.

The *rad* represents such a tremendous radiation dosage, in terms of permissible amounts for human beings, that another unit has been developed specifically for man. The term *roentgen-equivalent-man* (*rem*) is used. It corresponds to the amount of radiation that will produce an energy dissipation in the

Table 8-3 Half-lives of common radioactive isotopes

Atomic number	Nuclide	Half-life	Nature of radiation
1	^{3}H	12.3 yr	β
6	^{14}C	5730 yr	β
11	^{24}Na	15.0 h	β, γ
15	^{32}P	14.3 days	β
16	^{35}S	88 days	β
19	^{40}K	1.28×10^9 yr	β
27	^{60}Co	5.3 yr	β, γ
35	^{78}Br	6.4 min	β, γ
38	^{90}Sr	28.1 yr	β
53	^{131}I	8.0 days	β, γ
55	^{137}Cs	30 yr	β
88	^{226}Ra	1600 yr	α, γ
92	^{238}U	4.51×10^9 yr	α

human body that is biologically equivalent to one roentgen of radiation of X rays, or approximately 100 ergs/g. The recommended *Maximum Permissible Dose* (MPD) for radiation workers is $5\,rem$/year and for nonradiation workers $\frac{1}{2}$ *rem*/year.

Half-lives

Radioactive decomposition is a true unimolecular reaction. The rate is constant over a wide variety of environmental conditions. Half-lives of the radioactive elements vary from fractions of a second to about 10^{12} yr. The half-lives of a number of elements are given in Table 8-3. The kinetics of unimolecular or first-order reactions are discussed in Sec. 3-11.

8-4 ATOMIC TRANSMUTATIONS AND ARTIFICIAL RADIOACTIVITY

The experimental conversion of one element into another was accomplished by Rutherford in 1919. When alpha particles derived from radium C were passed through nitrogen gas, protons were detected. The collisions between alpha particles and nitrogen nuclei resulted in the formation of an isotope of oxygen and a proton as follows:

$$^{14}_{7}N + {}^{4}_{2}He \rightarrow {}^{17}_{8}O + {}^{1}_{1}H \tag{8-5}$$

By 1922 Rutherford and Chadwick had shown that all elements in the periodic table between boron and potassium, except carbon and oxygen, underwent similar transmutations when submitted to bombardment by alpha particles.

It was not until 1930 that radiation other than protons was detected when elements were subjected to alpha-particle bombardment. In that year, Bothe and Becker discovered a very penetrating, neutral, secondary radiation when beryllium or lithium was subjected to alpha particles from polonium. In 1932, Chadwick showed the particles to be neutrons, and the changes were described as follows:

$$^{7}_{3}Li + {}^{4}_{2}He \rightarrow {}^{10}_{5}B + {}^{1}_{0}n \tag{8-6}$$

$$^{9}_{4}Be + {}^{4}_{2}He \rightarrow {}^{12}_{6}C + {}^{1}_{0}n \tag{8-7}$$

where the neutron is presented by ${}^{1}_{0}n$.

The third important step in transmutation of elements involved the discovery that a third particle was found in certain instances. In 1934 I. Curie and Joliot noted that when boron, magnesium, or aluminum was bombarded with alpha particles, the expected transmutation with neutron release occurred and that *positrons*[1] (positive electrons) were also produced. In addition, they

---

[1] The existence of positrons had been predicted and confirmed previously in cosmic-ray studies.

found that positron emission continued after alpha bombardment was discontinued. The emission of positrons was shown to decrease in accordance with the decay law for radioactive materials. Through careful analysis of the materials produced, they were able to show that alpha bombardment of these elements had produced an atom with an unstable nucleus that underwent radioactive positron decay; thus the production of artificial radioactive materials by alpha bombardment was established.

$$^{10}_{5}B + ^{4}_{2}He \rightarrow ^{13}_{7}N + ^{1}_{0}n \tag{8-8}$$
$$\rightarrow ^{13}_{6}C + \beta^{+} \qquad t_{1/2} = 10.0 \text{ min}$$

$$^{27}_{13}Al + ^{4}_{2}He \rightarrow ^{30}_{15}P + ^{1}_{0}n \tag{8-9}$$
$$\rightarrow ^{30}_{14}Si + \beta^{+} \qquad t_{1/2} = 2.5 \text{ min}$$

It soon became apparent that there is no real distinction between a nuclear reaction leading to stable products and one leading to unstable products. According to the Bohr concept, all bombardments result in an absorption of the bombarding particle by the nucleus to produce an unstable compound nucleus. The life of the compound nucleus is extremely short (10^{-12} to 10^{-14} s), and decomposition occurs to a set of products. The products may be stable or they may be unstable.

The discovery of the neutron and the fact that radioactive elements could be produced artificially set the stage for the developments of nuclear energy.

8-5 NUCLEAR REACTIONS

Nuclear reactions may be induced by bombardment with a wide variety of particles. Most of the radioactive materials used in industry, research, and medicine today are produced by such bombardment, which may be done either in a nuclear reactor or in a particle accelerator.

Alpha-Induced Reactions

Because of the positive charge on the alpha particle, it has to overcome the repulsive forces of the positively charged nucleus of an atom before it can add to it. As the atomic number of elements increases, the repulsive force toward alpha particles increases. For this reason, alpha particles are unable to cause nuclear changes in elements of high atomic weight, and their use is restricted to action on the elements with light nuclei.

Alpha-induced reactions serve as the basis for the production of neutrons and therefore are extremely important.

Proton-Induced Reactions

Protons suffer from the same limitations as do alpha particles, even more so, because the ratio of mass to charge is only one-half that of the alpha particle. Therefore they are repelled more easily by the positively charged nuclei.

Deuteron-Induced Reactions

Deuterons $[(^2_1H)^+]$ are probably the most effective of the positively charged particles, since they have only one charge, and the ratio of mass to charge is the same as for the alpha particle. Certain deuteron-induced reactions are excellent sources of neutrons.

Gamma-Induced Reactions

Gamma and X rays are inefficient in producing nuclear reactions.

Neutron-Induced Reactions

Neutrons, being neutral, are extremely efficient particles for bombarding the nuclei of all elements. By their use, all elements, with the exception of helium, have been transmuted into other elements.

Two neutron-induced reactions are commonly utilized. One is the n, gamma (n,γ) reaction such as used to produce ^{60}Co by bombarding ^{59}Co with neutrons,

$$^{59}_{27}\text{Co} + ^1_0n \rightarrow ^{60}_{27}\text{Co} + \gamma \tag{8-10}$$

or
$$^{59}\text{Co}\,(n,\gamma)\,^{60}\text{Co}$$

The other reaction, termed the n, p reaction, can be illustrated by the bombardment of ^{14}N with neutrons to produce ^{14}C and a proton,

$$^{14}_{7}\text{N} + ^1_0n \rightarrow ^{14}_{6}\text{C} + ^1_1\text{H} \tag{8-11}$$

or
$$^{14}\text{N}(n,p)\,^{14}\text{C}$$

Activation Analysis

An analytical technique which employs induced radioactivity for quantitative analysis of elements is termed *activation analysis*. Here an unknown sample is subjected to bombardment, generally with neutrons, although charged particles and photons may also be used. The irradiation is conducted for chosen lengths of time, and the elements of interest are identified and assayed by measurement of the characteristic radionuclides formed. Usually the irradiation must be followed by chemical separation of the desired radionuclides after the addition of appropriate carriers. Such analyses are often sufficiently sensitive to measure trace elements in concentrations as low as parts per million or even parts per billion.

8-6 NUCLEAR FISSION

Shortly after the discovery of the neutron, Fermi found neutron bombardment of some heavy metals to be followed by beta activity. Bombardment of uranium produced beta particles of four distinct activities. The activities could not be correlated with those of any of the elements with a mass in the range of uranium. The answer was found by Hahn and Strassmann, who conducted chemical analyses of the products. They found isotopes of barium, lanthanum, strontium, and yttrium, as well as an inert gas (Xe or Kr) and an alkali metal (Cs or Rb) present. From this information it was concluded that neutron capture by the uranium atom was followed by a rupture of the nucleus to form several elements of lower atomic weight. This is termed *nuclear fission.*

Nuclear fission has become of interest because of the tremendous amounts of energy released as a result of the fission process. This release results from a conversion of some of the mass to energy. The energy released during the fission of one gram atom of ^{233}U, ^{235}U, or ^{239}Pu corresponds to 5.3×10^6 kwh. Since the gram atomic weights of these elements vary so little, the energy per pound of fissionable material is essentially as shown in Table 8-4.

Nuclear Explosions

Nuclear fission is initiated by neutron bombardment. During the fission of U^{235}, an average of 2.5 neutrons are released for each atom undergoing fission; thus the reaction can become self-perpetuating once fission is initiated. Fortunately this had not been the case in early laboratory studies. Probability considerations, however, indicated that, if the mass of fissionable material was large enough, a self-sustaining chain reaction would occur. This has been proven many times, beginning with the first atomic bomb test at Alamagordo, New Mexico, in 1945.

Nuclear explosions are accompanied by release of tremendous amounts of radioactivity. The effect on the surrounding ground area depends upon the distance from the ground at which the explosion occurs. In any event, a great deal of the radioactive matter is projected into the upper atmosphere where it is carried around the world. It is constantly returning to the earth, particularly at times of rain and snowfall, as *fallout.* There are over 200 fission products from a

Table 8-4 Energy liberated per pound of fissionable material

0.9×10^{13} cal
1.0×10^7 kwh
2.8×10^{13} ft-lb
3.6×10^{10} Btu

nuclear explosion, although only 17 of these contribute most of the radioactivity present in the mixture. Of particular concern are those having a relatively long half-life, such as strontium 90 and cesium 137 (see Table 8-3). Also of concern are fission products which can be taken into the body and deposited at critical locations: for example, strontium 90, which is a bone seeker, and iodine 131, a thyroid seeker.

Nuclear Power

Control of fission reactions so that the great amount of energy released can be utilized for beneficial purposes has been the objective of a great deal of research, and nuclear-fueled-power-generating plants are now in operation in many countries throughout the world. The reactors in which the fission occurs, of course, must contain the fissionable matter in excess of the "critical mass." To avoid an explosion, neutrons released in the fission process must be controlled in number to keep the chain reaction going at the desired speed. This is accomplished by means of neutron absorbers or moderators. Cadmium, graphite, and deuterium oxide are excellent neutron absorbers.

In nuclear reactors, ^{233}U, ^{235}U, or ^{239}Pu are generally used as fuel to maintain the controlled chain reaction. During decay they function to produce neutrons that can also be used to transmute nonradioactive elements into fissionable or radioactive forms. The reactor at Oak Ridge, Tennessee, has been used largely to produce a wide variety of radioactive isotopes for use in medical, biological, and industrial research.

Nuclear reactors require large quantities of water to dissipate the heat released by the nuclear fission that produces neutrons. Discharge of such cooling water to rivers has been of concern, not only because of possible thermal pollution effects, but also because of the induced radioactivity and possibly some radioactive fission products that it may contain. Concern has also been expressed over the safety of nuclear power plants from sabotage, earthquakes, and industrial accidents, and over the possible long-term fate of radioactive wastes removed during the reprocessing of fuel elements. The struggle between concern for uncertain risks and desire for potential benefits has been keenly evident in this area.

8-7 NUCLEAR FUSION

The fusion of two or more light atomic nuclei to form the nucleus of a heavier element is generally more productive of energy than the fission of heavy elements. The production of one atom of helium from the fusion of four atoms of hydrogen produces seven times as much energy per unit weight of material as the fission of ^{235}U or ^{239}Pu.

The hydrogen bomb depends upon *nuclear fusion* for its tremendous explosive power, resulting from the nuclear fusion of heavy hydrogen isotopes.

Temperature of over 50 000 000°C are needed to initiate the fusion process, and because of this the fusion process is known as a *thermonuclear reaction*. The high temperature required has been obtained by incorporating a fission device for ignition. Thus development of the atomic bomb made the hydrogen bomb a possibility.

The usual fuel for the fusion reaction is deuterium ($_1^2H$) and tritium ($_1^3H$) which combine to produce $_2^4He$ with the release of a neutron.

$$_1^2H + _1^3H \xrightarrow{\Delta\Delta\Delta} _2^4He + _0^1n \qquad (8\text{-}12)$$

$$\downarrow$$
$$\longrightarrow \text{energy}$$

Under the conditions, the neutron is converted to energy. In this reaction about 20 percent of the original mass of the hydrogen is converted to energy.

Nuclear fusion in itself does not release radioactive materials, and partly because of this a great deal of research is being conducted to produce a controlled fusion reaction for power production. At the International Conference on the Peaceful Uses of Atomic Energy, held in Geneva, Switzerland, in 1955, it was predicted that the fusion process would be harnessed to provide power for industrial uses within 20 yr. This period has passed, and now there is question whether this development can take place before the end of the century. Controlled fusion power is the hope of many who otherwise see a dwindling of the World's fuel supplies and a decreasing standard of living for the human race.

8-8 USE OF RADIOACTIVE MATERIALS AS TRACERS

Compounds containing radioactive elements, particularly ^{14}C and ^{125}I, have been used extensively by researchers in the fields of biology, chemistry, and medicine to determine the course of chemical and biochemical reactions. Tracers are being used in many industries to study various phenomena. Petroleum technologists have used tracers extensively to improve methods of processing crude oil and to evaluate the lubricating properties of oils and greases.

In environmental engineering considerable attention has been given to the use of tracers for determining the "flow-through time" of sedimentation tanks and of "reaches" or "stretches" in rivers. They are being used to determine direction and rate of flow of ground-water, and for evaluating the fate of organic and heavy metals in treatment processes and the environment.

The use of radioisotopes in chemical, biological, medical, and engineering research can create a problem for environmental engineers, since some of the materials contained in wastewaters reach sewers and rivers. In sewers and waste treatment plants, certain of the isotopes, such as radioiodine and radiophosphorus, can accumulate in biological slimes and sludges. In rivers, radioactive material may be concentrated by microscopic forms that serve as food for fish and other forms of life consumed by man. Many water supplies are derived from rivers; hence the disposal of radioactive wastes to rivers becomes a matter of concern to all the consumers. For this reason standards have been set for permissible levels of radioactive materials in water supplies and wastewater discharges. Some cities have programs for monitoring their water supplies for radioactivity as a protection to the public.

8-9 EFFECT OF RADIATION ON MAN

Radiation effects on man are classified as *somatic* or *genetic*. Somatic effects are those which cause damage to the individual and include anemia, fatigue, loss of hair, cataracts, skin rash, cancer, etc. Genetic effects include inheritable changes resulting from mutations in reproductive cells. It is widely held that even small dosages of radiation can have some adverse effects, genetic effects being of most concern. Man is exposed to varying levels of natural radiation, especially from extraterrestrial sources. It is generally felt that radiation created through the activities of man should be kept well within the bounds of the natural background radiation. What should be the upper acceptable levels within these bounds is a subject of much debate.

Different types of radiation produce different effects in man, and the effect is different for internal as compared with external exposure. External exposure to alpha particles represents a very small hazard since they have difficulty in penetrating the skin. However, alpha particles can be quite damaging if ingested since they can cause extensive ionization when they collide with matter making up the organs of the body. Beta particles are smaller and move faster than alpha particles and hence can penetrate to greater distances. They can penetrate from a few millimeters to a centimeter or so under the skin and so can be hazardous even with external exposure. Internally, beta particles are more hazardous than they are externally, but they are not as damaging as internal alpha particles. Gamma radiation is most dangerous since it has very high penetrating power and constitutes a hazard to the entire body. It can destroy tissue and inflict serious harm quite rapidly.

The interaction of radiation with biological material and with the water it contains results in the formation of a whole host of ionized species (such as H^+, H_2O^-, H_2O^+, e^-, e^+, H_3O^-, etc.), many of which are highly reactive. These go on to react with protein, to deactivate enzymes, to inhibit cell division, to disrupt cell membranes, and to otherwise damage cell performance. As knowledge of these effects has increased, so have efforts to prevent undue exposure to radiation. The risks versus benefits from use of radioactive materials should constantly be weighed in medical applications as well as in other uses which have already been discussed. With proper precautions, use of radioactive materials offer significant benefits to man.

PROBLEMS

8-1 What is the difference between alpha, beta, and gamma radiation?

8-2 What fraction of an initial amount of ^{14}C would be left after 1 year?

8-3 What is formed from the following disintegrations:

    (a) $^{230}_{90}Th$ emission of an alpha particle?

    (b) $^{40}_{19}K$ emission of a beta particle?

8-4 What is formed by the following induced reactions:

    (a) From ^{238}U by an n, gamma reaction?

    (b) From ^1_1H by an n, gamma reaction?

    (c) From ^6_3Li by an n, alpha reaction?

    (d) From $^{10}_5B$ by an n, alpha reaction?

8-5 The carbon in living plants and animals contains enough ^{14}C to yield about twelve ^{14}C disintegrations per minute per g C. Estimate the age of an entombed piece of wood which yields only 4 disintegrations per minute.

8-6 How many millions of electron volts (Mev) are released by the complete conversion of one gram of mass into energy?

8-7 Write balanced nuclear equations, in the same form as given in Sec. 8-4, for the noted decay of each of the following:

    (a) ^{14}C, beta

    (b) ^{22}Na, positron

    (c) ^{226}Ra, alpha

    (d) ^{32}P, beta

    (e) ^3H, beta

8-8 How long would a waste containing ^{24}Na have to be stored for the concentration to be reduced to 0.1 percent of its initial value?

REFERENCES

Blatz, H.: "Introduction to Radiological Health," McGraw-Hill, New York, 1964.

Etherington, H. (ed.): "Nuclear Engineering Handbook," McGraw-Hill, New York, 1958.

Friedlander, G., J. W. Kennedy, and J. M. Miller: "Nuclear and Radio-Chemistry," 2d ed., Wiley, New York, 1964.

Haïssinsky, M.: "Nuclear Chemistry and Its Applications," Addison-Wesley, Reading, Mass. 1964.

Harvey, B. G.: "Introduction to Nuclear Physics and Chemistry," 2d ed., Prentice-Hall, Englewood Cliffs, N.J., 1969.

Lefort, M.: "Nuclear Chemistry," Van Nostrand, London, 1968.

TWO

WATER AND WASTEWATER ANALYSIS

NINE

INTRODUCTION

9-1 IMPORTANCE OF QUANTITATIVE MEASUREMENTS IN ENVIRONMENTAL ENGINEERING PRACTICE

Quantitative measurements of one sort or another serve as the keystone of engineering practice. Environmental engineering is perhaps most demanding in this respect, for it requires the use of not only the conventional measuring devices employed by engineers but, in addition, many of the techniques and methods of measurement used by chemists, physicists and some of those used by biologists.

Every problem in environmental engineering must be approached initially in a manner that will define the problem. This approach necessitates the use of analytical methods and procedures, in the field and laboratory, that have been proved to yield reliable results in the hands of many people and on a wide variety of materials. Once the problem has been defined quantitatively, the engineer is usually in a position to design facilities that will provide a satisfactory solution.

After construction of the facilities has been completed and they have been placed in operation, usually constant supervision employing quantitative procedures is required to maintain economical and satisfactory performance. Records of performance are needed for reports that have to be made to supervisory personnel and regulatory agencies.

The increase in population density and new developments in industrial technology are constantly intensifying old problems and creating new ones. In addition, engineers are forever seeking more economical methods of solving old problems. Research is continuously under way to find answers to the new problems and better answers to old ones. Quantitative analysis will continue to serve as the basis for such studies.

9-2 CHARACTER OF ENVIRONMENTAL ENGINEERING PROBLEMS

Most problems in environmental engineering practice involve relationships between living organisms and their environment. Because of this, the analytical procedures needed to obtain quantitative information are often a strange mixture of chemical and biochemical methods, and interpretation of the data is usually related to the effect on microorganisms or human beings. Also, many of the determinations used fall into the realm of microanalysis because of the small amounts of contaminants present in the samples. For these reasons, the usual course in quantitative analysis offered by most schools of chemistry is of limited value to environmental engineers, other than to teach basic techniques. Specialized courses in environmental analytical chemistry have been developed to meet this particular need at nearly all schools that train environmental engineers.

9-3 STANDARD METHODS OF ANALYSIS

Concurrent with the evolution of environmental engineering practice, analytical methods have been developed to obtain the factual information required for the resolution and solution of problems. In many cases different methods were proposed for the same determination, and many of them were modified in some manner. As a result, analytical data obtained by analysts were often in disagreement. In cases involving litigation, judges often found it difficult to evaluate evidence based upon analytical methods. In an attempt to bring order out of chaos, the American Public Health Association appointed a committee to study the various analytical methods available and published the recommendations of the committee as "Standard Methods of Water Analysis" in 1905. Since that time, the scope of "Standard Methods" has been enlarged to include wastewaters, and the American Water Works Association and the Water Pollution Control Federation have become collaborators in its preparation. The fourteenth edition appeared in 1976.[1]

---

[1] "Standard Methods for the Examination of Water and Wastewater," American Public Health Association, Inc., New York.

"Standard Methods" as published today is the product of the untiring effort of hundreds of individuals who serve on committees and subcommittees, testing and improving analytical procedures for the purpose of selecting those best suited for inclusion in "Standard Methods." Evidence that is obtained by qualified analysts based upon methods recommended in "Standard Methods" is normally accepted in the courts of the United States without qualification.

9-4 SCOPE OF A COURSE IN WATER AND WASTEWATER ANALYSIS

It would be impossible and unwise to attempt to teach a course in analysis dealing with all the determinations described in "Standard Methods." In the first place, space does not permit such treatment, and in the second place, many of the determinations are highly specific for certain industrial wastes. On the other hand, it is important that a good foundation in analytical procedures be established so that any contingencies that may arise during one's career can be met and handled with a reasonable degree of confidence.

The choice of determinations and the order in which they are included in a particular course depend greatly upon the interests of the instructor. The selection of topics for discussion in the following chapters covers items that the authors have found essential for the basic training of environmental engineers. The order of presentation is a matter of personal opinion but is based upon a natural sequence of dependence and increasing complexity.

A major objective of a course in the analysis of water and wastewater should be to prepare the student for research, rather than as a technician. To this end, it is important that the student appreciate the nature and source of the materials under analysis, the limitations of the analytical methods, how to interpret the data, and how the information may be applied in environmental engineering practice.

9-5 EXPRESSION OF RESULTS

Most materials subjected to analysis in the fields of water and wastewater fall into the realm of dilute solutions, and it is impractical to express results in terms of percent, as is the usual practice in analytical chemistry. Ordinarily, the amounts determined are a few milligrams per liter and oftentimes only a few micrograms. Samples are usually measured by volume, using a volumetric pipet; therefore it is convenient to express results in terms of milligrams per liter (mg/l). Formerly, the term *parts per million* (ppm) was widely used but frequently led to misinterpretation.

Parts per Million

The term *parts per million* is a weight-to-weight ratio. Its use was more or less universal and unquestioned when analysis was principally concerned with water, because a liter of water weighs approximately 1000 g or 1 000 000 mg, and hence 1 mg/l was considered to be equal to 1 ppm. With the development and inclusion of methods for the analysis of polluted waters, such as domestic wastewater, the concept of the relation between parts per million and milligrams per liter did not change, because the specific gravity of domestic wastewater is essentially the same as that of water. As industrial wastes were included in the materials subjected to analysis, many of them were found to have specific gravities considerably different from that of water, and the close relationship between parts per million and milligrams per liter no longer applied. This discrepancy has led to abandonment of the term parts per million in water and wastewater analysis in favor of the use of milligrams per liter.

Milligrams per Liter

Milligrams per liter is a weight-volume relationship and, because environmental engineers deal largely with liquids, it offers a convenient basis for calculation. The expression

$$mg/l \times 8.34 = lb/million\ gal$$

is widely used and has general application. It replaces the original expression

$$ppm \times 8.34 = lb/million\ gal$$

which may be safely applied to problems involving water and wastewater and other liquids whose specific gravity is essentially 1.00, but may lead to serious errors with other liquids unless correction for specific gravity is applied.

The use of milligrams per liter eliminates any opportunity for misunderstanding and confusion. In the past many results have been reported in terms of parts per million with no reference to specific gravity. Obviously the results should have been reported in milligrams per liter. Furthermore, milligrams per liter is directly applicable to the metric system,

$$mg/l = g/m^3$$

or
$$mg/l \times 10^{-3} = kg/m^3$$

SI Units

In order to develop a uniform method of reporting results which is applicable worldwide, the International Organization for Standardization (ISU) published in 1973 a document listing a recommended International System of Units (SI).[2]

---

[2] "International Standard 1000," International Organization for Standardization (1973).

These recommendations were approved by Member Boards of ISU in 30 countries including the United States, and by the International Union of Pure and Applied Chemistry (IUPAC). Many engineering and scientific texts now use SI units exclusively, and many others present a portion of the problems and examples in SI units. In the United States increasing efforts are being made to familiarize the public with the metric system which is the cornerstone for the SI system. Thus, it is prudent for students and practitioners to develop a working knowledge of this system. Fortunately, the metric system has generally been used in environmental chemistry and so adoption to the SI system requires little basic change.

The SI system is founded upon seven base units as listed in Table 9-1. There are a series of units which are derived from the base units through multiplication or division; for example, the SI unit for concentration would be kilograms per cubic meter (kg/m^3). For some derived SI units, special names and symbols exist. Those appropriate to the subject of this text are given in Table 9-2. In addition, the SI system recognizes that there are units outside the SI system which should be retained because of their practical importance or because of their use in specialized fields. A listing of applicable alternative units is given in Table 9-3. Thus, the expression of concentration in milligrams per liter is acceptable under the SI system.

There is a current interest in environmental chemistry in trace organics and metals occurring in less than 1 mg/l amounts. For expressing such low concentrations, it is convenient to have a system of expression which allows for a wide range of possible contaminant levels. This is possible in the SI system through use of prefixes as listed in Table 9-4. The prefixes are used to form names and symbols for multiples of the SI units. For example, $\mu g/l$ is called micrograms per liter and is equivalent to 10^{-6} gram per liter. This is a concentration unit commonly used for trace materials in water. Measured pesticide concentrations in water are generally less than this and may be expressed in nanograms per liter (ng/l) which is equivalent to 10^{-9} gram per liter. Analytical instruments today are sufficiently sensitive so that picogram quantities of materials can often be detected. One pg/l of a pesticide is a concentration which would result from mixing about 4 mg of pesticide in the daily water supply for the City of Chicago. It is difficult to comprehend how such a small concentration of anything could have public health or environmental significance. Ability to measure such small concentrations does indicate the danger of arbitrarily setting "zero" concentration limits for materials.

Other Methods of Expression

In certain determinations, such as color and turbidity, reference is made to arbitrary standards. In these cases results are expressed in units without designation, 0, 1, 5, and so on.

Sludges and some industrial wastes contain enough suspended or dissolved solids so that results can be best expressed in terms of percent. A rule of thumb

Table 9-1 Base units in the SI system

Quantity	Name of base SI unit	Symbol
Length	Meter	m
Mass	Kilogram	kg
Time	Second	s
Electric current	Ampere	A
Thermodynamic temperature	Kelvin	K
Amount of substance	Mole	mol
Luminous intensity	Candela	cd

Table 9-2 Derived units in the SI system

Quantity	Name of derived SI unit	Symbol	Expressed in terms of base or supplementary SI units or in terms of other derived SI units
Force	Newton	N	$1 N = 1 kg \cdot m/s^2$
Pressure, stress	Pascal	Pa	$1 Pa = 1 N/m^2$
Energy, work, quantity of heat	Joule	J	$1 J = 1 N \cdot m$
Power	Watt	W	$1 W = 1 J/s$
Electric charge, quantity of electricity	Coulomb	C	$1 C = 1 A \cdot s$
Electric potential, potential difference, tension, electromotive force	Volt	V	$1 V = 1 J/C$
Electric capacitance	Farad	F	$1 F = 1 C/V$
Electric resistance	Ohm	Ω	$1 \Omega = 1 V/A$
Electric conductance	Siemens	S	$1 S = 1 \Omega^{-1}$

Table 9-3 Units outside the SI system which are acceptable

Quantity	Name	Symbol
Volume	Liter	l
Density	Grams per liter	g/l
Concentration	Moles per liter	mol/l
Time	Day	d
Time	Hour	h
Time	Minute	min

Table 9-4 Multiplication prefixes in SI system

Factor by which the unit is multiplied	Prefix	
	Name	Symbol
10^{12}	tera	T
10^9	giga	G
10^6	mega	M
10^3	kilo	k
10^2	hecto	h
10	deca	da
10^{-1}	deci	d
10^{-2}	centi	c
10^{-3}	milli	m
10^{-6}	micro	μ
10^{-9}	nano	n
10^{-12}	pico	p
10^{-15}	femto	f
10^{-18}	atto	a

often applied is this: When concentrations exceed 10 000 mg/l,[3] results are expressed as percent. Wherever practice has established a precedent in opposition to the rule, the rule is ignored.

Water chemists often prefer to express results in terms of milliequivalents (me) per liter. This allows them to translate results directly in terms of other chemicals. Milliequivalents per liter are obtained by dividing milligrams per liter of the element or ion by its equivalent weight in grams. Equivalent weights are discussed in Chap. 10. Moles per liter is also frequently used because of need when working with equilibrium relationships. Thus, the environmental engineer becomes exposed to many methods of expression, and to be competent in his field, he must develop a working knowledge of each. The need for a more uniform system of expression is evident!

9-6 OTHER ITEMS

It is expected that the environmental engineering student will become familiar with all the material in the Introduction to "Standard Methods." Particular attention should be directed to the sections on collection of samples; laboratory

---

[3] A figure of 10 000 mg/l is equivalent to 1 percent when the specific gravity is equal to 1.0.

apparatus, reagents, and techniques; statistics for the analyst; and significant figures.

Chapters 10 and 11 of this text contain discussions of some additional items of general importance and usefulness to water and wastewater analysis. It would be well for the student to become generally familiar with the contents of these chapters, and to refer back to them for details of specific importance to individual analyses which are discussed in subsequent chapters.

CHAPTER
TEN

BASIC CONCEPTS FROM
QUANTITATIVE CHEMISTRY

10-1 GENERAL OPERATIONS

Quantitative chemistry may be considered a keystone in the training of an
environmental engineer. It serves as the basis for most research work and
many field investigations. Unfortunately most college courses in quantitative
analysis, although they furnish excellent fundamental information, do not provide laboratory instruction and practice that are particularly valuable to environmental engineers. For this reason, specialized courses in quantitative analysis, often described as courses in "sanitary chemistry" or "water and waste
analysis," have been developed to meet the need.

At this writing, most leading schools concerned with the training of environmental engineers are using as a reference textbook "Standard Methods
for the Examination of Water and Wastewater."[1] This book is designed to be
used by trained analysts and therefore presumes a certain background of information. Since many environmental engineers do not have such a background,
the purpose of this chapter is to fill that need. It is recommended that each
environmental engineer add a copy of a standard, up-to-date text on quantitative analysis to his library for reference purposes. It is impossible and unnecessary in this chapter to cover in detail the many aspects of analytical
chemistry that are adequately dealt with in all standard textbooks.

[1] Published jointly by the American Public Health Association, American Water Works Association, and Water Pollution Control Federation.

Sampling

It is an axiom that the analytical results obtained in the laboratory can never be more reliable than the sample upon which the tests are performed. It may be safely stated that more results are in error because of inadequate sampling than because of faulty laboratory techniques. The subject of sample collection and care of samples is treated quite adequately in "Standard Methods," except for the subject of *grab* versus *composited* samples.

Grab samples are those taken more. or less instantaneously and analyzed separately. In general, most sampling in environmental engineering practice is of the grab variety. The major problem confronting the engineer is the decision of frequency of sampling. In this decision he must always balance the issues of a sufficient number of samples for reliability versus costs. The number of samples may vary from one to over a hundred per day, depending upon the nature of the material to be sampled. At this point a good deal of engineering judgment is involved. A few examples will serve to illustrate.

1. Consider a deep well which has been in service for some time. The water quality will be uniform, and a single grab sample will give a true picture of conditions.
2. Oftentimes changes occur slowly, as in large rivers, and once daily grab samples are adequate.
3. Where the character of a material changes considerably within a 24-h period, the use of grab samples at frequent intervals is dictated. In sampling industrial wastes, it may be necessary to obtain such samples every 10 or 15 min. The individual samples may be analyzed for certain characteristics such as pH, acidity, and alkalinity, and then pooled into 2-, 4-, 8-, 12-, or 24-h composites for more complete analysis.

Composited samples are used mainly in evaluating the efficiency of wastewater treatment facilities, where average results are adequate. Such samples are collected at regular intervals, usually every hour to two, and pooled into one large sample over a 24-h period. Under such conditions, detention times can be considered self-canceling, and the only requirement that must be met is that the amount of each individual sample be taken in proportion to the flow existing at the time.

When composited samples are taken for shorter periods than 24 h (8-h composited samples have been quite common in waste treatment practice), serious errors may be introduced by ignoring detention time. In conventional activated sludge plants, the theoretical detention time from influent to effluent is about 8 h; thus samples of influent collected from 8 A.M. to 4 P.M. measure the strength of daytime waste, but samples of effluent collected over the same period of time measure the treatment given to weak waste that entered the plant prior to 8 A.M. In taking short-term composited samples, sampling at downstream locations should be adjusted to detention times in the treatment units involved. In the average conventional activated sludge treatment plant with a total of 8 h of detention, sampling of the final effluent should not begin until 8 h after sampling of the influent is started; thus 16 h are needed to collect a set of 8-h composited samples.

Laboratory Apparatus and Reagents

A good grade of laboratory apparatus is adequate for all practical uses. Recalibrated or Bureau of Standards graduated glassware is very expensive and not required for routine work. Pyrex or a similar glassware of low solubility and low coefficient of expansion is highly recommended. Reagents should be of analytical-reagent grade or known to meet the specifications of purity established by the American Chemical Society. Lower grades of chemicals, even technical grade, may be used for some purposes, but the analyst must make sure that such grades do not contain undesirable amounts of certain impurities. It is not practical to buy a reagent grade of sodium hydroxide for purposes that require a high-purity sodium hydroxide. The best grades obtainable are too impure. It is just as easy to purify a cheaper grade in the laboratory, and it serves the purpose just as well, at a much lower cost.

Precipitation

Some analytical methods depend upon precipitation of an ion to allow its separation and measurement by actual weighing of the precipitated material. There are several requirements that must be met to make such procedures reliable. The major ones are as follows:

1. The precipitated material must have a very low solubility in water; i.e., its K_{sp} must be a very small number.
2. It must precipitate in a high state of purity or be capable of reprecipitation for further purification.
3. It must be capable of drying or of ignition at temperatures above 100°C to a definite compound of fixed composition.
4. It should not be hydroscopic at room temperatures.

   Precipitation may also be used as a method of purification. Under such conditions, the desired ion is precipitated from solution, filtered, placed into solution again by the use of special reagents, and measured by other means than weight analysis.

Filtration

Filtration of precipitates or solids is accomplished by means of paper filters or Gooch crucibles fitted with asbestos or glass fiber filter mats. Many materials that the environmental engineer wishes to measure cannot be subjected to temperatures above 103°C without losing chemically bound water. Gooch crucibles or similar filters are used for such measurements. Filtration through filter paper (analytical, ashless grade) is commonly used where ignition at temperatures above 600°C is required or permissible. At such temperatures the paper is destroyed, and the desired residue remains. An ashless paper or one whose ash weight is known should be used.

Filter papers vary greatly in their porosity. It is essential to use one that is fine enough to retain all the precipitate but as coarse as possible to obtain rapid filtration rates. Barium sulfate requires the use of a very fine paper. Ferric hydroxide is retained on very coarse fast-filtering paper.

Separation of a precipitate can be accomplished by the use of a proper filter. Quantitative results, however, depend upon two other important factors:

1. All the precipitate or solid material must be transferred from the original vessel to the filter.
2. The precipitate and filter must be washed with water or a suitable solvent to remove dissolved solids that remain in the precipitate and wetted filter. Three washings with complete drainage between washings are considered sufficient.

Drying or Ignition

The standard temperature for drying residues or solids is 103°C. This is stipulated because many of the residues or solids involved in environmental engineering practice are organic in nature and release water of composition in significant amounts at higher temperatures. Drying at 103°C ensures the removal of all free water, provided that the drying period is long enough, and minimizes the loss of other water.

Ignitions are commonly conducted at 600°C, unless specified otherwise, to ensure the destruction of all organic matter by oxidation to carbon dioxide and water while minimizing the loss of inorganic salts by volatilization or decomposition. Calcium carbonate is a major component of many residues and is stable at 600°C.

From these considerations, it should be obvious that much of the analytical work with which the environmental engineer is concerned involves measurement of organic materials; this is really the major reason why a conventional course in quantitative analysis falls short of supplying his needs.

Desiccation

Following drying or ignition operations, the residues and their containers (crucibles or evaporating dishes) *must be cooled to room temperature* before weighing on the analytical balance. If such cooling were allowed to take place in the open air, moisture would be picked up from the air by the residue and container. The amount of moisture pickup would depend upon the time of exposure and the relative humidity. To overcome this difficulty, residues and their containers are cooled in *desiccators* where the relative humidity is kept near zero percent by means of a *desiccant* such as anhydrous calcium chloride.

10-2 THE ANALYTICAL BALANCE

The construction of analytical balances and the theory of weighing are adequately covered in standard texts on quantitative analysis. The care of such instruments cannot be stressed too much, however, for engineers are accustomed to dealing in terms of pounds, tons, cast iron, concrete, and jackhammers. It is important that they recognize that analytical balances fall into the realm of *delicate* instruments and that great care and scrupulous cleanliness must be maintained with respect to both the balance and the weights used.

Automatic single-pan balances, like the one shown in Fig. 10-1, are by all odds the most practical devices for modern laboratories. They are about twice as expensive as other high-quality balances but, in commercial laboratories, will usually pay for themselves in time saved within a relatively short period.

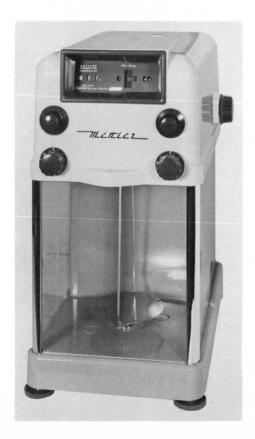

Figure 10-1 A single-pan automatic balance. *(Mettler Instrument Corporation.)*

The importance of having the objects to be weighed at room temperature cannot be overemphasized. The average student in quantitative analysis should not have difficulty in translating his knowledge of physics to explain how air currents cause an object having a temperature above room temperature to weigh less than it should and an object which is colder than room temperature to weigh more than it should.

10-3 GRAVIMETRIC ANALYSIS

Gravimetric analysis means analysis by weight and pertains to all determinations wherein the final results are obtained by means of the analytical balance. In the environmental engineering field, determination of total solids (residue on evaporation), suspended solids, and volatile solids is made by gravimetric procedures because there is no better way of gaining the desired information. In general, however, gravimetric procedures are avoided as much as possible because they are time-consuming. In other than solids determinations, the environmental engineer has little occasion to use the analytical balance except to make up standard solutions, and the like. An outstanding exception is the determination of sulfates. No other method, except possibly the turbidimetric method, approaches it in accuracy.

Because of the importance of solids determinations that must be done by gravimetric procedures, it is essential that the environmental engineer know the fundamentals of gravimetric analysis. Many of these have been discussed in Sec. 10-1; a few more follow.

Constant Weight

All gravimetric measurements require some sort of crucible or dish to hold the residue or precipitate. The weight of this container (tare) must be known and deducted from the gross weight to obtain the net weight of the material being measured.

Porcelain crucibles or dishes are commonly used for gravimetric work. Platinum ware has many advantages but is not ordinarily used, except for special determinations, because of its cost. All objects exposed to the air for any length of time have a film of dust and adsorbed moisture. Porcelain ware is manufactured with some of its surface unglazed; consequently, it will also adsorb moisture into its interior. To use such containers for gravimetric work without proper conditioning would lead to relatively large errors in the final results. Oftentimes negative results are obtained when small quantities are involved.

The conditioning of crucibles and dishes for gravimetric work involves pretreatment to eliminate the dirt and moisture. All containers should be thoroughly cleaned with water and then heat-treated under exactly the same conditions as those to which the container will be subjected in the actual determination. If the container is to be used to measure residue on evaporation at 103°C, it should be conditioned at that temperature. To heat the dish, particularly porcelain ware, at higher temperatures, such as at 600°C, will drive out more moisture than desired. The dishes should be heated or fired at the desired temperature, cooled in the desiccator and weighed, heated and fired, cooled and weighed, repeatedly, until the container reaches what is known as constant weight. For small dishes with a weight of less than 25 g, this is assumed to be ±0.0002 g. For heavier dishes the allowance is greater but never over ±0.0005 g. With a little experience, the beginner will learn that a preliminary heating or firing of less than 2 h is seldom worthwhile. This is particularly true of new dishes and of old dishes that have not been used for some time. After dishes have been brought to constant weight they should be kept in a desiccator to avoid collection of dust or adsorption of moisture prior to use.

Theoretically the principles outlined above for getting the tare weight of containers should be applied to the dish plus the residue to be measured. This is true when inorganic residues are involved; however, it has been demonstrated that when organic substances are concerned, such materials will usually continue to yield moisture in small amounts for long periods of time. Since free water is given off rather rapidly, barring physical interference to its release, the drying time for such solids is usually specified, and the weight obtained at the end of that time is accepted as being constant. The drying time for total solids in water (following evaporation to practical dryness on a steam bath) and for suspended solids by Gooch crucibles is 1 h. The drying time for total solids in sludges and many industrial wastes (following evaporation to dryness) is considered to be 10 h. Usually, drying overnight is more practical and is recommended for sludges and certain industrial wastes.

Preparation of Gooch Crucibles

Gooch crucibles require the use of a filter mat over the perforated bottom. Asbestos fibers are commonly used for this purpose.[2] Asbestos fiber suitable for use in Gooch crucibles should be specified when purchasing. It should be of fairly long fiber length and should be acid-washed with hydrochloric acid. A "soup" of the fiber can be made by placing some in a bottle, adding water so that the bottle is about one-half full, and then shaking vigorously. The finely divided material may be decanted and the procedure repeated. Coarse particles that will not disperse can be shredded by placing them with some water in a blender and running the agitator for about 10 s. The "soup" obtained should be thoroughly agitated by shaking, allowed to stand for 5 min, and then the fines that do not settle readily decanted to waste. The process should be repeated with addition of more water until a reasonably clear supernatant water is obtained following a 5-min settling period.

---

[2] Glass fiber filters are an excellent substitute for asbestos.

The Gooch crucible is prepared by adding asbestos "soup" so that it is about one-half full and then applying vacuum to withdraw the water. If all holes in the bottom are covered and the mat covers the bottom entirely, remove the crucible, hold it up to one eye, and look into a strong light. If none of the perforations in the crucible is discernible, the mat is of sufficient thickness. Replace the crucible in the suction filter, pour a small amount of the fines off the top of the unagitated "soup" into the crucible, and apply suction. This ensures a perfect filter mat. Filter about 250 ml of distilled water through the crucible. This will dislodge any fibers that may be in a vertical position in the perforations and give as nearly perfect a mat as possible. Remove any asbestos fibers that may be adhering to the rim or exterior of the crucible, place in the drying oven, and bring to constant weight. If the crucible is to be used for the determination of volatile solids as well as suspended solids, it should be dried at 103°C for a few minutes before placing in the furnace at 600°C to avoid disruption of the mat by explosions from steam formation.

Much of the dissatisfaction with the use of Gooch crucibles can be traced to improperly prepared filter mats.

10-4 VOLUMETRIC ANALYSIS

Volumetric analysis is a phase of quantitative analysis that depends upon the measurement of liquid reagent volumes of *standard solutions* needed to complete particular reactions in samples submitted to test. A standard solution is defined as follows: *a solution whose strength or reacting value per unit volume is known.* The facilities needed for conducting a simple volumetric analysis are (1) equipment to measure the sample accurately, either an analytical balance or volumetric glassware such as pipets; (2) a standard solution of suitable strength, (3) an indicator to show when the stoichiometric end point has been reached, and (4) a carefully calibrated buret for measuring the volume of standard solution needed to reach the stoichiometric end point as shown by the indicator.

Analysis by volumetric methods is very popular, as compared with gravimetric methods, because of the time that usually can be saved by such procedures. Environmental engineers use volumetric methods for many determinations such as dissolved oxygen, BOD, COD, and chlorides. There are a number of concepts and techniques involved in volumetric measurements that must be understood in order to obtain accurate results.

Calibrated Glassware

Calibrated glassware is of two types: (1) that which is calibrated to contain a definite volume, e.g., volumetric flasks and graduated cylinders; and (2) that which is calibrated to deliver prescribed volumes, e.g., pipets and burets. The former should never be used as a substitute for the latter when accuracy is required because some liquid will always remain in the container, and the delivered amount will be less than the calibrated amount. On the other hand, pipets are calibrated to deliver definite amounts under specified conditions and must always retain a small amount of the liquid after delivery.

Pipets are of two types, transfer and Mohr. Transfer or volumetric pipets have an enlarged section and only one graduation mark. They will deliver the specified amount provided that the pipet is clean, it is held in a near vertical position during delivery, contact is made between the tip of the pipet and the wall of the receiving vessel, and drainage is allowed to occur for 5 s after the level of the liquid in the tip appears to have reached a static condition. Transfer pipets should be used where volumes must be measured with a high degree of accuracy. Mohr-style pipets are made from glass tubing of uniform bore and have multiple graduation marks. The small sizes of 5-ml capacity or less may be used to deliver small quantities of liquids quite accurately, but with 10-ml and larger sizes the bore and calibrations are such that delivery of accurate amounts cannot be expected. The Mohr pipet is used largely for measuring fractional volumes of 1 ml or for measurement of volumes when accuracy is of little importance.

In general, it may be stated that transfer or volumetric pipets should be used for measurement of samples and of standard solutions. Mohr pipets may be used for the addition of nonstandardized reagents.

Calibrated glassware should be kept thoroughly clean. This is particularly true of transfer pipets and burets. Cleaning may be accomplished by means of some of the modern detergents designed for such purposes or by the use of chromic acid cleaning solution as described in "Standard Methods."

Equivalent or Normal Solutions

By definition, *a standard solution is one whose strength or reacting value per unit volume is known.* Volumetric analysis could be practiced solely with such solutions, but it is much more convenient to prepare and use solutions which are equivalent to one another in strength, so that 1.0 ml of reagent A will react with exactly 1.0 ml of reagent B, and so on.

We can establish the basis for a system of equivalent solutions for the measurement of acids and bases (acidimetry and alkalimetry) from the fundamental equation involved:

$$\underset{1.008}{H^+} + \underset{17.007}{OH^-} \rightarrow H_2O \tag{10-1}$$

From this equation it is seen that 1 g atomic weight or 1.008 g of hydrogen ion reacts with 1 g ionic weight or 17.007 g of hydroxyl ion. The figure 17 serves for all practical purposes in the latter case. Solutions containing these amounts of hydrogen ion or hydroxyl ion can be considered equivalent.

The *equivalent weight* of a compound is defined as *that weight of the compound which contains one gram atom of available hydrogen or its chemical equivalent.* The equivalent weight can be determined as follows:

$$\text{Equivalent weight} = \frac{MW}{Z} \qquad (10\text{-}2)$$

where MW is the molecular weight of the compound and Z is a positive integer whose value depends upon the chemical context.

For acids, the value of Z is equal to the number of moles of H^+ obtainable from one mole of the acid. For HCl, $Z = 1$; and for H_2SO_4, $Z = 2$. For acetic acid (CH_3COOH), $Z = 1$, since only one of the hydrogen atoms in the acetic acid molecule will ionize to yield available H^+ ions in solution.

$$CH_3COOH \rightarrow CH_3COO^- + H^+$$

For bases, the value of Z is equal to the number of moles of H^+ with which one mole of the base will react. For NaOH, $Z = 1$; for $Ca(OH)_2$, $Z = 2$.

A *normal solution* is defined as *one that contains one equivalent weight of a substance per liter of solution*. Thus, for the preparation of 1 liter of a normal solution of an acid or a base, all that is needed is enough of the compound that furnishes the ion to yield 1.008 g of H^+ or 17 g of OH^- and enough distilled water to make a solution having a volume of 1 liter. The *normality* of a solution is its relation to the normal solution, and the symbol N is used as the abbreviation for "normal." Thus a half-normal solution is expressed either as $0.5\,N$ or as $N/2$. The preparation and standardization of normal solutions will be discussed in Chap. 14.

Normal solutions are also used for volumetric measurements that do not involve acidimetry or alkalimetry. For example, chlorides are measured by titration with a reagent such as silver nitrate. The reaction involves precipitation of chloride ion as silver chloride in the following manner:

$$\underset{35.5}{Cl^-} + \underset{108}{Ag^+} \rightarrow \underline{AgCl} \qquad (10\text{-}3)$$

In this reaction 1 mole of Ag^+ is equivalent to 1 mole of Cl^-. From the equation

$$HCl \rightarrow H^+ + Cl^- \qquad (10\text{-}4)$$

it can be reasoned that 1 mole of Cl^- is equivalent to 1 mole of H^+. Since things equal to the same thing are equal to each other, it may be stated safely that 1 mole of Ag^+ is equivalent to 1 mole of H^+. Thus a normal solution of Ag^+ is one that contains one mole of Ag^+ per liter.

For all practical purposes, the equivalent weight of a compound involved in a precipitation reaction can be found from Eq. (10-2) by letting Z equal the oxidation number of the ion. The equivalent weight of $BaCl_2$ is MW/2. The equivalent weight of $Al_2(SO_4)_3$, considered either as an aluminum or as a sulfate salt, is MW/6. In certain cases, a salt may have more than one equivalent weight, depending on how it is used. Thus the equivalent weight of K_2HPO_4 is MW/2 when potassium is the reacting ion, and MW/3 when the reacting ion is phosphate.

A third form of normal solutions involves those used for their oxidizing or reducing values. Examples are solutions of potassium dichromate, ferrous ammonium sulfate, and sodium thiosulfate. The equivalent weight or that weight of a compound needed to prepare 1 liter of a normal oxidizing or reducing solution is derived from the following equation:

$$\underset{23}{Na^\circ} + \underset{1.008}{H^+} \rightarrow Na^+ + H^\circ$$

In this reaction one H^+ steals one electron from one atom of sodium, or 1 mole of hydrogen ions can be considered as equivalent to "1 mole of electrons." Sodium undergoes a change in oxidation number of 1; therefore the equivalent weight of a compound to be used as an oxidizing or reducing agent can be determined by letting Z equal the oxidation number change that the compound undergoes in the reaction involved. It is very important that this oxidation number change be known. For example, potassium permanganate undergoes an oxidation number change of 5 under acid conditions, and its equivalent weight for such use would be MW/5. Under alkaline conditions, it undergoes an oxidation number change of 3, and its equivalent weight for such use is MW/3. Table 10-1 shows the equivalent weights of some common oxidizing and reducing agents.

Primary Standards

The standardization or measurement of the exact strength of a normal or other solution depends upon the use of some standard material whose purity is

Table 10-1 Equivalent weights of some common oxidizing and reducing agents

Agent	Nature	Conditions	Oxidation number change	Equivalent weight
$KMnO_4$	Oxidizing	Acid	5	MW/5
$KMnO_4$	Oxidizing	Alkaline	3	MW/3
$K_2Cr_2O_7$	Oxidizing	Acid	2×3	MW/6
I	Oxidizing	Acid	1	AW/1
$KH(IO_3)_2$	Oxidizing	Acid	2×5	?*
$Na_2C_2O_4$	Reducing	Acid	2×1	MW/2
KI	Reducing	Acid	1	MW/1
As_2O_3	Reducing	Acid	2×2	MW/4
$Na_2S_2O_3$	Reducing	Acid	?	MW/1†
$Fe(NH_4)_2(SO_4)_2$	Reducing	Acid	1	MW/1

* In the case of $KH(IO_3)_2$, the reaction must be known in order to calculate the equivalent weight. For ordinary reactions it would be MW/10, but for the usual reaction with KI it is MW/12.

† The equivalent weight of $Na_2S_2O_3$ must be calculated indirectly because the oxidation number change of sulfur is not known. In the reaction with iodine, one $Na_2S_2O_3$ is equivalent to one atom of iodine. Therefore, the equivalent weight is MW/1.

known. Primary standards are usually salts or acid salts of high purity that can be dried at some convenient temperature without decomposing and that can be weighed with a high degree of accuracy. Examples are sodium carbonate and potassium acid phthalate, which are used to standardize acid and base solutions, respectively; potassium bi-iodate and potassium dichromate for reducing solutions; potassium oxalate for oxidizing solutions; and sodium chloride for solutions of silver ion. Analytical-reagent-grade chemicals are usually satisfactory for most purposes. For some research purposes and possibly for referee work, analyzed primary standards may be obtained from the U.S. Bureau of Standards.

Secondary Standards

Any solution that has been standardized against a primary standard is considered a secondary standard and may be used as such. In the preparation and use of solutions of a base, this practice is often followed and is a valuable time saver with all solutions that are not stable and therefore have to be restandardized frequently.

Choice of Indicators

Analysis by volumetric procedures requires that some method of indicating the stoichiometric end point be employed. Internal indicators are greatly preferred. In any event, they should herald the reaching of the completed reaction as closely as possible. Improper choice of indicators can introduce serious errors into volumetric work. Indicators in common use today are of several major types, viz., electrometric, acid-base, precipitation, adsorption, and oxidation-reduction. All types are used regularly in the practice of environmental engineering. Because of the variety, it is ordinarily best to discuss indicators according to type. For environmental engineers, however, it seems more appropriate to discuss applications to type reactions. The major type reactions are discussed in the following.

Acidimetry and Alkalimetry

Before proceeding with a discussion of the choice of indicators for measurements involved in acidimetry and alkalimetry, it is best to review some of the theoretical aspects developed in Sec. 5-5. In the first place, the titrating agent used is always a strong acid or a strong base, meaning, of course, highly ionized; therefore the combinations involved in titrations are strong plus strong, or weak plus strong. It is important that the pH changes occurring during acid and base titrations be understood. Several examples are presented in Sec. 5-5.

The pH of equivalence points involved in acidimetry and alkalimetry vary, depending upon the ionization constants and concentrations of the materials used. For this reason, the use of an electrometric indicator (pH meter) is superior for measuring end points in acid or base titrations, and will undoubtedly become standard practice in most laboratories. Such titrations are commonly referred to as *electrometric titrations* and require the use of a standard pH meter.

Color indicators A number of naturally occurring or synthetically prepared organic compounds undergo definite color changes in well-defined pH ranges. In general these compounds are weak acids or bases which change color when changed from the neutral to the ionized form. The pH at which the color change takes place depends upon the ionization constant for the particular indicator. A number of indicators which are useful for various pH ranges are listed in Table 10-2.

From the titration curves described in Sec. 5-5, it will be noted that the concept of neutrality at pH 7 has little application in acidimetry or alkalimetry. Therefore there is little common need for an indicator for this range. Only the curves for titration of strong versus strong can be said to have inflections at pH 7. It will be noted that the titration curves for all acids with ionization constants greater than 10^{-7} show inflection points at or below pH 8.3. Thus the stoichiometric end point for all such measurements can be said to have been reached at pH 8.3, and any indicator that gives a well-defined color change at such a pH is satisfactory for measuring acids. Phenolphthalein changes from colorless to pink in the pH range 8.2 to 8.4 and is the color indicator commonly used in environmental engineering practice. Phenolphthalein is also of great value in that it may be used to indicate the end point when strong bases (caustic alkalinity) are being measured, because it changes from pink to colorless at a pH of about 8.3. It is also used to measure carbonate ion (carbonate alkalinity) by indicating when the carbonate ion has been converted to bicarbonate ion,

Table 10-2 Properties of various acid–base indicators

Indicator	Acid color	Base color	pH Range
Methyl violet	Yellow	Violet	0–2
Malachite green (acidic)	Yellow	Blue-green	0–1.8
Thymol blue (acidic)	Red	Yellow	1.2–2.8
Methyl orange	Red	Yellow-orange	3.1–4.6
Bromcresol green	Yellow	Blue	3.8–5.4
Methyl red	Red	Yellow	4.4–6.2
Litmus	Red	Blue	4.5–8.3
Bromthymol blue	Yellow	Blue	6.0–7.6
Phenol red	Yellow	Red	6.8–8.4
Thymol blue (alkaline)	Yellow	Blue	8.0–9.6
Phenolphthalein	Colorless	Red	8.2–9.8
Thymolphthalein	Colorless	Blue	9.3–10.5
Alizarin yellow	Yellow	Lilac	10.1–11.1
Malachite green (alkaline)	Green	Colorless	11.4–13.0

and for carbon dioxide, showing when all the carbonic acid (CO_2) has been converted to bicarbonate.

The measurement of weakly basic substances requires the use of an indicator that will change color at pH levels of about 4.5. The indicator commonly used for such purposes has been methyl orange. It does not yield well-defined color changes that are easily detected by all people and is readily bleached by chlorine, but is still considered the standard for most analysis that the environmental engineer encounters. In using methyl orange indicator it is good practice to use a "blank" for comparison. Other indicators have been proposed as a substitute but most suffer from interference by carbon dioxide. This is particularly true of methyl red. The best substitute is electrometric measurement. Methyl orange is also used as an indicator to measure strong acids, commonly referred to as mineral acidity or *methyl orange acidity*.

In environmental engineering practice, methyl orange and phenolphthalein are commonly used for all acid and base titrations. The choice of which indicator to use is sometimes confusing to students, until they become acquainted with the chemistry involved. Choice of a methyl orange or phenolphthalein indicator can be simplified by applying the following steps:

1. Write the chemical equations for the reactions that are expected.
2. Consider the products of the reaction or reactions and decide whether the resulting solution will be acidic, basic, or neutral.
3. If acidic, use methyl orange. If basic, use phenolphthalein. If neutral, either indicator may be used.

Precipitation Methods

The best example of a volumetric method involving formation of a precipitate in environmental engineering practice is the determination of chloride (chloride ion) by titration with silver nitrate. The indicator commonly used is potassium chromate (K_2CrO_4). Like Cl^-, chromate ion (CrO_4^{2-}) also forms a precipitate with Ag^+.

$$2Ag^+ + CrO_4^{2-} \rightarrow \underline{Ag_2CrO_4} \qquad (10\text{-}5)$$

Silver chromate is red in color, and its appearance is used to show completion of the precipitation of Cl^-.

In order for CrO_4^{2-} to serve in this capacity the solubility of Ag_2CrO_4 must be sufficiently greater than that of AgCl so that essentially all chloride ions are precipitated before detectable amounts of Ag_2CrO_4 are formed. This means that the *effective* solubility product (K_{sp}) of Ag_2CrO_4 must be slightly greater than that of AgCl. Since indicators of this kind require an excess of reagent to form enough colored precipitate for visual detection, determination of this excess, commonly referred to as *indicator error* or blank, must be made and applied to all titrations.

Oxidation-Reduction Methods

Two types of indicators are commonly used: adsorption and those that change with oxidation-reduction potential. The former is illustrated by the use of starch solution to indicate the end point when solutions of iodine are titrated with sodium thiosulfate. Titration is "by eye" until the iodine concentration is near extinction, as shown by a pale yellow color. Upon the addition of a good starch indicator a blue solution results. This is due to adsorption of iodine upon the surface of the colloidal starch particles. As the titration proceeds, iodine is released from the starch, and disappearance of the blue color is taken as the end point. Other adsorption indicators show when an excess of titrant has been added. Starch acts in this manner when iodine solutions are used as the titrant.

Internal indicators that will change color with a change of oxidation-reduction potential (ORP) are commonly used (see Sec. 5-10). Such indicators are usually soluble organic substances which exist in two states of oxidation, the two forms being in equilibrium and of markedly different color. An example familiar to environmental engineers is Ferroin (ferrous 1,10-phenanthroline sulfate) which is used to indicate when sufficient ferrous ammonium sulfate titrant has been added to measure excess dichromate ion in the chemical oxygen demand (COD) test. It should be mentioned that selection of an ORP indicator depends upon the ORP at the stoichiometric end point for the particular reaction involved, just as the selection of an indicator for acidimetry or alkalimetry depends upon the pH of the solution resulting at the equivalence point.

The end point of an oxidation-reduction reaction can be determined electrometrically. A wide variety of electrode systems may be used. A rotating platinum electrode as employed in the Wallace & Tiernan amperometric titrator (Fig. 10-2) serves very well for many purposes.

Calculations

The data obtained during a titration must be translated into terms of weight to be of practical value. Since the unit of volume commonly used for measuring the amount of titrant added to a sample is the milliliter, the weight of active material per milliliter is important. For the normal system of standard solutions, one equivalent weight of a substance, in grams, in a liter of solution is a 1.0 normal solution. Each milliliter of such a solution contains one one-thousandth of an equivalent weight or what is more commonly referred to as 1 milliequivalent (me). Thus, when working with solutions of normality N, we find that

$$\text{ml titrant} \times N = \text{me of active material in titrant used}$$

The advantage of using equivalent solutions is that the me of active material in the titrant used is equal to the me of active material in the sample being titrated. In order to determine the concentration of active material in a sample, one must know the sample volume:

Figure 10-2 A Wallace & Tiernan amperometric titrator. *(Wallace & Tiernan Inc.).*

$$\text{me/liter active material in sample} = \frac{\text{ml titrant} \times N \times 1000}{\text{sample volume in ml}} \quad (10\text{-}6)$$

In environmental chemistry, it is normal practice to report the concentration of a material in weight units rather than in equivalent units. Also, the concentration of materials measured is usually so small that it is inconvenient to think in terms of grams. Therefore most thinking is done in terms of milligrams, and results are expressed in terms of milligrams per liter (mg/l). The number of milligrams of active material in a me is equal to the equivalent weight (EW) of the active material, so that

$$\text{mg/l of active material in sample} = \frac{\text{ml titrant} \times N \times \text{EW} \times 1000}{\text{sample volume in ml}}$$
$$(10\text{-}7)$$

Samples are commonly measured by volume, and to avoid unnecessary and repeated calculations, the sample size and the normality of titrant are often chosen so that the buret reading in milliliters times some whole number, such as 1, 10, 20, 50, or 100, gives the number of milligrams of material per liter. This requires the use of solutions whose normality is some exact value. The preparation, standardization, and use of such solutions will be discussed in Chap. 14.

10-5 COLORIMETRY

Analytical chemists as well as others who use analytical procedures are constantly striving to find faster, more economical, and convenient ways of obtaining quantitative data. To this end, colorimetric methods of analysis have been developed for many determinations of interest to the environmental engineer. Colorimetric methods are most applicable in the realm of dilute solutions. This is fortunate for the environmental engineer, because a large majority of materials with which he deals falls in this classification.

In order for a colorimetric method to be quantitative, it must form a compound with definite color characteristics and in amounts directly proportional to the concentration of the substance being measured. Solutions of the colored compound or complex must have properties that conform to Beer's law and to Lambert's law.

Lambert's Law

Lambert's law, sometimes referred to as Bouguer's law, relates the absorption of light to the depth or thickness of the colored liquid. This law states that each layer of equal thickness absorbs an equal fraction of the light which traverses it. Thus when a ray of monochromatic light passes through an absorbing medium, its intensity decreases exponentially as the length of the medium increases,

$$T = \frac{I}{I_0} = 10^{-k'l} \qquad (10\text{-}8)$$

or

$$A = \log \frac{I_0}{I} = k'l \qquad (10\text{-}9)$$

where I_0 = intensity of light entering solution
I = intensity of light leaving solution
l = length of absorbing layer
k' = constant for particular solution
T = transmittance of solution
$100T$ = percentage transmission of solution
A = absorbance, or optical density of solution

If light intensity decreases exponentially with increase in depth or thickness, the colored solution behaves in conformity with this law. There are no known exceptions to this law as long as homogeneous materials are involved. This law is discussed solely because a knowledge of it is germane when the depth or thickness of colored samples is varied to decrease a color to a level in the range of a series of prepared standards or of spectrophotometric equipment. Knowledge and application of this law can save much time and expense.

Beer's Law

Beer's law is concerned with light absorption in relation to solution concentration. It states that the intensity of a ray of monochromatic light decreases exponentially as the concentration of the absorbing medium increases,

$$T = \frac{I}{I_0} = 10^{-k''c} \tag{10-10}$$

or
$$A = \log \frac{I_0}{I} = k''c \tag{10-11}$$

where k'' = constant for particular solution
c = concentration of the solution

If light is absorbed exponentially with concentration over a reasonable and practical range of concentration, the colored material is said to conform to Beer's law. The best way to determine whether a colored compound or complex obeys Beer's law is to prepare a series of samples in the desired range of concentration and submit them to test on a photoelectric colorimeter or spectrophotometer. If the observations of percent light transmission plot along a straight line on a semilog graph, the material can be considered to obey Beer's law. Many colored systems do not conform to Beer's law, and therefore development of any new colorimetric method should involve such a test procedure.

By combining the two absorption laws, the Lambert-Beer or Bouguet-Beer law is obtained:

$$T = \frac{I}{I_0} = 10^{-kcl} \tag{10-12}$$

or
$$A = \log \frac{I_0}{I} = kcl \tag{10-13}$$

It follows from the Lambert-Beer law that if light of the same intensity enters two different solutions and adjustments of depths are made so that the emerging beams are of the same intensity, then the transmission are the same, and

$$c_1 l_1 = c_2 l_2 \tag{10-14}$$

Equation (10-14) indicates that if the depth of a sample is varied so that the intensity of color matches that of a standard, then the sample concentration is related to the standard concentration by the ratio of their depths.

Color-Comparison Tubes

Colorimetric measurements may be made in a wide range of equipment. The environmental engineer uses standard color-comparison tubes, photoelectric colorimeters, or spectrophotometers. Each has its place and particular application in water and wastewater analysis.

Color-comparison tubes, sometimes referred to as Nessler tubes, have been the standard equipment for making colorimetric measurements for many years. Tubes of this type are shown in Fig. 13-1. Their use has largely been replaced, however, because of the convenience of photoelectric and spectrophotometric methods. Precise work with color-comparison tubes requires that tubes of matched size or bore be used in order to comply with Lambert's law. The chief difficulty with their use is that standard color solutions are seldom stable, and every time a determination has to be made it becomes necessary to prepare a series of fresh standards. This adds greatly to the labor and time required. Another objection is that all comparisons are made by eye, and the "human error" involved is often considerable because sensitivity to different colors varies. Furthermore the analyst is required to interpolate values between standards.

Photoelectric Colorimeters

Photoelectric colorimeters have been used quite extensively in colorimetric work and are very satisfactory within their limitations. They make use of an electrometric device employing a photoelectric cell as the sensing element. The current developed by the photoelectric cell is translated into percent transmission or absorbance through a suitable galvanometer. The light source is an ordinary light bulb, and monochromatic light is obtained by allowing a beam of light to pass through a color filter. The monochromatic light is directed through a cell containing the sample, and the light that penetrates hits the photoelectric cell. The instrument is adjusted to yield a light transmission corresponding to 100 percent with the cell containing a "blank sample." The "blank sample" is a portion of distilled water that has been treated in the same manner as regular samples. A schematic diagram of the essentials of a photoelectric colorimeter is shown in Fig. 10-3, and a typical instrument is shown in Fig. 10-4.

Photoelectric colorimeters require a separate color filter for each determination on which they are to be used; thus the investment may become considerable, and the range of application is somewhat limited. They are not suitable for research purposes and should be considered for use principally where a very few well-established determinations are involved.

Spectrophotometers

The modern spectrophotometer employing either a glass prism or a diffraction grating to produce monochromatic light is an extremely valuable instrument for colorimetric analyses in the environmental engineering field. It has a wide range of adaptability that allows selection of monochromatic light of any wavelength in the visible spectrum. In addition, all instruments provide light in the ultraviolet and near-infrared regions. One filter usually suffices for the entire

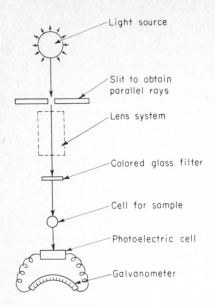

Light source

Slit to obtain
parallel rays

Lens system

Colored glass filter

Cell for sample

Photoelectric cell

Galvanometer

Figure 10-3 Schematic diagram of a photoelectric colorimeter.

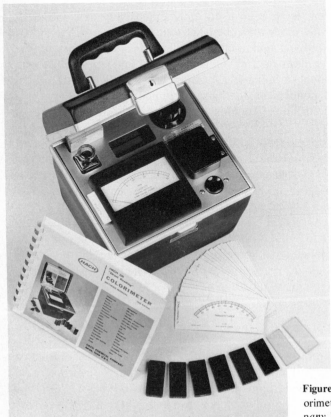

Figure 10-4 A photoelectric colorimeter. *(Hach Chemical Company.)*

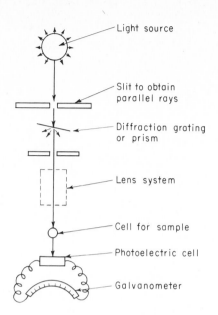

Light source

Slit to obtain
parallel rays

Diffraction grating
or prism

Lens system

Cell for sample

Photoelectric cell

Galvanometer

Figure 10-5 Schematic diagram of a spectrophotometer.

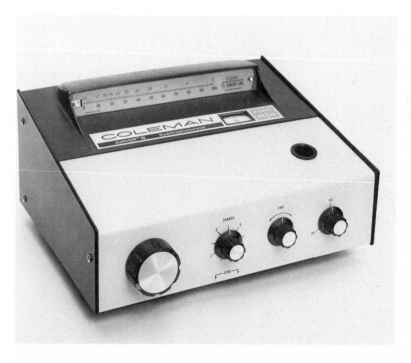

Figure 10-6 A spectrophotometer. *(Perkin-Elmer Corp., Coleman Instruments Div.)*

visible range of wavelengths. Separate filters are needed for the ultraviolet and for the infrared regions. The principle upon which spectrophotometers are based is the same as that for photoelectric colorimeters, except for the manner in which the monochromatic light is obtained. Figure 10-5 shows a schematic diagram of a spectrophotometer, and Fig. 10-6 shows a typical instrument.

A spectrophotometer is particularly recommended where a wide variety of determinations is made. Its versatility allows the best wavelength of light to be used at all times. The optimum wavelength can be determined at any time by establishing a spectral transmission curve. This is an essential part of research aimed at developing new methods of colorimetric analysis. The curve is established by making a series of observations of light transmission at several different wavelengths of light while using a typical colored solution in the cell. The results when plotted on linear coordinate paper yield a curve like that shown in

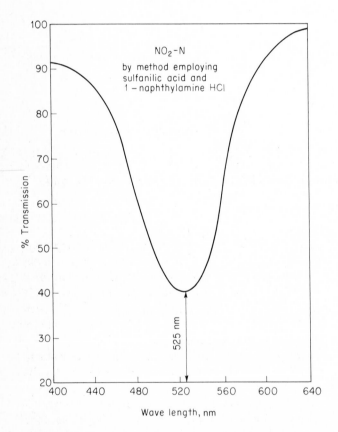

Figure 10-7 Spectral transmission curve for nitrites or nitrite nitrogen, showing optimum wavelength of light for photometric determination.

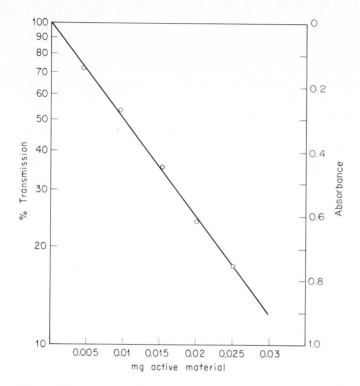

Figure 10-8 A typical calibration curve for photoelectric or spectrophotometric analysis.

Fig. 10-7 for nitrite. The wavelength that is absorbed to the greatest extent, in the case of nitrite 525 nanometers (nm), is the optimum wavelength to use.

Calibration and Use

Photoelectric colorimeters and spectrophotometers are calibrated for use in any particular determination by preparing a series of standards *in the same manner as regular tests are to be run* and by making observations on light transmission, using the wavelength specified for the determination. When such data are plotted on semilog paper, the curve should be essentially a straight line. The calibration curve, when carefully prepared, should serve for years and eliminate the need for the preparation of a series of standards. A typical calibration curve is shown in Fig. 10-8. It is usually good practice to include a standard in each set of samples, however, to make sure that unknown errors in reagents and the like do not lead to faulty results.

It should be emphasized that instrumental analysis does not necessarily ensure accurate results. The analyst must be constantly on guard to be sure that his instruments are in good working order, cells are kept scrupulously clean, and turbidity of samples is eliminated.

10-6 PHYSICAL METHODS OF ANALYSIS

The environmental engineer often employs other methods of analysis which are based upon physical measurements. Sensitive instruments are frequently employed to make these measurements. Such instruments not only allow rapid measurements to be made, but permit identification of both organic and inorganic materials occurring in extremely small concentration. Instrumental methods are also adaptable for the continuous monitoring of rivers and waste streams so that significant fluctuations in concentration can be recorded for evaluation. Because of their importance, a separate chapter (Chap. 11) is included in this book to describe the various instrumental methods of particular value to environmental engineers.

Some physical measurements commonly made in water and wastewater analysis do not require the use of expensive instruments. Two methods used to evaluate turbidity are discussed below.

Turbidimetry

The method in which ordinary white light transmitted through a finely divided suspension is compared with that transmitted by a standard suspension is known as turbidimetry. The environmental engineer employs turbidimetric analysis when he employs the Jackson candle turbidimeter or standard bottles for the measurement of turbidity. Some of the patented devices used for this purpose employ the same principle. Sulfates may also be determined in this manner.

Nephelometry

Nephelometric methods are also employed to measure turbidity. In this method light is allowed to strike a suspension at right angles to the eye of the observer or photoelectric cell of the instrument. The light reflected by the dispersed particles (Tyndall effect) is recorded. This principle is employed in measuring very low turbidities in filtered water. Bacteriologists often use nephelometry in following bacterial growth rates.

10-7 PRECISION, ACCURACY, AND STATISTICAL TREATMENT OF DATA

Every analyst should have a clear understanding of the differences between precision and accuracy insofar as analytical results are concerned. A knowledge of statistics is of considerable importance as an aid to establishing sampling programs so that data obtained may be subjected to statistical treatment when necessary. These subjects are treated quite adequately in "Standard Methods."

PROBLEMS

10-1 Explain the significance of the drying and ignition temperatures of 103°C and 600°C, respectively, in water quality analysis.

10-2 Explain why it is necessary (1) to cool a sample to room temperature before weighing, and (2) to allow the sample to cool in a desiccator.

10-3 Determine the equivalent weights of the following materials in grams:
(a) H_2SO_4; (b) HCl; (c) $Ca(OH)_2$; (d) CH_3COOH.
Answer: (a) 49; (b) 36.5; (c) 37; (d) 60

10-4 Determine the equivalent weights of the following materials in grams:
(a) $KMnO_4$ as an oxidizing agent in acid solutions.
(b) Ag_2SO_4 in a precipitation reaction involving silver.
(c) $K_2Cr_2O_7$ as an oxidizing agent.
(d) $BaCl_2$ in a precipitation reaction involving barium.

10-5 How many grams of $AgNO_3$ are required to prepare 500 ml of a 0.1 N solution to be used in a precipitation reaction?
Answer: 8.5 g

10-6 How many grams of $Fe(NH_4)_2(SO_4)_2 \cdot 6H_2O$ are required to prepare 1 liter of a 0.25 N solution to be used in a reduction reaction?

10-7 A 50-ml sample containing $Ca(OH)_2$ requires 10 ml of 0.02 N H_2SO_4 to reach the equivalence point in an acid–base titration. Determine the concentration in mg/l of $Ca(OH)_2$ in the sample.
Answer: 148 mg/l

10-8 A 100-ml sample containing chloride ion is titrated with $AgNO_3$ in a precipitation reaction. Calculate the concentration of chloride ion in mg/l if 10 ml of 0.01 N $AgNO_3$ is required to reach the equivalence point.

10-9 If the color in a sample of drinking water viewed through a depth of 10 cm corresponds in intensity to a standard having a color of 40 units and viewed through a depth of 30 cm, how many color units does the water sample contain?

10-10 A sample of drinking water is analyzed colorimetrically for ammonia nitrogen concentration. If the sample color when viewed through a depth of 5.5 cm corresponds in intensity to a 1 mg/l standard sample when viewed through a depth of 40 cm, what is the concentration of ammonia nitrogen in the sample?
Answer: 7.3 mg/l

10-11 Many spectrophotometers have two adjacent scales, one registering percent transmission and the other registering absorbance. What numerical value on the latter scale corresponds to 70 on the former?

10-12 If a sample containing 0.1 mg/l of nitrite-nitrogen gives 85 percent transmission when submitted to colorimetric analysis, what percent transmission should a sample containing 0.2 mg/l produce if the analysis follows Beer's law?

REFERENCES

Laitinen, H. A. and W. E. Harris: "Chemical Analysis," 2d ed., McGraw-Hill, New York, 1975.
Pierce, W. C., E. L. Haenisch, and D. T. Sawyer: "Quantitative Analysis," 4th ed., Wiley, New York, 1958.
Skoog, D. A. and D. M. West: "Fundamentals of Analytical Chemistry," 2d ed., Holt, Rinehart and Winston, New York, 1969.

ELEVEN

INSTRUMENTAL METHODS OF ANALYSIS

11-1 INTRODUCTION

A variety of sensitive instruments have been developed in recent years which have increased considerably the engineer's ability to measure and to cope with pollutional materials of increasing complexity. Instrumental methods of analysis are finding wide usage for routine monitoring of water quality in natural waterways, as well as during the course of water or waste treatment. Such methods have allowed analytical measurements to be made immediately at the source, and have permitted the recording of such measurements to be made at some distance from the place of actual measurement.

The pH meter and spectrophotometer are analytical instruments which were developed several decades ago, and have become indispensable in environmental control. They were described in Chap. 10. Some of the instruments of more recent origin will no doubt become of equal value in the future.

Nearly any physical property of an element or compound can be used as the basis for an instrumental measurement. The ability of a colored solution to absorb light, the capacity of a solution to carry a current, or the ability of a gas to conduct heat can all be used as the basis for an analytical method to measure the quantity of a material as well as to detect its presence. Most of the instrumental methods of analysis come under the heading either of optical methods or of electrical methods. There are methods, however, such as gas chromatography, which are particularly useful in environmental engineering, but are based on the measurement of other properties of materials. Some of the more common instrumental methods of analysis, their method of operation, and their major uses are briefly discussed in the following sections.

11-2 OPTICAL METHODS OF ANALYSIS

Optical methods of analysis measure the results of interactions between radiant energy and matter. The radiant energy used in such measurements may vary from X rays, through visible light, to radio waves. The parameter most frequently used to characterize radiant energy is the wavelength, which is the distance between adjacent crests of the wave in a beam of radiation. Wavelength is normally measured in angstroms (Å, 1 Å = 10^{-8} cm), or in nanometers (nm, 1 nm = 10^{-7} cm = 10 Å). Figure 11-1 indicates the normal range of wavelength for various types of radiant energy.

Radiant energy may also be considered to consist of *photons* or packets of energy which can interact with matter. The energy of a photon is related to its wavelength λ as follows:

$$E = \frac{hc}{\lambda} \tag{11-1}$$

where h is Planck's constant and is equal to 6.62 = 10^{-27} erg · s, c is the velocity of light and is equal to 3×10^{10} cm/s. This equation indicates that X rays, with short wavelength, are relatively high in energy and for this reason they can cause marked changes in matter. Microwaves and radio waves, on the other hand, have long wavelengths and are relatively low in energy. The changes which they cause upon interaction with matter are quite small and difficult to detect. Use of radiation of different energy content permits the determination of different properties of materials.

Optical methods of analysis may be designed to measure the ability of a material or solution to *absorb* radiant energy, to *emit* radiation when excited by an energy source, or to *disperse* or *scatter* radiation. Methods based upon these three principles will be described separately.

Absorption Methods

When a source of radiant energy, such as a beam of white light, is passed through a solution, the emergent beam will be lower in intensity than that

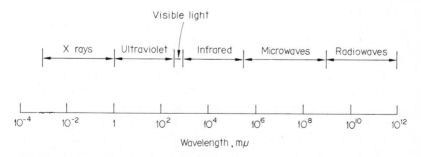

Figure 11-1 Range of wavelengths for different types of radiant energy.

entering. If the solution does not contain suspended particles which scatter light, then the reduction in intensity is due primarily to *absorption* by the solution. The extent of absorption of white light is generally greater for some colors than for others, with the result that the emerging beam is colored.

The use of a spectrophotometer to determine the extent of absorption of various wavelengths of visible light by a given solution is discussed in Sec. 10-5, and the result of such a measurement is illustrated in Fig. 10-7. Colorimetry, of course, refers to the visible region of the spectrum, but the same principles apply to other regions. Thus analytical methods based on absorption of ultraviolet or infrared radiation are also in common usage.

All instruments designed to measure the absorption of radiant energy have the basic components indicated in Fig. 11-2, which include an *energy source* to provide radiation of the desired wavelength, an *energy spreader* which permits separation of radiation of the desired wavelength from other radiation, and an *energy detector* which measures the portion of radiation which passes through the sample. Spectrophotometers for such measurements may vary from the simple and relatively inexpensive colorimeters discussed in Sec. 10-5 to highly complicated and expensive instruments which automatically scan the ability of a solution to absorb radiation over a wide range of wavelengths and automatically record the results of these measurements. One instrument cannot be used to measure absorbance at all wavelengths because a given energy source, energy spreader, and energy detector is suitable for use over only a limited range of wavelengths. Instruments which measure the absorption of visible light are widely used in water analysis. However, instruments measuring the absorption of ultraviolet or infrared radiation are finding increasing usage and will be briefly discussed.

Ultraviolet spectrophotometry When a molecule absorbs radiant energy in either the ultraviolet or the visible region, valence or bonding electrons in the molecule are raised to higher-energy orbits. Some smaller molecular changes can also take place, but these are usually masked by the above electronic

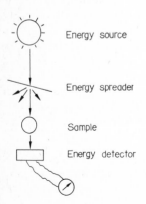

Energy source

Energy spreader

Sample

Energy detector

Figure 11-2 Basic components of instruments for measurement of absorption.

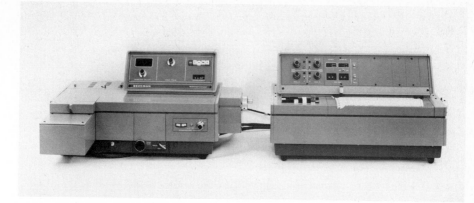

Figure 11-3 An ultraviolet and visible-range spectrophotometer with recorder. *(Beckman Instruments, Inc.)*

excitations. The result is that fairly broad absorption bands are usually observed in both the ultraviolet and the visible regions. Many instruments are designed to make measurements in both regions. A typical instrument of this type is shown in Fig. 11-3. The ultraviolet region is of more limited general usage, although it is particularly suitable for the selective measurement of low concentrations of organic compounds such as benzene-ring-containing compounds or unsaturated straight-chain compounds containing a series of double bonds. In general, ultraviolet spectrophotometry is used mainly as a research tool and only for highly specialized analyses.

Infrared spectrophotometry Nearly all organic chemical compounds show marked selective absorption in the infrared region. However, infrared spectra are exceedingly complex compared to ultraviolet or visible spectra. Infrared radiation is of low energy and its absorption by a molecule causes all sorts of subtle changes in the vibrational or rotational energy of the molecule. An understanding of these changes requires some knowledge of quantum mechanics, which is beyond the scope of this book. A person who is particularly knowledgeable in this area can use infrared spectra to identify particular atomic groupings which are present in an unknown molecule. To illustrate, the absorption of short-wavelength infrared radiation causes vibrations of hydrogen atoms in molecules, longer wavelengths cause vibrations of triple bonds, while still longer wavelengths cause vibrations of double bonds. Because of the complexity of infrared spectra, it is highly unlikely that any two different compounds will have identical curves. This fact has made infrared spectrophotometry a valuable aid in the identification of pesticides and other complex organic chemicals which have been extracted from waterways.

While infrared analysis gives much more information about a molecule than either ultraviolet or visible analysis, it is less sensitive, and thus a relatively high concentration (10^{-4} to 10^{-5} molar) of the substance to be analyzed is

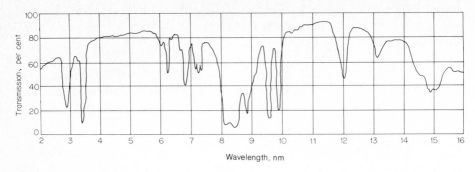

Figure 11-4 Infrared spectrogram of sodium tetrapropene benzene sulfonate.

needed. Another problem stemming from the complex infrared spectra produced by even simple molecules is that when many materials are present in a solution, interferences between their individual spectra make the use of infrared analysis for identification or quantification almost impossible. This is the reason a lengthy separation procedure is required in the "Stadard Methods" procedure for the infrared analysis of the alkyl benzene sulfonate type of detergents. A typical infrared spectum of a similar compound, tetrapropene benzene sulfonate, is shown in Fig. 11-4. An infrared spectrophotometer is shown in Fig. 11-5.

In principle, quantitative infrared analysis is the same as in the ultraviolet or visible regions. By examining the spectra of a pure substance, a wavelength may be found at which absorption is considerably greater than for other compounds present in the mixture. For analysis it is simply necessary to measure absorbance at the selected wavelength. Use is made of this principle in the "total carbon analyzer," which has been found useful for the rapid and continuous measurement of small quantities or organic carbon in water. Here a

Figure 11-5 An infrared spectrophotometer. *(Beckman Instruments, Inc.)*

liquid sample is injected into a furnace, where the water is evaporated and the organic carbon catalytically burned to carbon dioxide. The carbon dioxide is carried by a stream of oxygen through an infrared analyzer which is specifically designed to measure and record the concentration of carbon dioxide present. Actually total carbon is measured by this procedure, but if the sample is first prepared by acidification and aeration to remove inorganic carbon, then a selective measurement of the organic carbon present can be obtained.

Emission Methods

It has long been known that many metallic elements, when subjected to suitable excitation, will emit radiations of characteristic wavelengths. This is the basis for the familiar flame test for sodium (which emits a yellow light) and for other alkali and alkaline-earth metals. When a much more powerful method of excitation is used in place of the flame, most metallic and some nonmetallic elements will emit characteristic radiations. Under properly controlled conditions, the intensity of the emitted radiation at some particular wavelength can be correlated with the quantity of the element present. Thus both a quantitative and a qualitative determination can be made. The various analytical procedures which make use of emission spectra are characterized by the excitation method used, the nature of the sample (whether solid or liquid), and the method of detecting and recording the spectra produced. The more widely used methods are discussed below.

Flame photometry This method is used in water analysis for determining the concentration of alkali and alkaline-earth metals such as sodium, potassium, and calcium. A diagram showing the basic elements of a flame photometer is given in Fig. 11-6, and a typical instrument is shown in Fig. 11-7. A liquid sample to be analyzed is sprayed under controlled conditions into a flame where the water evaporates, leaving the inorganic salts behind as minute particles. The salts decompose into constituent atoms or radicals and may become vaporized. The vapors containing the metal atoms are excited by the thermal energy of the flame and this causes electrons of the metallic atoms to be raised to higher energy levels. When the electrons fall back to their original positions

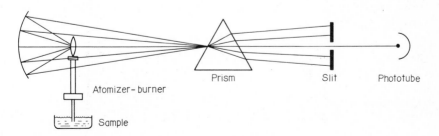

Figure 11-6 Schematic diagram of a flame photometer.

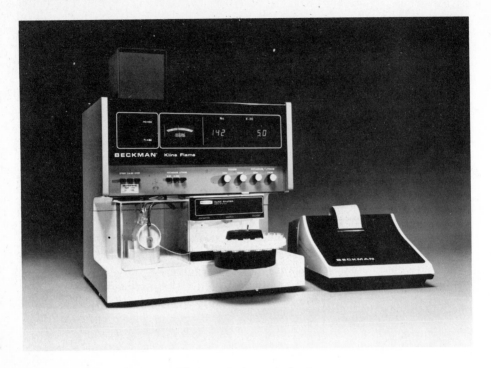

Figure 11-7 A flame photometer. *(Beckman Instruments, Inc.)*

or to a lower level they give off discrete amounts of radiant energy. The emitted radiation is passed through a prism, which separates the various wavelengths so that the desired region can be isolated. A photocell and some type of amplifier are then used to measure the intensity of the isolated radiation.

A number of fuel gases, such as acetylene, hydrogen, or the normal natural or manufactured gas used for heating in most laboratories can be used for the flame. The oxidant used is usually tank oxygen rather than air. Flames yielding higher temperatures are capable of exciting more elements and so are more versatile.

The emission spectrum for each metal is different and its intensity depends upon the concentration of atoms in the flame, the method of excitation used, and the after-history of the excited atoms. Sodium produces a characteristic yellow emission at 589 nm, lithium a red emission at 671 nm, and calcium a blue emission at 423 nm. Each also gives a less intense emission at shorter wavelengths. Concentrations of these elements can be measured down to 1 mg/l or less with some degree of accuracy, depending upon the sensitivity of the instrument used.

Atomic absorption spectrophotometry Although this is really an absorption method, it is included under emission spectroscopy because of its similarity to flame photometry. Atomic absorption spectrophotometry is a relatively new method of analysis, but has already gained wide interest in environmental en-

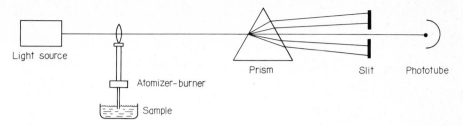

Figure 11-8 Schematic diagram of an atomic absorption spectrophotometer.

gineering because of its versatility for the measurement of trace quantities of most elements in water. Elements such as copper, iron, magnesium, nickel, and zinc can be measured accurately to a small fraction of a mg/l. Atomic absorption spectrophotometers are relatively expensive instruments, but because of their potential value to water quality research and investigations, it is well that environmental engineers be familiar with their uses and applications.

The similarity between atomic absorption spectrophotometry and flame photometry is indicated by a comparison of the illustrations in Figs. 11-6 and 11-8. Like the flame photometer, an atomic absorption system consists of a flame unit, a prism to disperse and isolate the emission lines, and a detector with appropriate amplifiers. In addition, the absorption system has a light source which emits a stable and intense light of a particular wavelength. Each element has characteristic wavelengths which it will readily absorb. A light source with wavelength readily absorbed by the element to be determined is directed through the flame and a measure of its intensity is made without the sample, and then with the sample introduced into the flame. The decrease in intensity observed with the sample is a measure of the concentration of the element. A disadvantage of this method is that a different light source must be used for each element. Fortunately sources such as hollow cathode lamps are available for most naturally occurring elements. A typical atomic absorption spectrophotometer is shown in Fig. 11-9.

The advantage of atomic absorption spectrophotometry is that it is quite specific for many elements. Absorption depends upon the presence of free unexcited atoms in the flame, and these are in much greater abundance than excited atoms. Thus elements such as zinc and magnesium, which are not easily excited in flames and so give poor results with the flame photometer, can be readily measured by the atomic absorption method.

A relatively new addition to atomic absorption spectrophotometry is the graphite furnace, which allows analysis for many heavy metals in the microgram per liter range. Here, the atomizer burner is replaced with a small cylindrical graphite tube which can be programmed through a series of different temperatures. The radiation from the cathode lamp source passes through the open ends of the horizontal cylinder and selected wavelengths are measured by the phototube as with the atomizer-burner method. A small quantity (5 to 100 μl) of a water sample is inserted into the cylinder through a hole in the side and

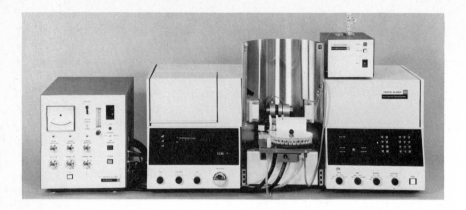

Figure 11-9 An atomic absorption spectrophotometer with graphite furnace. *(Perkin-Elmer Corp.)*

the temperature program is initiated. The temperature first rises to just over 100°C to allow the sample water to evaporate, leaving the metal containing salts behind. The temperature then increases to several hundred degrees Celcius which volatilizes the cations so they fill the cylindrical space, and the particular cation to be determined absorbs the characteristic radiation from the cathode tube. The graphite furnace, as it is called, allows the development of a greater density of atoms and thus effects greater sensitivity to the atomic absorption procedure. Use of the graphite furnace is rapidly becoming routine for water quality laboratories concerned with trace metal contamination.

Emission spectroscopy While the preceding methods are most widely used for water analysis, there are many other emission methods which employ more

Figure 11-10 An emission spectrophotometer. *(Jarrell-Ash Div., Fisher Scientific Co.)*

powerful excitation in place of the flame and can extend analysis to all metallic and many nonmetallic elements. In general, these instruments are highly complicated and exceedingly expensive, and are best used only by well-trained technicians. A typical emission spectrophotometer is shown in Fig. 11-10.

Emission instruments are particularly adapted to the analysis of solid as well as aqueous samples. Thus they are frequently employed for analysis of metals in sludges and other complex wastes. The basic components of most emission methods are the same as those of the flame photometer. However, higher energy excitation sources such as an a-c arc, a d-c arc, or a high-voltage spark are used to excite a greater variety of elements. The most important component of the spectrograph is the dispersing element, which can be either a prism or a grating. The detecting and recording components are designed to use the dispersive properties to the best advantage.

Dispersion and Scattering

The turbidity of a sample may be measured either by its effect on the transmission of light, which is termed *turbidimetry,* or by its effect on the scattering of light, which is termed *nephelometry.* As indicated in Chap. 12, these properties are used in the "Standard Methods" procedures for turbidity measurement. While these procedures use the human eye to detect the light emitted, methods employing common electric photometers can also be employed and have the advantage that continuous measurements of turbidity can be made and recorded. They can also overcome some of the human error in making turbidity observations. Nephelometric measurement is most sensitive for very dilute suspensions, but for moderately heavy turbidity, either this or turbidimetric measurements can be made.

In turbidimetry, as indicated in Fig. 11-11, the amount of light passing through a solution is measured. The higher the turbidity, the smaller the quan-

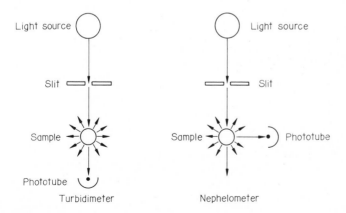

Figure 11-11 Schematic diagram of a turbidimeter and a nephelometer.

tity of light transmitted. In nephelometry, on the other hand, the detecting cell is placed at right angles to the light source to measure light scattered by the turbidity particles. Any spectrophotometer or photometer is satisfactory as a turbidimeter without modification. However, a special attachment is required for nephelometry.

Although turbidimetric analysis can be conducted at any wavelength of light, the "Standard Methods" procedure for determining sulfates by turbidimetric analysis recommends a wavelength of 420 nm. This results in a more sensitive analysis, because the blue light of this wavelength is scattered more strongly than red light at longer wavelengths.

Fluorimetry

Many organic and some inorganic compounds have the ability to absorb radiant energy of one wavelength and then to emit the energy as radiation at some longer wavelength. This phenomenon is known as *fluorescence* and it provides the basis for a very sensitive analytical tool. A schematic diagram of a *fluorimeter* which uses this principle is shown in Fig. 11-12 and a typical instrument is shown in Fig. 11-13. In many ways it is similar to a nephelometer. Predominantly ultraviolet light is passed first through a filter which excludes any visible light, and then through the sample. Fluorescent materials in the sample absorb the ultraviolet light and emit visible light, which passes through a second filter placed at right angles to the ultraviolet light source. This filter excludes any scattered ultraviolet light and permits the photocell to measure only the visible light emitted by the sample. This method is extremely sensitive for fluorescent materials.

One of the principal uses of fluorimetry in water quality studies has been to follow the movement of water and pollution. This is done by adding highly fluorescent dyes to the water and detecting their movement by fluoroscopic measurements. Rhodamine B and Pontacyl Pink B are dyes which have been used for this purpose and are detectable to concentrations of less than 1 μg/l. Their use is not always possible or practical with wastewaters, however, because of the high dosages needed to overcome background levels of fluorescent whitening agents from some detergents.

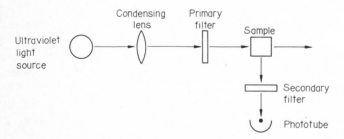

Figure 11-12 Schematic diagram of a fluorimeter.

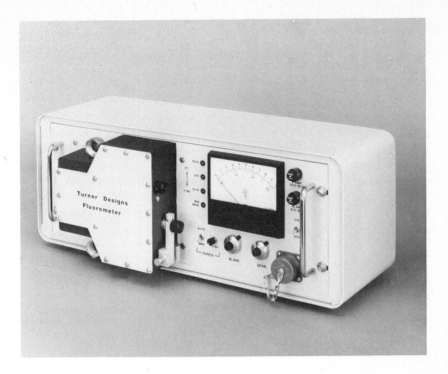

Figure 11-13 A fluorometer. *(Turner Designs.)*

11-3 ELECTRICAL METHODS OF ANALYSIS

Electrical methods of analysis make use of the relationships between electrical and chemical phenomena as outlined in Sec. 3-10. Such methods are particularly useful in water chemistry, as they lend themselves to continuous monitoring and recording. The pH meter is probably the most widely used electrical method of analysis. In this case a glass electrode and a reference electrode are inserted in a solution, and the electrical potential or voltage across these electrodes is a measure of the concentration of hydrogen ions in the solution. Methods based upon this principle are said to be *potentiometric*.

In other electrical methods, suitable electrodes are introduced into a solution and a small measured voltage is applied. The current which flows is dependent upon the composition of the solution and so may be used to make analytical measurements. Methods based upon this principle are said to be *polarographic*. There are many modifications of the different electrical methods of analysis, and the distinction between them is not always readily apparent. Other electroanalytical methods which have been adequately discussed in Sec. 3-10 are *conductimetry*, which measures the ability of a solution to carry a current, and *coulometry* which is a measure of the equivalence relationship between quantity of electricity and quantity of chemical change.

Potentiometric Analysis

The principles of the electrochemical cell were discussed in Sec. 3-10. Eq. (3-42) gives the relationship between the relative potential of an electrode and the concentration of a corresponding ionic species in solution. If use is made of this equation, the measurement of the potential of an electrode can permit the calculation of the activity or concentration of a component of the solution. A potentiometric analysis requires a special electrode designed for measurement of the specific ion of interest, a reference electrode, and a potential measuring device.

As an example, the chloride concentration in a solution can be measured, using the cell assembly diagramed in Fig. 11-14. The chloride measuring electrode is actually a silver–silver chloride electrode, at which the following reaction occurs:

$$AgCl + e^- \rightarrow Ag + Cl^- \qquad (11\text{-}2)$$

The potential of this half-cell, on the basis of Eq. (3-42) is given by the following:

$$E_{Cl} = E_{Cl}^\circ + \frac{RT}{zF} \ln [Cl^-] \qquad (11\text{-}3)$$

The measured potential of the complete cell thus becomes

$$E_{cell} = E_{Cl} - E_{SCE} = E_{Cl}^\circ - E_{SCE} + \frac{RT}{zF} \ln [Cl^-] \qquad (11\text{-}4)$$

This equation indicates that the potential of the whole cell varies with the log of the chloride activity or approximately with the log of the chloride concentration. Once this cell is calibrated with standard chloride solutions, it can be used to measure the chloride concentration in unknown solutions. This is the basis for all potentiometric methods of analysis. The major requirement is for an electrode which is specific for the ion of interest and which is also relatively free from interactions with other constituents of the solution.

A general description of different types of electrodes is given in Sec. 3-10. The above electrode for measuring chloride activity is termed a *metallic indi-*

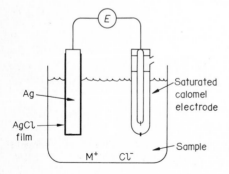

Figure 11-14 Cell assembly for potentiometric analysis for chlorides.

cator electrode. Generally, this type of electrode may be used for metals such as silver, copper, mercury, lead, and cadmium which show reversible electron transfer between the metal and its ions in solution. Other metals cannot be used which do not develop reproducible potentials, generally due to strain or crystal deformations in their structures or because of oxide coatings on their surfaces. Metals in this category include iron, nickel, cobalt, and chromium.

The chloride electrode is a specific application of a metal electrode used to measure an anion. Such an electrode is possible for anions which form slightly soluble precipitates with the cation of a metal.

Measurement of oxidation-reduction potential is another use for metallic indicator electrode systems in water quality control. The indicator electrode is most commonly an inert metal such as platinum or gold, and is used in conjunction with a reference electrode as illustrated in Fig. 11-15. As discussed in Sec. 5-10, oxidation-reduction reactions are very important in environmental engineering. The ability to measure the relative oxidation-reduction potential or pE of a system would be highly desirable. Unfortunately, in complex wastewater systems which are generally not in equilibrium and which contain many oxidants and reductants, meaningful measurement of the potential is often not possible. However, oxidation-reduction measurements with the electrode system has sometimes been useful in an empirical way to indicate whether a system is aerobic and oxidizing or anaerobic and reducing. In addition, electrode measurements are useful for showing end points in oxidation-reduction type titrations.

Another general class of electrodes which is important in water quality measurements is the *membrane electrode*. The development of membrane electrodes began around the turn of the century after an empirical observation by Cremer that a potential developed across a thin glass membrane when it was placed between two solutions with different hydrogen-ion concentration. This led to the development of the glass-membrane electrode which is now so widely used for potentiometric determination of pH. During the past twenty years membranes have been developed which are selectively sensitive to other ions.

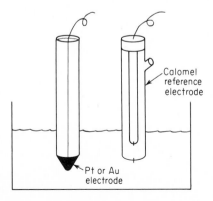

Calomel reference electrode

Pt or Au electrode

Figure 11-15 Electrode assembly for measurement of oxidation-reduction potentials (ORP).

Membrane electrodes are now available for direct potentiometric determination of K^+, Na^+, F^-, Ca^{2+}, NO_3^-, NH_4^+, and a number of others. Membrane electrodes are generally divided into three classes (1) glass electrodes, (2) liquid-membrane electrodes, and (3) solid-state or precipitate electrodes. The mechanism of potential development appears to be the same for the three classes. The glass electrode is most widely used and is discussed in more detail in the following, along with a general description of the other membrane electrode systems.

Glass electrodes The glass electrode is used almost universally for measurement of pH. For this purpose the electrode functions in highly colored solutions for which colorimetric methods are useless and in oxidizing media, reducing media, and colloidal systems for which other electrodes have failed almost completely. Earlier glass electrodes also developed a potential with Na^+ in alkaline solution, but electrodes are now available which can minimize this sodium interference. Advantage is taken of this interference, however, in the development of a sodium specific glass membrane electrode.

The action of the glass electrode is now fairly well understood. The glass used in the sensitive part of the electrode must have special characteristics with respect to composition and thickness. Construction of the electrode is essentially the same as that of the calomel reference electrode except that it does not have an opening or a wick to make direct electrical connection with the surrounding fluid. For pH measurements, the electrolyte within the glass electrode is an acid solution of definite strength rather than a KCl solution. The glass electrode for pH determination is used in conjunction with a standard calomel reference electrode, and the system may be described as shown in Fig. 11-16.

The single-electrode potential established on the pH electrode is determined by the concentration, or more specifically the activity, of hydrogen ions in the solution, $\{H_s^+\}$, in relation to the activity in the electrolyte within the electrode, $\{H_e^+\}$. It is generally considered that the potential is actually determined by the relative activities of adsorbed H^+ on the two sides of the glass membrane as follows:

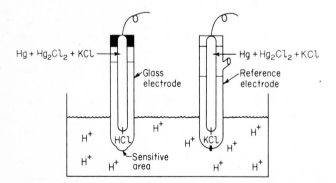

Figure 11-16 Electrode system employing the glass electrode for measurement of pH.

$$E_{obs} = k + 0.059 \log \frac{\{H_s^+\}}{\{H_e^+\}} \tag{11-5}$$

where k is a constant. The equation can be rewritten as follows:

$$E_{obs} = K + 0.059 \log \{H_s^+\}$$

or $\qquad\qquad E_{obs} = K - 0.059 \text{ pH} \tag{11-6}$

An electronic voltmeter (pH meter) is commonly employed for pH measurements and is calibrated to read pH directly.

Advantage has been taken of the observation that glasses with different compositions selectively develop potentials with different ions. Glass electrodes are commercially available for measuring concentration of the following ions: Na^+, K^+, NH_4^+, and Ag^+. The selectivity of the membrane to the various ions varies, however, as does the sensitivity. Before using glass electrodes for measuring cations other than H^+, the presence in the water and possible effects of interfering ions must be assessed.

Liquid membrane electrodes Liquid membrane electrodes are of more recent development than glass electrodes. They provide a means for direct potentiometric determination of the activities of several polyvalent cations such as Ca^{2+} and Mg^{2+}, and certain anions such as nitrate as well. In place of the glass membrane in the glass electrode, a liquid supported between porous glass or plastic disks is used. The liquid is an organic compound which is immiscible in water and contains functional groups which can exchange with ions from solution. By selecting functional groups which have strong affinity for particular cations, the liquid membrane electrode can become quite specific. When placed in a solution the ion for which the electrode is developed exchanges with ions within the water-liquid exchanger interface and establishes a potential which is measured. These electrodes must be used with the same precautions as other membrane electrodes.

Solid-state or precipitate electrodes Solid-state or precipitate electrodes contain a solid membrane (other than glass) between the sample and a reference solution which is selective for anions. The membrane generally consists of the anion of interest and a cation which can precipitate that anion from solution. A solid-state electrode for fluoride was developed within the last decade or so and is presently used in a "Standard Methods" procedure for water. The membrane consists of a crystal of lanthanum fluoride that has been treated with a rare earth to increase its electrical conductivity. It shows theoretical response to changes in fluoride ion activity, and shows a thousandfold selectivity for fluoride over other common anions.

Precautions in use of electrodes The fluoride specific electrode illustrates some of the limitations of all electrodes when used for direct measurement of concentration. Electrodes measure activities and not concentrations. For a given concentration, the activity varies with ionic strength as discussed in Secs. 2-12 and 5-3. Thus, to use an electrode for concentration measurements, the electrode must be calibrated in a standard solution of similar ionic strength. Also, electrodes measure specific ionic species and not complexes of the species. Fluoride forms complexes with Al(III), Fe(III), and Si(IV), and that portion of the fluoride associated with these cations will not be measured by the fluoride electrode. In order to overcome this interference, a method must be found for breaking up the complex or else appropriate corrections must be made. Change in solution pH or addition of materials which preferentially complex the interfering ions may be appropriate.

Ion-specific electrodes offer great promise for rapid, routine, and automatic monitoring of water quality. They have limitations, however, and these must be known and dealt with by the analyst for each particular ion. New electrode systems are rapidly evolving. Those concerned with water and wastewater analysis should keep abreast of these developments because of their potential for solving many difficult analytical problems.

Polarographic Analysis

A schematic diagram of an electrical circuit for polarographic analysis is given in Fig. 11-17, and a typical instrument is shown in Fig. 11-18. A voltage varying from 0 to 3 volts is impressed across the electrodes. The exact voltage at any setting of the control is recorded with a voltmeter. This voltage results in a flow of current through the solution, the amount being read with a microammeter. When the electrodes are immersed in a given solution and the applied voltage is gradually increased, the current will remain near zero until the potential reaches a point which is sufficient to bring about the reduction of some ion present in the solution. At this point, electrolytic reduction starts at the inert electrode, and an increased voltage causes a sharp increase in current in accordance with Ohm's law, $E = IR$. As reduction occurs, there is a depletion of ions next to the inert electrode, and fresh ions reach the electrode by diffusion. As

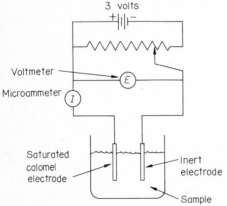

Figure 11-17 Electrical circuit for polarographic analysis.

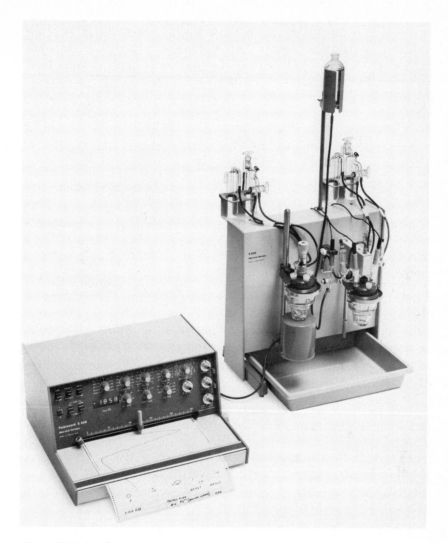

Figure 11-18 A polarograph. *(Brinkmann Instruments, Inc.)*

the voltage is continually increased, a point is reached at which the rate of diffusion of ions to the electrode limits the current. At this voltage, the rate of diffusion, and hence the current are proportional to the concentration of the ions in solution which are being reduced.

An illustration of a typical current-voltage curve for a particular ion is shown in Fig. 11-19. Point *A* on the curve is called the half-wave potential and is equal to the potential at which the current is one-half the diffusion current. This potential is related to the standard reduction potential of the ion involved and serves as a means of determining the nature of the substance being reduced. Thus the half-wave potential identifies the ion and the diffusion current determines the quantity present. When several reducible ions are present in the

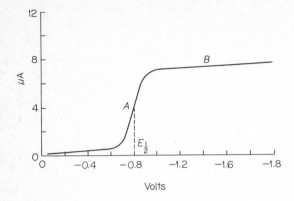

Figure 11-19 Typical current-voltage curve for polarographic analysis of a single ion.

solution, an increase in current will take place when the half-wave potential of each is reached, as illustrated in Fig. 11-20. If the respective individual curves are sufficiently separated so they do not interfere with one another, then the half-wave potential will allow identification of the particular ion, and the increase in diffusion current over that produced by the preceding ion will allow an estimate of the ion concentration.

The inert electrode used for polarographic analysis is usually a dropping-mercury electrode, which consists of metallic mercury dropping at a steady rate from a capillary tube. As the mercury drop grows, it continuously exposes fresh mercury to the solution so that unwanted poisoning by the ions being reduced can be avoided. Other electrodes such as streaming-mercury electrodes, rotating electrodes, or noble-metal electrodes have also been used, and under certain circumstances may have advantages over the dropping-mercury electrode.

Polarographic analysis is frequently used for measurement of the concentration of heavy metals in water or waste samples. It also forms the basis of a tentative procedure for nitrate measurement. "Standard Methods" gives an adequate description of procedures for such determinations and indicates that the lower limit of detection with ordinary equipment is about 0.1 mg/l.

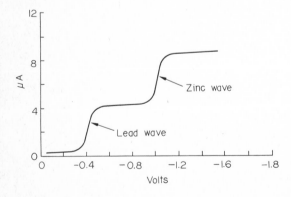

Figure 11-20 Current-voltage curve for lead and zinc in a single solution.

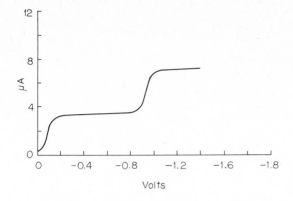

Figure 11-21 Typical current-voltage curve for oxygen.

Oxygen is one of the most significant interfering substances in polarographic analysis, as it is readily reduced at the inert electrode. This interference is normally removed by flushing the sample with a nonreducible gas such as nitrogen. However, advantage is taken of the reducibility of oxygen in the polarographic determination of dissolved oxygen. A typical oxygen polarogram is indicated in Fig. 11-21. Two waves are observed. The first, with a half-voltage of about -0.05 volt, is caused by the reduction of O_2 to H_2O_2, while the second at about -0.9 volt corresponds to the reduction of O_2 to H_2O. Normally the center of the current plateau between the two oxygen waves is used to make quantitative measurements.

Instruments specifically designed for dissolved oxygen determinations and using the dropping-mercury electrode have been on the market for some time.

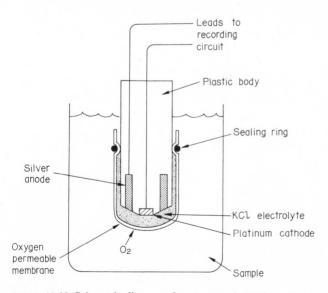

Figure 11-22 Schematic diagram of an oxygen electrode system.

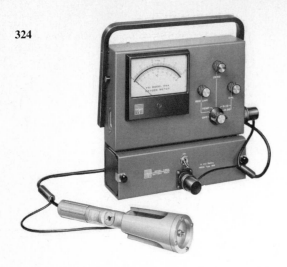

Figure 11-23 A dissolved oxygen analyzer. *(Yellow Springs Instruments Co.)*

Other electrodes which are more specific for dissolved oxygen have been developed and have proved quite useful for many water quality studies. A schematic diagram of such an electrode system is shown in Fig. 11-22. An inert metal such as gold or platinum serves as the cathode, and silver is used for the anode. These are electrically connected with a potassium chloride or other electrolytic solution. The complete cell is separated from the sample by means of a gas-permeable membrane, usually made of polyethylene or Teflon. The membrane shields the cathode and anode from contamination by interfering liquids and solids. When a potential of about 0.5 to 0.8 volt is applied across the anode and cathode, any oxygen which passes through the membrane will be reduced at the cathode, causing a current to flow. The magnitude of the current produced is proportional to the amount of oxygen in the sample. In another modification of this method, a lead cathode is used in place of the inert electrode, and a basic electrolyte such as potassium hydroxide is used. This, in combination with the silver anode, produces a galvanic cell having a sufficient potential to reduce the oxygen without the need for an externally applied voltage. An instrument equipped with this type of electrode is shown in Fig. 11-23. Instruments employing this type of electrode require that the sample be agitated to yield reliable results.

11-4 GAS CHROMATOGRAPHY

Gas chromatography is a highly versatile instrumental method of analysis which has been developed since 1951. It is currently used by environmental engineers for a variety of routine analyses which prior to its development were difficult and highly time-consuming. Gas chromatography entails the vaporization of a liquid sample followed by the separation of the various gaseous components formed so that they can be individually identified and quantitatively measured. A schematic diagram of the components of a typical gas chromato-

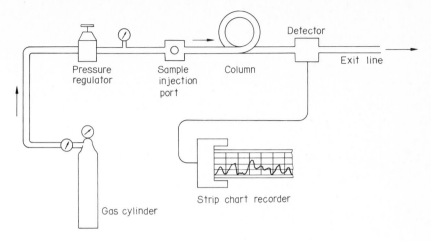

Figure 11-24 Schematic diagram of a gas chromatograph.

graph is shown in Fig. 11-24, and an instrument is illustrated in Fig. 11-25. The basic components are a gas cylinder with reducing valve, a constant-pressure regulator, a port for the injection of the sample, a chromatographic column, a detector, an exit line, and a strip chart recorder.

The gas cylinder contains a carrier gas such as hydrogen, helium, or nitrogen, which is continuously swept through the chromatographic column at a constant temperature and flow rate. A small sample for analysis is injected, usually with a syringe, into the sample port where it is flash-evaporated to convert its components into a gaseous state. The constantly flowing stream of carrier gas carries the gaseous constituents through the chromatographic column. The gases travel through at different rates so that they emerge from the column at different times. Their presence in the emerging carrier gas is detected by chemical or physical means, and the response of the detector is fed into a strip chart recorder. A typical chromatogram produced by this process is shown in Fig. 11-26. Each peak represents a specific chemical compound or a mixture of compounds with the same rate of movement through the column. The time for each compound to emerge from the column is a characteristic of the compound and is known as its *retention time*. The area under the peak is proportional to the concentration of the compound in the sample.

Chromatographic columns are generally glass or metal tubes varying from about 1 to 10 m in length and about 3 to 6 mm in diameter. The tube is packed with an inert solid impregnated with a nonvolatile liquid, such as silicone oil or polyethylene glycol. As a gaseous sample passes through the column, its components are partitioned between the stationary liquid phase of the column and the moving gas phase. Components which are relatively soluble in the liquid phase move through the column at slower rates than components which are not so soluble. Through the selection of suitable stationary phases and column lengths, most gaseous materials can be separated by this procedure. Recently,

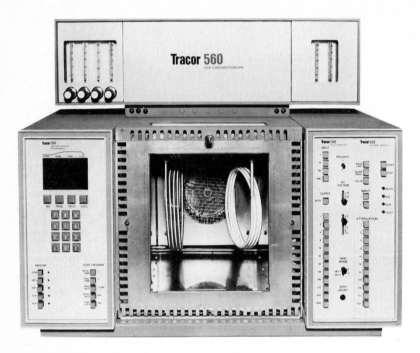

Figure 11-25 A gas chromatograph with oven door open to show columns. *(Tracor, Inc.)*

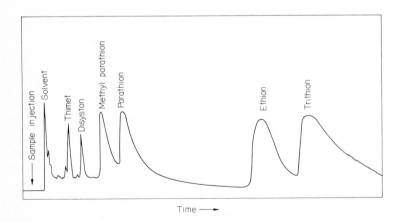

Figure 11-26 Gas chromatogram produced from pesticide analysis, using an electron capture detector.

capillary columns with diameter of 0.2 to 0.4 mm and length of 20 to 30 m have been developed and contain the solid phase on the inside wall of the capillary tube. The gas containing the organic components passes through the center of the capillary tube and the organics partition themselves between the gas and the stationary phase, emerging from the column at different times. The absence of packing minimizes the dispersion of the contaminants so that excellent resolution of the different compounds is obtained, even when there are twenty or more present. These columns have made possible the identification of a greater number of compounds in highly complex mixtures.

The temperature of the sample and column must be held sufficiently high during movement through the column and subsequent detection to maintain them in the gaseous state. Only materials which can be volatilized or which can be converted to volatile compounds can be detected by gas chromatography.

There are several different detectors in common usage. Each responds to some chemical or physical property of the eluting gases and converts this response to an electrical signal which can be recorded. One of the most widely used detectors responds to the difference in *thermal conductivity* of the eluting gases. In this detector, a heated wire is placed in the stream of gases emerging from the chromatographic column. Each gas has a different thermal conductivity or different ability to carry away heat from the wire, and thus, as they pass, the temperature of the wire changes. This results in a change in resistance of the wire and produces an electrical signal, which activates the recorder to draw a peak. Gas chromatographs using thermal-conductivity detectors are highly versatile because they respond to most gases. They are particularly useful for the analysis of gases resulting from anaerobic digestion.

Other detectors are available which are highly sensitive for specific organic or inorganic materials. A *flame ionization* detector is sensitive to organic compounds, but will not respond to water vapor. This permits the direct injection of water samples for analysis without prior separation. Organic compounds yield ions when burned in a flame, and the flame ionization detector makes use of this principle by measuring the difference in electrical conductivity of the flame as gases are eluted from the chromatographic column and burned. Organic acids in concentrations of only a few milligrams per liter can be separated and measured, using instruments equipped with this detector.

Pesticides in concentrations of a microgram per liter or less can be measured after extraction from water by use of *electron capture* or *coulometric* detectors. The electron capture detector measures the ability of compounds with halogen atoms or polar functional groups to capture beta particles emitted by a radioactive source. Coulometric detectors are not as sensitive, but are specific for organic compounds containing a specific element, such as chlorine. Here the effluent from the chromatographic column is burned, the products are absorbed in a titration cell, and the chloride ions released are titrated continuously with silver ions generated electrically. The electrical response activates the recorder pen.

Several other detectors measuring properties such as gas density, change in potential at electrodes, electrical conductivity, and mass spectrometery are being used today. The mass spectrometer has the ability not only to indicate the quantity of a given organic compound as it emerges from a gas chromatographic column, but also to make positive identification of each organic material. The use of a gas chromatograph in conjunction with a mass spectrometer (GC/MS) has led to the identification of hundreds of organic compounds in drinking water supplies throughout the country, and has resulted in increased research to answer questions about their public health significance. The possibilities for gas chromatography are extensive, and new developments are occurring rapidly. The environmental engineer would be well advised to keep abreast of these developments, as they may furnish significant aid in the solution of some of his more complex analytical problems.

11-5 OTHER INSTRUMENTAL METHODS

There are many other instrumental methods of analysis available. Most of these have not been used extensively in the past in environmental engineering, either because they are highly specific and not of sufficient usefulness, because the costs are excessively prohibitive, or else because they require operation by a highly trained specialist. Only a brief review of some of these methods will be given.

Mass Spectrometry

As indicated in Sec. 11-4, mass spectrometry when used in conjunction with gas chromatography (GC/MS) can give positive identification and quantification for a large number of individual organic compounds present in water and wastewater. A mass spectrometer is an instrument which will sort out charged gas molecules or ions according to their masses. The substance to be analyzed is vaporized and converted to positive ions by bombardment with rapidly moving electrons. The ions formed are pulled from the gas stream by an electrical field. The ions are accelerated in some fashion depending upon the type of instrument and are separated by their mass-to-charge ratio. A suitable detector can then record the particles of different mass either qualitatively, quantitatively, or both. Mass spectrometry is useful for tracer experiments using stable isotopes such as N^{15}, or for determining isotope ratios such as for oxygen when determining the age of water from different sources.

When used for the positive identification of organic materials, such as those emerging from a gas chromatograph, the bombardment of the organic molecule by the rapidly moving electrons breaks the organic molecules into a number of charged fragments. The mass-to-charge ratio of each fragment is measured as is the relative quantity of each fragment. Every organic molecule has its own pattern of fragmentation when bombarded under a given set of conditions. By comparing the mass-to-charge ratio and density of fragments of an unknown with that of known materials, positive identification can be made. This is a powerful tool which is rapidly increasing in usage, even though it is most expensive. It is helping to solve many difficult analytical problems and is greatly adding to our knowledge of the nature and fate of organic materials in the environment.

X-Ray Analysis

X rays are high-energy electromagnetic radiations with short wavelength. They are used for analytical purposes much in the same way as radiations of longer wavelength, such as visible light. *X-ray absorption* follows the same absorption laws as for other radiation, except that the phenomenon is on an atomic, rather than a molecular level. X-ray absorption is used for the measurement of the presence of heavy elements in substances composed primarily of low-atomic-weight materials. An example is the determination of the quantity of uranium in solution.

X-ray emission or *X-ray fluorescence* is used for the study of metals and other massive samples. A sample is bombarded with high-energy X rays which are absorbed by certain elements and re-emitted as X rays of lower energy. The bombarding X rays can dislodge an electron from an atom, leaving an unstable atom with a "hole" in one of the electron shells. Stability is regained by single or multiple transitions of electrons from outer shells to fill the hole. Each time an electron is transferred, the atom moves to a less energetic state and radiation is emitted at a wavelength corresponding to the energy difference between the initial and final states. The energy of the emitted radiation is characteristic for each emitting element and so can be used for analysis. This is similar to fluorimetry as previously discussed, except for the different energies of the respective radiations. The usefulness of X-ray fluorescence for rapid analysis for many heavy metals in water has been demonstrated. The heavy metals are first concentrated by passing a water sample through filter paper containing ion-exchange resins; the filter paper is then subjected to X-ray fluorescence. This technique is likely to see wider usage for water and wastewater analysis in the future.

X-ray diffraction is primarily of value for the study of crystalline material. X rays are reflected off the surfaces of crystals, and by studying the patterns of reflection as the crystalline material is rotated in the path of the X rays, much information about the structure of the material can be obtained.

Nuclear Magnetic Resonance

This analytical tool is used to detect and distinguish between the nuclear particles present in a sample. It measures the changes in the nuclei of materials when placed in a fixed magnetic field and then subjected to an alternating magnetic field. This method can be used to carry out specific chemical analyses or for ascertaining the structural formulas of organic compounds. It is a highly specialized research tool.

Radioactivity Measurements

Instrumental methods of analysis are routinely used for the measurement of radioactivity in the environment or in research studies which make use of radioactive tracers. Various instruments are available to measure particular types of radiations, the frequency of emissions, or both. A discussion of the many different instruments available is beyond the scope of this book.

REFERENCES

Ewing, G. W.: "Instrumental Methods of Chemical Analysis," 3d ed., McGraw-Hill, New York, 1969.

Willard, H. H., L. L. Merritt, and J. A. Dean: "Instrumental Methods of Analysis," 5th ed., D. Van Nostrand, New York, 1974.

TWELVE

TURBIDITY

12-1 GENERAL CONSIDERATIONS

The term *turbid* is applied to waters containing suspended matter that interferes with the passage of light through the water or in which visual depth is restricted. The *turbidity* may be caused by a wide variety of suspended materials, which range in size from colloidal to coarse dispersions, depending upon the degree of turbulence. In lake or other waters existing under relatively quiescent conditions, most of the turbidity will be due to colloidal and extremely fine dispersions. In rivers under flood conditions, most of the turbidity will be due to relatively coarse dispersions.

Turbidity may be caused by a wide variety of materials. In glacier-fed rivers and lakes most of the turbidity is due to colloidal rock particles produced by the grinding action of the glacier. The beautiful blues and greens of the lakes and rivers in Glacier National Park are typical examples. As rivers descend from mountain areas onto the plains, they receive contributions of turbidity from farming and other operations that disturb the soil. Under flood conditions, great amounts of topsoil are washed to receiving streams. Much of this material is inorganic in nature but considerable amounts of organic matter are included. As the rivers progress toward the ocean, they pass through urban areas where domestic and industrial wastewaters, treated or untreated, may be added. The domestic waste may add great quantities of organic and some inorganic materials that contribute turbidity. Certain industrial wastes may add large amounts of organic substances and others inorganic substances that produce turbidity. Street washings contribute much inorganic and some organic turbidity. Organic materials reaching rivers serve as food for bacteria, and the resulting bacterial growth and other microorganisms that feed upon the bacteria produce additional turbidity. Inorganic nutrients such as nitrogen and phosphorus present in wastewater discharges and agricultural runoff stimulate the growth of algae, which also contribute turbidity.

From the above considerations, it is safe to say that the materials causing turbidity may range from purely inorganic substances to those that are largely organic in nature. This disparity in the nature of the materials causing turbidity makes it impossible to establish hard and fast rules for its removal.

12-2 ENVIRONMENTAL SIGNIFICANCE

Turbidity is an important consideration in public water supplies for three major reasons.

Aesthetic

Consumers of public water supplies expect and have a right to demand turbidity-free water. Laymen are aware that domestic wastewater is highly turbid. Any turbidity in the drinking water is automatically associated with possible wastewater pollution and the health hazards occasioned by it. This fear has a sound basis historically, as anyone knows who is familiar with the water-borne epidemics that formerly plagued the water works industry.

Filterability

Filtration of water is rendered more difficult and costly when turbidity increases. The use of slow sand filters has become impractical in most areas because high turbidity shortens filter runs and increases cleaning costs. Satisfactory operation of rapid sand filters depends upon effective removal of turbidity by chemical coagulation before the water is admitted to the filters. Failure to do so results in short filter runs and production of an inferior-quality water, unless filters of special construction are used.

Disinfection

Disinfection of public water supplies is usually accomplished by means of chlorine or ozone. To be effective, there must be contact between the agent and the organisms that the disinfectant is to kill.

In turbid waters, most of the harmful organisms are exposed to the action of the disinfectant. However, in cases in which turbidity is caused by sewage solids, many of the pathogenic organisms may be encased in the particles and protected from the disinfectant. For this and aesthetic reasons the U.S. Environmental Protection Agency has placed a limit of 1 unit of turbidity as the maximum amount allowable in public water supplies.

12-3 STANDARD UNIT OF TURBIDITY

Because of the wide variety of materials that cause turbidity in natural waters, it has been necessary to use an arbitrary standard. The standard chosen was

$$1 \text{ mg } SiO_2/l = 1 \text{ unit of turbidity}$$

and the silica used must meet certain specifications as to particle size.

Standard suspensions of pure silica are not used in ordinary practice for measuring turbidity. They were used originally to calibrate the Jackson candle turbidimeter, which was chosen as the standard instrument for turbidity measurement. Today all instruments are copies of the original, and the glass tubes employed are calibrated in conformity with the original data obtained (see "Standard Methods," 14th ed., Table 214:I).

For routine work employing the Jackson turbidimeter no standards are required. Where other methods are employed, standard suspensions must be prepared for reference. The usual procedure is to use natural materials, where available, kaolin, or formazin polymer suspensions. The suspensions are standardized by means of the Jackson turbidimeter.

12-4 METHODS OF DETERMINATION

One instrumental and two visual methods are commonly employed. The instrumental method makes use of nephelometry to measure the intensity of light scattered by the turbidity particles, while the visual methods involve measurement of the interference to the passage of light caused by turbidity. Because of a basic difference in the phenomena measured, the instrumental method gives somewhat different results than the visual methods, the degree depending upon the nature of the turbidity particles. Because of this difference, values obtained by the instrumental procedure are expressed in nephelometric turbidity units (NTU), and those obtained by the visual methods in Jackson turbidity units (JTU).

The ranges of turbidity measured with reasonable accuracy by the different procedures are as follows:

Method	Range
Instrumental	0–40
Visual	
Jackson candle turbidimeter	100–1000 (short tube)
Jackson candle turbidimeter	25–1000 (long tube)
Bottle standards	5–100

It is obvious that the instrumental method must be used for measuring the low turbidity values appropriate for municipal water supplies.

Nephelometric Method

Measurement of turbidity by instrumentation involves employing the principles of nephelometry. In the instrument, a light source illuminates the sample and one or more photoelectric detectors are used with a readout device to indicate the intensity of scattered light at right angles to the path of the incident light. It is customary to use a particular formazin polymer suspension as a standard. Details are given in "Standard Methods." Results are easily obtained and are generally comparable to those obtained by visual methods. When using the formazin standard, 40 NTU are about equivalent to 40 JTU

Jackson Candle Turbidimeter

The Jackson turbidimeter is the standard instrument for measuring turbidity. It consists of three essential parts: a calibrated glass tube, a holder, and a candle, as shown in Fig. 12-1. Details of its construction and operation are given in

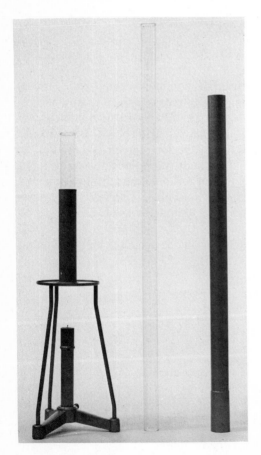

Figure 12-1 The Jackson candle turbidimeter, showing short- and long-form tubes.

"Standard Methods." With the tube in place over the lighted candle, portions of the sample are added until the outline of the candle flame is no longer discernible. Readings in terms of turbidity are then made directly from the calibrated tube. A series of readings should be taken on each sample to obtain reliable results until the operator becomes highly proficient in its use.

The Jackson candle turbidimeter is considered a rather crude instrument in this day of modern instrumentation.

Bottle Standards

Turbidity measurements in the range of 5 to 100 are usually made by reference to standard suspensions contained in glass bottles of similar size and color characteristics. Standard suspensions must be prepared by use of the Jackson turbidimeter. Usually a stock supply having a turbidity of 100 is prepared; then all suspensions of lower turbidity are prepared from it by dilution with distilled water. The standard suspensions must be replaced at frequent intervals because of changes in particle size and numbers that occur with time. Biological growths are controlled by use of mercuric chloride.

Measurement of turbidity is made by placing a suitable portion of the sample into a bottle of the same size and character as those used for the standards. Comparisons may be made by transmitted light or interference to visual perception. The latter is most common. Color of samples is often an interfering factor, but most analysts of experience can compensate for it without recourse to the addition of color to the standards.

12-5 APPLICATION OF TURBIDITY DATA

Turbidity measurements are of particular importance in the field of water supply. They have limited use in the field of domestic and industrial waste treatment.

Water Supply

Knowledge of the turbidity variation in raw-water supplies is of prime importance to the environmental engineer. He uses it in conjunction with other information to determine whether a supply requires special treatment by chemical coagulation and filtration before it may be used for a public water supply. Many large cities, such as New York, Boston, and Seattle, have upland or mountain supplies whose turbidities are so low that treatment other than chlorination is not required.

Water supplies obtained from rivers usually require chemical flocculation because of high turbidity. Turbidity measurements are used to determine the effectiveness of the treatment produced with different chemicals and the dosages needed. Thus they aid in selection of the most effective and economical chemical to use. Such information is necessary to design facilities for feeding the chemicals and for their storage.

Turbidity measurements help to gauge the amount of chemicals needed from day to day in the operation of treatment works. This is particularly important on "flashy" rivers where no impoundment is provided. Measurement of turbidity in settled water prior to filtration is useful in controlling chemical dosages so as to prevent excessive loading of rapid sand filters. Finally, turbidity measurements of the filtered water are needed to check on faulty filter operation.

Domestic and Industrial Waste Treatment

The suspended-solids determination is usually employed in waste treatment plants to determine the effectiveness of suspended-solids removal. The determination is slow and time-consuming, and in plants employing chemical treatment, changes in chemical dosages have to be made rather frequently. Turbidity measurements can be used to advantage, because of the speed with which they can be made, to gain the necessary information. By their use, chemical dosages can be adjusted to use the minimum amount of chemical while producing a high-quality effluent.

PROBLEMS

12-1 Discuss the nature of materials causing turbidity in
    (*a*) River water during a flash flood.
    (*b*) Polluted river water.
    (*c*) Domestic wastewater.

12-2 Give five precautions to be observed in operation of the Jackson candle turbidimeter (see "Standard Methods").

12-3 Describe how you would prepare a set of bottle standards having turbidities of 0, 5, 10, 15, and 20 units.

12-4 Discuss why turbidity in general cannot be correlated with the weight concentration of suspended matter in water samples (see "Standard Methods").

12-5 What limit is placed on turbidity in water supplies by the present U.S. EPA Standards and why has such a limit been set?

THIRTEEN

COLOR

13-1 GENERAL CONSIDERATIONS

Many surface waters, particularly those emanating from swampy areas, are often colored to the extent that they are not acceptable for domestic or some industrial uses without treatment to remove the color. The coloring material results from contact of the water with organic debris, such as leaves, needles of conifers, and wood, all in various stages of decomposition. It consists of vegetable extracts of a considerable variety. Tannins, humic acid, and humates, from the decomposition of lignin, are considered to be the principal color bodies.[1] Iron is sometimes present as ferric humate and produces a color of high potency.

Natural color exists in water primarily as negatively charged colloidal particles.[2] Because of this fact, its removal can usually be readily accomplished by coagulation with the aid of a salt having a trivalent metallic ion, such as aluminum or iron.

Surface waters may appear highly colored, because of colored suspended matter, when in reality they are not. Rivers which drain areas of red clay soils, such as those in the Piedmont area of the South Atlantic states, become highly colored during times of flood. Color caused by suspended matter is referred to as *apparent color* and is differentiated from color due to vegetable or organic extracts that are colloidal and which is called *true color*. In water analysis it is important to differentiate between apparent and true color. Color intensity generally increases with increase in pH. For this reason recording pH along with color is advised.

---

[1] T. R. J. McCrea, *J. Am. Water Works Assoc.*, **25:** 931 (1933).
[2] T. Saville, *J. New Engl. Water Works Assoc.*, **31:** 78 (1917).

Surface waters may become colored by pollution with highly colored wastewaters. Notable among these are wastes from dyeing operations in the textile industry and from pulping operations in the paper industry. Dye wastes may impart colors of wide variety that are readily recognized and traced. The pulping of wood produces considerable amounts of waste liquors containing lignin derivatives and other materials in dissolved form. The lignin derivatives are highly colored and quite resistant to biological attack. Much of this material is disposed of into natural watercourses, adding color which persists for great distances. Considerable research is under way to find an economical way of removing color from pulp-mill wastes.

13-2 PUBLIC HEALTH SIGNIFICANCE

Waters containing coloring matter derived from natural substances undergoing decay in swamps and forests are not considered to possess harmful or toxic properties. The natural coloring materials, however, give a yellow-brownish appearance to the water, somewhat like that of urine, and there is a natural reluctance on the part of water consumers to drink such waters because of the associations involved. Also, disinfection by chlorination of waters containing natural organics results in the formation of chloroform, a problem of current concern.

It is the responsibility of any water purveyor, public or private, to produce a product that is hygienically safe. Public health officials are aware of the fact that consumers will seek other sources of drinking water if the public water supply is not aesthetically acceptable, no matter how safe it may be from the hygienic viewpoint. Where waters are not aesthetically acceptable, consumers often shun safe domestic supplies and use waters from uncontrolled springs or private wells which may serve as foci for dissemination of pathogenic organisms. For this reason, waters intended for human use should not have a color exceeding 15 units.

13-3 METHODS OF DETERMINATION

Natural color, like turbidity, is due to a wide variety of substances, and it has been necessary to adopt an arbitrary standard for its measurement. This standard is employed directly and indirectly in the measurement of color. Many samples require pretreatment to remove suspended matter before true color can be determined. The method of pretreatment must be carefully selected to avoid introduction of errors.

Standard Color Solutions

Waters containing natural color are yellow-brownish in appearance. Through experience, it has been found that solutions of potassium chloroplatinate

Figure 13-1 Color-comparison tubes, commonly called Nessler tubes.

(K_2PtCl_6) tinted with small amounts of cobalt chloride yield colors that are very much like the natural colors. The shading of the color can be varied to match natural hues very closely by increasing or decreasing the amount of cobalt chloride.

The color produced by 1 mg/l of platinum (as K_2PtCl_6) is taken as the standard unit of color. The usual procedure is to prepare a stock solution of K_2PtCl_6 that contains 500 mg/l of platinum. Cobalt chloride is added to provide the proper tint. The stock solution has a color of 500 units, and a series of working standards may be prepared from it by dilution. Color-comparison tubes, commonly called Nessler tubes, as shown in Fig. 13-1, are usually used to contain the standards. A series ranging from 0 to 70 color units is employed and will serve for several months, provided that it is protected from dust and evaporation. The color-comparison tubes should be a matched set conforming to APHA standards as described in the introductory chapter of "Standard Methods."

Samples subjected to analysis may contain suspended matter which will interfere with the measurement of true color. Apparent color is determined on the sample "as is." Suspended matter must be removed to enable determination of true color. This can usually be accomplished by centrifuging the sample to separate the suspended solids. Analysis is performed on the clarified liquor. Filtration is not recommended because of possible adsorption of color on the filtering medium.

Samples with color less than 70 units are tested by direct comparison with the prepared standards. For samples with a color greater than 70 units, a dilution is made with distilled water to bring the resulting color within the range of the standards, and calculation of color is made, using a correction factor for the dilution employed.

Methods Employing Proprietary Devices

A number of instruments have been developed for the measurement of color to eliminate the need for renewing standard color solutions from time to time. Most of these instruments employ colored glass disks which simulate the various color standards when used in the particular instrument. One of the most popular is shown in Fig. 13-2.

Figure 13-2 The Hellige Aquatester. *(Hellige, Inc.)*

The proprietary devices find their greatest use in water works laboratories where trained chemists are not employed. They are not accepted as a standard procedure for measuring color because of variations in the color of the glass disks and their tendency to change characteristics, owing to fingerprints, dust, and so on. They should always be standardized against standard solutions of K_2PtCl_6 for highly important work.

Most of the proprietary devices suffer from the fact that replacement parts are rather expensive. This limits their use in laboratories designed for teaching, where breakage of glassware is normally rather great.

Field Methods

In some instances it becomes highly desirable to conduct color and other determinations in the field. The use of standard color solutions or of most proprietary devices is impractical because of possibilities of breakage, spillage, or lack of a suitable power source. The U.S. Geological Survey has developed a field kit. It employs aluminum tubes with glass windows on each end and a series of colored glass disks. Two tubes are used. One contains the sample and the other distilled water. The colored glass disks are placed over the end of the tube with distilled water until a combination is found that appears to have a color similar to that of the sample. The method is recognized as standard for field use.

Methods Applicable to Domestic and Industrial Wastewaters

Many industrial wastes are highly colored, and some contain colored substances that are quite resistant to biological destruction. Regulations concerning the color of effluents that may be discharged to streams are becoming more common. Evaluation of the color of yellow-brownish-hue wastes can be made by the standard procedures described above. Other systems of measurement have to be used to measure and describe colors that do not fall into this classification. A system adopted by the International Commission on Illumination utilized characterization and measurement of color by spectrophotometric means is recommended in "Standard Methods." A discussion is beyond the scope of this book.

13-4 INTERPRETATION AND APPLICATION OF COLOR DATA

The color of surface waters utilized for domestic supplies is of major concern for reasons mentioned above. Many industrial processes also require the use of color-free water. Removal of color is an expensive matter when capital investment and operating costs are considered. Therefore the water engineer, when developing or looking for new supplies, is always searching for a suitable supply with a color low enough so that chemical treatment will not be required. This "prospecting" may or may not be successful. If it is, he will use color data as one of the parameters to satisfy his client that expensive chemical treatment is not necessary. If it is not successful, he will use color data along with other information to prove that expensive chemical coagulation and sand filtration are needed to produce an acceptable supply.

Before a chemical treatment plant is designed, research should be conducted to ascertain the best chemicals to use and amounts required. In dealing with colored waters, color determinations serve as the basis of the decisions. Such data must be obtained for proper selection of chemical feeding machinery and the design of storage space.

Once operation of the treatment facilities has begun, color determinations on the raw and finished waters serve to govern the dosages of chemicals used, to ensure economical operation, and to produce a low-color water that is well within accepted limits.

PROBLEMS

13-1 Discuss briefly the causes of color in water.

13-2 Differentiate between "apparent" and "true" color.

13-3 What limit is generally placed on color and why are such standards set?

13-4 What is used as the standard unit of color?

13-5 What is the purpose of adding cobalt chloride to color standards?

CHAPTER

FOURTEEN

STANDARD SOLUTIONS

14-1 GENERAL CONSIDERATIONS

Chemists familiar with the analysis of water and wastewater involving volumetric procedures have learned that the use of standard reagents of definite normality saves a great deal of time in calculating results in terms of milligrams per liter. The preparation of solutions of a definite normality is not a tedious procedure if a logical system is used. Two or three important steps are involved, depending upon the nature of the materials.

Selection of the Proper Normality

In water and wastewater analysis it is usually desirable to report results in terms of milligrams per liter of some particular ion, element, or compound. As a rule, it is most convenient to have the standard titrating agent of such strength that 1 ml is equivalent to 1 mg of the material being measured. Thus 1 liter of a standard solution is ordinarily equivalent to 1000 mg or 1 g of the measured substance. The desired normality of the titrant is obtained by the relationship of 1 to the equivalent weight of the measured material. For example, the normality of acid solutions used to measure ammonia, ammonia nitrogen, and alkalinity (as $CaCO_3$) is as follows:

	Ammonia	Ammonia nitrogen	Alkalinity (as $CaCO_3$)
$\dfrac{1}{\text{eq. wt}}$	$= 1/17 = N/17$	$1/14 = N/14$	$1/50 = N/50$
	$= 0.0588\,N$	$0.0715\,N$	$0.02\,N$

344 WATER AND WASTEWATER ANALYSIS

The normality of basic solutions for measuring carbon dioxide (as CO_2) and mineral acidity (as $CaCO_3$) is

Acidity

CO_2 (as $CaCO_3$)

$$\frac{1}{eq.\ wt} = 1/44 = N/44 \qquad 1/50 = N/50$$

$$= 0.0227\ N \qquad\qquad 0.02\ N$$

The normality of silver nitrate solutions for measuring chlorides or sodium chloride is

Chlorides   *Sodium chloride*

$$\frac{1}{eq.\ wt} = 1/35.45 = N/35.45 \qquad 1/58.44 = N/58.44$$

The normality of reducing agents for measuring oxygen[1] is

Oxygen

$$\frac{1}{eq.\ wt} = 1/8 = N/8$$

The normality of oxidizing agents is obtained in a similar manner.

From the fact that standard solutions of titrating agents are of such strength that 1 ml is equivalent to 1 mg of the measured material, it will be readily apparent that when 1-liter samples are titrated the buret reading will give milligrams per liter directly. Usually it is inconvenient to use 1-liter samples, and calculations are easily made by the simple formula

$$ml\ titrant\ used \times \frac{1000}{ml\ sample} = mg/l$$

In some instances, as in the determination of dissolved oxygen where a fixed sample size is used, the strength of the titrant is adjusted so that each milliliter of titrant is equivalent to 1 mg/l.

Preparation of a Solution of Proper Normality

In cases in which a standard solution can be prepared from materials of known purity that can be accurately weighed on the analytical balance, the desired amount can be weighed, transferred to a volumetric flask, and diluted to the proper volume. Such solutions may be used without standardization against primary standards.

Many materials from which standard solutions are prepared are of such a character that their purity is not accurately known, or it may be impossible to weigh them exactly. In these cases, a solution is prepared that is known to be

---

[1] Actually a $N/40$ solution is used in practice, for reasons given in Sec. 21-4.

slightly stronger than desired, and it can be kept most conveniently in a graduated cylinder until standardization is completed. Standardization is accomplished by using a suitable primary standard.

Standardization of Solutions with Primary Standards

The procedure for standardizing solutions to an exact normality is somewhat peculiar to water and wastewater analysis and is not usually described in quantitative textbooks. Six fundamental steps are involved, as follows:

1. Calculate the weight of the primary standard that is exactly equivalent to 1 liter of the solution to be standardized.
2. Weigh three or four samples of the heat-dried primary standard that are sufficient to use about 20 ml of the solution. Weighings must be exact, and corrections must be made for percent purity if it differs significantly from 100 percent.
3. Calculate the volume of titrant of the desired normality needed to react with the corrected weight of each sample.
4. Add sufficient distilled water, and other reagents as needed, to each sample of primary standard to accomplish solution, and titrate two of them with the reagent to be standardized. The titrations should be less than the calculated amounts obtained in step 3. If so, the solution is stronger than desired, and the difference between the actual and calculated titrations represents the deficiency of water.
5. Calculate the amount of water to be added to the remaining solution. Drain the contents of the buret into the stock supply in the graduated cylinder and measure the total volume remaining. The amount of water to be added may be calculated from the following expression:

$$\frac{\text{Volume remaining}}{\text{Actual titration}} \text{ (calculated titration} - \text{actual titration)} = \text{ml}$$

   Make separate calculations for each of the titrations and add the least amount of water indicated.[2] *Mix thoroughly after adding the water, rinse, and fill the buret with the new solution.*
6. Titrate additional samples of primary standard and repeat steps 4 and 5 until the proper strength has been reached as shown by a correlation of actual titrations with calculated values. The solution may then be considered to have the desired normality.

14-2 PREPARATION OF *N*/1 AND *N*/50 H₂SO₄ SOLUTIONS

Standard solutions of sulfuric acid are used for the determination of alkalinity which is normally expressed in terms of $CaCO_3$, with an equivalent weight of

---

[2] Beginners should add slightly less water to avoid overdilution.

50; therefore $N/50$ solutions are required. Because large amounts of this reagent are used, it is most convenient to prepare a stock solution of $N/1$ acid and prepare the $N/50$ solution from it by simple dilution.

The purity of sulfuric acid as purchased usually varies from about 96 to 98 percent. In addition, it is very difficult to weigh accurately because of its hygroscopic properties. Solutions prepared from it must be standardized by means of some primary standard. Sodium carbonate is the primary standard usually used. The analytical grade is satisfactory, provided that it has been dried for 1 h at 140°C and kept in a desiccator prior to use.

Calculation of Concentrated H_2SO_4 Needed

In order to simplify calculations, they will be made on the basis of 1-liter amounts. By definition, 1 liter of $N/1$ acid contains 1.008 g of available hydrogen ion. Calculations are as follows:

$$1 \text{ GMW or } 98 \text{ g pure } H_2SO_4 = 2.016 \text{ g } H^+$$

$$\frac{\text{GMW}}{2} \text{ or } 49 \text{ g pure } H_2SO_4 = 1.008 \text{ g } H^+$$

Assume that concentrated acid is 96 percent pure.

Then $\qquad \dfrac{49}{0.96} = 51 \text{ g concentrated acid} = 1.008 \text{ g } H^+$

It is desirable that the solution be slightly stronger than $N/1$. To be sure of this, take 5 percent excess.

$$51 \times 1.05 = 53.5 \text{g}$$

Preparation of $N/1$ Acid Solution

Weigh approximately 53 g, ± 1 g, of concentrated acid into a small beaker on a trip balance. Place about 500 ml of distilled water in a 1-liter graduated cylinder and add the acid to it. Rinse the contents of the beaker into the cylinder with distilled water, and add water to the 1-liter mark. Mix thoroughly by stirring or pouring back and forth from the cylinder into a large beaker. Cool to room temperature before use.

Calculation of Primary Standard Needed

Sodium carbonate is a convenient primary standard. It has a molecular weight of 106 and an equivalent weight of 53 when reacting with H_2SO_4 to a pH of 4.2 to 4.4, the methyl orange end point.

$$53 \text{ g Na}_2CO_3 \simeq 1000 \text{ ml } N/1 \text{ H}_2SO_4$$

or $\qquad\qquad 1.06 \text{ g Na}_2CO_3 \simeq 20 \text{ ml } N/1 \text{ H}_2SO_4$

Weigh four samples of Na_2CO_3 ranging from 1.00 to 1.10 g and proceed as described in the third part of Sec. 14-1.

Preparation of *N*/50 Acid Solution

When a *N*/l acid solution is available, solution of any normality less than *N*/l can be prepared from it by dilution, provided that proper care is used in measuring the amount of *N*/1 acid needed and dilutions are made in volumetric flasks. The amount of acid of any normality needed to make a definite volume of an acid of another normality may be calculated from the relationship

$$ml \times N = ml \times N$$

If it is desired to make 1 liter of *N*/50 acid from a stock supply of *N*/1 acid, the calculation is as follows:

$$ml \times 1.0 = 1000 \times 0.02$$

$$ml = 20$$

Twenty milliliters of *N*/1 acid when diluted to 1000 ml with distilled water and *thoroughly mixed* yields a *N*/50 solution that is satisfactory for most purposes. For referee work, it would be advisable to check the normality of the *N*/50 acid against weighed samples of a primary standard.

14-3 PREPARATION OF *N*/1 AND *N*/50 NaOH SOLUTIONS

Standard solutions of sodium hydroxide are used to measure carbon dioxide and acidity. The equivalent weight of carbon dioxide when reacting with sodium hydroxide, to pH 8.2 or the phenolphthalein end point, is 44, as may be calculated from the equation

$$\underset{44}{CO_2} + Na^+ + \underset{17}{OH^-} \rightarrow Na^+ + HCO_3^- \qquad (14\text{-}1)$$

Therefore *N*/44 solutions of NaOH are best suited for determination of carbon dioxide. Mineral acidity is always expressed in terms of calcium carbonate, which has an equivalent weight of 50, and *N*/50 solutions of bases are used for its determination.

In practice, it is most convenient to prepare a *N*/1 solution of NaOH as a stock supply and make solutions of lower normality by dilution, as with sulfuric acid. In some laboratories, so few determinations of carbon dioxide are made that it is impractical to keep a supply of both *N*/44 and *N*/50 solutions on hand. In such cases, the *N*/50 solution is used for both determinations, and a factor of 0.88 is applied to correct buret readings when carbon dioxide is measured.

Crystal or pellet forms of sodium hydroxide cannot be purchased in a pure form. They are always contaminated with sodium carbonate as a result of reaction with carbon dioxide of the air during manufacture. Even the so-called analytical reagent grade contains several percent of sodium carbonate and is unfit for the preparation of standard solutions without purification. Several methods are used, but only two are commonly used in water quality analytical laboratories. Purified sodium hydroxide can be purchased in 50 percent solution form.

A primary standard is required in the preparation of standard solutions of sodium hydroxide. Potassium acid phthalate ($KHC_8H_4O_4$) is excellent because of its high equivalent weight and other desirable properties. It should be dried at 105 to 110°C for 1 h and kept in a desiccator prior to use.

Purification of NaOH

All sodium hydroxide must be subjected to a purification process to free it of sodium carbonate which also has basic properties. Two methods will be described in order of their preference.

1. Sodium carbonate is relatively insoluble in concentrated (approximately 50 percent or 18 M) solutions of sodium hydroxide. If 500 g of stick- or pellet-form sodium hydroxide is added to 500 ml of distilled water, the sodium hydroxide will dissolve, leaving the sodium carbonate undissolved. The resulting solution will be rather turbid because of the suspended sodium carbonate. On standing several days, the carbonate will float or settle, and a clear solution will result that is sufficiently free of carbonate for the preparation of standard solutions. Siphon the purified 50 percent solution into a Pyrex bottle, and use a rubber stopper to exclude the air. It is good practice to keep a considerable supply of purified 50 percent sodium hydroxide on hand because of the time involved in its preparation.
2. Dilute solutions of sodium hydroxide may be freed of carbonate by precipitation with barium hydroxide.

$$2Na^+ + CO_3^{2-} + Ba(OH)_2 \rightarrow \underline{BaCO_3\downarrow} + 2NaOH \qquad (14\text{-}2)$$

The resulting barium carbonate precipitate may be removed by filtration (if protected from carbon dioxide of air), or it may be allowed to settle and the clarified solution siphoned to another bottle. The purified solution must be standardized after this treatment, as barium hydroxide is used in excess and an amount of sodium hydroxide equivalent to the sodium carbonate originally present remains.

Calculation of NaOH Needed

For purposes of these calculations, it is assumed that a purified solution of sodium hydroxide, 50 percent, is available and that approximately 1 liter of a $N/1$ solution is to be prepared. By definition, 1 liter of $N/1$ base contains the equivalent of 1.008 g of H^+ or 17 g of OH^- per liter. Calculations are as follows:

$$1 \text{ GMW or } 40 \text{ g pure NaOH} = 17 \text{ g } OH^-$$

The stock solution of purified NaOH is 50 percent strength. Therefore

$$\frac{40}{0.5} = 80 \text{ g } 50\% \text{ NaOH} = 17 \text{ g OH}^-$$

Since the stock solution of NaOH may not contain exactly 50 percent and it is desirable to prepare a solution which is slightly stronger than $N/1$, take 10 percent excess.

$$80 \times 1.1 = 88 \text{ g}$$

Preparation of $N/1$ NaOH Solution

As rapidly as possible, so as to minimize absorption of carbon dioxide from the air, weigh 88 g, ± 1 g, of the purified 50 percent sodium hydroxide solution into a small Erlenmeyer flask on a trip balance. Place 500 ml of carbon-dioxide-free distilled water in a 1-liter graduated cylinder, and add the sodium hydroxide solution. Rinse the Erlenmeyer flask with carbon-dioxide-free water, and add rinsings to the cylinder. Dilute to approximately 1 liter with carbon-dioxide-free water and mix thoroughly with a plunger-type stirrer to minimize contact with the air. Protect the solution from the air by keeping an inverted beaker or some other form of cap over the top of the cylinder until standardization is completed.

Calculation of Primary Standard Needed

Potassium acid phthalate is an excellent primary standard. It may be obtained in essentially 100 percent pure form. Its equivalent weight is equal to its molecular weight, which is 204.

$$204 \text{ g KHC}_8\text{H}_4\text{O}_4 \simeq 1000 \text{ ml } N/1 \text{ NaOH}$$

or $\qquad 2.04 \text{ g KHC}_8\text{H}_4\text{O}_4 \simeq 10 \text{ ml } N/1 \text{ NaOH}$

Weigh four samples of $KHC_8H_4O_4$ ranging from 2.0 to 2.2 g, and proceed as described in the third part of Sec. 14-1. Titration should be to the phenolphthalein end point. When standardization is completed, the $N/1$ solution should be stored in a Pyrex bottle fitted with a rubber stopper to exclude air.

Preparation of $N/50$ NaOH Solution

Proceed in a manner similar to the instructions for preparing $N/50$ acid. The $N/1$ hydroxide should be diluted with carbon-dioxide-free water, stored in a Pyrex bottle, and protected from the carbon dioxide of the atmosphere.

Standardization with Secondary Standards

The $N/1$ and $N/50$ solutions of sodium hydroxide may be standardized against corresponding solutions of sulfuric acid. The acid solutions serve as secondary standards, and any error made in their preparation will be reflected in the hydroxide solutions. Methyl orange must be used as the indicator, and whether the acid or the hydroxide is used as the titrant is optional. It is usually most convenient to use the acid.

PROBLEMS

14-1 Define: molar solution, normal solution, standard solution, mole, equivalent weight, and milliequivalent.

14-2 How may carbonate-free sodium hydroxide solutions be prepared?

14-3 (*a*) What is the normality of a solution of H_2SO_4 if 16.2 ml were required to neutralize 1.22 g of Na_2CO_3 to the methyl orange end point?

(*b*) How many ml of water must be added to the above solution to make it exactly 1.0 N? Assume you have 927 ml of the solution.

14-4 How many ml of 1.0 N NaOH are required to neutralize 0.2 g of HCl?

14-5 A sample of potassium acid phthalate weighing 3.75 g required 15.0 ml of a NaOH solution for titration to the phenolphthalein end point. If 460 ml of NaOH remain, how much water should be added to it to make it exactly 1.00 N?

14-6 A sample of Na_2CO_3 weighing 1.50 g required 25.0 ml of a H_2SO_4 solution for titration to the methyl orange end point. If 980 ml of H_2SO_4 remain, how much water should be added to it to make it exactly 1.00 normal?

14-7 How many ml of 1.00 N NaOH are required to prepare 500 ml of 0.0227 N NaOH?

14-8 How many ml of 1.00 N H_2SO_4 are required to prepare 2 liters of 0.05 N H_2SO_4?

14-9 In the preparation of two liters of 1.0 N acid from 35 percent hydrochloric acid, what weight of the impure acid should be taken, assuming standardization in the recommended manner?

FIFTEEN

pH

15-1 GENERAL CONSIDERATIONS

pH is a term used rather universally to express the intensity of the acid or alkaline condition of a solution. It is a way of expressing the hydrogen-ion concentration, or more precisely, the hydrogen-ion activity. It is important in almost every phase of environmental engineering practice. In the field of water supplies, it is a factor that must be considered in chemical coagulation, disinfection, water softening, and corrosion control. In wastewater treatment employing biological processes, pH must be controlled within a range favorable to the particular organisms involved. Chemical processes used to coagulate wastewaters, dewater sludges, or oxidize certain substances, such as cyanide ion, require that the pH be controlled within rather narrow limits. For these reasons and because of the fundamental relationships that exist between pH, acidity, and alkalinity, it is very important to understand the theoretical as well as the practical aspects of pH.

15-2 THEORETICAL CONSIDERATIONS

The concept of pH evolved from a series of developments that led to a fuller understanding of acids and bases. Acids and bases were originally distinguished by their difference in taste and later by the manner in which they affected certain materials that came to be known as indicators. With the discovery of hydrogen by Cavendish in 1766, it soon became apparent that all acids contained the element hydrogen. Chemists soon found that neutralization reactions

351

between acids and bases always produced water. From this and other related information, it was concluded that bases contained hydroxyl groups.

In 1887 Arrhenius announced his theory of ionization. Since that time acids have been considered to be substances that dissociate to yield hydrogen ions, and bases have been considered to be substances that dissociate to yield hydroxyl ions. According to the concepts of Arrhenius, strong acids and bases are highly ionized and weak acids and bases are poorly ionized in aqueous solution. Proof of these claims had to await the development of suitable devices for the measurement of hydrogen-ion concentration.

Measurement of Hydrogen–Ion Concentration

The hydrogen electrode (Sec. 3-10) was found to be a suitable device for measuring hydrogen-ion concentration. With its use, it was found that pure water dissociates to yield a concentration of hydrogen ions equal to 10^{-7} mol/l.

$$H_2O \rightleftharpoons H^+ + OH^- \qquad (15\text{-}1)$$

Since water dissociates to produce one hydroxyl ion for each hydrogen ion, it is obvious that 10^{-7} mol of hydroxyl ion is produced simultaneously. By substitution into the equilibrium equation, we obtain

$$\frac{[H^+][OH^-]}{[H_2O]} = K \qquad (15\text{-}2)$$

but, since the concentration of water is so extremely large and is diminished so very little by the slight degree of ionization, it may be considered as constant, and Eq. (15-2) may be written.

$$[H^+][OH^-] = K_W \qquad (15\text{-}3)$$

and for pure water at about 25°C,

$$[H^+][OH^-] = 10^{-7} \times 10^{-7} = 10^{-14} \qquad (15\text{-}4)$$

This is known as the ion product or ionization constant for water.

When an acid is added to water, it ionizes in the water and the hydrogen-ion concentration increases; consequently the hydroxyl-ion concentration must decrease in conformity with the ionization constant. For example, if acid is added to increase the $[H^+]$ to 10^{-1}, the $[OH^-]$ must decrease to 10^{-13}.

$$10^{-1} \times 10^{-13} = 10^{-14}$$

Likewise, if a base is added to water to increase the $[OH^-]$ to 10^{-3}, the $[H^+]$ decreases to 10^{-11}. It is important to remember that the $[OH^-]$ or the $[H^+]$ can never be reduced to zero, no matter how acidic or basic the solution may be.

The pH Concept

Expression of hydrogen-ion concentrations in terms of molar concentrations is rather cumbersome. In order to overcome this difficulty, Sorenson (1909) proposed to express such values in terms of their negative logarithms and designated such values as p_H^+. His symbol has been superseded by the simple designation pH. The term may be represented by

$$pH = -\log [H^+] \quad \text{or} \quad pH = \log \frac{1}{[H^+]} \qquad (15\text{-}5)$$

and the pH scale is usually represented as ranging from 0 to 14, with pH 7 representing absolute neutrality.

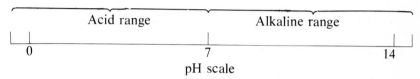

Acid conditions increase as pH values decrease, and alkaline conditions increase as the pH values increase. pH 7 has little significance as a reference point in water chemistry and therefore is of little importance in environmental engineering practice (see Sec. 5-5).

15-3 MEASUREMENT OF pH

The hydrogen electrode is the absolute standard for the measurement of pH. It is rather cumbersome and not well adapted for universal use, particularly in field studies or in solutions containing materials that are adsorbed on platinum black. A wide variety of indicators were calibrated with the hydrogen electrode to determine their color characteristics at various pH levels. From these studies it became possible to determine pH values fairly accurately by choosing an indicator that exhibited significant color changes in the particular range involved. With the use of about six to eight indicators it is possible to determine pH values in the range of interest to environmental engineers (see Sec. 10-4). Their use has been superseded by development of the glass electrode.

About 1925 it was discovered that an electrode could be constructed of glass (Sec. 11-3) which would develop a potential related to the hydrogen-ion concentration without interference from most other ions. Its use has become the standard method of measuring pH.

Measurement with the Glass Electrode

pH meters employing the glass electrode are manufactured by many concerns. They range from portable battery-operated units selling for about one hundred dollars to highly precise instruments selling for several hundred dollars. Units that could be operated on 110-volt alternating current were developed about 1940 and are highly satisfactory for most routine laboratory purposes, being capable of measuring pH within a ±0.1 pH unit. The small portable battery-operated

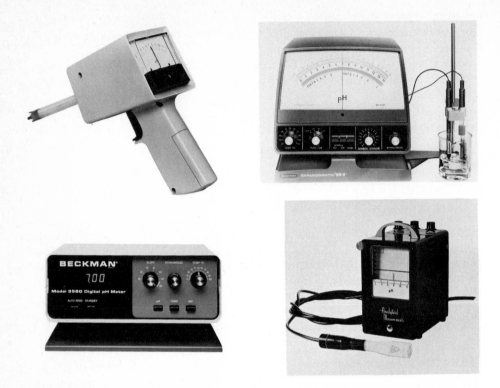

Figure 15-1 Commercial models of pH meters. *(a)* Battery-operated portable model. *(Beckman Instruments, Inc.)* *(b)* Line-operated laboratory model. *(Beckman Instruments, Inc.)* *(c)* Line-operated laboratory model with digital readout. *(Beckman Instruments, Inc.)* *(d)* Portable recording model. *(Analytical Measurements, Inc.)*

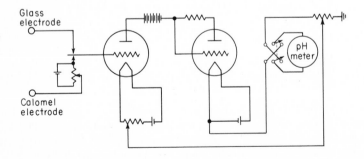

Figure 15-2 Simplified diagram of a pH circuit.

units are most suitable for field work. Figure 15-1 shows examples of four types of instruments available in the United States. Figure 15-2 is a simplified diagram of a pH meter circuit.

pH measurements can be made in a wide variety of materials and under extreme conditions, provided that attention is paid to the type of electrode used. Measurement of pH values above 10 and at high temperatures is best made with special glass electrodes designed for such service. The pH of semisolid substances can be made with a spear-type electrode. The instruments are normally standardized with buffer solutions of known pH values.

15-4 INTERPRETATION OF pH DATA

pH data should always be interpreted in terms of hydrogen-ion concentration, which, of course, is a measure of the intensity of acid or basic conditions. For all practical purposes the conversion is very simple, although in an exact sense it must be remembered that pH is a measure of ion activity and not concentration.

$$\text{at pH 2} \quad [H^+] = 10^{-2}$$

$$\text{at pH 10} \quad [H^+] = 10^{-10}$$

$$\text{at pH 4.5} \quad [H^+] = 10^{-4.5}$$

pH does not measure total acidity or total alkalinity. This can be illustrated by comparing the pH of $N/10$ solutions of sulfuric acid and acetic acids, which have the same neutralizing value. The pH of the former is approximately 1 because of its high degree of ionization, and the pH of the latter is about 3 because of its low degree of ionization.

In some instances the pOH, or hydroxyl-ion concentration, of a solution is of major interest. It is customary to calculate pOH from pH values, using the relationship given in Eq. (15-3). Approximations are often made from the relationship

$$pH + pOH = 14 \tag{15-6}$$

or
$$pOH = 14 - pH \tag{15-7}$$

It is just as important for the environmental engineer to remember that the $[OH^-]$ of a solution can never be reduced to zero, no matter how acid the solution is, as it is for him to remember that the $[H^+]$ can never be reduced to zero, no matter how alkaline a solution becomes.

Concepts of pOH, or hydroxyl-ion concentration, are of particular importance in precipitation reactions involving formation of hydroxides. Examples are the precipitation of Mg^{2+} in softening of water with lime and in chemical coagulation processes employing iron and aluminum salts.

PROBLEMS

15-1 What is the relationship (*a*) between pH and hydrogen-ion concentration and (*b*) between pH and hydroxide-ion concentration?

15-2 What would be the pH of a solution containing (*a*) 1.008 g of hydrogen ion per liter, (*b*) 0.1008 g of hydrogen ion per liter, and (*c*) 1.7×10^{-8} g of OH^- per liter?

15-3 One solution has a pH of 4.0 and another a pH of 6.0. What is (*a*) the hydrogen-ion concentration and (*b*) the hydroxide-ion concentration in each of the solutions?

15-4 A decrease in pH of one unit represents how much of an increase in hydrogen-ion concentration?

15-5 A 50 percent decrease in hydrogen-ion concentration represents how much of an increase in pH units?

15-6 Approximately what is the pH of a 2.0 *N* HCl solution?

15-7 Approximately what is the pH of a 0.02 *N* NaOH solution?

15-8 What is the hydroxide-ion concentration if the hydrogen-ion concentration is 3.0×10^{-2} mol/l?

SIXTEEN

ACIDITY

16-1 GENERAL CONSIDERATIONS

Most natural waters, domestic sewage, and many industrial wastes are buffered principally by a carbon dioxide–bicarbonate system. By reference to Fig. 5-2, which shows titration curves for several weak acids, it will be noted from the curve for carbonic acid that the stoichiometric end point is not reached until the pH has been raised to about 8.5. On the basis of this information, it is customary to consider that all waters having a pH lower than 8.5 contain acidity. Usually the phenolphthalein end point at pH 8.2 to 8.4 is taken as the reference point. Inspection of the curve for carbonic acid in Fig. 5-2 shows that at pH 7.0 considerable carbon dioxide remains to be neutralized. It also shows that carbon dioxide (carbonic acid) alone will not depress the pH below a value of about 4.5.

Figure 5-1 shows a titration curve for a strong acid, and from the nature of the curve, it may be concluded that neutralization of the acid is essentially complete at pH 4.5. Thus, from the nature of the titration curves for carbonic acid and for strong acids, it becomes obvious that the acidity of natural waters is caused by carbon dioxide or by strong mineral acids, the former being the effective agent in waters having pH values greater than 4.5 and the latter the effective agent in waters with pH values less than 4.5, as shown in Fig. 16-1.

16-2 SOURCES AND NATURE OF ACIDITY

Carbon dioxide is a normal component of all natural waters. It may enter surface waters by absorption from the atmosphere, but only when the partial

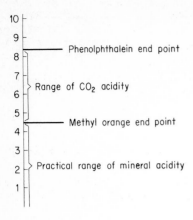

Figure 16-1 Types of acidity of importance in ordinary water and wastewater analysis, and the pH ranges in which they are significant.

pressure of carbon dioxide in the water is less than the partial pressure of the carbon dioxide in the atmosphere, in accordance with Henry's law. Carbon dioxide may also be produced in waters through biological oxidation of organic matter, particularly in polluted water. In such cases, if photosynthetic activity is limited, the partial pressure of carbon dioxide in the water may exceed that of the atmosphere and carbon dioxide will escape from the liquid. Thus it may be concluded that surface waters are constantly absorbing or giving up carbon dioxide to maintain an equilibrium with the atmosphere. The amount that can exist at equilibrium is very small because of the low partial pressure of carbon dioxide in the atmosphere.

Ground-waters and waters from the hypolimnion of stratified lakes and reservoirs often contain considerable amounts of carbon dioxide. This concentration results from bacterial oxidation of organic matter with which the water has been in contact, and under the conditions, the carbon dioxide is not free to escape to the atmosphere. Carbon dioxide is an end product of both aerobic and anaerobic bacterial oxidation; therefore its concentration is not limited by the amount of dissolved oxygen originally present. It is not uncommon to encounter ground-waters with 30 to 50 mg/l of carbon dioxide. This is particularly true of waters that have percolated through soils that do not contain enough calcium or magnesium carbonate to neutralize the carbon dioxide through formation of bicarbonates.

$$CO_2 + CaCO_3 + H_2O \rightarrow Ca(HCO_3)_2 \qquad (16\text{-}1)$$

Mineral acidity is present in many industrial wastes, particularly those of the metallurgical industry and some from the production of synthetic organic materials. Certain natural waters may also contain mineral acidity. The drainage from abandoned mines, lean ore dumps, and "gob" piles will contain significant amounts of sulfuric acid or salts of sulfuric acid if sulfur, sulfides, or iron pyrites are present. Conversion of these materials to sulfuric acid and sulfates is brought about by sulfur-oxidizing bacteria under aerobic conditions.

$$2S + 3O_2 + 2H_2O \xrightarrow{\text{bact.}} 2H_2SO_4 \qquad (16\text{-}2)$$

$$FeS_2 + 3\tfrac{1}{2}O_2 + H_2O \xrightarrow{\text{bact.}} FeSO_4 + H_2SO_4 \qquad (16\text{-}3)$$

Salts of heavy metals, particularly those with trivalent metal ions such as Fe^{3+} and Al^{3+}, hydrolyze in water to release mineral acidity.

$$FeCl_3 + 3H_2O \rightleftharpoons Fe(OH)_3 + 3H^+ + 3Cl^- \qquad (16\text{-}4)$$

Their presence is indicated by the formation of a precipitate as the pH of solutions containing them is increased during neutralization.

Many industrial wastes contain organic acids. Their presence and nature can be determined by use of electrometric titration curves or gas chromatography.

16-3 SIGNIFICANCE OF CARBON DIOXIDE AND MINERAL ACIDITY

Acidity is of little concern from a sanitary or public health viewpoint. Carbon dioxide is present in malt and carbonated beverages in concentrations greatly in excess of any concentrations known in natural waters, and no deleterious effects due to the carbon dioxide have been recognized. Waters that contain mineral acidity are so unpalatable that problems related to human consumption are nonexistent.

Acid waters are of concern because of their corrosive characteristics and the expense involved in removing or controlling the corrosion-producing substances. The corrosive factor in most waters is carbon dioxide, but in many industrial wastes it is mineral acidity. Carbon dioxide must be reckoned with in water-softening problems where the lime or lime–soda ash method is employed.

Where biological processes of treatment are used, the pH must ordinarily be maintained within the range of 6 to 9.5. This criterion often requires adjustment of pH to favorable levels, and calculation of the amount of chemicals needed is based upon acidity values in most cases.

16-4 METHOD OF MEASUREMENT

Both carbon dioxide and mineral acidity can be measured by means of standard solutions of alkaline reagents. Mineral acids are measured by titration to a pH of about 4.5. Since methyl orange was formerly used as the indicator in this titration, mineral acidity is also called *methyl orange acidity*. Titration of a sample to the phenolphthalein end point of pH 8.3 measures both mineral acidity plus acidity due to weak acids. This total acidity is also termed *phenolphthalein acidity*.

Carbon Dioxide

If reliable results are to be obtained, special precautions must be taken during the collection, handling, and analysis of samples for carbon dioxide, regardless of the method used. In waters where carbon dioxide is an important consideration, its partial pressure is greatly in excess of that in the atmosphere; therefore exposure to the air must be avoided or kept at a minimum. For this reason analysis can be accomplished to best advantage at the point of collection where exposure to the air and temperature change can be avoided.

The sample should be collected in the same manner that is used to obtain a sample for dissolved oxygen, i.e., by using a submerged tube or pipe inlet, to exclude air bubbles, and allowing the container to overflow in order to displace any water that has come in contact with the air. If the sample must be transported to the laboratory for analysis, the bottle should be filled completely and capped or stoppered so as to leave no air pocket. The temperature should be kept as near that at which it was collected as possible.

Titration method In order to minimize contact with the air, it is best to collect and titrate the sample in a graduated cylinder or a color-comparison tube. The tube or cylinder should be filled to overflowing and the excess siphoned off or removed with a pipet to get the proper sample size. After addition of the proper amount of phenolphthalein indicator, the titration is conducted in a manner to minimize loss of carbon dioxide. Ordinarily, appreciable amounts of carbon dioxide will be lost in the first titration because of the excessive stirring needed. Reliable results may be obtained by taking a second sample and adding the indicated amount of titrant to it before stirring. The titration may then be completed without significant loss of carbon dioxide. The final end point is reached somewhat slowly, and so it is recommended that the titration not be considered complete until a pinkish color persists for 30 s.

When sodium hydroxide is used as the standard reagent, it is important that it be free of sodium carbonate. The reaction involved in the neutralization may be considered to occur in two steps,

$$2NaOH + CO_2 \rightarrow Na_2CO_3 + H_2O \tag{16-5}$$

$$Na_2CO_3 + CO_2 + H_2O \rightarrow 2NaHCO_3 \tag{16-6}$$

and, from Eq. (16-6), it should be obvious that if sodium carbonate is originally present in the sodium hydroxide, it will cause erroneous results. In order to overcome this problem, a sodium carbonate solution is one of the standard titrants recommended for carbon dioxide measurements. Sodium carbonate can be used in this capacity because it reacts quantitatively with carbon dioxide, as shown in Eq. (16-6). It has a definite advantage in that it may be purchased in analytical-grade form.

Calculation from pH and alkalinity It is possible to calculate the amount of carbon dioxide in a water sample from ionization expressions for carbonic acid.

When the pH is less than about 8.5, the primary ionization constant for carbonic acid can be used, provided that the hydrogen-ion and bicarbonate-ion concentrations and the value for K_1 are known:

$$\frac{[H^+][HCO_3^-]}{[H_2CO_3]} = K_1 \tag{16-7}$$

In practice, $[H_2CO_3]$ in this expression is set equal to the sum of the molar concentrations of carbonic acid and free carbon dioxide because of the difficulty in distinguishing between these two forms. Since the free carbon dioxide represents about 99 percent of this total, the expression is only an approximation of a true equilibrium expression, although a fairly good one.

The use of Eq. (16-7) is illustrated in the following example. If $K_1 = 4.3 \times 10^{-7}$, $[H^+] = 10^{-7}$, and $[HCO_3^-] = 4.3 \times 10^{-3}$, then the CO_2 concentration would equal (10^{-7}) $(4.3 \times 10^{-3})/(4.3 \times 10^{-7}) = 10^{-3}$ mol/l or 44 mg/l. However, in order for such a calculation to be accurate, the effect of other ions, as discussed in Sec. 5-3, must be considered, as must the effect of temperature on K_1. Since these considerations can make the calculation of free carbon dioxide a complicated process, a nomographic chart is included in "Standard Methods" to facilitate the determination of free carbon dioxide from sample pH, alkalinity, dissolved solids, and temperature.

Determinations of carbon dioxide from pH and alkalinity measurements can result in highly accurate results, but not necessarily so. The method suffers from the fact that the total solids concentration must be known. This usually requires a separate determination by gravimetric or conductivity methods. Also, the pH must be measured very accurately, as small variations can introduce serious errors. For example, an inaccuracy of 0.1 in the pH determination causes a carbon dioxide error of about 25 percent. Therefore it is questionable whether results obtained by this method under ordinary laboratory or field conditions are more reliable than results obtained by the titration procedure, if proper attention is paid to details described for the titration method. Considering the difficulties with each procedure, it would appear that the titration procedure would normally be the method of choice for carbon dioxide concentrations greater than 2 mg/l, while for smaller concentrations the titration errors could be excessive, and thus the calculation procedure would be preferred.

Field method The titration procedure has many advantages and is sufficiently accurate for all practical purposes.

Methyl Orange Acidity

All natural waters and most industrial wastes that have a pH below 4 contain mineral or methyl orange acidity. Mineral acids are essentially neutralized by the time the pH has been raised to about 4.5 (see Fig. 5-1), and methyl orange is normally used as the indicator where a pH meter is not available. Results are reported in terms of methyl orange acidity expressed as $CaCO_3$. Since $CaCO_3$

has an equivalent weight of 50, $N/50$ NaOH is used as the titrating agent so that 1 ml is equivalent to 1 mg of acidity.

Phenolphthalein Acidity

Occasionally it is desirable to measure the total acidity resulting both from mineral acids and from weak acids in the sample. Since most weak acids are essentially neutralized by titration to pH 8.3, phenolphthalein indicator can be used for this titration. When heavy metal salts are present, it is usually desirable to heat the sample to boiling and then carry out the titration. The heat speeds the hydrolysis of the metal salts, allowing the titration to be completed more rapidly. Again, $N/50$ NaOH is used as the titrating agent, and results are reported in terms of phenolphthalein acidity expressed as $CaCO_3$.

16-5 APPLICATION OF ACIDITY DATA

Carbon dioxide determinations are particularly important in the field of public water supplies. In the development of new supplies, it is an important factor that must be considered in the treatment method and the facilities needed. Many underground supplies require treatment to overcome corrosive characteristics resulting from carbon dioxide. The amount present is an important factor in determining whether removal by aeration or simple neutralization with lime or sodium hydroxide will be chosen as the treatment method. The size of equipment, chemical requirements, storage space, and cost of treatment all depend upon amounts of carbon dioxide present. Carbon dioxide is an important consideration in estimating chemical requirements for lime or lime–soda ash softening.

Most industrial wastes containing mineral acidity must be neutralized before they may be discharged to rivers or sewers or subjected to treatment of any kind. Quantities of chemicals, size of chemical feeders, storage space, and costs are determined from laboratory data on acidity.

PROBLEMS

16-1 Can the pH of a water sample be calculated from a knowledge of its acidity? Why?

16-2 Can the carbon dioxide content of a waste sample known to contain a significant concentration of acetic acid be determined by the titration procedure? Why?

16-3 A sample of water collected in the field had a pH of 6.8. By the time the water sample reached the laboratory for analysis, the pH had increased to 7.5. Give a possible explanation for this change.

16-4 Estimate the carbon dioxide content of a natural water sample having a pH of 7.3 and a bicarbonate-ion concentration (as HCO_3^-) of 30 mg/l. Assume that the effect of the dissolved solids on ion activity is negligible and the water temperature is 25°C.

16-5 (*a*) A water supply was found to have a bicarbonate-ion concentration of 50 mg/l and a CO_2 content of 30 mg/l. Estimate the approximate pH of the water (temperature equals 25°C).

(*b*) If the CO_2 content of the above water were reduced to 3 mg/l by aeration, what would the pH then be?

16-6 Air contains an average of 0.03 percent by volume of carbon dioxide, and the Henry's law constant, α, for carbon dioxide in water at 20°C is 1740 mg/l-atm.

(*a*) Using Eq. (2-15), calculate the concentration of carbon dioxide in a water sample when in equilibrium with atmospheric carbon dioxide at sea level.

(*b*) On the basis of the above calculation, what will occur when a sample containing 10 mg/l of carbon dioxide is vigorously exposed to the air prior to titration for carbon dioxide?

16-7 A water sample has a methyl orange acidity of 60 mg/l. Calculate the quantity of lime in mg/l of $Ca(OH)_2$ required to raise the pH to 4.3.

SEVENTEEN

ALKALINITY

17-1 GENERAL CONSIDERATIONS

The *alkalinity* of a water is a measure of its capacity to neutralize acids. The alkalinity of natural waters is due primarily to the salts of weak acids, although weak or strong bases may also contribute. Bicarbonates represent the major form of alkalinity, since they are formed in considerable amounts from the action of carbon dioxide upon basic materials in the soil, as shown in Eq. (16-1). Other salts of weak acids, such as borates, silicates, and phosphates, may be present in small amounts. A few organic acids that are quite resistant to biological oxidation—for example, humic acid—form salts that add to the alkalinity of natural waters. In polluted or anaerobic waters, salts of weak acids such as acetic, propionic, and hydrosulfuric may be produced and would also contribute to alkalinity. In other cases, ammonia or hydroxides may make a contribution to the total alkalinity of a water.

Under certain conditions, natural waters may contain appreciable amounts of carbonate and hydroxide alkalinity. This condition is particularly true in surface waters where algae are flourishing. The algae remove carbon dioxide, free and combined, from the water to such an extent that pH values of 9 to 10 are often obtained. The chemistry involved is discussed in Sec. 17-7. Boiler waters always contain carbonate and hydroxide alkalinity. Chemically treated waters, particularly those produced in lime or lime–soda ash softening of water, contain carbonates and excess hydroxide.

Although many materials may contribute to the alkalinity of a water, the major portion of the alkalinity in natural waters is caused by three major classes of materials which may be ranked in order of their association with high pH values as follows: (1) hydroxide, (2) carbonates, and (3) bicarbonates. For most practical purposes, alkalinity due to other materials in natural waters is insignificant and may be ignored.

The alkalinity of waters is due principally to salts of weak acids and strong bases, and such substances act as buffers to resist a drop in pH resulting from acid addition, as discussed in Sec. 5-6. Alkalinity is thus a measure of the buffer capacity and in this sense is used to a great extent in wastewater treatment practice.

17-2 PUBLIC HEALTH SIGNIFICANCE

As far as is known, the alkalinity of a water has little public health significance. Highly alkaline waters are usually unpalatable, and consumers tend to seek other supplies. Chemically treated waters sometimes have rather high pH values which have met with some objection on the part of consumers. For these reasons, standards are sometimes established on chemically treated waters. Such standards, relating to phenolphthalein and total and excess alkalinity, are too detailed to summarize here.

17-3 METHOD OF DETERMINING ALKALINITY

Alkalinity is measured volumetrically by titration with $N/50$ H_2SO_4 and is reported in terms of equivalent $CaCO_3$. For samples whose initial pH is above 8.3, the titration is made in two steps. In the first step the titration is conducted until the pH is lowered to 8.2, the point at which phenolphthalein indicator turns from pink to colorless. The second phase of the titration is conducted until the pH is lowered to about 4.5, corresponding to the methyl orange end point. When the pH of a sample is less than 8.3, a single titration is made to a pH of 4.5.

The choice of pH 8.3 as the end point for the first step in the titration is in accord with the fundamentals of alkalimetry developed in Sec. 5-5. This value corresponds to the equivalence point for the conversion of carbonate ion to bicarbonate ion:

$$CO_3^{2-} + H^+ \rightarrow HCO_3^- \tag{17-1}$$

The use of a pH of about 4.5 for the end point for the second step of the titration corresponds approximately to the equivalence point for the conversion of bicarbonate ion to carbonic acid:

$$HCO_3^- + H^+ \rightarrow H_2CO_3 \tag{17-2}$$

On the basis of Eq. (5-31), the exact end point for this titration would be dependent upon the initial bicarbonate-ion concentration in the sample. Using

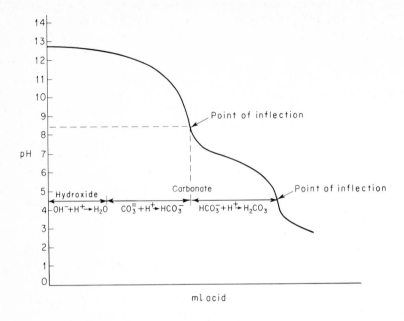

Figure 17-1 Titration curve for a hydroxide-carbonate mixture.

the ionization constant from Table 5-2 for the bicarbonate salt of carbonic acid, we see that Eq. (5-31) becomes

$$\text{pH (bicarbonate equivalence point)} = 3.2 - \tfrac{1}{2} \log [HCO_3^-] \qquad (17\text{-}3)$$

A $[HCO_3^-]$ of 0.01 M corresponds to an alkalinity of 500 mg/l as $CaCO_3$, for which the equivalence point would be 4.2. These considerations require that the carbonic acid or carbon dioxide formed from bicarbonate during the titration not be lost from solution. This is difficult to ensure, and for this reason the above considerations are largely of theoretical interest.

The actual pH of the stoichiometric end point in alkalinity determinations can be best determined by electrometric titration. This fact is particularly important in natural waters where the total alkalinity is a summation of the effects resulting from salts of weak acids of which bicarbonates are only one. The pH at which the inflection in the titration curve occurs (see Fig. 17-1) is taken as the true end point. The pH values given for the equivalence points for various alkalinities from Eq. (17-3) or in "Standard Methods" hold only for essentially pure bicarbonate solutions and should not be applied indiscriminately to domestic or industrial wastewaters, or even natural waters.

17-4 METHODS OF EXPRESSING ALKALINITY

Alkalinity measurements are made on a wide variety of materials. These range from relatively pure waters through polluted waters, such as municipal and industrial wastewaters, to digesting sludges. The methods of expressing alkalinity values vary considerably; therefore it is necesary to explain the methods in some detail and to indicate the areas where the various methods are employed.

Phenolphthalein and Total Alkalinity

Inspection of the titration curves for a strong base (hydroxide alkalinity), shown in Fig. 5-1, and for sodium carbonate, in Fig. 5-3, shows that essentially all the hydroxide has been neutralized by the time the pH has ben decreased to 10 and that the carbonate has been converted to bicarbonate by the time the pH has been lowered to about 8.3. In a mixture containing both hydroxide and carbonate, the carbonate modifies the titration curve to the extent that only the inflection at pH 8.3 occurs, as shown in Fig. 17-1. Because of this, it has become common practice to express the alkalinity measured to the phenolphthalein end point as *phenolphthalein alkalinity*. This term is quite widely used at present in the field of waste treatment and is still used to some extent in the area of water analysis.

If the titration of a sample that originally contained both carbonate and hydroxide alkalinity is continued beyond the phenolphthalein end point, the bicarbonates react with the acid and are converted to carbonic acid. The reaction is essentially complete when the pH has been lowered to about 4.5 (see Fig. 17-1). The amount of acid required to react with the hydroxide, carbonate, and bicarbonate represents the *total alkalinity*. It is customary to express alkalinity in terms of $CaCO_3$; therefore $N/50\ H_2SO_4$ is used in its measurement. Calculations are made as follows:

$$\text{Phenol. alk.} = (\text{ml } N/50\ H_2SO_4 \text{ to pH } 8.3)\ \frac{1000}{\text{ml sample}} \qquad (17\text{-}4)$$

$$\text{Total alk.} = \text{total ml } N/50\ H_2SO_4 \text{ to pH } \begin{cases} 5.0 \\ 4.8 \\ 4.6 \\ 4.0 \end{cases} \times \frac{1000}{\text{ml sample}} \qquad (17\text{-}5)$$

In the determination of total alkalinity, the pH at the stoichiometric end point is directly related to the amount of carbonate alkalinity originally present in the sample, as discussed in Sec. 17-3.

Hydroxide, Carbonate, and Bicarbonate Alkalinity

In water analysis it is often desirable to know the kinds and amounts of the various forms of alkalinity present. This information is especially needed in water-softening processes and in boiler-water analysis. It is customary to calculate hydroxide, carbonate, and bicarbonate alkalinities from the fundamental information given by the titration curves for strong bases and sodium carbonate. (See Figs. 5-1 and 5-3.) There are three procedures commonly used to make these calculations: (1) calculation from alkalinity measurements alone, (2) calculation from alkalinity plus pH measurements, and (3) calculation from equilibrium equations. The first is the classical method and is based on empirical relationships for the calculation of the various forms of alkalinity from phenolphthalein and total alkalinities. This is intended for use by technicians and others who do not have a knowledge of the fundamental chemistry involved. The results from this method are only approximate for samples with pH greater than 9. However, water chemists and engineers concerned with water softening, corrosion control, and prevention of scaling at elevated pH levels are vitally concerned with ionic species and concentrations. For these reasons it has become necessary to be able to calculate hydroxide-, carbonate-, and bicarbonate-ion concentrations at all pH levels with considerable accuracy. This can be done with either the second or the third procedure.

The second procedure gives sufficiently accurate estimates for most practical purposes, and also makes use of phenolphthalein and total alkalinity measurements. In addition, an accurate initial pH measurement is required for the direct calculation of hydroxide alkalinity. In the third procedure, the various equilibrium equations for carbonic acid are used to compute the concentrations of the various alkalinity forms. This method gives reasonably accurate results for constituents, even when present in the fractional mg/l range, provided that an accurate pH measurement is made. The concentration of constituents in low concentration is sometimes of importance. A total alkalinity, as well as a pH measurement, is required. In addition, a dissolved-solids measurement to correct for ion activity (see Sec. 5-3), and temperature measurement for the selection of a proper equilibrium constant must be made. It seems important that environmental engineers as well as chemists understand the basis for these procedures. This is presented in the following.

Calculation from Alkalinity Measurements Alone

In this procedure, phenolphthalein and total alkalinities are determined, and from these measurements the calculation of three types of alkalinity, hydroxide, carbonate, and bicarbonate, are made. This can be done by assuming (incorrectly) that hydroxide and bicarbonate alkalinity cannot exist together in the same sample. This permits only five possible situations to be present, which are as follows: (1) hydroxide only, (2) carbonate only, (3) hydroxide plus carbonate, (4) carbonate plus bicarbonate, and (5) bicarbonate only. Reference to Figs. 5-1 and 5-3 will demonstrate that neutralization of hydroxides is complete by the time enough acid has been added to decrease the pH to 8.3, and that a carbonate is exactly one-half neutralized when the pH has been decreased to the same degree. Upon continuation of the titration to reach a pH of about 4.5, a negligible amount of acid is needed in the case of the hydroxide, and an amount exactly equal to that needed to reach pH 8.3 is required for the carbonate. This is the fundamental information needed to determine which forms of alkalinity are present and the amounts of each.

A graphical representation of typical titrations obtained with the various combinations of alkalinity is shown in Fig. 17-2.

Hydroxide only Samples containing only hydroxide alkalinity have a high pH, usually well above 10. Titration is essentially complete at the phenolphthalein end point. In this case hydroxide alkalinity is equal to the phenolphthalein alkalinity.

Carbonate only Samples containing only carbonate alkalinity have a pH of 8.5 or higher. The titration to the phenolphthalein end point is exactly equal to one-half of the total titration. In this case carbonate alkalinity is equal to the total alkalinity.

Hydroxide-carbonate Samples containing hydroxide and carbonate alkalinity have a high pH, usually well above 10. The titration from the phenolphthalein to the methyl orange end point represents one-half of the carbonate alkalinity. Therefore carbonate alkalinity may be calculated as follows:

$$\text{Carbonate alk.} = 2 \text{ (titration from phenol. to methyl orange)} \times \frac{1000}{\text{ml sample}}$$

and
$$\text{Hydroxide alk.} = \text{total alk.} - \text{carbonate alk.}$$

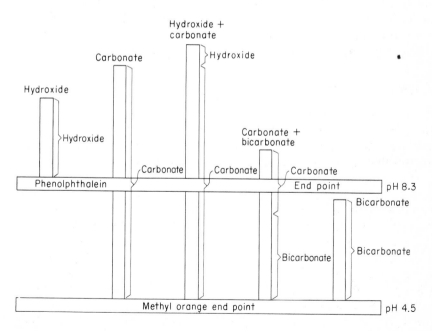

Figure 17-2 Graphical representation of titration of samples containing various forms of alkalinity.

Carbonate-bicarbonate Samples containing carbonate and bicarbonate alkalinity have a pH > 8.3 and usually less than 11. The titration to the phenolphthalein end point represents one-half of the carbonate. Carbonate alkalinity may be calculated as follows:

$$\text{Carbonate alk.} = 2 \text{ (titration to phenol. end point)} \times \frac{1000}{\text{ml sample}}$$

and $\qquad$ Bicarbonate alk. = total alk. − carbonate alk.

Bicarbonate only Samples containing only bicarbonate alkalinity have a pH of 8.3 or less, usually less. In this case bicarbonate alkalinity is equal to the total alkalinity.

The foregoing methods of approximate calculation have been superseded by the more precise methods described below.

Calculation from Alkalinity plus pH Measurements

In this procedure, measurements are made for pH, phenolphthalein, and total alkalinity. This will allow calculation of hydroxide, carbonate, and bicarbonate alkalinity.

Hydroxide First, the hydroxide alkalinity is calculated from the pH measurement, using the dissociation constant for water:

$$[OH^-] = \frac{K_W}{[H^+]} \tag{17-6}$$

This calculation requires a precise pH measurement for the determination of $[H^+]$. Since a hydroxide concentration of 1 mol/l is equivalent to 50,000 mg/l of alkalinity as $CaCO_3$, the above relationship can be expressed more conveniently as

$$\text{Hydroxide alk.} = 50,000 \times 10^{(\text{pH}-\text{p}K_w)} \tag{17-7}$$

At 24°C, $\text{p}K_W = 14.00$. However, it varies from 14.94 at 0°C to 13.53 at 40°C. Therefore it is important that a temperature measurement be made and the correct $\text{p}K_W$ be used. The relationship between pH, temperature, and hydroxide alkalinity is shown graphically in Fig. 17-3. To be more precise, a dissolved-solids measurement should be made to correct for ion activity, although in this case the correction is fairly negligible and not necessary for most practical purposes. A nomograph is available in "Standard Methods" which allows rapid calculation of hydroxide alkalinity, using pH, temperature, and dissolved-solids measurements.

Carbonate Once the hydroxide alkalinity is determined, use can be made of the principles from the preceding procedure to calculate the carbonate and bicarbonate alkalinity. The phenolphthalein alkalinity represents all the

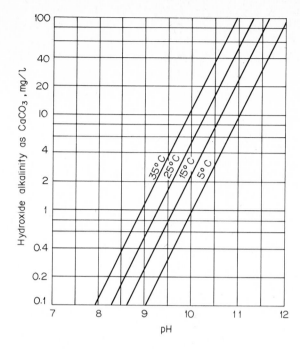

Figure 17-3 Relationship between hydroxide alkalinity and pH at various temperatures.

hydroxide alkalinity plus one-half the carbonate alkalinity. Therefore carbonate alkalinity may be calculated as follows:

$$\text{Carbonate alk.} = 2 (\text{phenol. alk.} - \text{hydroxide alk.}) \qquad (17\text{-}8)$$

Bicarbonate The titration from the phenolphthalein to the methyl orange end point measures the remaining one-half of the carbonate alkalinity plus all the bicarbonate alkalinity. It is also apparent that the bicarbonate alkalinity represents the remaining alkalinity after the hydroxide plus carbonate alkalinities are subtracted. From either standpoint, the bicarbonate alkalinity becomes

$$\text{Bicarbonate alk.} = \text{total alk.} - (\text{carbonate alk.} + \text{hydroxide alk.}) \quad (17\text{-}9)$$

Calculation from Equilibrium Equations[1]

The distribution of the various forms of alkalinity can be calculated from equilibrium equations plus a consideration of electroneutrality in solution. In order to preserve electroneutrality, the sum of the equivalent concentrations of the cations must equal that of the anions. Total alkalinity is a measure of the equivalent concentration of all cations associated with the alkalinity producing anions, except the hydrogen ion. Thus the balance of equivalent concentrations of alkalinity-associated cations and anions is given by

[1] E. W. Moore, *J. Am. Water Works Assoc.*, **31**: 51 (1939).

$$[H^+] + \frac{alkalinity}{50,000} = [HCO_3^-] + 2[CO_3^{2-}] + [OH^-] \qquad (17\text{-}10)$$

The equilibrium equations which must be considered are that for water [Eq. (17-6)] and that for the second ionization of carbonic acid,

$$\frac{[H^+][CO_3^{2-}]}{[HCO_3^-]} = K_2 \qquad (17\text{-}11)$$

From a pH measurement, $[H^+]$ and $[OH^-]$ can be determined, using Eq. (17-6). The only other unknowns are $[HCO_3^-]$ and $[CO_3^{2-}]$, and these can be determined from a simultaneous solution of Eqs. (17-10) and (17-11). The following equations result:

$$\text{Carbonate alkalinity} \atop \text{(mg/l as } CaCO_3) = \frac{50,000[(alkalinity/50,000) + [H^+] - (K_W/[H^+])]}{1 + ([H^+]/2K_2)}$$
$$(17\text{-}12)$$

$$\text{Bicarbonate alkalinity} \atop \text{(mg/l as } CaCO_3) = \frac{50,000[(alkalinity/50,000) + [H^+] - (K_W/[H^+])]}{1 + (2K_2/[H^+])}$$
$$(17\text{-}13)$$

At 25°C, K_W is 10^{-14} and K_2 is 4.7×10^{-11}. However, these values vary quite radically with temperature. Also, the activities of the ions vary considerably with ionic concentration, as indicated in Sec. 5-3. These corrections are rather tedious; consequently "Standard Methods" presents nomographs for the evaluation of carbonate and bicarbonate based on these considerations. The nomographs, as well as Eqs. (17-12) and (17-13), yield results in terms of alkalinity expressed as $CaCO_3$. At times the actual concentrations of carbonate or bicarbonate ion may be desired. Conversions to milligrams per liter of CO_3^{2-} or HCO_3^- are as follows:

$$\text{mg/l } CO_3^{2-} = \text{mg/l carbonate alk.} \times 0.6 \qquad (17\text{-}14)$$

$$\text{mg/l } HCO_3^- = \text{mg/l bicarbonate alk.} \times 1.22 \qquad (17\text{-}15)$$

Molar concentrations may be obtained by dividing milligrams per liter by the mole ionic weight in milligrams:

$$[CO_3^{2-}] = \frac{\text{mg/l } CO_3^{2-}}{60,000} \quad \text{and} \quad [HCO_3^-] = \frac{\text{mg/l } HCO_3^-}{61,000} \qquad (17\text{-}16)$$

17-5 CARBON DIOXIDE, ALKALINITY, AND pH RELATIONSHIPS IN NATURAL WATERS

From the equations

$$CO_2 + H_2O \rightleftharpoons H_2CO_3 \rightleftharpoons HCO_3^- + H^+ \qquad (17\text{-}17)$$

$$M(HCO_3)_2 \rightleftharpoons M^{2+} + 2HCO_3^- \qquad (17\text{-}18)$$

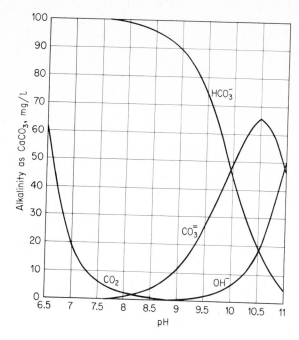

Figure 17-4 Relationship between carbon dioxide and the three forms of alkalinity at various pH levels (values calculated for water with a total alkalinity of 100 mg/l at 25°C).

$$HCO_3^- \rightleftharpoons CO_3^{2-} + H^+ \qquad (17\text{-}19)$$

$$CO_3^{2-} + H_2O \rightleftharpoons HCO_3^- + OH^- \qquad (17\text{-}20)$$

it is obvious that carbon dioxide and the three forms of alkalinity are all part of one system that exists in equilibrium, since all equations involve HCO_3^-. A change in concentration of any one member of the system will, of course, cause a shift in the equilibrium, alter the concentration of the other ions, and result in a change of pH. Conversely, a change in pH will shift the relationships. Figure 17-4 shows the relationship between carbon dioxide and the three forms of alkalinity in a water with 100 mg/l of total alkalinity, over the pH range of importance in environmental engineering practice. Use was made of Eqs. (16-7), (17-7), (17-12), and (17-13) for its construction. The information given in Fig. 17-4 is for illustrative purposes only, as the relationships differ with total alkalinity, temperature, and so on.

17-6 APPLICATION OF ALKALINITY DATA

Information concerning alkalinity is used in a variety of ways in environmental engineering practice.

Chemical Coagulation

Chemicals used for coagulation of water and wastewater react with water to form insoluble hydroxide precipitates. The hydrogen ions released react with alinity of the water. Thus the alkalinity acts to buffer the water in a pH where the coagulant can be effective. Alkalinity must be present in excess of that destroyed by the acid released by the coagulant for effective and complete coagulation to occur.

Water Softening

Alkalinity is a major item that must be considered in calculating the lime and soda-ash requirements in softening of water by precipitation methods. The alkalinity of softened water is a consideration in terms of whether such waters meet drinking water standards.

Corrosion Control

Alkalinity is an important parameter involved in corrosion control. It must be known in order to calculate the Langelier saturation index.[2]

Buffer Capacity

Alkalinity measurements are made as a means of evaluating the buffering capacity of wastewaters and sludges.

Industrial Wastes

Many regulatory agencies prohibit the discharge of wastes containing caustic (hydroxide) alkalinity to receiving waters. Municipal authorities usually prohibit the discharge of wastes containing caustic alkalinity to sewers. Alkalinity as well as pH is an important factor in determining the amenability of wastewaters to biological treatment.

17-7 OTHER CONSIDERATIONS OF INTEREST TO ENVIRONMENTAL ENGINEERS

The environmental engineer encounters a number of situations in practice which involve carbon dioxide–alkalinity–pH relationships that often require explanation.

pH Changes during Aeration of Water

It is common practice to aerate water to remove carbon dioxide. Since carbon dioxide is an acidic gas, its removal tends to raise the pH of the water in

[2] W. F. Langelier, *J. Am. Water Works Assoc.*, **28**: 1500 (1936); T. E. Larson and A. M. Buswell, *J. Am. Water Works Assoc.*, **34**: 1667 (1942).

accordance with Eq. (17-17). Normal air contains about 0.03 percent by volume of carbon dioxide. The Henry's law constant (see Eq. 2-15) for carbon dioxide at 25°C is about 1500 mg/l-atm; therefore the equilibrium concentration of carbon dioxide with air is 0.0003×1500 or about 0.45 mg/l. From Eq. (16-7) it can then be calculated that a water with an alkalinity of 100 mg/l, aerated until in equilibrium with the carbon dioxide in air, would have a pH of about 8.6. A water with higher alkalinity would tend to have a higher pH upon aeration, and one with lower alkalinity would tend to have a lower pH.

pH Changes in the Presence of Algal Blooms

Many surface waters support extensive algal blooms. pH values as high as 10 have been observed in areas where algae are growing rapidly, particularly in shallow water. Algae use carbon dioxide in their photosynthetic activity, and this removal is responsible for such high pH conditions. We have seen that aeration for removal of carbon dioxide tends to increase the pH to between 8 and 9 in waters with moderate alkalinity. Algae, however, can reduce the free carbon dioxide concentration below its equilibrium concentration with air and consequently can cause an even greater increase in pH. As the pH increases, the alkalinity forms change, with the result that carbon dioxide can also be extracted for algal growth both from bicarbonates and from carbonates in accordance with the following equilibrium equations:

$$2HCO_3^- \rightleftharpoons CO_3^{2-} + H_2O + CO_2 \qquad (17\text{-}21)$$

$$CO_3^{2-} + H_2O \rightleftharpoons 2OH^- + CO_2 \qquad (17\text{-}22)$$

Thus the removal of carbon dioxide by algae tends to cause a shift in the forms of alkalinity present from bicarbonate to carbonate, and from carbonate to hydroxide. It should be noted that during these changes the total alkalinity remains constant. Algae can continue to extract carbon dioxide from water until an inhibitory pH is reached, which is usually in the range from pH 10 to 11.

During the dark hours of the day, algae produce rather than consume carbon dioxide. This is because their respiratory processes in darkness exceed their photosynthetic processes. This carbon dioxide production has the opposite effect and tends to reduce the pH. Diurnal variations in pH due to algal photosynthesis and respiration are common in surface waters.

In natural waters containing appreciable amounts of Ca^{2+}, calcium carbonate precipitates when the carbonate-ion concentration, according to Eq. (17-21), becomes great enough so that the $CaCO_3$ solubility product is exceeded:

$$Ca^{2+} + CO_3^{2-} \rightleftharpoons CaCO_3 \qquad (17\text{-}23)$$

This precipitation usually happens before pH levels have exceeded 10, and it places a ceiling over the pH values obtainable. The calcium carbonate precipitated as a result of removal of carbon dioxide through algal action produces the marl deposits in lakes. Marl deposits are the precursors of limestone.

Alkalinity of Boiler Waters

Boiler waters contain both carbonate and hydroxide alkalinity. Both are derived from bicarbonates in the feed water. Carbon dioxide is insoluble in boiling water and so is removed with the steam. This causes an increase in pH and a shift in alkalinity forms from bicarbonate to carbonate and from carbonate to hydroxide, as indicated in Eqs. (17-21) and (17-22). Under these extreme conditions, pH levels in excess of 11.0 are often obtained.

PROBLEMS

17-1 On analysis, a series of samples was found to have the following pH values: 5.5, 3.0, 11.2, 8.5, 7.4, and 9.0. What can you conclude regarding the possible presence of a significant bicarbonate, carbonate, or hydroxide alkalinity in each sample?

17-2 Calculate the phenolphthalein and total alkalinities of the following samples:

(*a*) A 50-ml sample required 5.3 ml $N/50$ H_2SO_4 to reach the phenolphthalein end point, and a total of 15.2 ml to reach the methyl orange end point.

(*b*) A 100-ml sample required 20.2 ml of $N/50$ H_2SO_4 to reach the phenolphthalein end point, and a total of 25.6 ml to reach the methyl orange end point.

17-3 For the following, calculate the hydroxide, carbonate, and bicarbonate alkalinity by the procedures (1) using alkalinity measurements alone, and (2) using alkalinity plus pH measurements. The sample size is 100 ml, $N/50$ H_2SO_4 is used as the titrant, and the water temperature is 25°C.

Sample	Sample pH	Total ml titrant to reach end point	
		Phenol.	M.O.
A	11.0	10.0	15.5
B	10.0	14.4	38.6
C	11.2	8.2	8.4
D	7.0	0	12.7

17-4 Assuming that the effect of dissolved salts on ion activity is negligible and that the temperature of the water sample is 25°C, do the following:

(*a*) Estimate the bicarbonate-ion concentration (in mg/l as HCO_3^-) if the pH of the sample is 10.3 and the carbonate concentration is 120 mg/l (as CO_3^{2-}).

(*b*) Estimate the hydroxide, carbonate, bicarbonate, and total alkalinites of the above solution.

EIGHTEEN

HARDNESS

18-1 GENERAL CONSIDERATIONS

Hard waters are generally considered to be those waters that require considerable amounts of soap to produce a foam or lather and that also produce scale in hot-water pipes, heaters, boilers, and other units in which the temperature of water is increased materially. To the layman, the soap-consuming capacity is most important because of economic aspects and because of the difficulty encountered in obtaining suitable conditions for optimum cleansing; to the engineer, the scaling problem is the most challenging.

With the advent of synthetic detergents, many of the disadvantages of hard waters for household use have been diminished. However, soap is preferred for some types of laundering and for personal hygiene, and hard waters remain as objectionable as ever for these purposes. The scaling problem continues to be a consideration in spite of advances in our knowledge of water chemistry and the development of many proprietary devices which are claimed to prevent scaling through the application of principles not fully explainable.

Although there is less demand today, from the general public, for the removal of hardness through water-softening processes, the need is still great. The trend is toward private and industrial installations in preference to municipal softening plants, except where hardness is considered unreasonably high.

The hardness of waters varies considerably from place to place. In general, surface waters are softer than ground-waters. The hardness of water reflects the nature of the geological formations with which it has been in contact. Figure 18-1 shows the general character of the water supplies in the United States. The softest waters are found in the New England, South

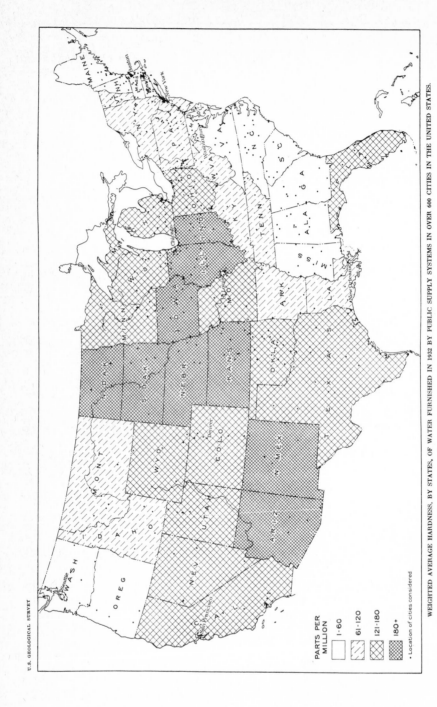

WEIGHTED AVERAGE HARDNESS, BY STATES, OF WATER FURNISHED IN 1932 BY PUBLIC SUPPLY SYSTEMS IN OVER 600 CITIES IN THE UNITED STATES.

Figure 18-1 Hardness characteristics of U.S. water supplies. *(From U.S. Geological Survey Paper 658.)*

Atlantic, and Pacific Northwest states. Iowa, Illinois, Indiana, Arizona, New Mexico, and the Great Plains states have the hardest waters.

Waters are commonly classified in terms of the degree of hardness, as follows:

mg/l	Degree of hardness
0–75	Soft
75–150	Moderately hard
150–300	Hard
300 up	Very hard

18-2 CAUSE AND SOURCE OF HARDNESS

Hardness is caused by divalent metallic cations. Such ions are capable of reacting with soap to form precipitates and with certain anions present in the water to form scale. The principal hardness-causing cations are calcium, magnesium, strontium, ferrous iron, and manganous ions. These cations, plus the most important anions with which they are associated, are shown in Table 18-1 in the order of their relative abundance in natural waters. Aluminum and ferric ions are sometimes considered as contributing to the hardness of water. However, their solubility is so limited at the pH values of natural waters that ionic concentrations are negligible

The hardness in water is derived largely from contact with the soil and rock formations. Rain water as it falls upon the earth is incapable of dissolving the tremendous amounts of solids found in many natural waters. The ability to dissolve is gained in the soil where carbon dioxide is released by bacterial action. The soil water becomes highly charged with carbon dioxide, which, of course, exists in equilibrium with carbonic acid. Under the low pH conditions that develop, basic materials, particularly limestone formations,

Table 18-1 Principal cations causing hardness in water and the major anions associated with them

Cations causing hardness	Anions
Ca^{2+}	HCO_3^-
Mg^{2+}	SO_4^{2-}
Sr^{2+}	Cl^-
Fe^{2+}	NO_3^-
Mn^{2+}	SiO_3^{2-}

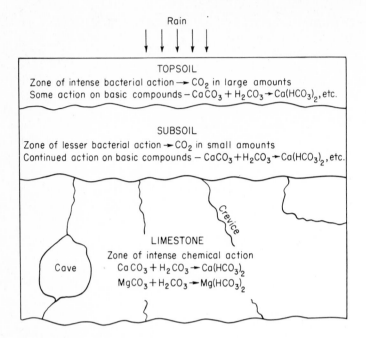

Figure 18-2 Source of carbon dioxide and the solution of substances causing hardness.

are dissolved. Figure 18-2 shows where the carbon dioxide originates and how it attacks the insoluble carbonates in the soil and in limestone formations to convert them to soluble bicarbonates. Since limestone is not pure carbonate but includes impurities such as sulfates, chlorides, and silicates, these materials become exposed to the solvent action of the water as the carbonates are dissolved, and they pass into solution too.

In general, hard waters originate in areas where the topsoil is thick and limestone formations are present. Soft waters originate in areas where the topsoil is thin and limestone formations are sparse or absent.

18-3 PUBLIC HEALTH SIGNIFICANCE

Hard waters are as satisfactory for human consumption as soft waters. Because of their adverse action with soap, however, their use for cleansing purposes is quite unsatisfactory, unless soap costs are disregarded.

18-4 METHODS OF DETERMINATION

Hardness is normally expressed in terms of $CaCO_3$. Many methods of determination have been proposed over the course of the years. Two are presently used.

Calculation Method

Perhaps the most accurate method of determining hardness is by calculation based upon the divalent ions found by a complete analysis. This method is to be preferred where complete analyses are available, but unfortunately such analyses are not made routinely. Seldom are complete analyses made except in exploratory work.

Recently it has been shown that some hard waters contain appreciable amounts of strontium. Unless it is analyzed for separately, it will be measured as calcium, and calculations of hardness based upon such limited analysis may be in considerable error.

Calculation of the hardness caused by each ion is performed by use of the general formula

$$\text{Hardness (in mg/l) as } CaCO_3 = M^{2+} \text{ (in mg/l)} \times \frac{50}{\text{eq. wt of } M^{2+}} \quad (18\text{-}1)$$

where M^{2+} represents any divalent metallic ion.

Example Calculate the hardness of a water with the following analysis:

mg/l	mg/l
Na^+—20	Cl^-—40
Ca^{2+}—15	SO_4^{2-}—16
Mg^{2+}—10	NO_3^-—1
Sr^{2+}—2	Alkalinity—50

Only the divalent cations, Ca^{2+}, Mg^{2+}, and Sr^{2+} cause hardness:

Cation	Equiv. wt.	Hardness, mg/l as $CaCO_3$
Ca^{2+}	20.0	(15)(50)/(20.0) = 37.5
Mg^{2+}	12.2	(10)(50)/(12.2) = 41.0
Sr^{2+}	43.8	(2)(50)/(43.8) = 2.3
		Total hardness = 80.8

EDTA Titrimetric Method

This method involves the use of solutions of ethylenediaminetetraacetic acid (EDTA) or its sodium salt as the titrating agent.

$$\begin{matrix} \text{HOOC—CH}_2 & & & \text{CH}_2\text{—COOH} \\ & \diagdown & \text{H H} & \diagup \\ & \text{N—C—C—N} & \\ & \diagup & \text{H H} & \diagdown \\ \text{HOOC—CH}_2 & & & \text{CH}_2\text{—COOH} \end{matrix}$$

Acid

$$\begin{matrix} \text{NaOOC—CH}_2 & & & \text{CH}_2\text{—COONa} \\ & \diagdown & \text{H H} & \diagup \\ & \text{N—C—C—N} & \\ & \diagup & \text{H H} & \diagdown \\ \text{NaOOC—CH}_2 & & & \text{CH}_2\text{—COONa} \end{matrix}$$

Sodium salt

These compounds, usually represented by EDTA, are chelating agents and form extremely stable complex ions with Ca^{2+}, Mg^{2+}, and other divalent ions causing hardness, as shown in the equation

$$M^{2+} + \text{EDTA} \rightarrow [M \cdot \text{EDTA}]_{\text{complex}} \qquad (18\text{-}2)$$

The successful use of EDTA for determining hardness depends upon having an indicator present to show when EDTA is present in excess, or when all the ions causing hardness have been complexed.

The dye known as Eriochrome Black T serves as an excellent indicator to show when all the hardness ions have been complexed. When a small amount of Eriochrome Black T, having a blue color, is added to a hard water with a pH of about 10.0, it combines with a few of the Ca^{2+} and Mg^{2+} ions to form a weak complex ion which is wine red in color, as shown in the equation

$$M^{2+} + \text{Eriochrome Black T} \rightarrow (M \text{ Eriochrome Black T}) \qquad (18\text{-}3)$$
Wine red complex

During the titration with EDTA, all free hardness ions are complexed according to Eq. (18-2). Finally the EDTA disrupts the red (M Eriochrome Black T) complex because it is capable of forming a more stable complex with the hardness ions. This action frees the Eriochrome Black T indicator, and the wine red color changes to a distinct blue color, heralding the end of the titration.

Although the EDTA method is subject to certain interferences, most of them can be overcome by proper modifications. The method yields very precise and accurate results. It is the method of choice in most laboratories at the present time.

18-5 TYPES OF HARDNESS

In addition to total hardness, it is desirable and sometimes necessary to know the types of hardness present. Hardness is classified in two ways: (1) with respect to the metallic ion and (2) with respect to the anions associated with the metallic ions.

Calcium and Magnesium Hardness

Calcium and magnesium cause by far the greatest portion of the hardness occurring in natural waters. In some considerations it is important to know the amounts of calcium and magnesium hardness in water. For example, it is necessary to know the magnesium hardness or the amount of Mg^{2+} in order to calculate lime requirements in lime–soda ash softening. The calcium and magnesium hardness may be calculated from the complete chemical analysis, as discussed in Sec. 18-4. Such information is not always available, and recourse is made to some method of analysis that allows separate measurement of calcium or magnesium hardness. If calcium hardness is determined, magnesium hardness is obtained by subtracting calcium hardness from total hardness, as follows:

$$\text{Total hardness} - \text{calcium hardness} = \text{magnesium hardness} \qquad (18\text{-}4)$$

This procedure yields reasonably reliable results because most of the hardness in natural waters is due to these two cations. Most methods for measuring calcium hardness will also include strontium hardness.

Carbonate and Noncarbonate Hardness

The part of the total hardness that is chemically equivalent to the bicarbonate plus carbonate alkalinities present in a water is considered to be *carbonate hardness*. Since alkalinity and hardness are both expressed in terms of $CaCO_3$, the carbonate hardness can be found as follows:

When alkalinity $<$ total hardness,

$$\text{Carbonate hardness (in mg/l)} = \text{alkalinity (in mg/l)} \qquad (18\text{-}5)$$

When alkalinity $\geq$ total hardness,

$$\text{Carbonate hardness (in mg/l)} = \text{total hardness (in mg/l)} \qquad (18\text{-}6)$$

Carbonate hardness is singled out for special recognition because the bicarbonate and carbonate ions with which it is associated tend to precipitate this portion of the hardness at elevated temperatures such as occur in boilers or during the softening process with lime.

$$Ca^{2+} + 2HCO_3^- \rightarrow \underline{CaCO_3} + CO_2 + H_2O \qquad (18\text{-}7)$$

$$Ca^{2+} + 2HCO_3^- + Ca(OH)_2 \rightarrow \underline{2CaCO_3} + 2H_2O \qquad (18\text{-}8)$$

It may also be considered as that part of the total hardness that originates from the action of carbonic acid on limestone, as illustrated in Fig. 18-2. Carbonate hardness was formerly called *temporary hardness* because it can be caused to precipitate by prolonged boiling [see Eq. (18-7)].

The amount of hardness which is in excess of the carbonate hardness is called *noncarbonate hardness*, and can be estimated as follows:

$$\text{Noncarbonate hardness (NCH)} = \text{total hardness} - \text{carbonate hardness} \qquad (18\text{-}9)$$

Since all forms of hardness as well as alkalinity are expressed in terms of $CaCO_3$, the calculations for Eqs. (18-5), (18-6), and (18-9) can be made directly. These are excellent examples of the reason why alkalinity and hardness are normally expressed in terms of $CaCO_3$. Noncarbonate hardness was formerly called *permanent hardness* because it cannot be removed or precipitated by boiling. Noncarbonate hardness cations are associated with sulfate, chloride, and nitrate anions.

Pseudo–Hardness

Sea, brackish, and other waters that contain appreciable amounts of Na^+ interfere with the normal behavior of soap because of the common ion effect. Sodium is not a hardness-causing cation, and so this action which it exhibits when present in high concentration is termed *pseudo-hardness*.

18-6 APPLICATION OF HARDNESS DATA IN ENVIRONMENTAL ENGINEERING PRACTICE

Hardness of a water is an important consideration in determining the suitability of a water for domestic and industrial uses. The engineer uses it as a basis for recommending the need for softening processes. The relative amounts of calcium and magnesium hardness and of carbonate and noncarbonate hardness present in a water are factors in determining the most economical type of softening process to use, and become important considerations in design. Determinations of hardness serve as a basis for routine control of softening processes.

PROBLEMS

18-1 What is hardness in water and by what is it caused?

18-2 A water has the following analysis:

mg/l	mg/l
Na^+—20	Cl^-—40
K^+—30	HCO_3^-—67
Ca^{2+}—5	CO_3^{2-}—0
Mg^{2+}—10	SO_4^{2-}—5
Sr^{2+}—2	NO_3^-—10

What is the total hardness, carbonate hardness, and noncarbonate hardness in mg/l as $CaCO_3$?

18-3 Discuss the principles involved in the EDTA titrimetric method of measuring hardness.

NINETEEN

RESIDUAL CHLORINE AND CHLORINE DEMAND

19-1 GENERAL CONSIDERATIONS

The prime purpose of chlorinating public water supplies and wastewater effluents is to prevent the spread of waterborne diseases. The practice of chlorination has become so widespread and generally accepted that the real reason is frequently taken very much for granted. It seems important that environmental engineers should be familiar with the history of the great plagues that have afflicted mankind and the developments that led to the proof that water is the major vehicle of transmission for some diseases. It is impossible to go into great detail here concerning the historical aspects that led to the practice of chlorination; hence supplementary reading is highly recommended.[1]

Early History of Diseases

Communicable diseases have been a curse of mankind since time immemorial. The intensity of the problem appears to have been magnified as the density of the population increased. During the fourteenth century a plague known as the "Black Death" swept over Europe, leaving about 25 percent of the people dead in its wake. An epidemic in London in the winter of 1664–1665 caused 70,000 deaths, equal to 14 percent of the population.

   With the development of the industrial revolution, which attracted people to urban areas and caused them to live under more crowded conditions, the

[1] M. J. Rosenau, "Preventive Medicine and Hygiene," Appleton-Century-Crofts, New York, 1935; S. C. Prescott and M. P. Horwood, "Sedgwick's Principles of Sanitary Science and Public Health," Macmillan, New York, 1946.

frequency of epidemics increased. Up until 1854, there had been a great deal of theorizing concerning the causes and modes of transmission of disease but no one had been able to prove his case. The science of bacteriology, upon which definite proof depended, was still unborn.

In 1854 a localized epidemic of Asiatic cholera broke out in London. Through the careful investigations of two men, John Snow and John York, it was demonstrated, as well as could be by the means available at that time, that the source of infection was water from the Broad Street Pump. It was further demonstrated that the well was contaminated by wastewater from a damaged sewer nearby and that the sewer carried wastewater from a home housing one suffering from the disease. The Broad Street Pump epidemic is a milestone in public health engineering practice, for it established without doubt that water was a major vehicle for the spread of Asiatic cholera, one of the greatest plagues of mankind. This discovery stimulated and gave real purpose to the practice of slow sand filtration which had been initiated about 1830.

The science of bacteriology is considered to have originated about 1870. Robert Koch, in 1875, was successful in growing a pure culture of the bacterium causing anthrax. This was another event of great significance, for within a few years the causative organisms of typhoid (1884), Asiatic cholera (1883), and many other diseases were grown in pure culture. These developments provided the means for absolute proof that water can serve as a major vehicle for disease transmission.

The cholera epidemic in Hamburg, Germany, in 1892 served as another milestone in the knowledge concerning waterborne disease. During the epidemic, cholera organisms were actually found in the river waters used for the water supply. In addition, the efficacy of slow sand filters for removing disease organisms was demonstrated.

The typhoid epidemic at Lausen, Switzerland, in 1872 was caused by contamination of a spring water supply. It is noteworthy because of the remoteness of the point of contamination and the considerable distance the water traveled underground without freeing itself of the disease organisms. This characteristic has been shown to be particularly true in limestone areas where cracks and crevices occur.

History of Chlorination Practice

Chlorination of water supplies on an emergency basis has been practiced since about 1850. With definite evidence at hand that certain diseases were transmitted by water, emergency treatment with hypochlorites became quite common during periods of epidemics.

It was not until 1904 that continuous chlorination of public water supply was attempted in England. Shortly thereafter, in 1908, George A. Johnson initiated treatment with calcium hypochlorite of the water at the Bubbly Creek filter plant of the Union Stock Yards in Chicago. In 1909, Jersey City, New Jersey, started hypochlorite treatment of its Boonton supply. This was the first attempt to chlorinate a public water supply in the United States and led to a celebrated court case in which a wise judge upheld the right of the city to chlorinate the water supply in the best interests of public health. From that day, chlorination of public water supplies has spread so that today it is almost routine practice.

The practice of chlorinating public water supplies did not spread rapidly at first because of the instability during storage of the calcium hypochlorites then available. The development of facilities for feeding gaseous chlorine occurred about 1912, and from that time chlorination practice has grown rapidly. With the increased use of chlorine for disinfecting purposes, there has been a corresponding decrease in the incidence of waterborne disease. The gross effect that modern sanitation practices, of which chlorination of water supplies and pasteurization of milk are major ones, have had upon the typhoid death rate in the United States is shown in Fig. 19-1. The continual decline in the importance of typhoid fever is illustrated in Fig. 19-2 by the number of waterborne outbreaks between 1946 and 1970. Of concern is the increase in cases of waterborne infectious hepatitis, a viral disease which results largely from direct

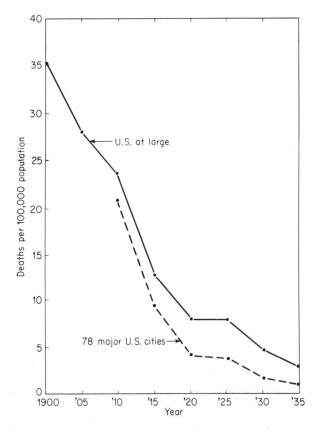

Figure 19-1 Typhoid and paratyphoid death rate in the United States and in 78 major U.S. cities. (*A.E. Gorman and A. Wolman, J. Am. Water Works Assoc.*, February 1939.)

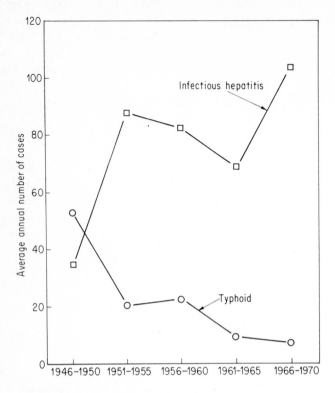

Figure 19-2 Average annual number of typhoid and hepatitis cases occurring in waterborne outbreaks—1946 to 1970. *(After G. F. Craun and L. J. McCabe, J. Am. Water Works Assoc.,* **65**, 74, 1973.)

contamination of underground water supplies and water distribution systems, the water reaching the consumer without disinfection.

19-2 CHEMISTRY OF CHLORINATION

Chlorine is used in the form of free chlorine or as hypochlorites. In either form it acts as a potent oxidizing agent and often dissipates itself in side reactions so rapidly that little disinfection is accomplished until amounts in excess of the chlorine demand have been added.

Reactions with Water

Chlorine combines with water to form hypochlorous and hydrochloric acids.

$$Cl_2 + H_2O \rightleftharpoons HOCl + H^+ + Cl^- \tag{19-1}$$

$$\frac{[H^+][Cl^-][HOCl]}{[Cl_2]} = 4 \times 10^{-4} \quad \text{(at 25°C)} \tag{19-2}$$

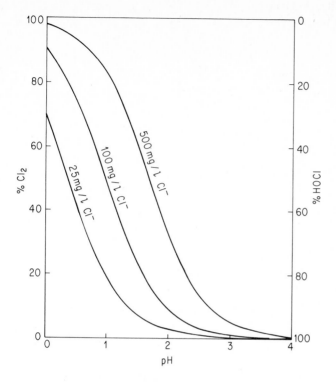

Figure 19-3 Effect of pH and chloride concentration on the distribution of chlorine and hypochlorous acid in water at 25°C.

This equilibrium is the dominant one in chlorine water from vacuum chlorinators with a pH of 2 to 3, and at the instant of contact when chlorine water is added to any water for disinfection (Fig. 19-3). The nature of the reactions is dominated by the free Cl_2. These often result in development of obnoxious compounds such as nitrogen trichloride, NCl_3. To minimize these effects, a high quality water is often used as chlorinator feed water and flash mixing is required at the point where the chlorine water is applied to prevent the development of localized low pH conditions. In dilute solution and at pH levels above about 4, the equilibrium shown above is displaced greatly to the right, and very little Cl_2 exists as such in solution. The hypochlorous acid formed is a weak acid and is very poorly dissociated at pH levels below 6.

$$HOCl \rightleftharpoons H^+ + OCl^- \qquad (19\text{-}3)$$

$$\frac{[H^+][OCl^-]}{[HOCl]} = 2.7 \times 10^{-8} \qquad (at\ 20°C) \qquad (19\text{-}4)$$

The relative amounts of OCl^- and $HOCl$ in solution as a function of pH and in accordance with the above equilibrium are shown in Fig. 19-4. This equilibrium is established very rapidly.

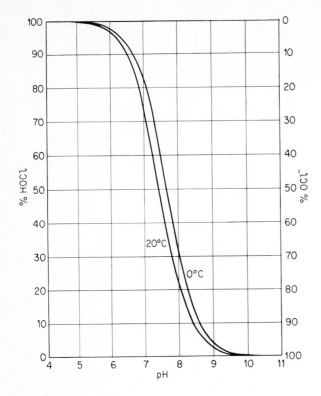

Figure 19-4 Effect of pH on the distribution of hypochlorous acid and hypochlorite ion in water.

Hypochlorites are used in the form of solutions of hypochlorite and high-test Ca hypochlorite in the dry form. The former is popular where large amounts are necessary such as in wastewater disinfection where local supplies are available or on-site generation from salt solutions is feasible. Ca hypochlorite is popular for situations where limited amounts are required or intermittent usage is dictated. Both compounds ionize in water to yield hypochlorite ion as illustrated below:

$$Ca(OCl)_2 \rightarrow Ca^{2+} + 2OCl^- \qquad (19\text{-}5)$$

$$NaOCl \rightarrow Na^+ + OCl^-$$

This ion, of course, establishes an equilibrium with hydrogen ions in accordance with Eq. (19-3). Thus it may be concluded that the same equilibria are established in water regardless of whether chlorine or hypochlorites are added. The significant difference would be in pH effects and its influence on the relative amounts of OCl^- and HOCl at equilibrium. Chlorine tends to decrease the pH, whereas hypochlorites tend to increase the pH.

Reactions with Impurities in Water

Chlorine and hypochlorous acid react with a wide variety of substances, including ammonia.

Reactions with ammonia Ammonium ion exists in equilibrium with ammonia and hydrogen ion [Eq. (2-45)]. The ammonia reacts with chlorine or hypochlorous acid to form monochloramines, dichloramines, and trichloramines, depending upon the relative amounts of each and to some extent upon the pH, as follows:

$$NH_3 + HOCl \rightarrow NH_2Cl + H_2O \qquad \text{(monochloramine)} \qquad (19\text{-}6)$$

$$NH_2Cl + HOCl \rightarrow NHCl_2 + H_2O \qquad \text{(dichloramine)} \qquad (19\text{-}7)$$

$$NHCl_2 + HOCl \rightarrow NCl_3 + H_2O \qquad \text{(trichloramine)} \qquad (19\text{-}8)$$

The mono- and dichloramines have significant disinfecting power and are therefore of interest in the measurement of chlorine residuals. The chemistry of the above reactions is discussed in more detail in Sec. 19-3.

Extraneous reactions Chlorine combines with a wide variety of materials, particularly reducing agents. Many of the reactions are very rapid, while others are much slower. These side reactions complicate the use of chlorine for disinfecting purposes. Their demand for chlorine must be satisfied before chlorine becomes available to accomplish disinfection. They result in the production of numerous chlorinated organic materials which has been of concern recently because of potential health significance.[2]

The reaction between hydrogen sulfide and chlorine will serve to illustrate the type of reaction that occurs with reducing agents:

$$H_2S + Cl_2 \rightarrow 2HCl + S \qquad (19\text{-}9)$$

Fe^{2+}, Mn^{2+}, and NO_2^- are examples of other inorganic reducing agents present in water supplies that will react with chlorine.

Organic compounds that possess unsaturated linkages will also add hypochlorous acid and increase the chlorine demand:

$$\underset{\text{H \quad H}}{-C{=}C-} + HOCl \rightarrow \underset{\text{H \quad H}}{\overset{\text{Cl \quad OH}}{-C{-}\!-C-}} \qquad (19\text{-}10)$$

Chlorine can react with phenols to produce mono-, di-, or trichlorophenols, which can impart tastes and odors to waters. Chlorine also reacts with humic substances present in most water supplies, forming a variety of chlorinated products.[3] One of much concern is chloroform ($CHCl_3$) which is carcinogenic, at least to animals. A survey of 80 water supplies in the United States indicated the average concentration of chloroform resulting from disinfection with chlorine was 21 $\mu g/l$, and the maximum was over 300 $\mu g/l$.[4] This has led to concern over the public health significance of the materials, but has not led to

---

[2] R. L. Jolley, ed., "The Environmental Impact of Water Chlorination," CONF-751096, UC-11,41,48, Oak Ridge National Laboratory (July 1976).

[3] J. J. Rook, "Water Treat. Exam." **23**, 234 (1974).

[4] *Ibid.*, A. A. Stevens, *et al.*

an effort to abandon the practice of chlorination. The benefits from chlorine disinfection are immense, as already discussed. Further, it is the only recognized method of disinfection that is capable of providing protective residuals within the distribution system to guard against inadvertent bacterial contamination, from back siphonage or cross connections. Efforts are being directed toward minimizing the production of chlorinated organics from chlorination.

Chlorine also reacts with other halogens in water. For example, hypochlorous acid reacts with bromide to form hypobromous acid:

$$Br^- + HOCl \rightarrow HOBr + Cl^- \tag{19-11}$$

HOBr is also a disinfectant, but reacts more rapidly than chlorine. When bromide is present in water, the chlorine appears to be more reactive for this reason. HOBr also reacts with organics. It is generally noted that chlorination of a water supply results in the formation of some brominated organics. The above conversion helps to explain this phenomena.

19-3 PUBLIC HEALTH SIGNIFICANCE OF CHLORINE RESIDUALS

Disinfection is a process designed to kill harmful organisms, and it does not ordinarily produce a sterile water. These generalizations hold for disinfection with chlorine. Two factors are extremely important in disinfection: time of contact and concentration of the disinfecting agent. Where other factors are constant, the disinfecting action may be represented by

$$\text{Kill} \propto C^n \times t \quad (n > 0) \tag{19-12}$$

The important point is that with long contact times a low concentration of disinfectant suffices, whereas short contact times require high concentration to accomplish equivalent kills.

It has become common practice to refer to chlorine, hypochlorous acid, and hypochlorite ion as *free chlorine residuals,* and the chloramines are called *combined chlorine residuals.* Research has shown that with free chlorine residuals, a lower pH, which favors the formation of HOCl over OCl⁻, is more effective for disinfection. Research has also shown that a greater concentration of combined chlorine residual than of free chlorine residual is required to accomplish a given kill in a specified time. For these reasons it is important to know both the concentration and the kind of residual chlorine acting.

The rate of the reaction between ammonia and hypochlorous acid varies considerably, depending upon the pH and temperature. The reaction rate is most rapid at pH 8.3 and decreases rapidly as the pH is decreased or increased. For this reason, it is common to find free chlorine and combined chlorine residuals coexisting after contact periods of 10, 15, or even 60 min.

The action of excess chlorine on waters containing ammonia merits special consideration. With mole ratios of chlorine to ammonia up to 1:1, both monochloramine and dichloramine are formed. The relative amounts of each are a function of the pH, which affects the relative rate of formation of the two species as well as the thermodynamic equilibrium between them. Greater proportions of dichloramine appear at lower pH values.

Further increases in the mole ratio of chlorine to ammonia result in formation of some trichloramine and oxidation of part of the ammonia to N_2 or NO_3^-. With chlorination within the pH range of 6 to 7, these reactions are essentially complete when 1.5 mol of chlorine has been added for each mole of ammonia nitrogen originally present in the water. Chloramine residuals usually reach a maximum when about 1 mol of chlorine has been added for each mole of ammonia, and then decline to a minimum value at a chlorine-to-ammonia ratio of 1.5. Further additions of chlorine produce free chlorine residuals. Chlorination of a water to the extent that all the ammonia is converted to N_2 or a higher oxidation state is referred to as "break-point chlorination" because of the peculiar character of the chlorine residual curve, as illustrated in Fig. 19-5.

Theoretically it would require 3 mol of chlorine for the complete conversion of ammonia to nitrogen trichloride (trichloramine), and 4 mol for the complete oxidation to nitrate. The fact that in the pH range of 6 to 7, only 1.5 mol is required can be accounted for if the following overall reaction were predominant:

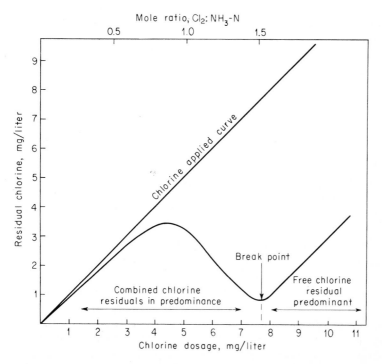

Figure 19-5 A residual-chlorine curve showing a typical break point. Ammonia-nitrogen content of water, 1.0 mg/l.

$$2NH_3 + 3Cl_2 \rightarrow N_2 + 6H^+ + 6Cl^- \qquad (19.13)$$

At lower pH, a higher ratio of chlorine to ammonia is required, presumably because some trichloramine is formed, and at higher pH, a higher ratio is required, presumably because some nitrate is formed.[5] The nature and rates of the intermediate reactions involved have been investigated recently.[6] Equation (19.13) indicates that 7.6 g of chlorine per g of ammonia nitrogen would be required to reach the breakpoint.

Breakpoint chlorination is required to obtain a free chlorine residual for better disinfection if ammonia is present in a water supply. Breakpoint chlorination has also been proposed as one method for removal of ammonia when required in wastewater treatment for reasons discussed in Sec. 24-4.

While free chlorine residuals have good disinfecting powers, they are usually dissipated quickly in the distribution system. For this reason final treatment with ammonia is often practiced to convert free chlorine residuals to longer lasting combined chlorine residuals.

19-4 METHODS OF DETERMINATION

It was not until about 1940 that the difference in disinfecting power of chloramine and free chlorine residuals was demonstrated. Prior to that time no attempt had been made to develop analytical procedures for differentiation. Therefore methods for measuring chlorine residuals may be classed into old methods that measure total chlorine and new methods that allow measurement of free and combined forms.

Total Chlorine Residual

All common methods of measuring chlorine residuals depend upon their oxidizing power; consequently any other oxidizing agents present may interfere with the test. Manganese in valences above 2 and nitrites are the most common interferences.

Starch-iodide method The starch-iodide method is of interest, historically, because it served as the basis of controlling chlorination until about 1913 when the orthotolidine test was developed. The method depends upon the oxidizing

---

[5] Pressly, et al., "Nitrogen Removal by Breakpoint Chlorination," U.S. Dept. of Interior Report (Sept. 1970).

[6] Wel, I. W., "Chlorine-Ammonia Breakpoint Reactions; Kinetics and Mechanisms," Ph.D. thesis, Harvard University, Cambridge, Mass., May, 1972; Savnier, B., and R. E. Selleck, "Kinetics of Breakpoint Chlorination and Disinfection," SERL Rept. 76-2, University of California, Berkeley, May, 1976.

power of free and combined chlorine residuals to convert iodide ion to free iodine, the reactions being represented as

$$Cl_2^0 + 2I^- \rightarrow I_2^0 + 2Cl^- \tag{19.14}$$

$$I_2 + starch \rightarrow blue\ color \quad (qualitative\ test) \tag{19.15}$$

In the presence of starch, the iodine produced a blue color, which was accepted as evidence of the presence of residual chlorine but, of course, did not indicate the amount of residual present, except as people were able to judge the intensity of the blue color.

The starch-iodide method provides a means of quantitative measurement of total residual if the iodine released is titrated with a standard solution of a reducing agent. The usual reagent is sodium thiosulfate, and the end point is indicated by the disappearance of the blue color.

$$I_2 + 2Na_2S_2O_3 \rightarrow Na_2S_4O_6 + 2NaI \tag{19.16}$$

or

$$I_2 + 2S_2O_3^{2-} \rightarrow S_4O_6^{2-} + 2I^- \tag{19.17}$$

Orthotolidine method In 1909 Phelps proposed the use of orthotolidine as a colorimetric indicator for chlorine residuals. Ellms and Hauser (1913) incorporated the use of color standards and thereby made the test quantitative. Shortly thereafter, proprietary devices employing colored glass disks or sealed colored liquid standards, suitable for field as well as laboratory use, were developed. The introduction of a simple testing procedure, which allowed close control of chlorination, was an important factor in the widespread acceptance of chlorination of public water supplies. It made chlorination a practical method of disinfection on even the smallest supplies because of the simplicity of the test.

Orthotolidine is an aromatic organic compound that is oxidized in acid solution by chlorine, chloramines, and other oxidizing agents to produce a yellow-colored compound. The reaction is represented as follows:

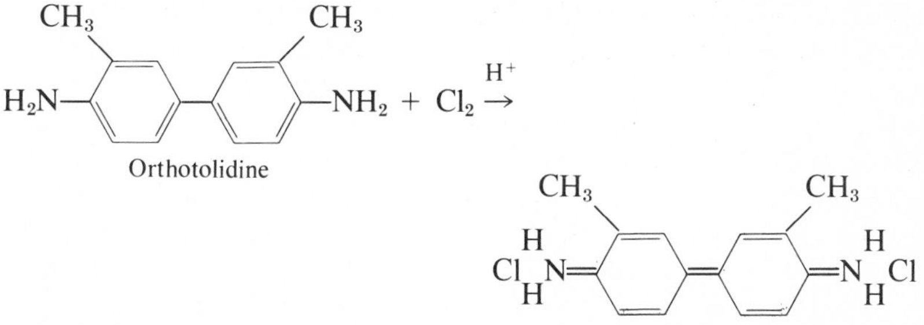

The holoquinone produced is yellow at pH values less than 1.8, and the intensity of the yellow color is proportional to the amount present. It meets the requirements of Beer's law (Sec. 10-5) and is suitable for quantitative measurement.

Nitrites and oxidized forms of manganese both oxidize orthotolidine to produce holoquinones, thus producing false indications of chlorine residuals. Also, orthotolidine is not as accurate for measuring free and combined chlorine residuals as are some of the newer methods. Largely for these reasons, the "acid" orthotolidine procedure is no longer considered a "standard method" for residual chlorine determination.

Free and Combined Chlorine Residuals

With the development of knowledge concerning the relative disinfecting powers of free and combined chlorine residuals, it became important to have ways of differentiating and measuring them. There are several methods which are now available.

Amperometric titration method Oxidation-reduction titration procedures were among the first methods for measuring free and combined chlorine residuals. One device used for such titrations is the amperometric titrator which contains an internal indicator or electrometric device to show when the reactions are completed. Such an instrument with a rotating platinum electrode is shown in Fig. 10-2.

Phenylarseneoxide (C_6H_5AsO) is the reducing agent normally used as the titrating agent. It reacts with free chlorine residuals at pH 6.5 to 7.5 in a quantitative manner. The reaction becomes sluggish at pH greater than 7.5. Also, chloramines are reduced at pH levels below 6.0, provided that iodide ion is present. The chloramines oxidize iodide to free iodine, and the phenylarseneoxide reduces the free iodine, thereby measuring the amount of chloramines present. By conducting a two-stage titration, with the pH adjusted at about 7 and then at about 4, it is possible to measure separately free chlorine residuals and combined chlorine residuals. Interferences from nitrites and oxidized forms of manganese are eliminated by conducting the titrations at pH levels above 3.5. The amperimetric titration procedure is not subject to interference from color or turbidity, which is a particular advantage when measuring chlorine residuals in some wastewaters.

SNORT method In the stabilized neutral orthotolidine method (SNORT), advantage is taken of the slow rate of reaction of orthotolidine at neutral pH with combined chlorine and interferences such as iron and nitrite. In order to measure free chlorine residual, a neutral orthotolidine solution is added to the sample together with a buffer-stabilizer solution which helps maintain the pH within the 6.5 to 7.5 range required, and stabilizes the oxidized orthotolidine which would otherwise decompose. At neutral pH, a blue rather than a yellow color is produced by the reaction of orthotolidine with chlorine indicated in Eq. (19.18). In order to measure the concentration of monochloramine, a potassium iodide solution is then added to the sample. The monochloramine oxidizes the iodide to produce free iodine which in turn oxidizes the orthotolidine to quantitatively produce more blue color. Colorimetric analysis then yields a value for free chlorine plus monochloramine residuals. If the sample is then acidified, dichloramine then oxidizes iodide to iodine, and upon neutralization, the total chloramine residual can be measured. From the three values obtained by the different steps in this overall procedure, concentrations of each of the three different chlorine residuals can be calculated.

DPD method With the DPD method, the principle is similar to that for the SNORT and amperimetric methods. When N,N-diethyl-p-phenylenediamine (DPD) is added to a sample containing free chlorine residual, an instantaneous reaction occurs, producing a red color. If a small amount of iodide is then added to the sample, monochloramine reacts to produce iodine, which in turn oxidizes more DPD to form additional red color. If a large quantity of iodide is then added, dichloramines will react to form still more red color. By measuring the intensity of the red color produced in the neutral pH range of 6.2 to 6.5 after each of the three steps outlined above, then free chlorine, monochloramine, and dichloramine residuals can be determined. The intensity of the red color produced can be determined either by titration with ferrous ion until the red color disappears, or directly by colorimetric analysis. This is a relatively easy procedure to use.

Leuco crystal violet method This is another variation of the same principle outlined for the other methods. This colorimetric procedure permits measurement of either free or total chlorine residual. An organic material, N,N-dimethylaniline, which has the common name of leuco crystal violet, is added to a sample and free chlorine oxidizes it to form a bluish material, which can be determined colorimetrically. Iodide is then added which reacts with chloramines producing additional bluish color which is also measured colorimetrically. This procedure tends to be somewhat more complex than the other procedures, and the steps in the analysis must be followed exactly.

19-5 MEASUREMENT OF CHLORINE DEMAND

The *chlorine demand* of a water is the difference between the amount of chlorine applied and the amount of free, combined, or total available chlorine remaining at the end of the contact period. The chlorine demand is different with different waters, and even with a given water will vary with the amount of chlorine applied, the desired residual, time of contact, pH, and temperature. The test should be conducted with chlorine or with hypochlorites, depending upon the form that will be used in practice.

Measurement of chlorine demand can be readily made by treating a series of samples of the water in question with known but varying dosages of chlorine or hypochlorite. The water samples should be at a temperature within the range of interest, and after the desired contact period, determination of residual chlorine in the samples will demonstrate which dosage satisfied the requirements of the chlorine demand, in terms of the desired residual.

19-6 DISINFECTION WITH CHLORINE DIOXIDE

A discussion of chlorination should also include chlorine dioxide, as it has certain advantages over free chlorine and hypochlorites. It is about as effective an oxidizing agent as hypochlorous acid and is a particularly good disinfectant at high pH values. It does not combine with ammonia to produce chloramines nor with natural organics to form chloroform or other trihalomethanes. Another advantage of chlorine dioxide is that it is very effective in the destruction of phenolic compounds that combine with other types of chlorine to create the undesirable taste-producing chlorinated phenols.

Chlorine dioxide is an unstable gas and so is usually produced at the plant site by mixing a solution of sodium chlorite ($NaClO_2$) with a strong chlorine solution:

$$2NaClO_2 + Cl_2 \rightarrow 2ClO_2 + 2NaCl \qquad (19.19)$$

The yield of chlorine dioxide is increased if the pH is depressed to less than 4 by the addition of a small excess of chlorine. The more extensive use of chlorine dioxide in water treatment has been hampered both by the lack of a suitable test for residual measurement and by its high cost compared to other forms of chlorine.

19-7 APPLICATION OF CHLORINE DEMAND AND CHLORINE RESIDUAL DATA

Determination of the chlorine demand of a water or wastewater is an important consideration in design. It serves as the basis for determining the number and capacity of chlorinators required, the amount of chlorine needed, the type of shipping containers, and all appurtenances required for handling and storage.

Chlorine residuals are used universally in disinfection practice to control addition of chlorine so as to ensure effective disinfection without waste of chlorine. They are also used to control chlorination of domestic and industrial wastes and usually are the sole criteria immediately available to determine whether or not desired objectives are being attained.

PROBLEMS

19-1 Why is it important to determine chlorine residual in water treatment practice?

19-2 Compare the relative significance of free chlorine residuals and chloramine residuals in water treatment practice.

19-3 Discuss applications of the chlorine demand test.

19-4 Write three equations to illustrate type reactions which cause chlorine demand.

19-5 Discuss the significance of contact time, chlorine residual, and pH as factors influencing the degree of disinfection obtained by chlorination.

19-6 What general class of substances interfere with the orthotolidine test for residual chlorine? Give three specific examples.

19-7 Outline one procedure for determining residual chlorine to allow differentiation between free chlorine, chloramines, and interferences.

19-8 By use of appropriate equilibrium equations, show why the addition of chlorine tends to decrease the pH of water, while hypochlorite tends to increase the pH.

19-9 Compute the relative proportions of free chlorine occurring as $HOCl$ and as OCl^- at a pH of 6.8 and a temperature of 20°C.

19-10 According to Chick's law, the rate of kill of bacteria by chlorination follows first-order reaction kinetics. Assuming this to be true, how much contact time is required to kill 99 percent of the bacteria with a chlorine residual of 0.1 mg/l, if 80 percent are killed in 2 minutes with this residual? (See Sec. 3-11.)

TWENTY

CHLORIDES

20-1 GENERAL CONSIDERATIONS

Chlorides occur in all natural waters in widely varying concentration. The chloride content normally increases as the mineral content increases. Upland and mountain supplies usually are quite low in chlorides, whereas river and ground-waters usually have a considerable amount. Sea and ocean waters represent the residues resulting from partial evaporation of natural waters that flow into them, and chloride levels are very high.

Chlorides gain access to natural waters in many ways. The solvent power of water dissolves chlorides from topsoil and deeper formations. Spray from the ocean is carried inland as droplets or as minute salt crystals, which result from evaporation of the water in the droplets. These sources constantly replenish the chlorides in inland areas where they fall. Ocean and sea waters invade the rivers that drain into them, particularly the deeper rivers. The salt water, being more dense, flows upstream under the fresh water which is flowing downstream. There is a constant intermixing of the salt water with the fresh water above. In the case of the Hudson River, which has a deep channel and rather slight gradient, seawater invades for a distance of about 50 miles upstream. This invasion has been a major factor in preventing New York City from developing the Hudson River as a source of water supply. Ground-waters in areas adjacent to the ocean are in hydrostatic balance with seawater. Over-pumping of ground-waters produces a difference in hydrostatic head in favor of the seawater, and it intrudes into the fresh-water area. Such intrusion has occurred at many locations in Florida and California.

Human excreta, particularly the urine, contain chloride in an amount about equal to the chlorides consumed with food and water. This amount averages about 6 g of chlorides per person per day and increases the amount of Cl^- in sewage about 15 mg/l above that of the carriage water. Thus sewage effluents add considerable chlorides to receiving streams. Many industrial wastes contain appreciable amounts of chlorides. Control of contamination of surface waters by chlorides contained in industrial wastes is a major consideration in the Ohio River valley and in all areas where oil-field brines and other salt brines are allowed to reach receiving streams.

20-2 PUBLIC HEALTH SIGNIFICANCE OF CHLORIDES

Chlorides in reasonable concentrations are not harmful to humans. At concentrations above 250 mg/l they give a salty taste to water, which is objectionable to many people. For this reason chlorides are generally limited to 250 mg/l in supplies intended for public use. In many areas of the world where water supplies are scarce, sources containing as much as 2000 mg/l are used for domestic purposes without the development of adverse effects, once the human system becomes adapted to the water.

Before the development of bacteriological testing procedures, chemical tests for chloride and for nitrogen, in its various forms, served as the basis of detecting contamination of ground-waters by wastewater. The chloride test was of special value in areas where the level of chlorides was low. In Massachusetts, for example, a survey of the chloride content of ground-waters was made and maps were prepared showing normal levels. Locations with similar chloride levels were connected by lines and a map with chloride contours (isochlors) was obtained. Such maps provided reference information of considerable value in sanitary surveys where wastewater contamination was suspected. A great deal of judgment and caution was needed for proper interpretation. The chemical tests have been largely replaced by the more sensitive bacteriological tests used today.

Chlorides are used to some extent as tracers in environmental engineering practice. They are inconvenient to use in many instances because of the quantities required to produce significant increases in chloride level and because of their tendency to produce density currents. Their use as tracers has been superseded to a great extent by dyes, nitrites, and radioactive materials.

20-3 METHODS OF DETERMINATION

Chlorides may be readily measured by means of volumetric procedures employing internal indicators. For most purposes the Mohr method employing silver nitrate as the titrant and potassium chromate as the indicator is satisfactory. However, the procedure using mercuric nitrate as the titrant and diphenylcarbazone as the indicator has certain inherent advantages over the Mohr or argentometric method.

Mohr Method (Argentometric)

The Mohr method employs a solution of silver nitrate for titration, and "Standard Methods" recommends the use of a 0.0141 N solution. This corresponds to a $N/71$ solution or one in which each milliliter is equivalent to 0.5 mg of chloride ion. The silver nitrate solution can be standardized against standard chloride solutions prepared from pure sodium chloride. In the titration the chloride ion is precipitated as white silver chloride.

$$Ag^+ + Cl^- \rightleftharpoons AgCl \qquad (K_{sp} = 3 \times 10^{-10}) \qquad (20\text{-}1)$$

The end point cannot be detected by eye unless an indicator capable of demonstrating the presence of excess Ag^+ is present. The indicator normally used is potassium chromate, which supplies chromate ions. As the concentration of chloride ions approaches extinction, the silver-ion concentration increases to a level at which the solubility product of silver chromate is exceeded and it begins to form a reddish-brown precipitate.

$$2Ag^+ + CrO_4^{2-} \rightleftharpoons Ag_2CrO_4 \qquad (K_{sp} = 5 \times 10^{-12}) \qquad (20\text{-}2)$$

This is taken as evidence that all the chloride has been precipitated. Since an excess of Ag^+ is needed to produce a visible amount of Ag_2CrO_4, the indicator error or blank must be determined and subtracted from all titrations.

Several precautions must be observed in this determination if accurate results are to be obtained:

1. A uniform sample size must be used, preferably 100 ml, so that ionic concentrations needed to indicate the end point will be constant.
2. The pH must be in the range of 7 to 8 because Ag^+ is precipitated as AgOH at high pH levels and the CrO_4^{2-} is converted to $Cr_2O_7^{2-}$ at low pH levels.
3. A definite amount of indicator must be used to provide a certain concentration of CrO_4^{2-}; otherwise Ag_2CrO_4 may form too soon or not soon enough.

The indicator error or blank varies somewhat with the ability of individuals to detect a noticeable color change. The usual range is 0.2 to 0.4 ml of titrant.

If the silver nitrate solution used for titration is exactly 0.0141 N, the calculation for chlorides as given in "Standard Methods" may be simplified as

$$Cl^- \text{ (in mg/l)} = \frac{(\text{ml } AgNO_3 - \text{blank}) \times 0.5 \times 1000}{\text{ml sample}} \qquad (20\text{-}3)$$

since $0.0141 \times 35.46 = 0.5$.

For routine control work, it is most convenient to prepare a $N/35.46$ (0.0282 N) silver nitrate solution of which each milliliter is equivalent to 1.0 mg of Cl^-, and the factor 0.5 can be eliminated from the calculation.

Mercuric Nitrate Method

The mercuric nitrate method of determining chlorides is much less subject to interferences than the Mohr method because the titration is performed in a sample whose pH is adjusted to a value of about 2.5. Under these conditions, Hg^{2+} ion combines with Cl^- to form the $HgCl_2$ complex which is soluble,

$$Hg^{2+} + 2Cl^- \rightleftharpoons HgCl_2 \qquad (K = 2.6 \times 10^{-15}) \qquad (20\text{-}4)$$

therefore making end-point detection easier than with the Mohr procedure. As the Cl^- concentration approaches zero, the Hg^{2+} concentration increases to a level where it becomes significant as the mercuric nitrate is added.

Diphenylcarbazone is the indicator used to show the presence of excess Hg^{2+} ions. It combines with them to form a distinct purple color. A blank correction is needed as with the Mohr procedure, although the value is usually less. Nitric acid is added to the indicator to reduce the sample pH to 2.5, a value which must be maintained uniformly in unknown samples, standards, and blanks. A pH indicator, xylene cyanol FF, which is blue-green at pH 2.5 is also included and improves the end point by masking the pale color developed by diphenylcarbazone during the titration.

"Standard Methods" recommends that a 0.0141 N solution of mercuric nitrate be used as the titrant in determining chlorides. Each milliliter of such a solution is equivalent to 0.5 mg of Cl^-, and therefore the comments made above concerning calculations with similar strength solutions of $AgNO_3$ also apply in this case.

The use of $N/35.46$ (0.0282 N) solutions of the mercuric nitrate titrant is more convenient for routine determinations, since each milliliter is equivalent to 1.0 mg of Cl^-.

20-4 APPLICATION OF CHLORIDE DATA

In many areas the level of chlorides in natural waters is an important consideration in the selection of supplies for human use. Where brackish waters must be used for domestic purposes, the amount of chlorides present is an important factor in determining the type of desalting apparatus to be used. The chloride determination is used to control pumping of ground-water from locations where intrusion of seawater is a problem.

In areas where the discharge of salt-water brines and industrial wastes containing high concentrations of chlorides must be controlled to safeguard receiving waters, the chloride determination serves to excellent advantage for regulatory purposes.

Chlorides interfere in the determination of chemical oxygen demand (COD).

A correction must be made on the basis of the amounts present or else a complexing agent such as $HgSO_4$ can be added.

Sodium chloride has a considerable history as a tracer. One of its principal applications has been in tracing pollution of wells. It is admirably suited for such purposes for five reasons:

1. Its presence is not detectable by eye.
2. It is a normal constituent of water and has no toxic effects.
3. The chloride ion is not adsorbed by soil formations.
4. It is not altered or changed in amount by biological processes.
5. The chloride ion is easily measured.

It is to be expected that chlorides will continue in limited use as tracers where other methods are not applicable.

PROBLEMS

20-1 Discuss the significance of the presence of high chloride concentrations in water supplies.

20-2 Explain why a blank correction must be applied to the titration values in both the Mohr and mercuric nitrate methods in the calculation of chloride content.

20-3 Would the analytical results by the Mohr method for chlorides be higher, lower, or the same as the true value if an excess of indicator were accidentally added to the sample? Why?

20-4 Why must the sample pH be neither high nor low in the Mohr method for chlorides?

20-5 What purpose is served by the nitric acid added to the indicator in the mercuric nitrate method for chlorides?

20-6 (a) In the determination of chlorides by the Mohr method, what will be the equilibrium concentration of silver ions in mg/l, on the basis of the solubility-product principle, when the chloride concentration has been reduced to 0.2 mg/l?

(b) If the concentration of chromate indicator used is 5×10^{-3} mol/l, how much excess silver ion in mg/l must be present before the formation of a red precipitate will begin?

20-7 Estimate the concentration of mercuric ions in the mercuric nitrate method of determining chlorides under the same conditions as given in Prob. 20-6(a).

CHAPTER
TWENTY-ONE

DISSOLVED OXYGEN

21-1 GENERAL CONSIDERATIONS

All living organisms are dependent upon oxygen in one form or another to maintain the metabolic processes that produce energy for growth and reproduction. Aerobic processes are the subject of greatest interest because of their need for free oxygen. Man is vitally concerned with the oxygen content of the air that he breathes, since he knows from experience that an appreciable reduction in oxygen content will lead to discomfort and possibly death. For this reason, he is careful to restrict the number of occupants within enclosures to the ventilating capacity.

The environmental engineer is, of course, interested in atmospheric conditions in relation to man, but in addition, he is vitally concerned with the "atmospheric conditions" that exist in liquids, water being the liquid in greatest abundance and importance.

All the gases of the atmosphere are soluble in water to some degree. Both nitrogen and oxygen are classed as poorly soluble, and since they do not react with water chemically, their solubility is directly proportional to their partial pressures. Hence Henry's law may be used to calculate the amounts present at saturation at any given temperature. The solubility of both nitrogen and oxygen varies greatly with the temperature over the range of interest to environmental engineers. Figure 21-1 shows solubility curves for the two gases in distilled or low-solids-content water in equilibrium with air at 760-mm pressure. The solubility is less in saline waters. It will be noted that under the partial-pressure conditions that exist in the atmosphere, more nitrogen than oxygen dissolves in water. At saturation, the dissolved gases contain about 38 percent oxygen, or nearly twice as much oxygen as in the normal atmosphere.

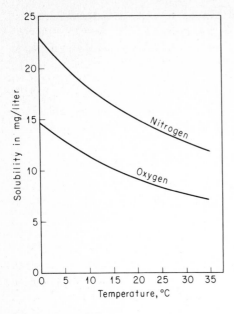

Figure 21-1 Solubility of oxygen and nitrogen in distilled water saturated with air at 760 mm Hg.

The solubility of atmospheric oxygen in fresh waters ranges from 14.6 mg/l at 0°C to about 7 mg/l at 35°C under 1 atm of pressure. Since it is a poorly soluble gas, its solubility varies directly with the atmospheric pressure at any given temperature. This is an important consideration at high altitudes. Because rates of biological oxidation increase with temperature, and oxygen demand increases accordingly, high-temperature conditions, where dissolved oxygen is least soluble, are of greatest concern to environmental engineers. Most of the critical conditions related to dissolved-oxygen deficiency in environmental engineering practice occur during the summer months when temperatures are high and solubility of oxygen is at a minimum. For this reason it is customary to think of dissolved-oxygen levels of about 8 mg/l as being the maximum available under critical conditions.

The low solubility of oxygen is the major factor that limits the purification capacity of natural waters and necessitates treatment of wastes to remove pollutional matter before discharge to receiving streams. In aerobic biological treatment processes, the limited solubility of oxygen is of great importance because it governs the rate at which oxygen will be absorbed by the medium and therefore the cost of aeration.

The solubility of oxygen is less in salt-containing water than it is in clean water. For this reason the solubility for a given temperature decreases as one progresses from fresh water to estuary water to the ocean. The extent of this effect is indicated in Table 21-1, which contains a listing of oxygen solubility as a function of temperature and chloride content. Chloride concentration is used as a measure of the seawater–fresh water mix in a sample. The chloride content of seawater is about 19,000 mg/l.

**Table 21-1 Solubility of dissolved oxygen in water in
equilibrium with dry air at 760 mm Hg and containing
20.9 percent oxygen[1]**

Tempera- ture, °C	Chloride concentration, mg/l				
	0	5000	10,000	15,000	20,000
0	14.6	13.8	13.0	12.1	11.3
1	14.2	13.4	12.6	11.8	11.0
2	13.8	13.1	12.3	11.5	10.8
3	13.5	12.7	12.0	11.2	10.5
4	13.1	12.4	11.7	11.0	10.3
5	12.8	12.1	11.4	10.7	10.0
6	12.5	11.8	11.1	10.5	9.8
7	12.2	11.5	10.9	10.2	9.6
8	11.9	11.2	10.6	10.0	9.4
9	11.6	11.0	10.4	9.8	9.2
10	11.3	10.7	10.1	9.6	9.0
11	11.1	10.5	9.9	9.4	8.8
12	10.8	10.3	9.7	9.2	8.6
13	10.6	10.1	9.5	9.0	8.5
14	10.4	9.9	9.3	8.8	8.3
15	10.2	9.7	9.1	8.6	8.1
16	10.0	9.5	9.0	8.5	8.0
17	9.7	9.3	8.8	8.3	7.8
18	9.5	9.1	8.6	8.2	7.7
19	9.4	8.9	8.5	8.0	7.6
20	9.2	8.7	8.3	7.9	7.4
21	9.0	8.6	8.1	7.7	7.3
22	8.8	8.4	8.0	7.6	7.1
23	8.7	8.3	7.9	7.4	7.0
24	8.5	8.1	7.7	7.3	6.9
25	8.4	8.0	7.6	7.2	6.7
26	8.2	7.8	7.4	7.0	6.6
27	8.1	7.7	7.3	6.9	6.5
28	7.9	7.5	7.1	6.8	6.4
29	7.8	7.4	7.0	6.6	6.3
30	7.6	7.3	6.9	6.5	6.1

[1] After G. C. Whipple and M. C. Whipple, "Solubility of Oxygen in Sea Water," *J. Am. Chem. Soc.*, 33: 362 (1911).

In polluted waters the saturation value is also less than that of clean water. The ratio of the value in polluted water to that in clean water is referred to as the β value. The rate of solution of oxygen in polluted waters is normally less than in clean water and the ratio is referred to as the α value. They may range as low as 0.8 for β and 0.4 for α in some wastewaters, and both α and β values are important design factors in selection of aeration equipment.

21-2 ENVIRONMENTAL SIGNIFICANCE OF DISSOLVED OXYGEN

In liquid wastes, dissolved oxygen is the factor that determines whether the biological changes are brought about by aerobic or by anaerobic organisms. The former use free oxygen for oxidation of organic and inorganic matter and produce innocuous end products, whereas the latter bring about such oxidations through the reduction of certain inorganic salts such as sulfates, and the end products are often very obnoxious. Since both types of organisms are ubiquitous in nature, it is highly important that conditions favorable to the aerobic organisms (aerobic conditions) be maintained; otherwise the anaerobic organisms will take over, and development of nuisance conditions will result. Thus dissolved-oxygen measurements are vital for maintaining aerobic conditions in natural waters that receive pollutional matter and in aerobic treatment processes intended to purify domestic and industrial wastewaters.

Dissolved-oxygen determinations are used for a wide variety of other purposes. It is one of the most important single tests that the environmental engineer uses. In most instances involving the control of stream pollution, it is desirable to maintain conditions favorable for the growth and reproduction of a normal population of fish and other aquatic organisms. This condition requires the maintenance of dissolved-oxygen levels that will support the desired aquatic life in a healthy condition at all times.

Determinations of dissolved oxygen serve as the basis of the BOD test; thus they are the foundation of the most important determination used to evaluate the pollutional strength of domestic and industrial wastes. The rate of biochemical oxidation can be measured by determining residual dissolved oxygen in a system at various intervals of time.

All aerobic treatment processes depend upon the presence of dissolved oxygen, and tests for it are indispensable as a means of controlling the rate of aeration to make sure that adequate amounts of air are supplied to maintain aerobic conditions and also to prevent excessive use of air.

Oxygen is a significant factor in the corrosion of iron and steel, particularly in water distribution systems and in steam boilers. Removal of oxygen from boiler-feed waters by physical and chemical means is common practice in the power industry. The dissolved-oxygen test serves as the means of control.

21-3 COLLECTION OF SAMPLES FOR DETERMINATION OF DISSOLVED OXYGEN

A certain amount of care must be exercised in the collection of samples to be used for dissolved-oxygen determinations. In most cases of interest, the dissolved-oxygen level will be below saturation, and exposure to the air will lead to erroneous results. For this reason, a special sampling device similar to the one described in "Standard Methods" is needed. All such instruments are designed on the principle that contact with air cannot be avoided during the time the sample bottles are being filled. However, if space is available to allow the bottles to overflow, a sample of water that is representative of the mixture being sampled can be obtained. Most samplers are designed to provide an overflow of two or three times the bottle volume to ensure collection of representative samples.

Most samples for dissolved oxygen are collected in the "field," where it is not convenient to perform the entire determination. Since oxygen values may change radically with time because of biological activity, it is customary to "fix" the samples immediately after collection. The usual procedure is to treat the samples with the conventional reagents used in the dissolved-oxygen test and then perform the titration when the samples are brought to the laboratory. This procedure will give low results for samples with a high iodine demand, and in this case it is better to preserve the sample by addition of 0.7 ml concentrated sulfuric acid and 0.02 g sodium azide. When this is done, it is necessary to add 3 ml of alkali-iodide reagent rather than the usual 2 ml because of the extra acid the sample contains. Better results are also obtained if the "fixed" samples are stored in the dark and on ice until the analyses can be completed.[1] The chemical treatment employed in "fixing" is radical enough to arrest all biological action, and the final titration may be delayed up to six hours.

21-4 CHOICE OF STANDARD REAGENT FOR MEASURING DISSOLVED OXYGEN

Most modern methods of determining dissolved oxygen depend upon reactions that release an amount of iodine equivalent to the amount of oxygen originally present, with subsequent measurement of the amount of iodine released by means of a standard solution of a reducing agent. Sodium thiosulfate is the reducing agent normally used, and starch solution is used to determine the end point. All reactions in the determination of oxygen involve oxidation and reduction. The starch indicator, however, acts in the capacity of an adsorption indicator. It adsorbs iodine from dilute solutions to produce a brilliant blue color and returns to a colorless form when the iodine is all reduced to iodide ion.

Selection of $N/40$ Thiosulfate Solution

The equivalent weight of oxygen is 8. Since the normality of most titrating agents used in water and wastewater analysis is adjusted so that each milliliter is equivalent to 1.0 mg of the measured material, it would follow that a $N/8$ solution of thiosulfate should be used. However, such a solution is too concentrated to allow accurate determinations of dissolved oxygen unless unreason-

[1] R. Porges, "Dissolved Oxygen Determinations for Field Surveys," *Journal Water Pollution Control Federation* **36:** 1247 (1964).

ably large samples are titrated. It has become standard practice to use 200-ml samples for titration. This is one-fifth of a liter. By using a titrating agent which is one-fifth as strong as conventionally used, the results obtained on 200-ml samples, in terms of milliliters of titrant used, are the same as if 1-liter samples had been treated with the $N/8$ reagent. Thus when a $N/40$ ($N/8 \times \frac{1}{5}$) solution of thiosulfate is used to titrate 200-ml samples, dissolved-oxygen values in milligrams per liter are equal to the titration. This eliminates the need for calculations.

Preparation and Standardization of $N/40$ ($0.025\ N$) Thiosulfate

Sodium thiosulfate ($Na_2S_2O_3 \cdot 5H_2O$) can be obtained in relatively pure form. However, because of its water of hydration, it cannot be dried to a compound of definite composition, and even loses water at room temperature under conditions of low humidity. It is necessary, therefore, to prepare solutions that are slightly stronger than desired and to standardize them against a primary standard.

The equivalent weight of sodium thiosulfate cannot be calculated from its formula and anticipated valence change, as is the case with most reducing agents (see Table 10-1). Rather, it must be calculated from its reaction with the oxidizing agent, in this case iodine.

$$2Na_2S_2O_3 \cdot 5H_2O + I_2^{\circ} \rightarrow Na_2S_4O_6 + 2NaI + 10H_2O \qquad (21\text{-}1)$$

From Eq. (21-1) it may be concluded that each molecule of sodium thiosulfate is equivalent to one atom of iodine. Since each atom of iodine gains one electron in the conversion to iodide ion, it is obvious that each molecule of thiosulfate supplies one electron when oxidized to the tetrathionate, or

$$2S_2O_3^{2-} + I_2^{\circ} \rightarrow S_4O_6^{2-} + 2I^- \qquad (21\text{-}2)$$

From these considerations, it may be concluded that the equivalent weight of sodium thiosulfate is equal to the molecular weight; and something in excess of 1/40 of the molecular weight, 6.205 g, should be taken to prepare 1 liter of a solution that is slightly stronger than $N/40$. Usually 6.5 g is sufficient.

The thiosulfate solution may be standardized with either of two primary standards, potassium dichromate or potassium bi-iodate. Both can be obtained in essentially 100 percent pure form. It is customary to prepare $N/40$ solutions of either by weighing out exact amounts on the analytical balance and diluting to the proper volume in volumetric flasks. Both primary standards react with iodide ion in acid solution to release free iodine:

$$Cr_2O_7^{2-} + 6I^- + 14H^+ \rightarrow 2Cr^{3+} + 3I_2^{\circ} + 7H_2O \qquad (21\text{-}3)$$

$$2IO_3^- + 10I^- + 12H^+ \rightarrow 6I_2^{\circ} + 6H_2O \qquad (21\text{-}4)$$

The iodine released is chemically equivalent to the oxidizing agent used. Thus if 20 ml of $N/40$ $K_2Cr_2O_7$ or $N/40$ $KIO_3 \cdot HIO_3$ (see Table 10-1) is used in the standardization procedure, exactly 20 ml of $N/40$ thiosulfate solution should be used in the titration. If the solution of thiosulfate is stronger than desired, it may be brought to proper strength by diluting with water in accordance with the principles set forth in Sec. 14-1.

The reaction between $Cr_2O_7^{2-}$ and I^- is not instantaneous under the conditions prescribed for the determination. Five minutes are normally needed for completion. The Cr^{3+} produces a greenish-blue color which interferes slightly with the normal starch end point. This difficulty may be overcome to a considerable extent by dilution before titration. Standardization with potassium biiodate is usually preferred because it does not suffer from these limitations.

Thiosulfate solutions are subject to attack by bacterial action and by carbon dioxide. Sulfur bacteria oxidize thiosulfates to sulfates under aerobic conditions, and carbon dioxide can depress the pH sufficiently to cause the thiosulfate ion to decompose into SO_3^{2-} and $S°$. The SO_3^{2-} is converted to SO_4^{2-} by dissolved oxygen. Thiosulfate solutions may be protected from bacterial and CO_2 attack by adding 0.4 g of NaOH per liter. The resulting high pH prevents bacterial growth and keeps the pH from falling as a result of small amounts of carbon dioxide that may gain access to the solution. Excess NaOH must be avoided, as this too can make the solution less stable. Chloroform may be used as a preservative, but it does not protect the solution from carbon dioxide that may be absorbed from the air. In addition, solutions preserved with it tend to dissolve stopcock lubricant.

21-5 METHODS OF DETERMINING DISSOLVED OXYGEN

Originally, measurement of dissolved oxygen was made by heating samples to drive out the dissolved gases and analyzing the collected gases for oxygen by methods applied in gas analysis. Such methods require large samples and are very cumbersome and time-consuming.

The Winkler or iodometric method and its modifications are the standard procedures for determining dissolved oxygen at the present time. The test depends upon the fact that oxygen oxidizes Mn^{2+} to a higher state of valence under alkaline conditions and that manganese in higher states of valence is capable of oxidizing I^- to free $I_2°$ under acid conditions. Thus the amount of free iodine released is equivalent to the dissolved oxygen originally present. The iodine is measured with standard sodium thiosulfate solution and interpreted in terms of dissolved oxygen.

The Winkler Method

The unmodified Winkler method is subject to interference from a great many substances. Certain oxidizing agents such as nitrite and Fe^{3+} are capable of oxidizing I^- to $I_2°$ and produce results that are too high. Reducing agents such as Fe^{2+}, SO_3^{2-}, S^{2-}, and polythionates, reduce $I_2°$ to I^- and produce results that are too low. The unmodified Winkler method is applicable only to relatively pure waters.

The reactions involved in the Winkler procedure are as follows:

$$Mn^{2+} + 2OH^- \rightarrow \underline{Mn(OH)_2} \quad \text{(white precipitate)} \quad (21\text{-}5)$$

If no oxygen is present, a pure white precipitate of $Mn(OH)_2$ forms when $MnSO_4$ and the alkali-iodide reagent (NaOH + KI) are added to the sample. If oxygen is present in the sample, then some of the Mn^{2+} is oxidized to Mn^{4+} and precipitates as a brown hydrated oxide. The reaction is usually represented as follows:

$$Mn^{2+} + 2OH^- + \tfrac{1}{2}O_2 \rightarrow \underline{MnO_2} + H_2O \quad (21\text{-}6)$$

or

$$Mn(OH)_2 + \tfrac{1}{2}O_2 \rightarrow \underline{MnO_2} + H_2O \quad (21\text{-}7)$$

The oxidation of Mn^{2+} to MnO_2, sometimes called fixation of the oxygen, occurs slowly, particularly at low temperatures. Furthermore, it is necessary to move the flocculated material throughout the solution to enable all the oxygen to react. Vigorous shaking of the samples for at least 20 seconds is needed. In the case of brackish or seawaters, much longer contact times are required.

After shaking the samples for a time sufficient to allow all oxygen to react, the floc is allowed to settle so as to leave at least 5 cm of clear liquid below the stopper; then sulfuric acid is added. Under the low pH conditions that result, the MnO_2 oxidizes I^- to produce free I_2.

$$MnO_2 + 2I^- + 4H^+ \rightarrow Mn^{2+} + I_2 + 2H_2O \quad (21\text{-}8)$$

The sample should be stoppered and shaken for at least 10 seconds to allow the reaction to go to completion and to distribute the iodine uniformly throughout the sample.

The sample is now ready for titration with $N/40$ thiosulfate. The use of $N/40$ thiosulfate is based upon the premise that a 200-ml sample will be used for titration. In adding the reagents used for the Winkler test, a certain amount of dilution of the sample occurs; therefore it is necessary to take a sample somewhat greater than 200 ml for the titration. When 300-ml bottles are used in the test, 2 ml of $MnSO_4$ and 2 ml of alkali-KI solutions are used. These are added in such a manner as to displace approximately 4 ml of sample from the bottle, and a correction should be made. When the 2 ml of acid is added, none of the oxidized floc is displaced; thus no correction need be made for its addition. To correct for the addition of the first two reagents, 203 ml of the treated sample is taken for titration.

Titration of a sample of a size equivalent to 200 ml of the original sample with $N/40$ thiosulfate solution yields results in milliliters, which can be interpreted directly in terms of milligrams per liter of dissolved oxygen.

The Azide Modification of the Winkler Method

The nitrite ion is one of the most frequent interferences encountered in the dissolved-oxygen determination. It occurs principally in effluents from sewage treatment plants employing biological processes, in river waters, and in incubated BOD samples. It does not oxidize Mn^{2+} but does oxidize I^- to free I_2 under acid conditions. It is particularly obnoxious because its reduced form, N_2O_2, is oxidized by oxygen, which enters the sample during the titration procedure, and is converted to NO_2^- again, establishing a cyclic reaction that can lead to erroneously high results, far in excess of amounts that would be expected. The reactions involved may be represented as follows:

$$2NO_2^- + 2I^- + 4H^+ \rightarrow I_2 + N_2O_2 + 2H_2O \qquad (21\text{-}9)$$

and
$$N_2O_2 + \tfrac{1}{2}O_2 + H_2O \rightarrow 2NO_2^- + 2H^+ \qquad (21\text{-}10)$$

When interference from nitrites is present, it is impossible to obtain a permanent end point. As soon as the blue color of the starch indicator has been discharged, the nitrites formed by the reaction in Eq. (21-10) will react with more I^- to produce I_2 and the blue color of the starch indicator will return.

Nitrite interference may be easily overcome by the use of sodium azide (NaN_3). It is most convenient to incorporate the azide in the alkali-KI reagent. When sulfuric acid is added, the following reactions occur and the NO_2^- is destroyed.

$$NaN_3 + H^+ \rightarrow HN_3 + Na^+ \qquad (21\text{-}11)$$

$$HN_3 + NO_2^- + H^+ \rightarrow N_2 + N_2O + H_2O \qquad (21\text{-}12)$$

By this procedure, nitrite interference is eliminated and the method of determination retains the simplicity of the original Winkler procedure.

Rideal–Stewart Modification of the Winkler Method

The Rideal-Stewart or permanganate modification is designed to overcome the effects of a wide variety of interferences caused by reducing substances, including nitrites. It involves pretreatment of the sample with potassium permanganate under acid conditions. The permanganate is added in excess and oxidizes the reducing agents present. The excess is destroyed by adding a reducing agent, potassium oxalate, which in slight excess does not react with free iodine. Some of the reactions involved are as follows:

$$5NO_2^- + 2MnO_4^- + 6H^+ \rightarrow 5NO_3^- + 2Mn^{2+} + 3H_2O \qquad (21\text{-}13)$$

$$5Fe^{2+} + MnO_4^- + 8H^+ \rightarrow 5Fe^{3+} + Mn^{2+} + 4H_2O \qquad (21\text{-}14)$$

$$\text{Aldehydes} + MnO_4^- + H^+ \rightarrow \text{acids} + Mn^{2+} + H_2O \qquad (21\text{-}15)$$

$$\underset{\text{H H}}{-\overset{\displaystyle |}{C}=\overset{\displaystyle |}{C}-} + MnO_4^- + H^+ \rightarrow \text{acids} + Mn^{2+} + H_2O \qquad (21\text{-}16)$$

The NO_3^- formed in Eq. (21-13) does not oxidize I^- under the conditions of the test. Fe^{3+}, in concentrations below 10 mg/l, does not interfere. At levels above 10 mg/l it must be treated to lower its ionic concentration to avoid interference. Potassium fluoride is usually added for this purpose, since it supplies F^-, which combines with Fe^{3+} to form poorly ionized FeF_3.

$$Fe^{3+} + 3F^- \rightarrow FeF_3 \tag{21-17}$$

Excess $KMnO_4$ is destroyed by adding potassium oxalate.

$$5(COO^-)_2 + 2MnO_4^- + 16H^+ \rightarrow 10CO_2 + 2Mn^{2+} + 8H_2O \tag{21-18}$$

After the excess permanganate has been destroyed, the regular Winkler procedure is followed, except that additional amounts of alkali-KI are needed to overcome the effects of the acid added originally to facilitate the action of the permanganate. Proper corrections must be made for the volumes of reagents added, in order to calculate the volume of sample required for the titration with thiosulfate.

21-6 DISSOLVED-OXYGEN ELECTRODES

The use of dissolved-oxygen electrodes, which allow *in situ* measurements to be made, has increased significantly since their development. These electrodes, described under polarographic analysis in Sec. 11-3, are especially useful for taking dissolved-oxygen profiles of reservoirs and streams. The electrodes can be lowered to various depths, and the dissolved-oxygen concentration can be read from a connecting microammeter located at the surface. They can also be suspended in biological waste treatment tanks to monitor the dissolved-oxygen level at any point. The rate of biological oxygen utilization in such a tank can be determined by placing a sample of the mixed liquor in a BOD bottle and then inserting a dissolved-oxygen electrode to observe the rate at which oxygen is depleted.

Dissolved-oxygen electrodes are usually calibrated by making measurements in water samples which have been analyzed by the Winkler procedure. Thus any errors in the Winkler analysis will be carried over to the electrode calibration. During dissolved-oxygen measurements, it is important that sufficient movement of the sample by the electrode be maintained to prevent low readings which result if oxygen is depleted at the membrane as it is reduced at the cathode. Dissolved-oxygen electrodes are very sensitive to temperature, and thus either accurate temperature measurements must be made along with dissolved-oxygen measurements so that a correction can be applied, or else instruments which are equipped with a thermistor or other device to compensate automatically for temperature changes must be used. Because of the versatility of dissolved-oxygen electrodes, their use should become more widespread in the future, especially when the occasional problem of getting the electrodes to operate properly is overcome.

21-7 APPLICATION OF DISSOLVED-OXYGEN DATA

Dissolved-oxygen data are used in a wide variety of applications. Many of these have been discussed under general conditions in Sec. 21-1.

PROBLEMS

21-1 Discuss why it is desirable to maintain a significant dissolved-oxygen concentration in rivers and streams.

21-2 Write chemical equations summarizing all the essential reactions involved in the unmodified Winkler method for dissolved oxygen.

21-3 Prepare a table showing five substances which interfere with the Winkler method, and indicate which modification would be used to overcome each interference.

21-4 What is the function of the NaOH sometimes used in preparing the thiosulfate solution used for dissolved-oxygen determinations?

21-5 Give two reasons why fixation of dissolved oxygen should be performed in the field if at all feasible.

21-6 Two samples were collected simultaneously at the same spot in the river for dissolved-oxygen analysis. One sample was "fixed" immediately after collection, and the other was treated later in the laboratory. Indicate two possible factors that could cause lower results to be obtained in the second sample.

21-7 Calculate the percent saturation of dissolved oxygen in a water sample with a temperature of 22°C and a dissolved-oxygen concentration of 5.3 mg/l when the atmospheric pressure is 740 mm Hg. Assume the sample salinity is less than 100 mg/l.

21-8 What advantages do dissolved-oxygen electrodes have over the Winkler test for dissolved-oxygen measurements? What are some disadvantages?

TWENTY-TWO

BIOCHEMICAL OXYGEN DEMAND

22-1 GENERAL CONSIDERATIONS

Biochemical oxygen demand (BOD) is usually defined as the amount of oxygen required by bacteria while stabilizing decomposable organic matter under aerobic conditions. The term "decomposable" may be interpreted as meaning that the organic matter can serve as food for the bacteria, and energy is derived from its oxidation.

The BOD test is widely used to determine the pollutional strength of domestic and industrial wastes in terms of the oxygen that they will require if discharged into natural watercourses in which aerobic conditions exist. The test is one of the most important in stream-pollution-control activities. This test is of prime importance in regulatory work and in studies designed to evaluate the purification capacity of receiving bodies of water.

The BOD test is essentially a bioassay procedure involving the measurement of oxygen consumed by living organisms (mainly bacteria) while utilizing the organic matter present in a waste, under conditions as similar as possible to those that occur in nature. In order to make the test quantitative, the samples must be protected from the air to prevent reaeration as the dissolved-oxygen level diminishes. In addition, because of the limited solubility of oxygen in water, about 9 mg/l at 20°C, strong wastes must be diluted to levels of demand in keeping with this value to ensure that dissolved oxygen will be present throughout the period of the test. Since this is a bioassay procedure, it is extremely important that environmental conditions be suitable for the living organisms to function in an unhindered manner at all times. This condition means that toxic substances must be absent and that all accessory nutrients needed for bacterial growth, such as nitrogen, phosphorus, and certain trace elements, must be present. Biological degradation of organic matter under natural conditions is brought about by a diverse group of organisms that carry the oxidation essentially to completion, i.e., almost entirely to carbon dioxide and water. Therefore it is important that a mixed group of organisms, commonly called "seed," be present in the test.

The BOD test may be considered as a wet oxidation procedure in which the living organisms serve as the medium for oxidation of the organic matter to carbon dioxide and water. A quantitative relationship exists between the amount of oxygen required to convert a definite amount of any given organic compound to carbon dioxide, water, and ammonia, and this can be represented by the following generalized equation:

$$C_nH_aO_bN_c + \left(n + \frac{a}{4} - \frac{b}{2} - \frac{3}{4}c \right) O_2 \rightarrow n\,CO_2 + \left(\frac{a}{2} - \frac{3}{2}c \right) H_2O + cNH_3$$

$$(22\text{-}1)$$

On the basis of the above relationship, it is possible to interpret BOD data in terms of organic matter, as well as the amount of oxygen used during its oxidation. This concept is fundamental to an understanding of the rate at which BOD is exerted.

The oxidative reactions involved in the BOD test are a result of biological activity, and the rate at which the reactions proceed is governed to a major extent by population numbers and temperature. Temperature effects are held constant by performing the test at 20°C, which is, more or less, a median value as far as natural bodies of water are concerned. The predominant organisms responsible for the stabilization of organic matter in natural waters are forms native to the soil. The rate of their metabolic processes at 20°C and under the conditions of the test is such that time must be reckoned in days. Theoretically an infinite time is required for complete biological oxidation of organic matter, but for all practical purposes, the reaction may be considered complete in 20 days. However, a 20-day period is too long to wait for results in most instances. It has been found by experience that a reasonably large percentage of the total BOD is exerted in 5 days; consequently the test has been developed on the basis of a 5-day incubation period. It should be remembered, therefore, that 5-day BOD values represent only a portion of the total BOD. The exact percentage depends upon the character of the "seed" and the nature of the organic matter, and can be determined only by experiment. In the case of domestic and many industrial wastewaters, it has been found that the 5-day BOD value is about 70 to 80 percent of the total BOD. This is a large enough percentage of the total so that 5-day values are used for many considerations. The 5-day incubation period was selected also to minimize interferences from oxidation of ammonia as discussed later.

22-2 THE NATURE OF THE BOD REACTION

Studies of the kinetics of BOD reactions have established that they are for most practical purposes "first-order" in character (see Sec. 3-11), or the rate of the reaction is proportional to the amount of oxidizable organic matter remaining at any time, as modified by the population of active organisms. Once the population of organisms has reached a level at which only minor variations occur, the reaction rate is controlled by the amount of food available to the organisms and may be expressed as follows:

$$\frac{-dC}{dt} \propto C \quad \text{or} \quad \frac{-dC}{dt} = k'C \tag{22-2}$$

where C represents the concentration of oxidizable organic matter (pollutants) at the start of the time interval t, and k' is the rate constant for the reaction. This means that the rate of the reaction gradually decreases as the concentration C of food or organic matter decreases.

In BOD considerations it is customary to use L in place of C, where L represents the ultimate demand, and the expression

$$\frac{-dL}{dt} = k'L \tag{22-3}$$

represents the rate at which organic polluting matter is destroyed. Since oxygen is used in stabilizing the organic matter in direct ratio to the amount of organic matter oxidized, it is possible to interpret L in terms of organic polluting matter or in terms of oxygen used, as preferred.

Upon integration of Eq. (22-3), the expression

$$\frac{L_t}{L} = e^{-k't} = 10^{-kt} \tag{22-4}$$

where $k = k'/2.303$ is obtained. This formula says that the amount of pollutants remaining after any time t has elapsed is a fraction of L corresponding to 10^{-kt}, or the BOD that has not been exerted is a percentage of L corresponding to 10^{-kt}. This expression is used widely in engineering practice, particularly in stream-pollution-control studies.

In many cases the analyst and the environmental engineer are interested in the BOD exerted. This value is usually determined by actual test through dissolved-oxygen measurements. Often it is desirable to translate 5-day results to total BOD (L) or the BOD at some other time. This is done by a modification of Eq. (22-4) to

$$y = L(1 - 10^{-kt}) \tag{22-5}$$

In this expression $y = $ BOD at any time t, and L is the total or ultimate BOD. The value of k must be determined by experiment.

Since the BOD reaction is closely related to a first-order type of reaction, a plot of the amount of organic matter remaining versus time yields a parabolic

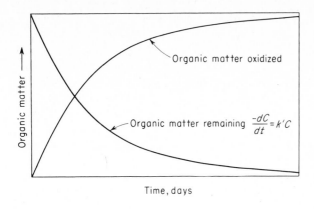

Figure 22-1 Changes in organic matter during biological oxidation of polluted waters under aerobic conditions.

curve similar to the decay curve for a radioactive element. Likewise, if a plot is made showing the amount of organic matter oxidized versus time, another parabolic curve is obtained that is the reciprocal of the first. Curves illustrating these changes are shown in Fig. 22-1.

Because oxygen is used in direct ratio to the amount of organic matter oxidized in biochemical oxidations, a plot of oxygen used versus time should produce a parabolic type of curve like the one for organic matter oxidized in Fig. 22-1. A typical BOD or oxygen-used curve is shown in Fig. 22-2. It will be noted that the curve has characteristics similar to those for the curve for

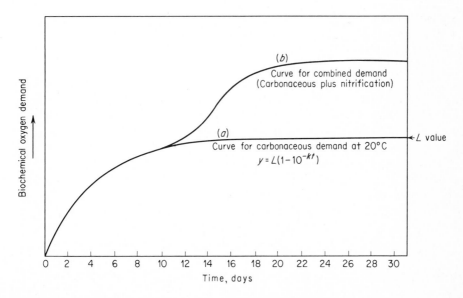

Figure 22-2 The BOD curve. *(a)* Normal curve for oxidation of organic matter. *(b)* The influence of nitrification.

organic matter oxidized in Fig. 22-1 during the first 8 to 10 days. Following that, the BOD curve digresses radically from the course it would be expected to follow as a unimolecular or first-order reaction.

The importance of having a mixed culture of organisms corresponding to those in the soil, for proper measurement of BOD, has been mentioned. Such cultures, when derived from the soil or domestic wastewater, contain large numbers of saprophytic bacteria and other organisms that utilize the carbonaceous matter present in the samples subjected to BOD analysis, and use oxygen in a corresponding amount. In addition, they normally contain certain autotrophic bacteria, particularly nitrifying bacteria, which oxidize noncarbonaceous matter for energy. The nitrifying bacteria are usually present in relatively small numbers in untreated domestic wastewater, and fortunately, their reproductive rate at 20°C is such that their populations do not become sufficiently large to exert an appreciable demand for oxygen until about 8 to 10 days have elapsed in the regular BOD test. Once the organisms become established, they oxidize nitrogen in the form of ammonia to nitrous and nitric acids in amounts that introduce serious error into BOD work.

$$2NH_3 + 3O_2 \xrightarrow[\text{bacteria}]{\text{nitrite-forming}} 2NO_2^- + 2H^+ + 2H_2O \qquad (22\text{-}6)$$

$$2NO_2^- + O_2 + 2H^+ \xrightarrow[\text{bacteria}]{\text{nitrate-forming}} 2NO_3^- + 2H^+ \qquad (22\text{-}7)$$

It is true that the oxidation of inorganic nitrogen can deplete the dissolved oxygen in streams, and this effect must be taken into account by the engineer. However, it is not desirable to use normal BOD measurements for such estimates, because ammonia nitrogen is added to BOD dilution water as a required nutrient and its oxidation could lead to erroneous conclusions about the waste. The potential dissolved oxygen utilization by nitrification is best evaluated by an analysis of the waste for the different forms of nitrogen present and use of the stoichiometric relationships between oxygen and nitrogen given by Eqs. (22-6) and (22-7).

The interference caused by nitrifying organisms makes the actual measurement of total carbonaceous BOD impossible unless provision is made to eliminate them. The interference caused by the nitrifying bacteria was a major reason for selecting a 5-day incubation period for the regular BOD test.

In cases in which the effluent from biological treatment units, such as trickling filters and activated sludges, are to be analyzed for BOD the effluents often contain populations of nitrifying organisms sufficient to utilize significant amounts of oxygen during the regular 5-day incubation period. It is important to know the amount of residual carbonaceous BOD in such cases in order to be able to measure plant efficiency.[1] The action of the nitrifying bacteria can be arrested by the use of specific inhibiting agents such as methylene blue or

[1] C. N. Sawyer and L. Bradney, *Sewage Works J.*, **18**: 1113 (1946).

allylthiourea (ATU),[2] or the nitrifying populations can be reduced to insignificant levels by pretreatment of the sample by pasteurization, chlorination, or acid treatment, thus allowing measurement of residual carbonaceous BOD without interference from nitrification.

Samples from rivers and estuaries often contain significant populations of nitrifying organisms. No standard procedure has been advanced for measuring carbonaceous BOD in such waters. Furthermore, algal growths, when present, introduce another variable which makes BOD data on rivers and estuaries difficult to interpret.[3]

22-3 METHOD OF MEASURING BOD

The BOD test is based upon determinations of dissolved oxygen; consequently the accuracy of the results is influenced greatly by the care given to its measurement. BOD may be measured directly in a few samples, but in general, a dilution procedure is required.

Direct Method

With samples whose 5-day BOD does not exceed 7 mg/l, it is not necessary to dilute them, provided that they are aerated to bring the dissolved-oxygen level nearly to saturation at the start of the test. Many river waters fall into this category.

The usual procedure is to adjust the sample to about 20°C and aerate with diffused air to increase or decrease the dissolved gas content of the sample to near saturation. Two or more BOD bottles are then filled with the sample; at least one is analyzed for dissolved oxygen immediately, and the others are incubated for 5 days at 20°C. After 5 days, the amount of dissolved oxygen remaining in the incubated samples is determined, and the 5-day BOD is calculated by subtraction of the 5-day results from those obtained on 0 day.

The direct method of measuring BOD involves no modification of the sample, and therefore produces results under conditions as nearly similar as possible to the natural environment. Unfortunately the BOD of very few samples falls within the range of dissolved oxygen available in this test.

Dilution Method

The dilution method of measuring BOD is based upon the fundamental concept that the rate of biochemical degradation of organic matter is directly pro-

[2] J. C. Young, *J. Water Pollution Control Fed.*, **45**: 639 (1973).

[3] T. F. Wisniewski, "Oxygen Relationships in Streams," U.S. Public Health Service Seminar, *R. A. Taft Sanitary Eng. Center Tech. Rept.* W58-2, Cincinnati, Ohio (March, 1958); G. P. Fitzgerald, "The Effect of Algae on BOD Measurements," *J. Water Pollution Control Federation*, **36**: 1524 (1964).

portional to the amount of unoxidized material existing at the time, as discussed in Sec. 22-2. According to this concept, the rate at which oxygen is used in dilutions of the waste is in direct ratio to the percent of waste in the dilution, provided that all other factors are equal. For instance, a 10 percent dilution uses oxygen at one-tenth the rate of a 100 percent sample. Experience has served as the basis for the mathematical development of the BOD reaction; therefore it is safe to assume the validity of the concept.

In any bioassay work, it is important to control all environmental and nutritional factors in a manner that will not interfere with the desired action. In the BOD test, this means that everything influencing the rate at which organic matter is biologically stabilized must be kept under close control and highly reproducible from test to test. The major items of importance are (1) freedom from toxic materials, (2) favorable pH and osmotic conditions, (3) presence of available accessory nutrient elements, (4) standard temperature, and (5) presence of a significant population of mixed organisms of soil origin.

A wide variety of waste materials are subject to the BOD test. These range from industrial wastes that may be free of microorganisms to domestic wastewater with an abundance of organisms. Many industrial wastes have extremely high BOD values, and very high dilutions must be made to meet the requirements imposed by the limited solubility of oxygen. Domestic wastewater has an ample supply of accessory nutrient elements, such as nitrogen and phosphorus, but many industrial wastes are deficient in one and sometimes both of these elements. Because of these limitations, the dilution water used in BOD work must compensate for the limitations imposed by any sample subjected to analysis. Since these limitations are not always known, it is safe practice to use a dilution water that will provide for all contingencies. This is not necessary and may be undesirable, however, when domestic wastewater is the sole consideration.

The dilution water A wide variety of waters have been used for BOD work. Natural surface waters would appear to be ideal, but they have a number of disadvantages, including variable BOD, variable microorganism population (often including algae and significant populations of nitrifying bacteria), and variable mineral content. Tap water has been used, but it suffers from most of the limitations found in surface waters plus the possibility of toxicity from chlorine residuals. Through long experience, it has developed that a synthetic dilution water prepared from distilled or demineralized water is best for BOD testing because most of the variables mentioned above can be kept under control.

The quality of the distilled water used for the preparation of dilution water is of prime importance. It must be free from toxic substances. Chlorine or chloramines and copper are the two most commonly found. In many cases it is necessary to dechlorinate the water fed to the still to obtain a chlorine-free distillate. Copper contamination is normally due to exposed copper in the condenser. The BOD of distilled waters prepared from potable supplies is usually sufficiently low to allow use of the water without storage other than that needed to bring its temperature into a favorable range.

The pH of dilution water may range anywhere from 6.5 to 8.5 without affecting the action of the saprophytic bacteria. It is customary to buffer the solution by means of a phosphate system at about pH 7.0. The buffer is essential to maintain favorable pH conditions at all times.

The proper osmotic conditions are maintained by the potassium and sodium phosphates added to provide buffering capacity. In addition, calcium and magnesium salts are added which contribute to the total salt content.

The potassium, sodium, calcium, and magnesium salts added to give buffering capacity and proper osmotic conditions also serve to provide the microorganisms with any of these elements that are needed in growth and metabolism. Ferric chloride, magnesium sulfate, and ammonium chloride supply the requirements for iron, sulfur, and nitrogen. The phosphate buffer furnishes any phosphorus that may be needed. The nitrogen should be eliminated in cases where nitrogenous oxygen demand is being measured.

The dilution water now contains all the essential materials for the measurement of BOD except the necessary microorganisms. A wide variety of materials have been used for "seeding" purposes. Experience has shown that domestic wastewater, particularly from combined sewer systems, provides about as well balanced a population of mixed organisms as anything, and usually 2 ml of wastewater per liter of dilution water is sufficient. Some river waters are satisfactory, but care must be taken to avoid using waters that contain algae or nitrifying bacteria in significant amounts.

The dilution water should always be "seeded" with wastewater or other material to ensure a uniform population of organisms in various dilutions and to provide an opportunity for any organic matter present in the dilution water blanks to be exposed to the same type of organisms as those involved in the stabilization of the waste. The latter is a point that is often ignored; this has led to erroneously high results in many cases.

Finally, the dilution water should be aerated to saturate it with oxygen before use.

The need for blanks In the determination of BOD by the dilution technique, it is safe to assume that the dilution water containing the "seeding" material will contain organic matter and that addition of the diluting water to the sample will increase the amount of oxidizable organic matter; therefore a correction must be applied. Usually the correction is made without special calculations by letting the 5-day dissolved-oxygen value of the blank represent the 0-day corrected value. It is not necessary to determine the dissolved oxygen of the dilution water on 0 day, unless it is desired to have some measure of the amount of organic matter in the dilution water. "Standard Methods" recommends setting up a separate series of seed dilutions in order to make a seed correction. However, this extra effort to correct a minor variation would not appear warranted in most instances because of other larger biological variations which are normal to the test.

At least three blanks should be included with each set of BOD samples. In any bioassay test there is a certain amount of biological variation. Since the blanks serve as the reference value from which all calculations of BOD are made, it is important that it have some statistical reliability. Usually three blanks provide such reliability, but each analyst should satisfy his particular requirements.

Dilutions of waste The analyst has a real responsibility in deciding what dilutions should be set for determination of BOD. Usually it is best to set three different dilutions. When the strength of a sample is known with some assurance, two dilutions may suffice. Where samples of unknown strength are involved, the dilutions should cover a considerable range, and in some instances it may be necessary to set as many as four dilutions. In any case there should be an overlapping of the BOD measurable by successive dilutions.

It has been demonstrated that BOD is not influenced by oxygen concentrations as low as 0.5 mg/l. It has also been learned that it is not statistically reliable to base BOD values upon dilutions that produce a depletion of oxygen less than 2 mg/l. Therefore it has become customary to base calculations of BOD on samples that produce a depletion of at least 2 mg/l and have at least 0.5 mg/l of dissolved oxygen remaining at the end of the incubation period. This restriction usually means a range of 2 to 7 mg/l. With this information at hand it is possible to construct a table showing the range of BOD measurable by various dilutions. Table 22-1 presents such information for dilutions prepared

Table 22-1 BOD measurable with various dilutions of samples

Using percent mixtures		By direct pipeting into 300-ml bottles	
% mixture	Range of BOD	ml	Range of BOD
0.01	20,000–70,000	0.02	30,000–105,000
0.02	10,000–35,000	0.05	12,000– 42,000
0.05	4,000–14,000	0.10	6,000– 21,000
0.1	2,000– 7,000	0.20	3,000– 10,500
0.2	1,000– 3,500	0.50	1,200– 4,200
0.5	400– 1,400	1.0	600– 2,100
1.0	200– 700	2.0	300– 1,050
2.0	100– 350	5.0	120– 420
5.0	40– 140	10.0	60– 210
10.0	20– 70	20.0	30– 105
20.0	10– 35	50.0	12– 42
50.0	4– 14	100	6– 21
100	0– 7	300	0– 7

on a percentage basis and also for dilutions prepared by direct pipeting into bottles of about 300-ml capacity. It is customary to estimate the BOD of a sample and set one dilution based upon the estimate. Two other dilutions, one higher and one lower, are also set up. For example, a sample is estimated to have a BOD of 1000 mg/l. Reference to Table 22-1 will show that a 0.5 percent mixture should be used. If a 0.2 and a 1.0 percent mixture are included, the range of measurable BOD is extended from 200 to 3500 mg/l and should compensate for any errors in the original estimate.

In the direct-pipeting technique, preliminary dilutions should be made of all samples that require less than 0.5 ml of the sample, so that amounts added to the bottles can be measured without serious error. The volumes of all bottles must be known in order to allow calculation of the BOD when this method is used.

Incubation bottles The bottles used for BOD analysis should be equipped with glass stoppers that are ground to a point to prevent trapping of air when the stopper is inserted. The bottles should be equipped with some form of water seal to prevent air from entering the bottle during the incubation period.

It is extremely important that bottles used for BOD work be free of organic matter. Cleaning can be best accomplished by use of a chromic acid solution or a good grade of detergent.[4] If the latter cleaning agent is used, bottles should be rinsed with hot water to kill nitrifying organisms which tend to develop on the walls of the bottles. Care must be exercised to make sure that all the cleaning agent is removed from the bottle before use. This assurance can usually be accomplished by four rinses with tap water and a final rinse with distilled or demineralized water.

Initial dissolved oxygen With samples whose BOD is below 200 mg/l, it is necessary to use amounts of sample in excess of 1.0 percent. Serious errors may be introduced in results if the dissolved oxygen of the sample differs materially from that of the dilution water and corrections are not made. With sample dilutions of less than 20 percent, it is usually sufficient to adjust the samples to 20°C, aerate to saturation, and then assume the dissolved-oxygen concentration is the same as in the dilution water. This eliminates the need for measuring dissolved oxygen on such samples, and also satisfies any immediate oxygen demand. With sample dilutions greater than 20 percent, the dissolved oxygen of the sample should be determined separately.

Calculation of BOD For all practical purposes, the calculation of BOD by the dilution method can be made by the use of either of two simple formulas.

---

[4] J. P. Mascarenhas and G. Klein, "Evaluation of BOD Bottle Cleaning Techniques," *Sewage and Ind. Wastes,* **30:** 976 (1958).

For percent mixtures:

$$\text{BOD (in mg/l)} = \left[(DO_b - DO_i)\frac{100}{\%}\right] - (DO_b - DO_s) \qquad (22\text{-}8)$$

For direct pipeting:

$$\text{BOD (in mg/l)} = \left[(DO_b - DO_i)\frac{\text{vol. of bottle}}{\text{ml sample}}\right] - (DO_b - DO_s) \qquad (22\text{-}9)$$

In these calculations, DO_b and DO_i are the dissolved-oxygen values found in the blanks and the dilutions of the sample, respectively, at the end of the incubation period, and DO_s is the dissolved oxygen originally present in the undiluted sample. From Eqs. (22-8) and (22-9), it becomes obvious that when DO_s approaches the value of DO_b it may be ignored. Further, it becomes unnecessary to correct for the dissolved oxygen of the sample when the BOD exceeds 200 mg/l because the value of $DO_b - DO_s$ seldom exceeds a value of 8. The BOD test is considered to have an accuracy of ± 5 percent, and 8 is within the expected error at and above such levels of BOD.

In BOD analysis it often happens that more than one dilution of the sample yields results that show a depletion of oxygen greater than 2 mg/l and have a residual of more than 0.5 mg/l. Thus it is possible to make more than one calculation of BOD. According to the fundamental concepts upon which the dilution method is based, the calculated values should check within the normal experimental variation of about ± 5 percent. This is not always the case, however. When discrepancies occur, the question naturally arises: Which value is most reliable? In general, the acceptable sample with the greatest oxygen depletion is statistically the best, because with this sample possible errors in the blank dissolved-oxygen determination are minimized.

22-4 RATE OF BIOCHEMICAL OXIDATIONS

For a great many years the BOD reaction was considered to have a rate constant k equal to 0.10 per day at 20°C. This value was established by extensive studies on polluted river waters and domestic wastes in the United States and England. As application of the BOD test spread to the analysis of industrial wastes, and the use of synthetic dilution waters became established, it was soon noted that k values considerably in excess of 0.10 per day were involved and that an appreciable variation occurred for different waste materials. In addition, it was found that k values for domestic wastes varied considerably from day to day and averaged about 0.17 per day, rather than 0.10 per day as originally determined. Also, the k values for effluents from biological waste treatment plants were found to be significantly lower than that of the raw wastes. Another factor affecting BOD rates is temperature. The magnitude of this effect and a method for correction when projecting laboratory determined rates to field conditions were discussed in Sec. 3-11. The importance of the reaction rate k with respect to the BOD developed at any time is shown in Table 22-2.

Table 22-2 Significance of reaction rate k upon BOD

Time, days	Percent of total BOD exerted				
	$k = 0.05$	$k = 0.10$	$k = 0.15$	$k = 0.20$	$k = 0.25$
1	10.9	20.6	29.2	36.9	43.8
2	20.6	37	50	60	68
3	29	50	64	75	82
4	37	60	75	84	90
5	44	68	82	90	94
6	50	75	87	94	97
7	55	80	91	96	98
10	68	90	97	99	99+
20	90	99	99+	99+	99+

From the values shown in Table 22-2, it will be noted that the course of the BOD reaction varies greatly, depending upon the reaction rate. The 5-day BOD values represent about 68 percent of the total BOD when $k = 0.1$ per day and as much as 94 percent when $k = 0.25$ per day. From this it may be concluded that k values must be known if a proper evaluation of ultimate BOD, or L, is to be obtained from 5-day values.

The significance of k in determining the course of the BOD reaction is illustrated in Fig. 22-3. For a waste having a given L value, the BOD values on any given day will vary widely until about 15 days have elapsed. In the past it

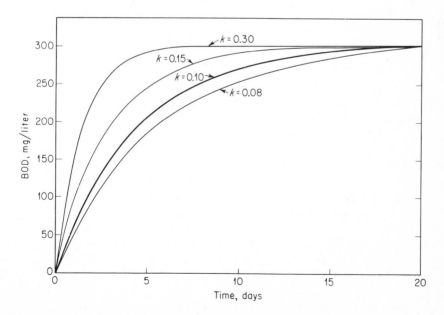

Figure 22-3 Effect of velocity constant k on BOD (for a given L value).

was common practice to interpret 5-day BOD in terms of L values by assuming a k value of 0.10 per day. Figure 22-4 shows how the L value of a sample with a 5-day BOD of 200 varies with the value of k.

The variation in k values leaves considerable room for speculation as to why such differences in rates of reaction occur. Two factors of major importance are involved: (1) the nature of the organic matter and (2) the ability of the organisms present to utilize the organic matter.

Organic matter occurring in domestic and industrial wastes varies greatly in chemical character and availability to microorganisms. That part which exists in true solution is readily available, but that part which occurs in colloidal and coarse suspension must await hydrolytic action before it can diffuse into the bacterial cells where oxidation can occur. The rate of hydrolysis and diffusion are probably the most important factors in controlling the rate of the reaction. It is well known that simple substrates, such as glucose, are removed from solution at very rapid rates, and k values are correspondingly high. More complex materials are removed much more slowly, and k values are lower. In a complex material such as domestic waste, reaction rates are modified greatly by the more complex substances, whereas in an industrial waste containing soluble compounds of simple character, the reaction rate is usually very rapid. Certain organic compounds, such as lignin, are very slowly attacked by bacteria. Some of the synthetic detergents also fall into this category.

A lag period is often noted in the BOD reaction with some industrial wastes, particularly those containing organic compounds of synthetic origin and of

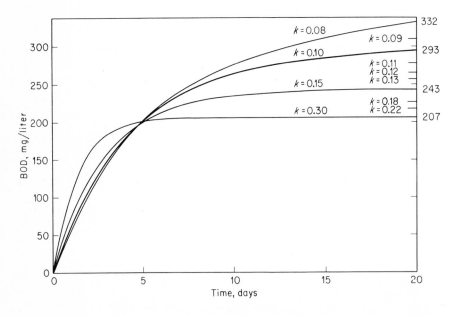

Figure 22-4 Effect of rate k on ultimate BOD (based on assumed 5-day BOD of 200 mg/l).

chemical structure not found in natural materials. The ''seeding'' organisms used in the BOD test may or may not have specific bacteria that can utilize the material as food. If not, the substance will not exert a BOD. Oftentimes only a few bacteria are present that can oxidize the substance, and the rate of oxidation is so slow for a period, possibly several days, that a measurable BOD cannot be detected. With sufficient time, however, the population of the specific bacteria will increase to levels at which the oxidation progresses at normal rates. In cases of this kind, the lag period can usually be overcome by using for ''seeding'' purposes water from a river into which such wastes are discharged. The water should be taken well downstream from the point of discharge. Growths attached to rocks downstream from the point of waste discharge sometimes will furnish an adequate seed. A properly acclimated seed may also be developed in the laboratory by aerating for several days a mixture composed of the neutralized waste and a small portion of domestic wastewater.

The phenomenon of the lag period is sometimes explained on the basis that the ''seeding'' organisms do not have the proper enzyme systems to utilize the organic matter. With time, however, they adapt themselves to the new food supply and furnish the necessary enzymes.

The rate of biochemical reactions, or rate constant k, can be evaluated in a number of ways. All are dependent upon making BOD observations at two or more time intervals so as to establish the ''trajectory'' of the reaction.[5] A treatment of the various methods is beyond the scope of this discussion.

22-5 DISCREPANCY BETWEEN L VALUES AND THEORETICAL OXYGEN DEMAND VALUES

For many years the total BOD, or L, value of organic substances was considered to be equal to the theoretical oxygen demand as calculated from the chemical equation involved [see (Eq. 22-1)]. For example, oxidation of glucose to carbon dioxide and water requires 192 g of oxygen per mole or 1.065 mg of oxygen per milligram of glucose.

$$\underset{180}{C_6H_{12}O_6} + \underset{192}{6O_2} \rightarrow 6CO_2 + 6H_2O \qquad (22\text{-}10)$$

A great deal of BOD work has been done with glucose solutions in a concentration of 300 mg/l. Such a solution has a theoretical oxygen demand of 320 mg/l. Actual BOD measurements made upon such solutions have yielded 20-day[6] and calculated L values in the range of 250 to 285. Thus it is evident that

[5] E. W. Moore, H. A. Thomas, and W. B. Snow, ''Simplified Method for Analysis of BOD Data,'' *Sewage and Ind. Wastes,* **22**: 1343 (1950); H. A. Thomas, ''Graphical Determination of BOD Curve Constants,'' *Water and Sewage Works,* **97**: 123 (1950).

[6] A. Q. Y. Tom, ''Investigations on Improving the BOD Test,'' Sc.D. Thesis, Massachusetts Institute of Technology, 1951.

not all the glucose is completely converted to carbon dioxide and water. The explanation of the discrepancy involves an understanding of the transformations that organic matter undergoes when subjected to biological attack.

In order for organic matter to be oxidized by bacteria, it must serve as food material from which the organisms can derive energy for growth and reproduction. This means that part of the organic matter is converted to cell tissue. The part that is converted to cell tissue will remain unoxidized until such time as the organisms must draw upon cell tissue to derive energy to maintain life (endogenous respiration). When bacteria die they become food material for other bacteria, and a further transformation to carbon dioxide, water, and cell tissue occurs. Living bacteria, as well as dead ones, serve as food material for higher organisms such as protozoans. In each transformation further oxidation occurs, but in the final analysis there remains a certain amount of organic matter that is quite resistant to further biological attack. This is commonly referred to as humus and represents an amount of organic matter corresponding to the discrepancy between the total BOD and the theoretical oxygen demand.

22-6 DISCREPANCY BETWEEN OBSERVED RATES AND FIRST-ORDER RATES

BOD studies[7] using soluble organic materials have shown certain limitations of the assumption that the reaction follows "first-order" kinetics. In many cases the exertion of carbonaceous BOD has been observed to occur in two phases similar to that pictured in Fig. 22-2. The second phase, however, is not due to nitrification as in the illustration, but is speculated to result from the secondary action of protozoa. The sequence of biochemical action can be explained as follows. During the first one or two days of incubation, the soluble organic material is rapidly consumed by the bacteria, about 30 to 50 percent is oxidized, and the remainder is converted to bacterial cells as discussed in Sec. 22-5. When this conversion is completed, a "plateau" representing a reduced rate of oxidation occurs and is attributed to the endogenous respiration phase of bacterial metabolism. Within a day or two a secondary rise in the rate of oxidation occurs and is attributed to a rise in the population of protozoa which are predators and consume the bacteria for food. The occurrence and duration of a noticeable plateau between these phases depend upon the interval of time between the peak of the bacterial population and that of the protozoan population. These observations indicate protozoa can play a significant role in the BOD test.

While these observations indicate caution should be used in the interpretation of BOD data, they do not invalidate the use of "first-order" kinetics for

---

[7] A. W. Busch, "BOD Progression in Soluble Substrates," *Sewage and Ind. Wastes,* **30:** 1336 (1958); M. N. Bhatla and A. F. Gaudy, "Role of Protozoa in the Diphasic Exertion of BOD," *J. Sanit. Eng. Div., Am. Soc. Civil Engrs.,* **91:** SA3, 63 (1965).

solving practical stream pollution problems involving complex wastes. The BOD exertion curves for most such wastes, which contain both soluble and particulate organic matter, tend to be "composite" curves representing the summation of the oxidations for each individual compound. Since the rates of breakdown of different compounds vary so widely, the bumps and valleys in the curve tend to be leveled out, with the result that a relatively smooth "first-order" type of curve is obtained.

22-7 APPLICATION OF BOD DATA

BOD data have wide application in environmental engineering practice. It is the principal test applied to domestic and industrial wastes to determine strength in terms of oxygen required for stabilization. It is the only test applied that gives a measure of the amount of biologically oxidizable organic matter present that can be used to determine the rates at which oxidation will occur, or BOD will be exerted, in receiving bodies of water. BOD is therefore the major criterion used in stream pollution control where organic loading must be restricted to maintain desired dissolved-oxygen levels. The determination is used in studies to measure the purification capacity of streams and serves regulatory authorities as a means of checking on the quality of effluents discharged to such waters.

Information concerning the BOD of wastes is an important consideration in the design of treatment facilities. It is a factor in the choice of treatment method and is used to determine the size of certain units, particularly trickling filters and activated sludge units. After treatment plants are placed in operation, the test is used to evaluate the efficiency of various units.

Many municipalities and sewer authorities finance wastewater treatment operations through sewer rental charges. Public Law 92-500 requires that industries contributing wastes to municipal systems financed by Federal funds contribute a fair share of the operation and maintenance costs. BOD is one of the factors normally used in calculating such charges, particularly where secondary treatment employing biological processes is employed.

PROBLEMS

22-1 What use is made of the BOD test in water pollution control?

22-2 List five requirements which must be complied with in order to obtain reliable BOD data.

22-3 List five requirements of a satisfactory dilution water for BOD work.

22-4 What purpose or purposes are served by each of the following in BOD dilution water: (a) $FeCl_3$, (b) $MgSO_4$, (c) K_2HPO_4, (d) NH_4Cl, and (e) $CaCl_2$?

22-5 Explain how a sample of river water having a temperature below 20°C should be pretreated in preparation for BOD analysis.

22-6 Explain how a proper seed might be obtained in order to determine the BOD of an industrial waste which is not readily oxidized biologically.

22-7 Why is it best not to use the normal BOD test to estimate the possible nitrogenous oxygen demand potential of a waste?

22-8 What three methods can be used to control nitrification in the 5-day BOD test?

22-9 What justification does the engineer have for using first-order reaction kinetics to describe the complex biochemical processes occurring in the BOD test?

22-10 What factors affect the rate of biochemical oxidation in the BOD test?

22-11 What significant part do protozoa play in the BOD test?

22-12 The following data were obtained in the analysis of an industrial waste: After 5 days' incubation at 20°C, the residual dissolved oxygen in blanks was 7.80 mg/l, and in a 0.1 percent dilution of the waste was 2.80 mg/l.

(a) What was the 5-day BOD of the waste?

(b) How many pounds of 5-day BOD are contained in 10,000 gallons of the waste?

22-13 The following dissolved-oxygen values were found after 5 days of incubation in 310-ml BOD bottles: 7.7, 7.9, and 7.9 in three blank samples; 6.5, 4.0, and 0.5 mg/l in bottles containing 2, 5, and 10 ml of sample, respectively. The 0-day dissolved oxygen of the sample was 0.0 mg/l. What was the most probable 5-day BOD of the waste?

22-14 A sample of wastewater was incubated for seven days at 20°C and showed a BOD of 208 mg/l.

(Assume k equals 0.15/day.)

(a) Calculate its 5-day BOD.

(b) Calculate its 10-day BOD.

(c) Calculate the ultimate BOD.

22-15 Two wastes of equal volume have identical 5-day BOD values. One has a k of 0.19/day and the other a k of 0.25/day. Draw a graph to show how the oxygen sag curves would differ in the receiving streams, assuming that stream flows and other conditions are identical.

22-16 Approximately what would be the ultimate and 5-day carbonaceous BOD values for samples containing 200 mg/l of the following materials, assuming that they were readily oxidized in the BOD test?

(Assume that k equals 0.20/day.)

(a) Acetic acid

(b) Butanol

(c) Glucose

(d) Benzoic acid

(e) Alanine

TWENTY-THREE

CHEMICAL OXYGEN DEMAND

23-1 GENERAL CONSIDERATIONS

The chemical oxygen demand (COD) test is widely used as a means of measuring the pollutional strength of domestic and industrial wastes. This test allows measurement of a waste in terms of the total quantity of oxygen required for oxidation to carbon dioxide and water in accordance with Eq. (22-1). It is based upon the fact that all organic compounds, with a few exceptions, can be oxidized by the action of strong oxidizing agents under acid conditions. The amino nitrogen will be converted to ammonia nitrogen as indicated in Eq. (22-1). However, organic nitrogen in higher oxidation states will be converted to nitrates.

During the determination of COD, organic matter is converted to carbon dioxide and water regardless of the biological assimilability of the substances. For example, glucose and lignin are both oxidized completely. As a result, COD values are greater than BOD values and may be much greater when significant amounts of biologically resistant organic matter is present. Wood-pulping wastes are excellent examples because of their high lignin content. For reasons presented in Sec. 22-5, the COD of materials such as glucose is always greater than the L value.

One of the chief limitations of the COD test is its inability to differentiate between biologically oxidizable and biologically inert organic matter. In addition, it does not provide any evidence of the rate at which the biologically active material would be stabilized under conditions that exist in nature.

The major advantage of the COD test is the short time required for evaluation. The determination can be made in about 3 h rather than the 5 days required for the measurement of BOD. For this reason it is used as a substitute for the BOD test in many instances. COD data can often be interpreted in terms of BOD values after sufficient experience has been accumulated to establish reliable correlation factors.

23-2 HISTORY OF THE COD TEST

Chemical oxidizing agents have long been used for measuring the oxygen demand of polluted waters. Potassium permanganate solutions were used for many years, and the results were referred to as *oxygen consumed* from permanganate. The oxidation caused by permanganate was highly variable with respect to various types of compounds, and the degree of oxidation varied considerably with the strength of reagent used. Oxygen-consumed values were always considerably less than 5-day BOD values. This fact demonstrated the inability of permanganate to carry the oxidation to any particular end point.

Ceric sulfate, potassium iodate, and potassium dichromate are other oxidizing agents that have been studied extensively for the determination of chemical oxygen demand. Potassium dichromate has been found to be the most practical of all, since it is capable of oxidizing a wide variety of organic substances almost completely to carbon dioxide and water. Because all oxidizing agents must be used in excess, it is necessary to measure the amount of excess remaining at the end of the reaction period in order to calculate the amount actually used in the oxidation of the organic matter. It is relatively easy to measure any excess of potassium dichromate, an important point in its favor.

In order for potassium dichromate to oxidize organic matter completely, the solution must be strongly acidic and at an elevated temperature. As a result, volatile materials originally present and those formed during the digestion period are lost unless provision is made to prevent their escape. Reflux condensers are ordinarily used for this purpose and allow the sample to be boiled without significant loss of volatile organic compounds.

Certain organic compounds, particularly low-molecular-weight fatty acids, are not oxidized by dichromate unless a catalyst is present. It has been found that silver ion acts effectively in this capacity. Aromatic hydrocarbons and pyridine are not oxidized under any circumstances.

23-3 CHEMICAL OXYGEN DEMAND BY DICHROMATE

Potassium dichromate is a relatively cheap compound which can be obtained in a high state of purity. The analytical-reagent grade, after drying at $103°C$, can be used to prepare solutions of an exact normality by direct weighing and dilution to the proper volume. The dichromate ion is a very potent oxidizing agent in solutions that are strongly acid. The reaction involved may be represented in a general way as follows:

$$C_nH_aO_b + c\,Cr_2O_7^{2-} + 8cH^+ \xrightarrow{\Delta} nCO_2 + \frac{a + 8c}{2}\,H_2O + 2c\,Cr^{3+} \quad (23\text{-}1)$$

where $c = \frac{2}{3} n + \frac{a}{6} - \frac{b}{3}$

For these and other reasons mentioned previously dichromate approaches an ideal reagent for the measurement of COD.

Selection of Normality

COD results are reported in terms of milligrams of oxygen. Since the equivalent weight of oxygen is 8, it would seem logical to use a $N/8$ or $0.125 N$ solution of oxidizing agent in the determination, so that results can be calculated in accordance with the general procedure described in Sec. 14-1. Experience with the test has shown that it has sufficient sensitivity to allow the use of a stronger solution of dichromate, and a $N/4$ or $0.25 N$ solution is recommended. This allows the use of larger samples by doubling the range of COD that can be measured in the test procedure, since each milliliter of a $0.25 N$ solution of dichromate is equivalent to 2 mg of oxygen.

Measurement of Excess Oxidizing Agent

In any method of measuring COD, an excess of oxidizing agent must be present to ensure that all organic matter is oxidized as completely as is within the power of the reagent. This requires that a reasonable excess be present in all samples. It is necessary, of course, to measure the excess in some manner so that the actual amount reduced can be determined. A solution of a reducing agent is ordinarily used.

Nearly all solutions of reducing agents are gradually oxidized by oxygen dissolved from the air unless special care is taken to protect them from oxygen. Ferrous ion is an excellent reducing agent for dichromate. Solutions of it can be best prepared from ferrous ammonium sulfate which is obtainable in rather pure and stable form. In solution, however, it is slowly oxidized by oxygen, and standardization is required each time the reagent is to be used. The standardization is made with the $0.25 N$ solution of dichromate. The reaction between ferrous ammonium sulfate and dichromate may be represented as follows:

$$6Fe^{2+} + Cr_2O_7^{2-} + 14H^+ \rightarrow 6Fe^{3+} + 2Cr^{3+} + 7H_2O \qquad (23\text{-}2)$$

Blanks

Both the COD and BOD tests are designed to measure oxygen requirements by oxidation of organic matter present in the samples. It is important, therefore, that no organic matter from outside sources be present if a true measure of the amount present in the sample is to be obtained. Since it is impossible to exclude extraneous organic matter in the BOD test and impractical to do so in the COD test, blank samples are required in both determinations.

Indicator

A very marked change in oxidation-reduction potential (ORP) occurs at the end point of all oxidation-reduction reactions. Such changes may be readily detected by electrometric means if the necessary equipment is available. Oxidation-reduction indicators may also be used; Ferroin (ferrous 1,10-phenanthroline sulfate) is an excellent one to indicate when all dichromate has been reduced by ferrous ion. It gives a very sharp color change that is easily detected in spite of the green color produced by the Cr^{3+} formed on reduction of the dichromate.

Calculations

Although an oxidizing agent is used in the measurement of COD, it does not figure directly in the calculation of COD. This is because a solution of a reducing agent must be used to determine how much of the oxidizing agent was used, and it is simpler to relate everything to the reducing agent in this case, because its strength varies from day to day and its normality is seldom, if ever, exactly equal to 0.25 N.

Calculation of COD is made using the following formula:

$$\text{COD (mg/l)} = \frac{8000 \text{ (blank titr.} - \text{sample titr.)[norm. Fe(NH}_4)_2(\text{SO}_4)_2]}{\text{ml sample}} \quad (23\text{-}3)$$

Alternate Procedures

The COD test is precise and accurate for samples with a COD of 50 mg/l or greater. For more dilute samples it is preferred that a more dilute dichromate solution be used so that a significant relative difference between the quantity of dichromate added and that remaining after refluxing results. With dilute samples, care must be exercised to avoid sample contamination, and good analytical techniques must be used if reasonably accurate results are to be obtained. It is also important in any modification that the volume of concentrated sulfuric acid to volume of sample plus dichromate solution be maintained at a 1:1 ratio. If it is smaller, the oxidizing power of the solution will decrease significantly, while if it is larger, the blank consumption of dichromate becomes excessive.

23-4 INORGANIC INTERFERENCES

Certain reduced inorganic ions can be oxidized under the conditions of the COD test and thus can cause erroneously high results to be obtained. Chlorides cause the most serious problem because of their normally high concentration in most wastewaters,

$$6Cl^- + Cr_2O_7^{2-} + 14H^+ \rightarrow 3Cl_2 + 2Cr^{3+} + 7H_2O \quad (23\text{-}4)$$

Fortunately this interference can be eliminated by the addition of mercuric sulfate to the sample prior to the addition of the other reagents. The mercuric ion combines with the chloride ions to form a poorly ionized mercuric chloride complex (see Sec. 2-13).

$$Hg^{2+} + 2Cl^- \rightleftharpoons HgCl_2 \qquad (\beta_2 = 1.7 \times 10^{13}) \qquad (23\text{-}5)$$

In the presence of excess mercuric ions the chloride ion concentration is so small that it is not oxidized to any extent by dichromate. Exception to the use of mercuric salts can be raised, however, unless provisions are made to prevent discharge of the mercury containing samples to the sewer.[1]

Nitrites are oxidized to nitrates and this interference can be overcome by the addition of sulfamic acid to the dichromate solution. However, significant amounts of nitrite seldom occur in wastes or in natural waters. This also holds true for other possible interferences such as ferrous iron and sulfides.

23-5 APPLICATION OF COD DATA

The COD test is used extensively in the analysis of industrial wastes. It is particularly valuable in surveys designed to determine and control losses to sewer systems. Results may be obtained within a relatively short time and measures taken to correct errors on the day they occur. In conjunction with the BOD test, the COD test is helpful in indicating toxic conditions and the presence of biologically resistant organic substances. The test is widely used in the operation of treatment facilities because of the speed with which results can be obtained.

PROBLEMS

23-1 Give four different applications for the COD analysis in environmental engineering practice.

23-2 What general groups of organic compounds are not oxidized in the COD test?

23-3 Indicate whether COD results would probably be higher, lower, or the same as the true value under the following conditions, and briefly explain why: (a) mercuric sulfate was not added; (b) silver sulfate was not added; (c) the ferrous ammonium sulfate was assumed to have the same normality as it did two weeks prior to the current analysis.

23-4 Why do the COD analysis and BOD analysis usually give different results for the same waste?

23-5 What is the theoretical COD of samples containing 300 mg/l of (a) ethyl alcohol (b) phenol, and (c) leucine?

23-6 (a) Estimate the COD of a solution containing 500 mg/l of butanol.

(b) If this compound were readily degradable biologically, about what would you expect the 5-day BOD to be?

[1] R. B. Dean, R. T. Williams and R. H. Wise, *Environmental Sci. and Technol.* **5:** 1044 (1971).

23-7 What could be inferred from the following analytical results concerning the relative ease of biodegradability of each waste?

	mg/l	
Waste	5-day BOD	COD
A	240	300
B	100	500
C	120	240

TWENTY-FOUR

NITROGEN

24-1 GENERAL CONSIDERATIONS

The compounds of nitrogen are of great interest to environmental engineers because of the importance of nitrogen compounds in the atmosphere and in the life processes of all plants and animals. The chemistry of nitrogen is complex because of the several oxidation states that nitrogen can assume and the fact that changes in oxidation state can be brought about by living organisms. To add even more interest, the oxidation state changes wrought by bacteria can be either positive or negative, depending upon whether aerobic or anaerobic conditions prevail.

From the viewpoint of the inorganic chemist, nitrogen can exist in seven oxidation states, and compounds in all are of interest to him.

$$\overset{-III}{NH_3} - \overset{O}{N_2} - \overset{I}{N_2O} - \overset{II}{NO} - \overset{III}{N_2O_3} - \overset{IV}{NO_2} - \overset{V}{N_2O_5}$$

As far as is known, compounds of nitrogen in I, II, and IV oxidation states have little if any significance in biological processes. All other forms are important, and the chemistry of nitrogen of interest to environmental engineering may be summarized as follows:

$$\overset{-III}{NH_3} - \overset{O}{N_2} - \overset{III}{N_2O_3} - \overset{V}{N_2O_5}$$
$$\uparrow$$
Organic
derivatives

N_2O_3 and N_2O_5 are the acid anhydrides of nitrous and nitric acids.

The relationships that exist between the various forms of nitrogen compounds and the changes that can occur in nature are best illustrated by a diagram known as the nitrogen cycle, shown in Fig. 24-1. From it, it will be seen that the atmosphere serves as a reservoir from which nitrogen is constantly removed by the action of electrical discharge and nitrogen-fixing bacteria and algae. During electrical storms large amounts of nitrogen are oxidized to N_2O_5, and its union with water produces HNO_3 which is carried to the earth in the rain. Nitrates are also produced by direct oxidation of nitrogen or of ammonia in the production of commercial fertilizers. The nitrates serve to fertilize plant life and are converted to proteins.

$$NO_3^- + CO_2 + \text{green plants} + \text{sunlight} \rightarrow \text{protein} \qquad (24\text{-}1)$$

Atmospheric nitrogen is also converted to proteins by nitrogen-"fixing" bacteria and certain algae.

$$N_2 + \text{special bacteria or certain algae} \rightarrow \text{protein} \qquad (24\text{-}2)$$

In addition, ammonia and ammonium compounds are applied to soils to supply plants with ammonia for further production of proteins. Urea is one of the popular ammonium compounds because it releases ammonia gradually.

$$NH_3 + CO_2 + \text{green plants} + \text{sunlight} \rightarrow \text{protein} \qquad (24\text{-}3)$$

Animals and human beings are incapable of utilizing nitrogen from the atmosphere or from inorganic compounds to produce proteins. They are de-

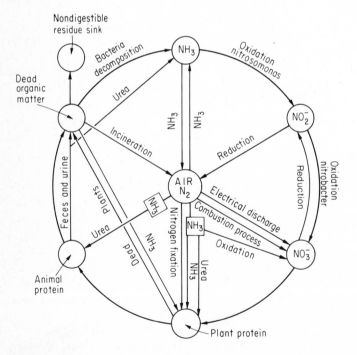

Figure 24-1 The nitrogen cycle.

pendent upon plants, or other animals that feed upon plants, to provide protein, with the exception of ruminants. The multiple-stomached animals are capable of producing part of their protein requirement from carbohydrate matter and urea through bacterial action. Within the animal body, protein matter is used largely for growth and repair of muscle tissue. Some may be used for energy purposes. In any event, nitrogen compounds are released in the waste products of the body during life. At death the proteins stored in the body become waste matter for disposal. The urine contains the nitrogen resulting from the metabolic breakdown of proteins. The nitrogen exists in urine principally as urea which is hydrolyzed rather rapidly by the enzyme urease to ammonium carbonate:

$$\begin{matrix} NH_2 \\ / \\ C{=}O \\ \backslash \\ NH_2 \end{matrix} + 2H_2O \xrightarrow[\text{urease}]{\text{enzyme}} (NH_4)_2CO_3 \qquad (24\text{-}4)$$

The feces of animals contain appreciable amounts of unassimilated protein matter (organic nitrogen). It and the protein matter remaining in the bodies of dead animals and plants are converted in large measure to ammonia by the action of saprophytic bacteria, under aerobic or anaerobic conditions:

$$\text{Protein (organic N)} + \text{bacteria} \rightarrow NH_3 \qquad (24\text{-}5)$$

Some nitrogen always remains in nondigestible matter and becomes part of the nondigestible residue sink. As such it becomes part of the *detritus* in water or sediments, or the *humus* in soils.

The ammonia released by bacterial action on urea and proteins may be used by plants directly to produce plant protein. If it is released in excess of plant requirements, the excess is oxidized by autotrophic nitrifying bacteria. The *Nitrosomonas* group, known as the nitrite formers, convert ammonia under aerobic conditions to nitrites and derive energy from the oxidation:

$$2NH_3 + 3O_2 \xrightarrow{\text{bacteria}} 2NO_2^- + 2H^+ + 2H_2O \qquad (24\text{-}6)$$

The nitrites are oxidized by the *Nitrobacter* group of nitrifying bacteria, which are also called the nitrate formers.

$$2NO_2^- + O_2 \xrightarrow{\text{bacteria}} 2NO_3^- \qquad (24\text{-}7)$$

The nitrates formed may serve as fertilizer for plants. Nitrates produced in excess of the needs of plant life are carried away in water percolating through the soil because the soil does not have the ability to hold them. This frequently results in relatively high concentrations of nitrates in ground-waters, and is an extensive problem in Illinois and other midwestern states.

Under anaerobic conditions nitrates and nitrites are both reduced by a process called denitrification. Presumably nitrates are reduced to nitrites, and then reduction of nitrites occurs. Reduction of nitrites is carried all the way to ammonia by a few bacteria, but most of them carry the reduction to nitrogen gas, which escapes to the atmosphere. This constitutes a serious loss of fertilizing matter in soils when anaerobic conditions develop. The formation of nitrogen gas by reduction of nitrates is sometimes a problem in the activated sludge process of wastewater treatment. Prolonged detention of activated sludge in final settling tanks allows formation of sufficient nitrogen gas to buoy the sludge, if nitrates are present in adequate amounts. This is often referred to as the "rising" sludge problem.

Advantage is taken of denitrification in one proposed scheme for removing nitrogen from wastes where this is required to prevent undesirable growths of algae and other aquatic plants in receiving waters. Ammonia and organic nitrogen are first biologically converted to nitrites and nitrates by aerobic treatment. The waste is then placed under anoxic conditions, where denitrification converts the nitrites and nitrates to nitrogen gas, which escapes to the atmosphere. For denitrification to occur, organic matter must be present, and is oxidized for energy while the nitrogen is being reduced. Methyl alcohol is a favorite form of organic matter but, due to its increased cost, other materials are being extensively investigated.

24-2 ENVIRONMENTAL SIGNIFICANCE OF NITROGEN DATA

Analyses for nitrogen in its various forms have been performed on potable and polluted waters ever since man became convinced that water was a vehicle for the transmission of disease. The determinations served as one basis of judging the sanitary quality of water for a great many years. Today nitrogen analyses are performed largely for other reasons.

An Indicator of Sanitary Quality

It has long been known that polluted waters will purify themselves, provided that they are allowed to age for sufficient periods of time. The hazard to health or the possibility of contracting disease by drinking such waters decreases markedly with time and temperature increase, as shown in Fig. 24-2.

Prior to the development of bacteriological tests for determining the sanitary quality of water (about 1893), environmental engineers and others concerned with the public health were largely dependent upon chemical tests to provide circumstantial evidence of the presence of contamination. The chloride test was one of these (see Sec. 20-2), but it gave no evidence of how recently the contamination had occurred. Chemists working with wastes and freshly polluted waters learned that most of the nitrogen is originally present in the form of organic (protein) nitrogen and ammonia. As time progresses, the organic nitrogen is gradually converted to ammonia nitrogen, and later on, if aerobic conditions are present, oxidation of ammonia to nitrites and nitrates occurs. The

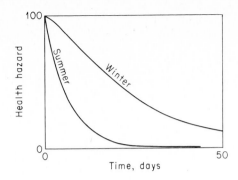

Figure 24-2 Surface waters; health hazard in relation to age of pollution.

progression of events was found to occur somewhat as shown in Fig. 24-3, and more refined interpretations of the sanitary quality of water were based upon this knowledge. For example, waters that contained mostly organic and ammonia nitrogen were considered to have been recently polluted and therefore of great potential danger. Waters in which most of the nitrogen was in the form of nitrates were considered to have been polluted a long time previously and therefore offered little threat to the public health. Waters with appreciable amounts of nitrites were of highly questionable character. The bacteriological

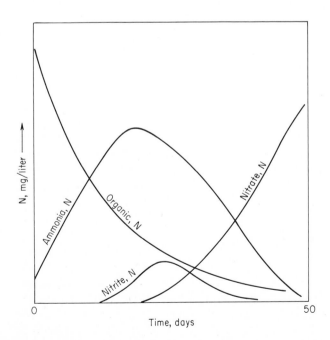

Figure 24-3 Changes occurring in forms of nitrogen present in polluted water under aerobic conditions.

test for coliform organisms provides circumstantial evidence of much greater reliability concerning the hygienic safety of water, and it has eliminated the need for extended nitrogen analysis in most water supplies.

In 1940 it was found that drinking waters with high nitrate content often caused methemoglobinemia in infants. From extended investigations in Iowa, Minnesota, and Ohio, where the problem has been most acute, it has been concluded that the nitrate content should be limited.[1] The proposed Environmental Protection Agency drinking water regulations require that the nitrate concentration in terms of nitrogen not exceed 10 mg/l in public water supplies.

Nutritional and Related Problems

All biological treatment processes employed by environmental engineers are dependent upon reproduction of the organisms employed, as discussed in Sec. 7-6. In planning waste treatment facilities it becomes important to know whether the waste contains sufficient nitrogen for the organisms. If not, any deficiency must be supplied from outside sources. Determinations of ammonia and organic nitrogen are normally made to obtain such data.

Nitrogen is one of the fertilizing elements essential to the growth of algae. Such growth is often stimulated to an undesirable extent in bodies of water that receive either treated or untreated effluents, because of the nitrogen and other fertilizing matter contributed by them. Nitrogen analyses are an important means of gaining information on this problem.

Oxidation in Rivers and Estuaries

The autotrophic conversion of ammonia to nitrites and nitrates requires oxygen, as indicated in Eqs. (24-6) and (24-7), and so the discharge of ammonia nitrogen and its subsequent oxidation can seriously reduce the dissolved-oxygen levels in rivers and estuaries, especially where long residence times required for the growth of the slow-growing nitrifying bacteria are available. Also, these organisms are produced in large numbers by highly efficient aerobic biological waste treatment systems, and their discharge with the treated effluent can cause rapid nitrification to occur in waterways. Disinfection of effluents with chlorine has minimized this problem. Nitrogen analyses are important in assessing the possible significance of the problem.

Control of Biological Treatment Processes

Determinations of nitrogen are often made to control the degree of purification produced in biological treatment. With the use of the BOD test, it has been

---

[1] K. F. Maxcy, Report on Relation of Nitrate Nitrogen Concentration in Well Waters to the Occurrence of Methemoglobinemia in Infants, *Natl. Acad. Sci.-Research Council Sanit. Eng. and Environment Bull.* 264, 1950.

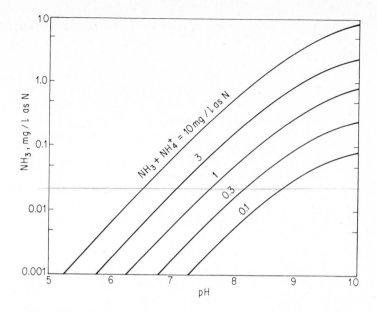

Figure 24-4 The effect of pH and ammonia nitrogen concentration ($NH_3 + NH_4^+$) on the concentration of free ammonia in water.

learned that effective stabilization of organic matter can be accomplished without carrying the oxidation into the nitrification stage. This results in a material saving of time and air required.

Nitrogen control has become an important consideration in the design and operation of wastewater treatment plants for reasons cited above. In some states limitations have been imposed because of suspected toxic effects upon fish life. It is well known that molecular ammonia is toxic but that the ammonium ion is not. Since the relationship between the two is pH-dependent,

$$NH_3 + H^+ \rightleftharpoons NH_4^+$$

a discussion is in order. Figure 24-4 shows the relationship between free ammonia and ammonium ion that exists for several concentrations of ammonia nitrogen over the pH range of interest in most natural waters. Free ammonia in concentrations above about 0.2 mg/l can cause fatalities in several species of fish. Applying the usual safety factor, a National Academy of Sciences–National Academy of Engineering Committee has recommended that no more than 0.02 mg/l free ammonia be permitted in receiving waters.[2] From this and the data presented in Fig. 24-4, it is safe to conclude that ammonia toxicity will

[2] Committee on Water Quality Criteria, *Water Quality Criteria, 1972,* Superintendent of Documents, Washington, D.C. (1972).

not be a problem in receiving waters with pH below 8 and ammonia nitrogen concentrations less than about 1 mg/l.

Control of ammonia for the above reasons can be accomplished by actual removal of ammonia or by nitrification. In some cases, limitations are placed upon the amount of total nitrogen that can be present in an effluent. This condition can be met by removal of ammonia in some instances but more often it requires nitrification and denitrification. Because of these new requirements, methods of measuring all forms of nitrogen have become extremely important.

24-3 METHODS OF ANALYSIS

Since nitrogen exists in four forms that are of interest to environmental engineers, a discussion of the determination of each form is required. Because of differences in the amounts present in potable and polluted waters, methods applied to water may differ somewhat from those applied to wastes. It is customary to report all results in terms of nitrogen so that values may be interpreted from one form to another without use of a factor.

Ammonia Nitrogen

All nitrogen that exists as ammonium ion or in the equilibrium

$$NH_4^+ \rightleftharpoons NH_3 + H^+ \tag{24-8}$$

is considered to be ammonia nitrogen.

By direct Nesslerization In samples that have been properly clarified by a pretreatment method using zinc sulfate and sodium hydroxide, it is possible to obtain a measure of the amount of ammonia nitrogen by treatment with Nessler's reagent, which is a strongly alkaline solution of potassium mercuric iodide. Its formula may be represented as K_2HgI_4 or simply $2KI \cdot HgI_2$. It combines with NH_3 in alkaline solution to form a yellowish-brown colloidal dispersion whose intensity of color is directly proportional to the amount of NH_3 originally present. The reaction may be represented by the equation

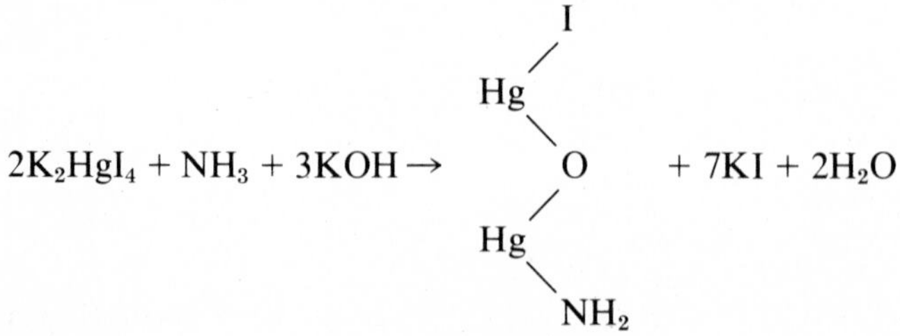

$$2K_2HgI_4 + NH_3 + 3KOH \rightarrow \begin{matrix} & I \\ & / \\ & Hg \\ & \backslash \\ & O \\ & / \\ & Hg \\ & \backslash \\ & NH_2 \end{matrix} + 7KI + 2H_2O \tag{24-9}$$

Yellowish-brown colloid

The color developed with Nessler's reagent is easily matched by eye, and many people prefer to make visual comparisons rather than to depend upon photometric methods. Nessler's reagent prepared by different procedures and on different occasions varies somewhat in its sensitivity to ammonia. It is usually good technique to prepare fresh visual standards or a new calibration curve each time a change is made in Nessler's reagent.

By distillation The direct Nesslerization procedure is subject to serious error from extraneous color and turbidity. It is impractical to use it on many samples. The distillation procedure is used to separate the ammonia from interfering substances, and measurement of ammonia nitrogen can then be made in a number of ways.

Ammonium ion exists in equilibrium with ammonia and hydrogen ion, as shown in Eq. (24-8). At pH levels above 7, the equilibrium is displaced far enough to the right so that ammonia is liberated as a gas along with the steam produced when a sample is boiled, as follows:

$$NH_4^+ \xrightarrow{\Delta} NH_3\uparrow + H^+ \tag{24-10}$$

If provision is made to condense the steam, ammonia will be absorbed in the condensate. Removal of ammonia allows the hydrogen ions released in the decomposition of ammonium ion to accumulate in the residue, and a decrease in pH will result unless a buffer is present to combine with the hydrogen ions. A phosphate buffer is added to maintain the pH in a range of 7.2 to 7.4. Higher pH levels are not recommended because of the danger of some ammonia being released from organic sources at the temperature of boiling water. Experience has shown that essentially all the free ammonia will be expelled by the time 200 ml of water has been distilled from solutions whose pH is maintained between 7.2 to 7.4, when samples of 500 ml or less are used.

For water samples that contain small amounts of ammonia nitrogen, the usual procedure is to measure the amount of nitrogen in the distillate by use of Nessler's reagent, as described above. The calculation of ammonia nitrogen in terms of milligrams per liter must take into consideration the volume of distillate as well as the sample size. Many analysts find it convenient to distill an amount that exceeds 200 ml to avoid close attention at the end of the distillation. The calculation can be made, provided that the volume of distillate is known, by the following method:

$$\text{mg/l } NH_3\text{–N} = \frac{V_D}{V_{DN}} \times N \times \frac{1000}{s} \tag{24-11}$$

where V_D is the ml of distillate, V_{DN} is the ml of distillate actually Nesslerized, N is the mg of NH_3–N found in the Nesslerized portion of the distillate, and s is the ml of sample used for the distillation.

When samples contain more than 0.2 mg/l of ammonia nitrogen, as is the case with domestic and many industrial wastes, it is best to absorb the ammonia in boric acid. The chemistry involved may be represented as follows: Boric acid is an excellent buffer. It combines with ammonia to form ammonium and borate ions, as shown in the equation.

$$NH_3 + H_3BO_3 \rightarrow NH_4^+ + H_2BO_3^- \tag{24-12}$$

This causes the pH to increase somewhat, as was discussed under buffer action in Sec. 5-6, but the pH is held in a favorable range for absorption of ammonia by the use of an excess of boric acid. The ammonia may then be measured by Nesslerization or by a back titration with a strong acid. Actually the acid measures the amount of borate ion present in the solution as follows:

$$H_2BO_3^- + H^+ \rightarrow H_3BO_3 \tag{24-13}$$

When the pH of the boric acid solution has been decreased to its original value, an amount of strong acid equivalent to the ammonia has been added. The titration is most easily conducted by electrometric methods which eliminate the need for internal indicators. The proper pH for the end point is best determined by diluting the specified volume of boric acid solution with ammonia-free distilled water in an amount equal to the volume of distillate desired and measuring the pH of the mixture.

Organic Nitrogen

All nitrogen present in organic compounds may be considered organic nitrogen. This includes the nitrogen in amino acids, amines, amides, imides, nitro derivatives, and a number of other compounds. Most of these have very little significance in water analysis unless specific industrial wastes are involved.

Most of the organic nitrogen that occurs in domestic wastes is in the form of proteins or their degradation products: polypeptides and amino acids. Therefore the methods employed in water analysis have been designed to ensure measurement of these forms without particular regard to other organic forms. Actually most forms except the nitrogen in nitro compounds are measured by the method used.

Most organic compounds containing nitrogen are derivatives of ammonia, and destruction of the organic portion of the molecule by oxidation frees the nitrogen as ammonia. The Kjeldahl method employing sulfuric acid as the oxidizing agent is standard procedure. A catalyst is ordinarily needed to hasten the oxidation of some of the more resistant organic materials. The reaction that occurs may be illustrated by the oxidation of alanine (α-aminopropionic acid). In the reaction, carbon and hydrogen are oxidized to carbon dioxide and water, while the sulfate ion is reduced to sulfur dioxide. The amino group is released as ammonia but, of course, cannot escape from the acid environment and is held as an ammonium salt.

$$\underset{\text{Alanine}}{CH_3CHNH_2COOH} + 7H_2SO_4 \xrightarrow[\text{cat}]{\Delta} 3CO_2 + 6SO_2 + 8H_2O + NH_4HSO_4 \tag{24-14}$$

The oxidation proceeds rapidly at temperatures slightly above the boiling point of sulfuric acid (340°C). The boiling point of the acid is increased by addition of sodium or potassium sulfate.

The complete digestion of organic matter is essential if all organic nitrogen is to be released as ammonia. A misinterpretation often occurs as to what conditions exist when digestion is complete, and some explanation seems in order. This can best be done by listing the changes that samples undergo during digestion.

1. Excess water is expelled, leaving concentrated sulfuric acid to attack the organic matter.
2. Copious white fumes form in the flask at the time sulfuric acid reaches its boiling point. Digestion is just beginning at this stage.
3. The mixture turns black, owing to the dehydrating action of the sulfuric acid on the organic matter.
4. Oxidation of carbon occurs. Boiling during this period is characterized by extremely small bubble formation due to the release of carbon dioxide and sulfur dioxide.
5. Complete destruction of organic matter is indicated by a clearing of the sample to a "water-clear" solution.
6. Digestion should be continued for at least 20 min after the samples appear clear to ensure complete destruction of all organic matter.

Once the organic nitrogen has been released as ammonia nitrogen, it may be measured in a manner similar to that described above in the discussion of ammonia nitrogen. The excess sulfuric acid must be neutralized and the pH of the sample adjusted to 7 or above. Usually phenolphthalein indicator is used and the pH raised well above 8. Under such conditions, the equilibrium shown in Eq. (24-8) is displaced greatly to the right, and ammonia can be distilled with ease. The ammonia nitrogen in the distillate may be measured by Nesslerization or by absorption in boric acid and back titration with standard acid. Calculation of organic nitrogen is made in the same manner as for ammonia nitrogen.

Many water chemists prefer to use a $N/14$ solution of sulfuric acid for the measurement of ammonia and organic nitrogen when ammonia is distilled into boric acid. Since each milliliter of $N/14$ acid is equivalent to 1.0 mg of nitrogen, its use eliminates the need for the 0.28 factor required when $N/50$ acid is used. Furthermore much smaller volumes of $N/14$ acid are needed for titration, an important factor in conserving reagent and preventing undue dilution of the sample during titration.

Albuminoid Nitrogen

The albuminoid nitrogen determination is mainly of historic interest. It was used extensively at one time to provide supplementary information on the sanitary quality of water, particularly when it was not convenient to measure organic nitrogen. It measures certain forms of organic nitrogen that are released as ammonia nitrogen from alkaline solutions containing potassium permanganate.

Nitrite Nitrogen

Nitrite nitrogen seldom appears in concentrations greater than 1 mg/l, even in waste-treatment-plant effluents. Its concentration in surface and ground-waters is normally much below 0.1 mg/l. For this reason sensitive colorimetric methods are needed for its measurement. A modification of the Griess-Ilosvay diazotization method is used. This employs the use of two organic reagents: sulfanilic acid and N-(1-napththyl)-ethylenediamine dihydrochloride. The reactions involved may be represented as follows:

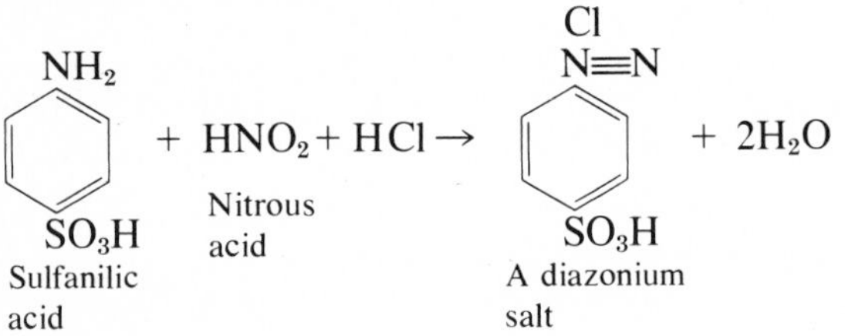

$$(24\text{-}15)$$

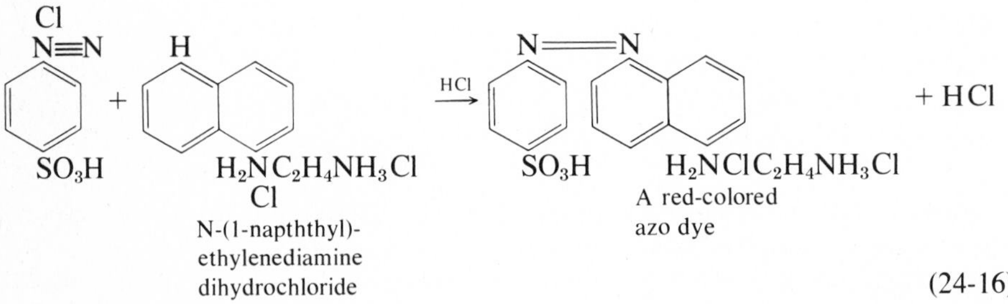

$$(24\text{-}16)$$

Under acid conditions nitrite ion as nitrous acid reacts with the amino group of sulfanilic acid to form a diazonium salt that combines with N-(1-napththyl)-ethylenediamine dihydrochloride to form a bright-colored pinkish-red azo dye. The color produced is directly proportional to the amount of nitrite nitrogen present in the sample, and determination of the amount can be made by comparison with color standards or by means of photometric measurement. Photometric measurement is preferred because standards for visual comparison are not permanent and must be prepared each time analyses are performed.

Nitrate Nitrogen

The determination of nitrate nitrogen is probably one of the most difficult an analyst has to perform in order to obtain results in which he has real confidence.

Six tentative procedures are given in the 14th edition of "Standard Methods." All have severe limitations, and results obtained on natural samples can best be classed as semiquantitative. The need is great for a more refined and exact method of analysis.

Ultraviolet spectrophotometry Nitrates in water absorb ultraviolet radiation with a wavelength of 220 nm. For this reason optical methods of analysis as described in Sec. 11-2 can be used to measure nitrates. The method is quite sensitive but requires a spectrophotometer which can be operated in the ultraviolet range. Any materials which absorb 220 nm radiation will interfere in this analysis. This includes nitrites, hexavalent chromium, and many different organic compounds. Since interfering organic matter is always present in wastewater and is usually present in low concentration in most potable water supplies, caution in the use of this method is required.

Nitrate electrode method The nitrate electrode is a liquid membrane electrode as described in Sec. 11-3 and can detect the presence of nitrate nitrogen down to concentrations of about 1 mg/l. The advantage of the electrode method is that once calibrated, analysis for nitrate is rapid. Also, the electrode is readily adapted to continuous monitoring and to automatic process control. Disadvantages are that several common ions such as chloride and bicarbonate cause interference, and the electrode is not sensitive in the lower nitrate concentrations frequently encountered.

Cadmium reduction method The cadmium reduction method offers a highly sensitive method for nitrate analysis. A filtered sample with added NH_4Cl is passed through a specially prepared column containing amalgamated cadmium filings. During passage, nitrates are quantitatively reduced by the cadmium to nitrite. The nitrite produced is determined by the diazotization method previously described. Since nitrites originally present in the sample are also measured, this procedure in effect measures the sum of nitrate plus nitrite nitrogen. In order to determine the nitrate concentration, a separate analysis of nitrite alone is required, and this value is subtracted from the results of the cadmium reduction procedure. Because of the sensitivity of the diazotization method for nitrites, nitrate concentrations as low as 0.01 mg/l can be detected. Samples with nitrate nitrogen concentrations greater than 1 mg/l can also be measured if the samples are diluted prior to passage through the cadmium column. A standard nitrate solution should be passed through the column on occasion to establish that quantitative conversion of nitrate to nitrite is being obtained. If not, then the column should be reactivated or the sample flow rate should be adjusted to obtain quantitative conversion. Few substances in natural waters interfere with this procedure, making it a likely candidate to be raised from a "tentative" to a "standard" method in the future. This procedure has successfully been used in an automated method of analysis.

Brucine method Brucine is a complex naturally occurring organic compound which reacts with nitrates under acid conditions and at an elevated temperature to produce a yellow color. Unfortunately, the color development does not obey Beer's law, and thus a curved rather than a straight line is obtained. In order to minimize this problem, samples should be diluted so that the nitrate nitrogen concentration is in a range of 0.1 to 1 mg/l. Another problem is that the intensity of the color developed is a function of both time and temperature; hence both factors must be closely controlled. For this reason it is good practice to analyze a series of samples and standards simultaneously so that a suitable curve can be developed for each test.

Chromotrophic acid method This is another colorimetric method in which a yellow color is formed from the reaction between two moles of nitrate and one mole of chromotropic acid. Certain heavy metals interfere with this method and possible interferences should be evaluated before use.

Devarda's alloy reduction method This method is a new addition to the "tentative" list of procedures outlined in "Standard Methods." It is not particularly sensitive for nitrates and so is generally applicable for nitrate-nitrogen concentrations above 2 mg/l. It involves the addition of a special heavy metal alloy which results in the reduction of nitrate to ammonia under hot alkaline conditions. The ammonia distills as it is formed, and the distillate is collected and analyzed as in the test for ammonia previously described. Procedures to eliminate or correct for interferences from ammonia present in the original sample, and nitrite which is also reduced to ammonia by the procedure, must be used.

24-4 APPLICATIONS OF NITROGEN DATA

At the present time, data concerning the nitrogen compounds that exist in water supplies are used largely in connection with disinfection practice. The amount of ammonia nitrogen present in a water determines to a great extent the chlorine needed to obtain free chlorine residuals in breakpoint chlorination and determines to some extent the ratio of monochloramines to dichloramines when combined chlorine residuals are involved. Nitrate determinations are important in determining whether water supplies meet Environmental Protection Agency recommendations for the control of methemoglobinemia in infants.

Nitrogen data are extremely important in connection with waste treatment. By controlling nitrification, aerobic-treatment costs can be kept at a minimum. Ammonia and organic nitrogen determinations are important in determining whether sufficient available nitrogen is present for aerobic biological treatment. If not, they are needed to calculate the amounts that must be supplied from outside sources, an important economic consideration in many instances.

Where wastewater sludges are sold for their fertilizing value, the nitrogen content of the sludge is a major factor in determining their value for such purposes.

The productivity of natural waters in terms of algal growths is related to the fertilizing matter that gains entrance to them. Nitrogen in its various forms is a major consideration. Also, reduced forms of nitrogen are oxidized in natural waters, thereby affecting the dissolved-oxygen resources. For these reasons nitrogen data are often part of the information needed in stream-pollution-control programs.

PROBLEMS

24-1 In what forms does nitrogen normally occur in natural waters?

24-2 Discuss the significance of nitrogen analysis in water pollution control.

24-3 Analyses for various forms of nitrogen were made at three points in a stream as follows:

			Nitrogen concentration mg/l			
Point	Location	DO mg/l	org-N	NH_3–N	NO_2-N	NO_3-N
1	Point of waste discharge	7	3	4	0	0
2	5 miles downstream	2	1	2	1	3
3	10 miles downstream	0	1	0	0	2

On the basis of your knowledge of the nitrogen cycle, explain the relative change in each nitrogen form, as well as the decrease in total nitrogen in moving downstream from point 1 to point 3.

24-4 Would you expect to find the highest concentration of each of the following in raw domestic wastewater or in the effluent from an aerobic biological waste treatment plant? Why? (a) Organic-N, (b) NH_3-N, (c) NO_2-N, (d) NO_3-N.

SOLIDS

25-1 GENERAL CONSIDERATIONS

The environmental engineer is concerned with the measurement of solid matter in a wide variety of liquid and semiliquid materials ranging from potable waters through polluted waters, domestic and industrial wastes, and sludges produced in treatment processes. Strictly speaking, all matter except the water contained in liquid materials is classed as solid matter. The usual definition of *solids,* however, refers to the matter that remains as *residue upon evaporation* and drying at 103 to 105°C. All materials that exert significant vapor pressure at such temperatures are, of course, lost during the evaporation and drying procedures. The residue, or solids, remaining represent only those materials present in a sample that have a negligible vapor pressure at 105°C.

Because of the wide variety of inorganic and organic materials encountered in the analyses for solids, the tests are empirical in character and relatively simple to perform. Gravimetric methods are used in almost all cases, and reference should be made to Sec. 10-3 in preparation for such determinations. Exceptions are the measurement of settleable solids and the estimation of dissolved solids by specific conductance measurements. The major problems in the analyses for solids are concerned with specific tests designed to gain information on the amounts of various kinds of solids present, e.g., dissolved, suspended, volatile, and fixed.

Dissolved and Undissolved Solids

The amount and nature of dissolved and undissolved matter occurring in liquid materials vary greatly. In potable waters, most of the matter is in dissolved form and consists mainly of inorganic salts, small amounts of organic matter, and dissolved gases. The total solids content of potable waters usually ranges from 20 to 1000 mg/l, and as a rule, hardness increases with total solids. In all other liquid materials, the amounts of undissolved colloidal and suspended matter increase with the degree of pollution. Sludges represent an extreme case in which most of the solid matter is undissolved, and the dissolved fraction is of minor importance. Determination of the amounts of dissolved and undissolved matter is accomplished by making tests upon filtered and unfiltered portions of samples. The undissolved substances are usually referred to as *suspended matter* or *suspended solids*.

Volatile and Fixed Solids

One of the major objectives of performing solids determinations upon domestic wastes, industrial wastes, and sludge samples is to obtain a measure of the amount of organic matter present. This test is accomplished by a combustion procedure in which organic matter is converted to carbon dioxide and water, while the temperature is controlled to prevent decomposition and volatilization of inorganic substances as much as is consistent with complete oxidation of the organic matter. *The loss in weight is interpreted in terms of organic matter.*

The standard procedure is to conduct ignitions at 600°C. It is about the lowest temperature at which organic matter, particularly carbon residues resulting from pyrolysis of carbohydrates and other organic matter, as shown in Eq. (25-1), can be oxidized at reasonable speed.

$$C_x(H_2O)_y \xrightarrow{\Delta\Delta} xC + yH_2O\uparrow \qquad (25\text{-}1)$$

$$C + O_2 \xrightarrow{\Delta} CO_2 \qquad (25\text{-}2)$$

Also, at 600°C decomposition of inorganic salts is minimized. Any ammonium compounds not released during drying are volatilized, but most other inorganic salts are relatively stable, with the exception of magnesium carbonate, as shown in the equation

$$MgCO_3 \xrightarrow[350°C]{\Delta} MgO + CO_2\uparrow \qquad (25\text{-}3)$$

In the determination of the volatile content of suspended solids, dissolved inorganic salts are not a consideration because they are removed during the filtration procedure. In sludge analysis, the ammonium compounds exist mainly as ammonium bicarbonate, are completely volatilized during the evaporation and drying procedures, and are not present to interfere in a volatile-solids determination.

$$NH_4HCO_3 \xrightarrow{\Delta} NH_3\uparrow + H_2O\uparrow + CO_2\uparrow \qquad (25\text{-}4)$$

Other unstable inorganic salts present in sludge are normally in such small amounts in relation to the amount of total solids that their influence is usually ignored.

Serious errors can be introduced in volatile-solids determinations by conducting ignitions at uncontrolled temperatures. For this reason it is standard practice to conduct combustions in a muffle furnace where the temperature can be accurately controlled. Calcium carbonate is decomposed at temperatures above 825°C, and since it is a major component of the inorganic salts normally present in samples subjected to volatile-solids analysis, its decomposition can introduce significant errors.

Unless care is exercised in the initial stages, the determination of volatile solids in sludges is often subject to serious error because of physical or mechanical losses due to decrepitation during the procedure. Decrepitation may be eliminated by a preliminary controlled firing of samples with a bunsen burner to destroy all flammable materials before placing the samples in the muffle furnace. If ignitions are properly performed, the weight loss incurred is a reasonably accurate measure of organic matter, and the residue remaining represents the ash or fixed solids.

Settleable Solids

The term *settleable solids* is applied to solids in suspension that will settle, under quiescent conditions, because of the influence of gravity. Only the coarser suspended solids with a specific gravity greater than that of water will settle. Sludges are accumulations of settleable solids. Their measurement is important in engineering practice to determine the need for sedimentation units and the physical behavior of waste streams entering natural bodies of water.

25-2 ENVIRONMENTAL SIGNIFICANCE OF SOLIDS DETERMINATIONS

The amount of dissolved solids present in water is a consideration in its suitability for domestic use. In general, waters with a total solids content of less than 500 mg/l are most desirable for such purposes. Waters with higher solids content often have a laxative and sometimes the reverse effect upon people whose bodies are not adjusted to them. This is important to people who travel and to the transportation companies who are interested in the welfare of their passengers. In many areas, it is impossible to find natural waters with a solids content under 500 mg/l; consequently it is impossible to meet this desired level without some form of treatment. In many instances, treatment to reduce the solids content is not practiced, and residents who regularly use such waters appear to suffer no ill effects. Standards generally recommend an upper limit of 1000 mg/l on potable waters.

25-3 DETERMINATION OF SOLIDS IN WATER SUPPLIES

Because of the wide variety of materials subjected to solids determinations, the tests applied vary somewhat, and it is best to discuss them in terms of water, polluted waters, and sludges. Dissolved solids are the major concern in water supplies; therefore the total-solids determination and the specific conductance measurement are of greatest interest. Suspended-solids tests are seldom made because of the small amounts present. They are more easily evaluated by measurement of turbidity.

Total Solids or Residue on Evaporation

The determination of total solids is easily made by evaporation and drying of a measured sample in a tared container. The use of platinum dishes is highly recommended because of the ease with which they can be brought to constant weight before use. Vycor ware is a good substitute. The use of porcelain dishes is to be avoided because of their tendency to change weight.

The determination of volatile solids (organic content) is ordinarily not a consideration in waters intended for domestic use. The results obtained cannot be interpreted in terms of organic matter with any degree of reliability. In cases in which the organic content of a water is important, it is usually best to obtain such information by means of a COD, BOD, or organic carbon determination.

Specific Conductance

A rapid estimation of the dissolved-solids content of a water supply can be obtained by specific-conductance measurements. Such measurements indicate the capacity of a sample to carry an electrical current, which in turn is related to the concentration of ionized substances in the water. Most dissolved inorganic substances in water supplies are in the ionized form and so contribute to the specific conductance. Although this measurement is affected by the nature of the various ions, their relative concentrations, and the ionic strength of the water, such measurements can give a practical estimate of the variations in dissolved mineral content of a given water supply. Also, by the use of an empirical factor, specific conductance can allow a rough estimate to be made of the dissolved mineral content of water samples. The use of specific conductance and the factors affecting this measurement are discussed in Sec. 3-10.

Dissolved and Suspended Matter

In cases in which turbidity measurements are not deemed adequate to provide the necessary information, the suspended solids may be determined by filtration through a Gooch crucible in the usual manner. Because of the small amounts of suspended solids usually present in water supplies, the determination is subject to considerable error unless an abnormally large sample is filtered. The preferred technique is to filter a sample of water through filter paper and determine total solids in the filtrate. The difference between total solids in unfiltered and filtered samples is a measure of the suspended solids present.

25-4 DETERMINATIONS APPLICABLE TO POLLUTED WATERS AND DOMESTIC WASTEWATERS

The settleable- and suspended-solids determinations are of greatest value in assessing the strength of domestic wastes and lightly polluted waters.

Settleable Solids

The determination of settleable solids is of particular importance in the analysis of wastewaters. The test is ordinarily conducted in an Imhoff cone (see Fig. 25-1), allowing 1 h settling time under quiescent conditions. Samples should be

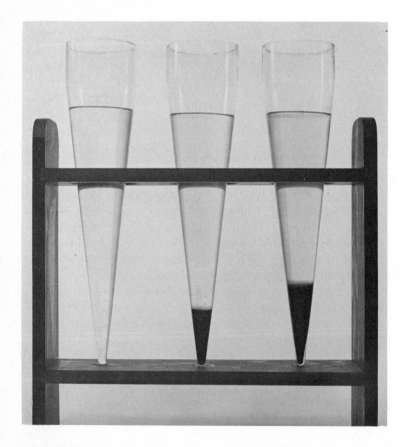

Figure 25-1 Imhoff cones used to measure settleable solids.

adjusted to near room temperature and the test conducted in a location where direct sunlight does not interfere with normal settlement of the solids. Results are measured and reported in terms of milliliters per liter.

Total Solids or Residue on Evaporation

Total-solids determinations are ordinarily of little value in the analysis of polluted waters and domestic wastewaters because they are difficult to interpret with any degree of accuracy. In most instances the dissolved solids present in the original water represent such a large and variable percentage of the total amount found in the polluted wastewaters that it is impossible to evaluate data, unless the amount of solids originally present in the carriage water is known. The test was originally devised as a means of evaluating the amount of polluting matter present in wastewaters. Since the BOD and COD tests are capable of evaluating the strength of such materials much more exactly, there is little justification for running total-solids tests for such purposes.

Some waste treatment processes, particularly those involving sedimentation, are adversely affected by radical changes in the density of wastewater. In coastal cities, seawater often gains access to sewer systems at times of high tide, and in industrial cities, intermittent discharge of highly mineralized wastes may occur. Both can cause significant changes in density. The total-solids test can be used to good advantage to detect such changes, although such contamination can be detected more easily by the chloride determination or by specific conductance measurements.

Suspended Solids

The suspended-solids determination is extremely valuable in the analysis of polluted waters. It is one of the major parameters used to evaluate the strength of domestic wastewaters and to determine the efficiency of treatment units. In stream-pollution-control work, all suspended solids are considered to be settleable solids, as time is not a limiting factor. Deposition is expected to occur through biological and chemical flocculation; therefore measurement of suspended solids is considered fully as significant as BOD.

The suspended-solids determination is subject to considerable error if proper precautions are not taken, as discussed in Secs. 10-1 and 10-3. Usually the sample size is limited to 50 ml or less because of difficulties encountered in filtration of larger samples. The weight of solids removed seldom exceeds 20 mg and is often less than 10 mg. Small errors in weighing or losses of filter mat can be quite significant. It is extremely important that the Gooch crucibles be carefully prepared and brought to constant weight before use. Sufficient sample should be filtered, if possible, to yield an increase in weight of about 10 mg. This often requires the filtration of 500 ml or more of samples of biologically treated wastewaters or lightly polluted waters.

The volatile content of suspended solids can be determined by direct ignition in the muffle furnace because the amount of solids involved is too small for decrepitation to occur. Suspended solids often contain as much as 80 percent of volatile matter. The fixed solids remaining frequently weigh less than 2 mg. This illustrates why it is so very important to use Gooch crucibles that have been brought to constant weight. It often happens that a crucible plus fixed solids weighs less than the original tare weight. This indicates that the crucible used had not been brought to constant weight or that loss of filter mat occurred during the filtration process. A similar error can be caused by weighing crucibles before they have returned to room temperature.

An alternate procedure for measurement of suspended solids makes use of a glass-fiber filter instead of a Gooch crucible.[1] This results in a considerable saving in time, because prior filter preparation is not required, and both filtration and weighing can be more rapidly performed. Most important, experience has shown that students can learn to conduct the analysis much more rapidly, and the results of their analyses are quite consistent compared with results obtained by using Gooch crucibles. The volatile content of suspended solids can be determined by placing the glass-fiber filter in a muffle furnace. The filter is not destroyed by this procedure, provided that certain precautions are taken. A temperature of 600°C is near the melting point of the filter, and for this reason it should not be left in the furnace for more than 5 to 10 minutes. This problem can be minimized by using a combustion temperature of 580°C. A slight weight loss occurs in the filter during drying and combustion, and for this reason a blank filter should be subjected to the same treatment as the sample filter so that an appropriate correction can be made.

Suspended solids are reported in terms of milligrams per liter, and volatile suspended solids are normally reported in terms of percent of the suspended solids.

25-5 DETERMINATIONS APPLICABLE TO INDUSTRIAL WASTES

Industrial wastes include such a wide variety of materials that analyses for exploratory purposes should include all determinations that can possibly provide significant information. For this reason all the solids tests commonly applied to domestic wastewaters are important. The settleable-solids test is particularly important, as it serves as the principal basis of determining whether primary sedimentation facilities are required for treatment. In addition, the total-solids determination has special significance. Many industrial wastes contain unusual amounts of dissolved inorganic salts, and their presence is easily detected by the total-solids test. Their concentration and nature are factors in

---

[1] B. M. Wyckoff, "Rapid Solids Determination Using Glass Fiber Filters," *Water and Sewage Works,* **III:** 277 (June 1964).

determining the susceptibility of wastes to anaerobic treatment. Before the development of the COD test, the volatile content of total solids was used extensively to measure the amount of organic matter present. It was very helpful in assessing the amount of biologically inert organic matter, such as lignin in the case of wood-pulping waste liquors.

25-6 DETERMINATION OF SOLIDS IN SLUDGES

The total- and volatile-solids determinations are important in the analysis of raw and digested sludges. Both are subject to some error because of the loss of volatile organic compounds during the drying process. This factor is not particularly significant in samples of raw or well-digested sludges but it is important when partially digested sludges containing appreciable amounts of volatile acids are being analyzed. This loss often leads to faulty interpretation of volatile-solids destruction.

Because it is impossible to pipet samples of raw and digested sludges, it is common practice to weigh the samples in previously tared dishes. It is customary to use small procelain evaporating dishes about 3 in. in diameter. It is important that the dishes be previously ignited to constant weight in order to provide reliable results for volatile solids. Because of the rather nonuniform character of sludges, it is necessary to use relatively large samples of 25 to 50 g, unless some method of homogenization has been employed. As a result, considerable amounts of residue are obtained upon evaporation of the samples. It is usually necessary to dry the samples at 103°C for several hours to be sure that all the moisture has had a chance to escape. Most analysts prefer to dry the samples overnight.

The volatile-solids content of sludges is an extremely important determination. Accurate results can be obtained, provided that care is taken to control decrepitation, as described in Sec. 25-1.

The steps involved in the collection of data for total- and volatile-solids determinations are not always understood. A typical set of data and calculations are as follows:

Wt of dish on analytical balance		30.160 g
Wt of dish on trip balance	30.5 g	
Wt of dish + sample	70.8	
Wt of sample	40.3	
Wt of dish + dry sludge solids		32.780
Wt of sludge solids		2.620

Percent solids in sludge $\dfrac{2.62}{40.3} \times 100 = 6.50\%$

Wt of dish + ash	30.720
Wt of volatile matter	2.06

Percent volatile matter $\dfrac{2.06}{2.62} \times 100 = 78.7\%$

It is unnecessary to report the percent of fixed solids, as it can be readily reckoned from the percent of volatile solids.

The measurement of solids in activated sludges is a special case. Because of the relatively low concentrations usually involved, serious errors can be introduced by measuring total solids, which would also include dissolved solids. It is customary to measure activated sludge by procedures used to determine suspended solids. This allows the dissolved solids to pass into the filtrate. A number of modifications are used to measure activated sludge solids, but if volatile solids are desired, as is often the case, the test is best performed with the standard Gooch-crucible or the newer glass-fiber-filter technique. The sample should be limited to 5 or possibly 10 ml because of filtration problems.

25-7 APPLICATIONS OF SOLIDS DATA IN ENVIRONMENTAL ENGINEERING PRACTICE

In the realm of public and industrial water supplies, the total-solids determination is the only one of importance. It is used to determine the suitability of potential supplies for development. In cases in which water softening is needed, the type of softening procedure used may be dictated by the total-solids content, since precipitation methods decrease, and exchange methods increase, the solids. Corrosion control is frequently accomplished by the production of stabilized waters through pH adjustment. The pH at stabilization depends to some extent upon the total solids present as well as the alkalinity and temperature.

The settleable-solids determination has two very important applications. First, it is used extensively in the analysis of industrial wastes to determine the need for and design of primary settling tanks in plants employing biological treatment processes. The test is also widely used in waste-treatment-plant operation to determine the efficiency of sedimentation units. It is fully as important in the operation of large treatment plants as in the smaller.

The suspended- and volatile-suspended-solids determinations are used to evaluate the strength of domestic and industrial wastes. The tests are particularly valuable in determining the amount of suspended solids remaining after settleable solids have been removed in primary settling units, for the purpose of determining the loading of remaining materials on secondary biological treatment units. In the larger treatment plants, suspended-solids determinations are used routinely as a measure of the effectiveness of treatment units. From the viewpoint of stream pollution control, the removal of suspended solids is usually as important as BOD removal. Both suspended- and volatile-suspended-solids determinations are used to control the aeration solids in the activated sludge process.

The total- and volatile-solids tests are the only solids determinations normally applied to sludges. They are indispensable in the design and operation of sludge-digestion, vacuum-filter, and incineration units.

PROBLEMS

25-1 Why are each of the following solids analyses of interest in water quality control?
(*a*) Total dissolved solids in municipal water supplies.
(*b*) Total and volatile suspended solids in domestic wastewater.
(*c*) Total and volatile solids in sludge.
(*d*) Settleable solids in domestic wastewater.

25-2 What significant information is furnished by the determination of volatile solids?

25-3 What precautions must be taken in the determination of volatile solids? Why?

25-4 List the steps involved in preparation of a Gooch crucible which must be observed to ensure reliable results from its use.

25-5 Would you expect the analytical results to be higher than, lower than, or the same as the true value under the following conditions, and why?
(*a*) Weighing a warm crucible.
(*b*) Estimating organic content by combustion at 600°C of a sludge sample with a high magnesium carbonate content.
(*c*) Estimating the organic content by volatile solids analysis of a sample containing a large quantity of organic materials having a high vapor pressure.
(*d*) Estimating the organic content of a sample by combustion at 800°C rather than at 600°C.

25-6 A domestic wastewater contains 350 mg/l of suspended solids. Primary sedimentation facilities remove 65 percent. Approximately how many gallons of sludge containing 5.0 percent solids will be produced per million gallons of wastewater settled?

IRON AND MANGANESE

26-1 GENERAL CONSIDERATIONS

Both iron and manganese create serious problems in public water supplies. The problems are most extensive and critical with underground waters, but difficulties are encountered at certain seasons of the year in waters drawn from some rivers and some impounded surface supplies. Why some underground supplies are relatively free of iron and manganese and others contain so much has always been something of an enigma that defied explanations when viewed solely from the viewpoint of inorganic chemistry. More recent developments and experiences have indicated that biochemical changes, or more exactly, changes in environmental conditions brought about by biological reactions, are major considerations. Since both manganese and iron are present in insoluble forms in significant amounts in nearly all soils, any explanation of how appreciable amounts can gain entrance to water flowing through or coming in contact with the soil must consider how the iron and manganese are converted to soluble forms.

Iron exists in soils and minerals mainly as insoluble ferric oxide and iron sulfide (pyrite). It occurs in some areas also as ferrous carbonate (siderite), which is very slightly soluble. Since ground-waters usually contain significant amounts of carbon dioxide (see Sec. 16-2), appreciable amounts of ferrous carbonate may be dissolved by the reaction shown in the equation

$$FeCO_3 + CO_2 + H_2O \rightarrow Fe^{2+} + 2HCO_3^- \qquad (26\text{-}1)$$

in the same manner that calcium and magnesium carbonates are dissolved. However, iron problems are prevalent where it is present in the soil as insoluble ferric compounds. Solution of measurable amounts of iron from such solids

464

does not occur, even in the presence of appreciable amounts of carbon dioxide, as long as dissolved oxygen is present. Under reducing (anaerobic) conditions, however, the ferric iron is reduced to ferrous iron, and solution occurs without difficulty.

Manganese exists in the soil principally as manganese dioxide, which is very insoluble in water containing carbon dioxide. Under reducing (anaerobic) conditions, the manganese in the dioxide form is reduced from an oxidation state of IV to II, and solution occurs, as with ferric oxides.

Evidence to indicate that iron and manganese gain entrance to water supplies through changes produced in environmental conditions as a result of biological reactions has stemmed from four sources, as follows:

1. Ground-waters that contain appreciable amounts of iron or manganese or both are always devoid of dissolved oxygen and are high in carbon dioxide content. The iron and manganese are present as Fe^{2+} and Mn^{2+}. The high carbon dioxide content indicates that bacterial oxidation of organic matter has been extensive, and the absence of dissolved oxygen shows that anaerobic conditions were developed.
2. Wells producing good-quality water, low in iron and manganese, for many years have been known to produce poor-quality water when organic wastes have been discharged on the soil around or near the well, thereby creating anaerobic conditions in the soil.
3. The iron and manganese problem in impounded surface supplies has been correlated with reservoirs that stratify, but occurs only in those in which anaerobic conditions develop in the hypolimnion. The soluble iron and manganese released from the bottom muds are contained in the waters of the hypolimnion until the fall overturn occurs. At that time they are distributed throughout the reservoir and cause trouble in the water supply until sufficient time has elapsed for oxidation and sedimentation to occur under natural conditions.
4. It has been shown, on the basis of thermodynamic considerations, that Mn(IV) and Fe(III) are the only stable oxidation states for iron and manganese in oxygen-containing waters.[1] Thus these forms can be reduced to the soluble Mn(II) and Fe(II) only under highly anaerobic reducing conditions.

When oxygen bearing water is injected into the ground for recharge of the ground-water aquifer, it is sometimes noted that the soluble iron content of the water increases, an observation which seems to contradict the above stated need for anaerobic conditions. The explanation is that the oxygen is consumed

[1] J. J. Morgan and W. Stumm, The Role of Multivalent Metal Oxides in Limnological Transformations, as Exemplified by Iron and Manganese, *Proc. Intern. Conf. on Water Pollution Research,* Pergamon Press, New York, 1964.

through the oxidation of insoluble pyrite (FeS_2), leading to anaerobic conditions and the formation of soluble iron sulfate.

$$2FeS_2 + 7O_2 + 2H_2O \rightarrow 2Fe^{2+} + 4SO_4^{2-} + 4H^+ \qquad (26\text{-}2)$$

In summary, the evidence seems clear that the development of anaerobic conditions is essential for appreciable amounts of iron and manganese to gain entrance to a water supply. Only under anaerobic conditions are the soluble forms of iron, Fe(II), and manganese, Mn(II), thermodynamically stable.

26-2 ENVIRONMENTAL SIGNIFICANCE OF IRON AND MANGANESE

As far as is known, humans suffer no harmful effects from drinking waters containing iron and manganese. Such waters, when exposed to the air so that oxygen can enter, become turbid and highly unacceptable from the aesthetic viewpoint, owing to the oxidation of iron and manganese to the Fe (III) and Mn (IV) states which form colloidal precipitates. The rates of oxidation are not rapid, and thus reduced forms can persist for some time in aerated waters. This is especially true when the pH is below 6 with iron oxidation and below 9 with manganese oxidation. The rates may be increased by the presence of certain inorganic catalysts or through the action of microorganisms, Both iron and manganese interfere with laundering operations, impart objectionable stains to plumbing fixtures, and cause difficulties in distribution systems by supporting growths of iron bacteria. Iron also imparts a taste to water which is detectable at very low concentrations. For these reasons public water supplies ought not to contain more than 0.3 mg/l of iron or 0.05 mg/l of manganese.

26-3 METHODS OF DETERMINING IRON

A great many methods of determining iron have been developed. Precipitation methods are commonly used where quantities are relatively large, as in some industrial wastes. However, in water supplies the amounts present are normally so small that colorimetric procedures are more satisfactory. The colorimetric procedures have a major advantage in that they are usually highly specific for the ion involved, and a minimum of pretreatment is required. Iron may also be determined by atomic absorption spectrophotometry (Sec. 11-2).

Phenanthroline Method

The phenanthroline method is the preferred standard procedure for the measurement of iron in water at the present time, except when phosphate or heavy metal interferences are present. The method depends upon the fact that 1,10-phenanthroline combines with Fe^{2+} to form a complex ion which is orange-red in color. The color produced conforms to Beer's law and is readily measured by visual or photometric comparison.

Water samples subjected to analysis have usually been exposed to the atmosphere; consequently some oxidation of Fe(II) to Fe(III) and precipitation of ferric hydroxide may have occurred. It is necessary to make sure that all the iron is in a soluble condition. This is done by treating a portion of the sample with hydrochloric acid to dissolve the ferric hydroxide:

$$Fe(OH)_3 + 3H^+ \rightarrow Fe^{3+} + 3H_2O \qquad (26\text{-}3)$$

Since the reagent 1,10-phenanthroline is specific for measuring Fe(II), all iron in the form of Fe(III) must be reduced to the ferrous condition. This is most readily accomplished by using hydroxylamine as the reducing agent. The reaction involved may be represented as follows:

$$4Fe^{3+} + 2NH_2OH \rightarrow 4Fe^{2+} + N_2O + H_2O + 4H^+ \qquad (26\text{-}4)$$

Three molecules of 1,10-phenanthroline are required to sequester or form a complex ion with each Fe^{2+}. The reaction may be represented as shown in the equation

1,10-phenanthroline   Orange-red complex   (26-5)

By proper modifications of the test procedure, measurements of total, dissolved, and suspended iron can be made. These considerations are not normal, however, and, whenever they are, special precautions must be taken in sampling and transportation of samples to ensure that no changes occur before analyses are performed. Because of the possible errors that may result, it is best that the analyst assume full responsibility for sampling as well as for analysis. When interfering materials such as phosphates and heavy metals are present, satisfactory results can be obtained by use of a procedure which involves acidification of the sample with HCl and extraction of the iron into diisopropyl ether prior to the addition of the phenanthroline solution.

26-4 METHODS OF DETERMINING MANGANESE

In environmental engineering practice, manganese is principally of concern in water supplies. Its concentration seldom exceeds a few milligrams per liter; therefore colorimetric methods are most applicable. Two colorimetric methods are recommended in "Standard Methods"; both depend upon oxidation of the manganese from its lower oxidation state to VII, where it forms the highly colored permanganate ion. The color produced is directly proportional to the concentration of manganese present over a considerable range of concentration in accordance with Beer's law, and it is easily measured by eye or photometric means. Chlorides interfere because of their reducing action in an acid medium, and so provisions must be made to overcome their influence. Most other reducing agents are rendered inactive by the strong oxidizing agents used to form the permanganate ion. Manganese may also be determined by atomic absorption spectrophotometry (Sec. 11-2).

Persulfate Method

The persulfate method is best suited for routine determinations of manganese because pretreatment of samples is not needed to overcome chloride interference. Ammonium persulfate is commonly used as the oxidizing agent. It is subject to deterioration during prolonged storage; for this reason, it is always good practice where samples are not run routinely to include a standard sample with each set of samples to verify the potency of the persulfate used.

Chloride interference is overcome in the persulfate method by adding Hg^{2+} to form poorly ionized $HgCl_2$. Since the ionization constant of $HgCl_2$ is about 5×10^{-14}, the concentration of chloride ion is decreased to such a low level that it cannot reduce the permanganate ions formed.

The oxidation of manganese in lower valences to permanganate by persulfate requires the presence of Ag^+ as a catalyst. The reaction involved in the oxidation may be represented as follows:

$$2Mn^{2+} + 5S_2O_8^{2-} + 8H_2O \xrightarrow{Ag^+} 2MnO_4^- + 10SO_4^{2-} + 16H^+ \qquad (26\text{-}6)$$

The color produced by the permanganate ion is stable for several hours, provided that a good-quality distilled water is used for dilution purposes and reasonable care is taken to protect the sample from contamination by dust of the atmosphere.

Periodate Method

The periodate method is somewhat more sensitive to small amounts of manganese than the persulfate method, and the colored solutions produced are stable for longer periods of time. It is especially applicable where manganese concentrations are below 0.1 mg/l. Chlorides interfere, and it is often necessary to expel them as HCl by evaporating the sample with sulfuric acid to the point at which the sulfuric acid begins to distill. This is recognized by the formation of white fumes resulting from the condensation of water vapor in the atmosphere by the sulfuric acid as it distills.

The oxidation of manganese from its lower oxidation states to permanganate by periodate is normally accomplished without the aid of a catalyst. However, where small amounts of manganese are involved, the use of Ag^+ as a catalyst is recommended. The reaction involved may be represented as follows:

$$2Mn^{2+} + 5IO_4^- + 3H_2O \rightarrow 2MnO_4^- + 5IO_3^- + 6H^+ \qquad (26\text{-}7)$$

In practice an excess of periodate is used, and in its presence the permanganate ion is stable.

26-5 APPLICATIONS OF IRON AND MANGANESE DATA IN ENVIRONMENTAL ENGINEERING PRACTICE

In explorations for new water supplies, particularly from underground sources, iron and manganese determinations are an important consideration. Supplies may be rejected on this basis alone. When supplies containing amounts in excess of 0.3 mg/l iron or 0.05 mg/l manganese are developed, the engineer must decide whether treatment is justified and, if so, the best method of treatment. The ratio of iron to manganese is a factor that determines the type of treatment used, as well as the amount of organic matter present in the water. The efficiency of treatment units is determined by routine tests for iron and manganese. They are also used to aid in the solution of problems in distribution systems where iron-fixing bacteria are troublesome.

Corrosion of cast-iron and steel pipelines often produces "red-water" troubles in distribution systems. The iron determination is helpful in assessing the extent of corrosion and aiding in the solution of these problems. Research on corrosion and methods of corrosion control require the use of many types of tests to evaluate the extent of metal loss. The iron determination is one of them.

PROBLEMS

26-1 What is the environmental significance of iron and manganese in water supplies?

26-2 Discuss the different oxidation states of iron and manganese occurring in natural water supplies, and discuss the conditions under which each form is stable.

26-3 Discuss briefly how iron and manganese get into underground water supplies.

26-4 In what form must iron exist in order to be measured by the phenanthroline method, and how is iron converted to this form?

26-5 In the determination of manganese:
    (a) What is the function of ammonium persulfate?
    (b) What is the function of Ag^+?
    (c) What is the function of $HgSO_4$?
    (d) In what oxidation state must the manganese be for colorimetric measurement?

TWENTY-SEVEN

FLUORIDE

27-1 GENERAL CONSIDERATIONS

The water engineer has at least a dual interest in the determination of fluorides. It is his responsibility to design and operate units for the removal of fluorides from water supplies that contain excessive amounts and, on the other hand, it becomes his responsibility to supervise the addition of fluorides to optimum levels in water supplies that are deemed to be deficient in fluorides by local health agencies. In some areas, particularly in the neighborhood of aluminum-processing plants, contamination of the atmosphere and vegetation by fluorides has been a serious problem. Control methods have had to be employed to protect cattle and other herbivorous animals from damage to bones and teeth.

Significance of High Fluorides in Water Supplies

A disfigurement in the teeth of humans known as *mottled enamel* (dental fluorosis) has been recognized for many years. United States immigration authorities, at an early date, noticed that people arriving from certain areas of Europe were severely afflicted, whereas people from other areas showed little or no evidence of mottling. This led dental authorities to believe that the diseased condition was due to a local factor. Shortly after this information became known, reports of mottled enamel among people native to the United States began to appear. These cases came largely from cities in the Great Plains and Rocky Mountain states, but no real clue was offered to explain the cause of the defective teeth until about 20 years later.

Substantial evidence that fluorides are the cause of mottled enamel was obtained by Churchill of the Aluminum Co. of America in 1930.[1] The city of Bauxite, Arkansas, was one of the communities that had reported a high incidence of mottled enamel among its people. Churchill, through spectrographic analysis, found appreciable amounts of fluoride ion in the Bauxite water supply. In collaboration with McKay, a dentist of Colorado Springs, Colorado, he studied waters from five areas where mottling was endemic and from 40 areas where it was not a problem. From these studies it was concluded that excessive fluoride levels in drinking water are the cause of mottled enamel. Their data showed that mottling did not appear unless the fluoride-ion concentration was in excess of 1.0 mg/l and that the degree and severity of mottling increased as the fluoride level rose.

As soon as excessive amounts of fluorides in water supplies had been established as the cause of dental fluorosis, research on methods of removal were initiated. The passing of water through various types of defluoridation media such as tricalcium phosphate, bone char, bone meal, and activated alumina was found to accomplish fluoride removal by a combination of ion exchange and sorption. Although such methods are presently the most widely used in practice, they are relatively expensive, and few plants providing facilities for fluoride removal are presently in operation. The numbers appear to be slowly on the increase. California in 1965 had seven facilities in operation with two more planned.[2] Development of cheaper methods would make fluoride removal a practical matter for many cities.

Significance of Low Fluorides in Water Supplies

As a result of the great interest focused on the fluoride content of public water supplies in relation to the dental fluorosis problem, a great deal of information became available on fluorides. It was natural that the dental profession would use this information to determine whether fluorides were correlated with other dental disease.

In 1938, Dean[3] presented information which demonstrated that dental caries is less prevalent when mottled enamel occurs. This led to extensive correlation studies on dental caries versus fluoride levels in public waters at many places in the United States. The results obtained, as summarized by Dean,[4] are presented in Fig. 27-1. From this information, a dental caries–fluoride hypothesis evolved: Approximately 1 mg/l of fluoride ion is desirable in public waters for optimal dental health. At decreasing levels, dental caries

---

[1] H. V. Churchill, *Ind. Eng. Chem.*, **23**: 996 (1931).

[2] J. A. Harmon and S. G. Kalishman, "Defluoridation of Drinking Water in Southern California," *J. Am. Water Works Assoc.*, **57**: 245 (1965).

[3] H. T. Dean, *Public Health Repts. (U.S.)*, **53**: 1443 (1938).

[4] H. T. Dean, *J. Am. Water Works Assoc.*, **35**: 1161 (1943).

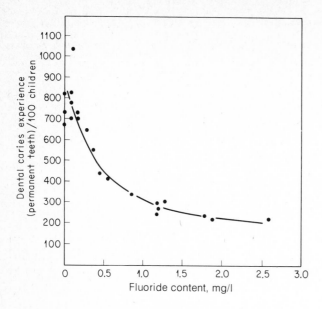

Figure 27-1 Relationship between dental caries and fluoride level in drinking water. *(After Dean.)*

becomes a serious problem, and at increasing levels, dental fluorosis becomes a problem.

The dental caries–fluoride hypothesis has served as the basis for programs of supplementing public water supplies having low fluoride levels with fluorides to bring the concentration up to about 1 mg/l. Because there was some question about the physiological effects of supplemental fluorides, as well as their efficacy as a substitute for "natural" fluorides for controlling dental caries, extensive 10-year pilot studies were conducted at Newburgh, New York; Grand Rapids, Michigan; and several other cities. The results of the investigations have been unanimous in demonstrating the safety of supplemental fluoridation and showing that added fluorides are as effective in controlling dental caries as so-called "natural" fluorides. The program for fluoridation of public water supplies deficient in natural fluorides has been sponsored by many organizations interested in the public health, including the American Dental and Medical Associations. There has been considerable opposition to the program from several quarters, but the program has advanced steadily. In 1970 over 4500 community water supplies in the United States serving over 84 million people were being supplemented with fluorides.[5] Control of additions is based upon the fluoride determination.

[5] F. J. Maier, *J. Am. Water Works Assoc.*, **62**, 3 (1970).

Fluorides and Air Pollution

Cryolite (Na_3AlF_6) is used as a solvent for Al_2O_3 in the electrolytic method of producing aluminum. At operating temperatures, the cryolite is molten and exerts a considerable vapor pressure. As a result, appreciable amounts of fluorides escape to the atmosphere through exhaust systems. The fluorides condense to form a smoke, and much of the particulate matter settles on vegetation and the soil in the immediate area. Considerable damage to livestock has occurred in certain areas of the United States. The problem has been solved in most instances by use of electrostatic precipitation units.

27-2 CHEMISTRY OF FLUORINE AND ITS COMPOUNDS

Fluorine is the most active element known and is not used in the elemental form in environmental engineering practice. It forms simple fluoride compounds and many complex ions. The principal forms in which fluorides are added to public water supplies are

NaF	Na_2SiF_6 (sodium silicofluoride)
CaF_2	H_2SiF_6 (hydrofluosilicic acid)
H_2F_2	$(NH_4)_2SiF_6$ (ammonium fluosilicate)

All the compounds and complex ions containing fluorine dissociate to yield fluoride ion. At the concentrations of about 1 mg/l involved in water treatment practice, it is generally considered that hydrolysis of the fluosilicate ion is essentially complete, as shown in the equation

$$SiF_6^{2-} + 3H_2O \rightleftharpoons 6F^- + 6H^+ + SiO_3^{2-} \qquad (27\text{-}1)$$

On this basis, the fluoride in silicofluorides can be determined by any method that is sensitive to fluoride ion.

27-3 METHODS OF DETERMINING FLUORIDES

There are three standard methods for determination of fluoride at the present time: the electrode method and two colorimetric procedures. The electrode method is the simplest but does require use of an expanded-scale pH meter and a special electrode. The principles behind and limitations of the electrode method are described under Solid State or Precipitate Electrodes in Sec. 11-3. All three methods are subject to interferences from other ions, and it is often necessary to separate the fluorides from the interfering ions before making the colorimetric test. The separation is accomplished by a distillation procedure such as that used in the determination of ammonia nitrogen. In the case of fluorides, however, the distillation is performed from acidified solutions.

Under these conditions, all fluoride ion is converted to poorly ionized hydrogen fluoride, according to the equation

$$F^- + H^+ \xrightarrow{\Delta} HF\uparrow \tag{27-2}$$

The hydrogen fluoride is evolved with the steam and held in the condensate.

Samples of water that do not contain significant amounts of interfering ions, and distillates obtained from the purification procedure may be analyzed by any one of three methods. The colorimetric methods involve the bleaching of a preformed color by the fluoride ion. The preformed color is the result of the action between zirconium ion and either alizarin dye or SPADNS dye, depending upon the method used. The color produced is commonly referred to as a "lake," and the intensity of the color produced is reduced if the amount of zirconium present is decreased. Fluoride ion combines with zirconium ion to form a stable complex ion ZrF_6^{2-}, and the intensity of the color lake decreases accordingly. The reaction involved may be represented as follows when the alizarin dye is used:

$$\text{(Zr–alizarin lake)} + 6F^- \rightarrow \text{alizarin} + ZrF_6^{2-} \tag{27-3}$$

Reddish color   Yellow

The bleaching action is a function of the fluoride-ion concentration and is directly proportional to it. Thus Beer's law is satisfied in an inverse manner. Comparisons may be made visually or photometrically. The bleaching action of fluorides is slow in the alizarin dye procedure, and a 1-h contact period is recommended before comparisons are made. Because temperature and time are important variables, it is necessary that standards be prepared for making comparisons each time analyses are to be made. When photometric measurements are used, care must be exercised to keep contact time and temperature the same as employed in developing the calibration curve. Good practice requires that at least one standard be included with samples each time photometric measurements are made.

The SPADNS method overcomes the strict time limitations of the alizarin procedure and so is particularly useful when photometric methods of analysis are preferred. Here use is made of the almost instantaneous reaction rate between fluoride and zirconium ions which result when more acid is added to the reaction mixture. By this procedure, readings can be made immediately after the sample and reagents are mixed. Under the somewhat different conditions of this modification, less interference is encountered when SPADNS dye rather than alizarin dye is used.

27-4 APPLICATION OF FLUORIDE DATA

Because of the public health significance of fluorides in water supplies intended for human use, determination of fluorides has become extremely important. In situations where fluorides are added to provide an optimum level for the control of dental caries, it is necessary to know the amount of natural fluorides present so that proper amounts of supplemental fluoride can be added. Wherever supplementation is practiced, it is necessary to maintain surveillance on the finished water to be sure that proper amounts of chemicals are being fed. The usual practice is to collect samples on the distribution system as well as at the treatment plant.

In areas where natural fluorides exceed the U.S. Environmental Protection Agency recommended upper limits, the fluoride content of a water may determine the suitability of a supply for development. These upper limits vary from 0.8 to 1.7 mg/l, depending upon the climatic conditions of the area of use. The lower values pertain to areas with highest maximum temperatures where water consumption by children would be the greatest. In cases where high-fluoride waters must be used, local authorities, in cooperation with the engineer, must decide whether fluoride-removal facilities are to be installed. The size and design of such units will depend upon the level of fluorides present in the water. Fluoride determinations, of course, serve as the basis of determining when removal units require regeneration.

PROBLEMS

27-1 Describe three situations in water works practice in which the engineer would require fluoride determinations.

27-2 How can the engineer remove excessive fluorides from drinking water supplies?

27-3 What is the purpose of sample acidification prior to distillation to separate fluorides from interfering ions?

TWENTY-EIGHT

SULFATE

28-1 GENERAL CONSIDERATIONS

The sulfate ion is one of the major anions occurring in natural waters. It is of importance in public water supplies because of its cathartic effect upon humans when it is present in excessive amounts. For this reason the recommended upper limit is 250 mg/l in waters intended for human consumption. Sulfates are important in both public and industrial water supplies because of the tendency of waters containing appreciable amounts to form hard scales in boilers and heat exchangers.

Sulfates are of considerable concern because they are indirectly responsible for two serious problems often associated with the handling and treatment of wastewaters. These are odor and sewer-corrosion problems resulting from the reduction of sulfates to hydrogen sulfide under anaerobic conditions, as shown in the following equations:

$$SO_4^{2-} + \text{organic matter} \xrightarrow[\text{bacteria}]{\text{anaerobic}} S^{2-} + H_2O + CO_2 \qquad (28\text{-}1)$$

$$S^{2-} + 2H^+ \rightleftharpoons H_2S \qquad (28\text{-}2)$$

A knowledge of the sulfur cycle, as represented in Fig. 28-1, is essential to an understanding of the transformations that occur.

Odor Problems

In the absence of dissolved oxygen and nitrates, sulfates serve as a source of oxygen (or more correctly as an electron acceptor) for biochemical oxidations

476

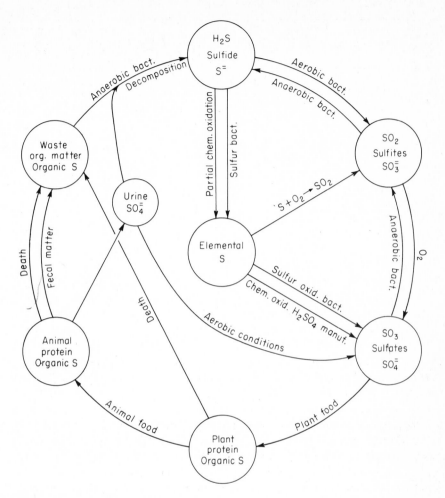

Figure 28-1 The sulfur cycle.

produced by anaerobic bacteria. Under anaerobic conditions, the sulfate ion is reduced to sulfide ion, which establishes an equilibrium with hydrogen ion to form hydrogen sulfide in accordance with its primary ionization constant $K_1 = 9.1 \times 10^{-8}$. The relationships existing between H_2S, HS^-, and S^{2-} at various pH levels in a 10^{-3} molar solution are shown in Fig. 28-2. At pH values of 8 and above, most of the reduced sulfur exists in solution as HS^- and S^{2-} ions, and the amount of free H_2S is so small that its partial pressure is insignificant, and odor problems do not occur. At pH levels below 8, the equilibrium shifts rapidly toward the formation of un-ionized H_2S and is about 80 percent complete at pH 7. Under such conditions the partial pressure of hydrogen sulfide becomes great enough to cause serious odor problems whenever sulfate reduction yields a significant amount of sulfide ion.

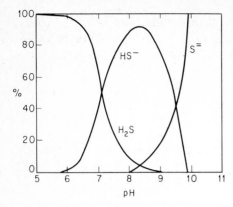

Figure 28-2 Effect of pH on hydrogen sulfide-sulfide equilibrium (10^{-3} molar solution, 32 mg H_2S/l).

Corrosion of Sewers

In many areas of the United States—particularly in the southern states where domestic wastewater temperatures are high, detention times in the sewers are long, and sulfate concentrations are appreciable—"crown" corrosion of concrete sewers has been an important problem. The difficulty is always associated with reduction of sulfates to hydrogen sulfide, and the hydrogen sulfide is often blamed for the corrosion. Actually H_2S, or hydrosulfuric acid as its aqueous solutions are called, is a weaker acid than carbonic acid and has little effect on good concrete. Nevertheless, "crown" corrosion of gravity-type sewers does occur, and hydrogen sulfide is indirectly responsible.

Gravity-type sewers provide an unusual environment for biological changes in the sulfur compounds present in wastewaters. Sewers are really part of a treatment system, for biological changes are constantly occurring during transportation. These changes require oxygen, and if sufficient amounts are not supplied through natural reaeration from air in the sewer, reduction of sulfates occurs, and sulfide ion is formed. At the usual pH level of domestic wastewaters, most of the sulfide is converted to hydrogen sulfide and some of it escapes into the atmosphere above the wastewater. Here it does no damage if the sewer is well ventilated and the walls and crown are dry. In poorly ventilated sewers, however, moisture collects on the walls and crown. Hydrogen sulfide dissolves in this water in accordance with its partial pressure in the sewer atmosphere. As such it does no harm.

Bacteria capable of oxidizing hydrogen sulfide to sulfuric acid are ubiquitous in nature and are always present in domestic sewage. It is natural that some of these organisms should infect the walls and crown of sewers at times of high flows or in some other manner. Because of the aerobic conditions normally prevailing in sewers above the wastewater, these bacteria oxidize the hydrogen sulfide to sulfuric acid,

$$H_2S + 2O_2 \xrightarrow{\text{bacteria}} H_2SO_4 \qquad (28\text{-}3)$$

and the latter, being a strong acid, attacks the concrete. This effect is particularly serious in the crown, where drainage is at a minimum. Figure 28-3 summarizes the important aspects of odor and corrosion problems in sewer systems.

28-2 METHODS OF ANALYSIS

Two methods of determining sulfates, employing gravimetric and turbidimetric procedures, are currently considered standard. The choice of method depends to a considerable extent upon the purpose for which the determination is being made and the concentration of sulfates in the sample.

Gravimetric

The gravimetric method is considered to yield the most accurate results and is the recommended standard procedure for sulfate concentrations above 10 mg/l. The quantitative aspects of this method depend upon the fact that barium ion combines with sulfate ion to form poorly soluble barium sulfate as follows:

$$Ba^{2+} + SO_4^{2-} \rightarrow \underline{BaSO_4} \qquad (28\text{-}4)$$

The precipitation is normally accomplished by adding barium chloride in slight excess to samples of water acidified with hydrochloric acid and kept near the boiling point. The samples are acidified to eliminate the possibility of precipita-

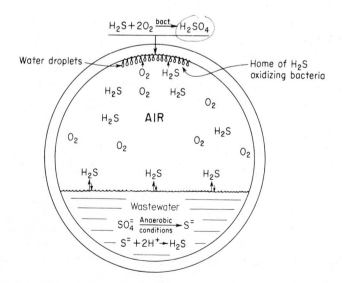

Figure 28-3 Formation of hydrogen sulfide in sewers, and "crown" corrosion resulting from oxidation of hydrogen sulfide to sulfuric acid.

tion of $BaCO_3$, which might occur in highly alkaline waters maintained near the boiling temperature. Excess barium chloride is used to produce sufficient common ion to precipitate sulfate ion as completely as possible.

Because of the great insolubility of barium sulfate ($K_{sp} = 1 \times 10^{-10}$), there is a considerable tendency for much of the precipitate to form in a colloidal condition that cannot be removed by ordinary filtration procedures. Digestion of the samples at temperatures near the boiling point for a few hours usually results in a transfer of the colloidal to crystalline forms, in accordance with the principle discussed in Sec. 3-6, and filtration can then be accomplished. The crystals of barium sulfate are usually quite small; for this reason, a special grade of filter paper (suitable for sulfate determinations) should be used. Some analysts prefer to use Gooch crucibles with asbestos mats. With reasonable care to make sure that all crystals have been transferred to the filter and with sufficient washing to remove all excess barium chloride and other salts, this method is capable of measuring sulfates with a high order of accuracy. Its major limitation is the time required.

Turbidimetric

The turbidimetric method of measuring sulfates is based upon the fact that barium sulfate tends to precipitate in a colloidal form and that this tendency is enhanced in the presence of a sodium chloride—hydrochloric acid solution containing glycerol and other organic compounds. By standardizing the procedure used to produce the colloidal sol of barium sulfate, it is possible to obtain results that are quantitative and acceptable for a great many purposes. The method is very rapid and has wide application, since samples with sulfate concentrations greater than 10 mg/l can be analyzed by taking smaller portions and diluting them to the recommended 50-ml sample size. Because of the variables that enter into a determination of this sort, it is recommended that at least one standard sample of sulfate ion be included in each set of samples to verify that conditions used in the test are comparable to those used in establishing the calibration curve.

Volumetric

A number of volumetric methods have been proposed for the determination of sulfates. None is considered standard at the present time. All of them suffer from a lack of sensitivity and precision and are not applicable to natural waters of potable character. They are used in analysis of boiler and other saline waters for control purposes.

28-3 APPLICATIONS OF SULFATE DATA

The sulfate content of natural waters is an important consideration in determining their suitability for public and industrial water supplies. The amount of sulfate in wastewaters is a factor of concern in determining the magnitude of problems that can arise from reduction of sulfates to hydrogen sulfide. In anaerobic digestion of sludges and industrial wastes, the sulfates are reduced to hydrogen sulfide, which is evolved with methane and carbon dioxide. If the gas is to be used in gas engines, the hydrogen sulfide content should not exceed 50 grains/100 ft^3. A knowledge of the sulfate content of the sludge or waste fed to digestion units provides a means of estimating the hydrogen sulfide content of the gas produced. From this information, the designing engineer can determine whether scrubbing facilities will be needed to remove hydrogen sulfide, and the size of the units required.

Many organic compounds contain sulfur as sulfates, sulfonates, or sulfides. During aerobic treatment of such wastes, complete utilization or dissimilation results in release of the organically bound sulfur as sulfate ion.

PROBLEMS

28-1 What is the significance of a high sulfate concentration in water supplies and in wastewater disposal?

28-2 List three conditions which must occur more or less simultaneously for "crown" corrosion of a sewer to take place.

28-3 List four precautions which must be observed to ensure an accurate gravimetric determination of sulfate concentration.

28-4 What is the purpose of digestion of the sample in the gravimetric analysis for sulfates?

28-5 In the determination of sulfate concentration by the gravimetric procedure, a 400-ml sample yielded 0.0360 g of BaSO$_4$. How many mg/l of sulfates were in the sample?

28-6 Why must exactly the same procedure be followed each time an analysis is made for sulfate concentration by the turbidimetric method?

28-7 What are two purposes for the conditioning reagent used in the turbidimetric determination of sulfate concentration?

28-8 From equilibrium considerations, calculate the relative proportions of sulfide in the H$_2$S, HS$^-$, and S^{2-} forms at (a) a pH of 6.0; (b) a pH of 7.5; (c) a pH of 10.0.

TWENTY-NINE

PHOSPHORUS AND PHOSPHATE

29-1 GENERAL CONSIDERATIONS

The phosphate determination has grown rapidly in importance in environmental engineering practice as engineers have realized the many ways in which phosphorus compounds affect phenomena with which they are concerned. The only inorganic compounds of phosphorus of significance in engineering practice are the phosphates or their molecularly dehydrated forms, usually referred to as polyphosphates or condensed phosphates. Organically bound phosphorus is usually a minor consideration.

Water Supplies

Polyphosphates are used in some public water supplies as a means of controlling corrosion. They are also used in some softened waters for stabilization of calcium carbonate to eliminate the need for recarbonation.

All surface water supplies support growths of minute aquatic organisms. The free swimming and floating organisms are called *plankton* and are of great interest to environmental engineers. The plankton are composed of animals, *zooplankton,* and plants, *phytoplankton*. The latter are predominantly algae, and since they are chlorophyll-bearing organisms, their growth is influenced greatly by the amount of fertilizing elements in the water. Research has shown that nitrogen and phosphorus are both essential for the growth of algae and that limitation in amounts of these elements is usually the factor that controls their rate of growth. Where both nitrogen and phosphorus are plentiful, algal blooms occur which may produce a variety of nuisance conditions. Experience has shown that such blooms do not occur when nitrogen or phosphorus or both are present in very limited amounts. The critical level for inorganic phosphorus has been established as somewhere near 0.005 mg/l or 5 μg/l under summer growing conditions.

Wastewater Treatment

Domestic wastewater is relatively rich in phosphorus compounds. Prior to the development of synthetic detergents, the content of inorganic phosphorus usually ranged from 2 to 3 mg/l and organic forms varied from 0.5 to 1.0 mg/l. Most of the inorganic phosphorus was contributed by human wastes as a result of the metabolic breakdown of proteins and elimination of the liberated phosphates in the urine. The amount of phosphorus released is a function of protein intake and, for the average person in the United States, this release is considered to be about 1.5 g/day.[1]

Most heavy-duty synthetic detergent formulations designed for the household market contain large amounts of polyphosphates as "builders." Many of them contain from 12 to 13 percent phosphorus or over 50 percent of polyphosphates. The use of these materials as a substitute for soap has greatly increased the phosphorus content of domestic wastewater. It has been estimated from sales of polyphosphates to the detergent industry that domestic wastewater probably contains from two to three times as much inorganic phosphorus at the present time as it did before synthetic detergents became widely used, unless local ordinances limit the use of phosphate-based detergents.

The organisms involved in biological processes of wastewater treatment all require phosphorus for reproduction and synthesis of new cell tissue. Domestic wastewater contains amounts of phosphorus far in excess of the amount needed to stabilize the limited quantity of organic matter present. This fact is demonstrated by the presence of appreciable amounts in effluents from treatment plants. Many industrial wastes, however, do not contain sufficient quantities of phosphorus for optimum growth of the organisms used in treatment. In such cases, deficiencies may be supplied by the addition of inorganic phosphates.

Fertilizing Value of Sludges

A major problem in wastewater treatment practice is the disposal of sludges remaining from aerobic and anaerobic treatment processes. All sludges contain nitrogen and phosphorus in significant amounts and have value for fertilizing purposes. The total phosphorus content of digested sludges is ordinarily about 1 percent and that of heat-dried activated sludge about 1.5 percent. In the United States, where phosphate fertilizers are relatively abundant and cheap, most sludges are sold on the basis of their nitrogen content, and little or no credit is given for the phosphorus.

[1] C. N. Sawyer, Factors Involved in Disposal of Sewage Effluents to Lakes, *Sewage and Ind. Wastes,* **26:** 317 (1954).

Boiler Waters

Phosphate compounds are widely used in steam power plants to control scaling in boilers. If complex phosphates are used, they are rapidly hydrolyzed to orthophosphate at the high temperatures involved. Control of phosphate levels is accomplished through determinations of orthophosphate.

29-2 PHOSPHORUS COMPOUNDS OF IMPORTANCE

Phosphorus compounds of wide variety are encountered in environmental engineering practice. A list of the more important ones is given in Table 29-1.

All the polyphosphates (molecularly dehydrated phosphates) gradually hydrolyze in aqueous solution and revert to the ortho form from which they were derived. The rate of reversion is a function of temperature and increases rapidly as the temperature approaches the boiling point. The rate is also increased by lowering the pH, and advantage is taken of this fact in the preparation of samples for the determination of complex phosphates. The hydrolysis of complex phosphates is also influenced by bacterial enzymes. The rate of reversion is very slow in pure waters but is more rapid in wastewaters. Experiments have shown that pyrophosphates are hydrolyzed more rapidly than tripolyphosphates in some waters, while in others the reverse is true. A matter of several hours and possibly days is required for the complete reversion of polyphosphate to orthophosphate, particularly at low temperatures or at high pH values.

From the considerations given above, it should be obvious that determinations for phosphorus or phosphates must employ procedures to measure polyphosphates if a true measure of the total inorganic forms present is to be obtained.

Table 29-1 Phosphorus compounds commonly encountered in environmental engineering practice

Name	Formula
Orthophosphates:	
Trisodium phosphate	Na_3PO_4
Disodium phosphate	Na_2HPO_4
Monosodium phosphate	NaH_2PO_4
Diammonium phosphate	$(NH_4)_2HPO_4$
Polyphosphates:	
Sodium hexametaphosphate	$Na_3(PO_3)_6$
Sodium tripolyphosphate	$Na_5P_3O_{10}$
Tetrasodium pyrophosphate	$Na_4P_2O_7$

29-3 METHODS OF DETERMINING PHOSPHORUS OR PHOSPHATE

The engineer is often interested in knowing the amounts of ortho, poly, and organic phosphorus present. Fortunately it is possible to measure orthophosphate with very little interference from polyphosphates because of their stability under the conditions of pH, time, and temperature used in the test. Both poly and organic forms of phosphorus must be converted to orthophosphate for measurement.

Orthophosphate

Phosphorus occurring as orthophosphate ($H_2PO_4^-$, HPO_4^{2-}, PO_4^{3-}) can be measured quantitatively by gravimetric, volumetric, or colorimetric methods. The gravimetric method is applicable where large amounts of phosphorus are present, but such situations do not occur in ordinary engineering practice. The volumetric method is applicable when phosphate concentrations exceed 50 mg/l, but such concentrations are seldom encountered except in boiler waters and digester supernatant liquors. The method involves formation of a precipitate, filtration, careful washing of the precipitate, and titration. The procedure is time-consuming, and most analysts prefer to use a colorimetric method, possibly at some sacrifice of accuracy.

Three colorimetric methods are used for measuring orthophosphate. They are essentially the same in principle but differ in the nature of the agent added for final color development. The chemistry involved is essentially as follows: Phosphate ion combines with ammonium molybdate under acid conditions to form a complex compound known as ammonium phosphomolybdate.

$$PO_4^{3-} + 12(NH_4)_2MoO_4 + 24H^+ \rightarrow (NH_4)_3PO_4 \cdot 12MoO_3$$
$$+ 21NH_4^+ + 12H_2O \quad (29\text{-}1)$$

When large amounts of phosphates are present, the phosphomolybdate forms a yellow precipitate that can be filtered and used for volumetric determination. At lower concentrations of phosphates, a yellow colloidal sol is formed which has been proposed as a basis of colorimetric measurement of intermediate concentrations. With concentrations of phosphates under 30 mg/l, the usual range in water analysis, the yellow color of the colloidal sol is not discernible, and other means of color development are necessary. In one modification, vanadium is added and forms a vanadomolydophosphoric acid complex which yields a much more intense yellow color, permitting analysis for phosphorus down to the mg/l or lower range.

The molybdenum contained in ammonium phosphomolybdate is also readily reduced to produce a blue-colored sol that is proportional to the amount of phosphate present. Excess ammonium molybdate is not reduced and therefore does not interfere. Either ascorbic acid or stannous chloride may be used as the reducing agent. The colored compound formed has never been isolated, and its formula is unknown. It is referred to as molybdenum or heteropoly blue in most cases. The chemistry involved with stannous chloride as the reducing agent may be represented in a qualitative manner as follows:

$$(NH_4)_3PO_4 \cdot 12MoO_3 + Sn^{2+} \rightarrow (\text{molybdenum blue}) + Sn^{4+} \qquad (29\text{-}2)$$

With the stannous chloride procedure an extraction procedure can be used both for increased sensitivity and to obtain accurate results when excessive interferences are present in the sample. In this case, the phosphomolybdate is extracted from the sample into a benzene-isobutanol solution prior to addition of the stannous chloride.

Polyphosphates

Polyphosphates may be converted to orthophosphates by boiling samples that have been acidified with sulfuric acid for at least 90 min. The hydrolysis may be hastened by heating in an autoclave at 20 psi. The excess acid added to speed the hydrolysis must first be neutralized before proceeding with the addition of the ammonium molybdate solution. The orthophosphate formed from the polyphosphate is measured in the presence of orthophosphates originally present in the sample by one of the methods applicable to orthophosphates. The amount of polyphosphate is obtained by difference as follows:

$$\text{Total inorganic phosphate} - \text{orthophosphate} = \text{polyphosphate} \qquad (29\text{-}3)$$

Organic Phosphorus

Engineers are often interested in measuring the amount of organic phosphorus present in industrial wastes or in sludges. This analysis requires that the organic matter be destroyed so that the phosphorus is released as phosphate ion. The organic matter may be destroyed by any of the three wet oxidation or digestion procedures listed in "Standard Methods." The oxidant used differs with the procedure and may be perchloric acid, sulfuric acid–nitric acid, or persulfate. Perchloric acid is the most rigorous oxidant, but is also the most hazardous to use. In order to avoid and prevent damage from explosions, special hoods for the digestion should be used, and care in the order of adding chemicals is essential. For these reasons, perchloric acid digestion should be carried out only by an experienced chemist and only if the added rigor of this procedure is necessary. Persulfate digestion is recommended if experience proves the results obtained are suitable for the need.

Once digestion has been accomplished, measurement of the phosphorus released can be made by any of the methods applied to orthophosphate. All forms of phosphorus (total) are measured in an organic phosphorus determination. Therefore the organic phosphorus is obtained as follows:

$$\text{Total phosphorus} - \text{inorganic phosphorus} = \text{Org-P} \qquad (29\text{-}4)$$

29-4 APPLICATIONS OF PHOSPHORUS DATA

Phosphorus data are becoming more and more important in environmental engineering practice as engineers appreciate their significance as a vital factor in life processes. In the past, the data have been used principally to control phosphate dosages in water systems for corrosion prevention and in boilers for control of scale. Phosphorus determinations are extremely important in assessing the potential biological productivity of surface waters, and in many areas limits are being established on amounts of phosphorus that may be discharged to receiving bodies of water, particularly lakes and reservoirs. Phosphorus determinations are routine in the operation of wastewater treatment plants and in stream pollution studies in many areas. Because of the importance of phosphorus as a nutrient in biological methods of wastewater treatment, its determination is essential with many industrial wastes and in the operation of waste treatment plants.

PROBLEMS

29-1 Discuss the significance of phosphorus in water pollution control.

29-2 What is the difference between orthophosphates, polyphosphates, and organic phosphorus? In which form must the phosphorus be for colorimetric analysis?

29-3 How is the analysis for phosphorus conducted to differentiate between the three forms of phosphorus?

29-4 Would you expect the analytical results for orthophosphate to be higher than, lower than, or the same as the original value in a sample of domestic wastewater which had been acidified to prevent bacterial action and stored for several days prior to analysis?

THIRTY

GREASE

30-1 GENERAL CONSIDERATIONS

The grease content of domestic and certain industrial wastes, and of sludges, is an important consideration in the handling and treatment of these materials for ultimate disposal. Grease is singled out for special attention because of its poor solubility in water and its tendency to separate from the aqueous phase. Although this characteristic is advantageous in facilitating the separation of grease by use of flotation devices, it does complicate the transportation of wastes through pipelines, their destruction in biological treatment units, and their disposal into receiving waters.

Wastes from the meatpacking industry, particularly where hard fats from the slaughtering of sheep and cattle are involved, have resulted in serious decreases in the carrying capacity of sewers. Such experiences, and other factors related to treatment or ultimate disposal, have served as the basis for ordinances and regulations governing the discharge of greasy materials to sewer systems or receiving waters and have forced the installation of preliminary treatment facilities by many industries for the recovery of grease or oil before discharge is permitted.

A number of problems are caused by grease in waste treatment practice. Very few plants have provisions for the separate disposal of grease to scavengers or by incineration; consequently that which separates as scum in primary settling tanks is normally transferred with the settled solids to disposal units. In sludge digestion tanks, the grease tends to separate and float to the surface to form dense scum layers, because of its poor solubility in water and its low specific gravity. Scum problems have been particularly severe where high-grease-content wastes, such as those from the meatpacking and oil and fat industries, have been admitted to public sewer systems. The vacuum filtration of sludge is also complicated by high grease content.

Not all the grease is removed from sewage by primary settling units. Appreciable amounts remain in the clarified wastewater in a finely divided emulsified form. During subsequent biological attack in secondary treatment units or in the receiving stream, the emulsifying agents are usually destroyed, and the finely divided grease particles become free to coalesce into larger particles which separate from the water. In activated sludge plants, the grease often accumulates into "grease balls" which give an unsightly appearance to the surface of final settling tanks. Both trickling filters and the activated sludge process are adversely affected by unreasonable amounts of grease which seems to coat the biological forms sufficiently to interfere with oxygen transfer from the liquid to the interior of the living cells. This is sometimes described as a "smothering" action.

Separation of floating grease in final settling tanks has been a problem in some treatment plants employing high-rate processes. This has been attributed to short-term contact of the waste with limited amounts of biological growths which destroy the emulsifying agents present but do not have sufficient adsorptive powers to hold the grease that is released, nor time to oxidize it. As a result, the grease is free to separate under quiescent conditions such as occur in final settling tanks or receiving waters.

30-2 GREASE AND ITS MEASUREMENT

The term *grease* applies to a wide variety of organic substances that are extracted from aqueous solution or suspension by hexane or trichloro-trifluoroethane (Freon). Hydrocarbons, esters, oils, fats, waxes, and high-molecular-weight fatty acids are the major materials dissolved by these solvents. All these materials have a "greasy feel" and are associated with the problems in waste treatment related to grease.

Hexane and Freon have been selected for the solvent in grease determinations because they are good solvents for all the materials normally associated with the term "grease" and have a minimum solvent power for other organic compounds. In the current 14th edition of "Standard Methods," Freon only is recommended as it has less explosion hazard than hexane. Chloroform, diethyl ether, and other solvents have been used in the past but are less desirable in one or more respects. Chloroform, for example, dissolves carbohydrates to a limited extent.

The method of determining grease by means of solvent extraction does not measure low-molecular-weight hydrocarbons such as gasoline. Preparation of the sample for extraction requires that it be dried at 103°C. As a result, all materials with boiling points below this temperature are lost, as well as significant amounts of all other materials that have appreciable vapor pressures at 103°C. Such compounds, except in unusual cases, are normally present in relatively small amounts in domestic wastewaters and are of little concern except in wastes from the petroleum industry. Most of the materials generally classed as "grease" have very low vapor pressures at 103°C and can be recovered essentially 100 percent by hexane or Freon extraction. In cases in which drying oils are present, some oxidation occurs at the unsaturated linkages during the drying procedure and may render them insoluble. Such oils, however, do not normally occur in domestic waste to any great extent.

Although the methods employed for the determination of grease may seem highly unrefined and inaccurate, they are the result of years of effort to obtain a reasonable measure of those things in water, domestic and industrial wastes, and sludges that tend to separate fróm the aqueous phase and create special problems. Once the engineer becomes fully acquainted with the purposes of the grease determination and understands the relative significance of volatile versus nonvolatile materials, he can usually adjust his thinking to the terms of the test method and its limitations.

30-3 METHODS OF ANALYSIS

All the common methods of determining grease depend upon a preferential solution of the greasy materials using extractions with hexane or Freon and are subject in some degree to the limitations discussed above. The methods employed for water, polluted waters, and sludges differ somewhat, and separate discussions are needed.

Water

The oil or grease content of relatively clean waters is not a routine determination and is seldom performed except in special cases in which accidental contamination has occurred. The choice of method of analysis depends upon the volatility of the contaminants. High-boiling-point materials may be measured by the extraction method given in "Standard Methods," but all materials with appreciable vapor pressures at 70°C must be measured by a special distillation procedure or by the use of infrared or other such analysis.

Wastewaters

Oils, fats, waxes, and fatty acids are the principal substances classed as grease in domestic wastewaters. Industrial wastewaters may contain simple esters and, possibly, a few other compounds in the same category.

The term "oil" represents a wide variety of substances ranging from low- to high-molecular-weight hydrocarbons of mineral origin, spanning the range from gasoline through heavy fuel and lubricating oils. In addition, it includes all glycerides of animal and vegetable origin that are liquid at ordinary temperatures. The fatty acids occur principally in a precipitated form as calcium and magnesium soaps. As such, they are insoluble in the solvents. Samples are acidified with hydrochloric acid to a pH of about 1.0 to release the free fatty acids. The reaction involved may be represented by the equation

$$Ca(C_{17}H_{35}COO)_2 + 2H^+ \rightarrow 2C_{17}H_{35}COOH + Ca^{2+} \qquad (30\text{-}1)$$

The high-molecular-weight fatty acids are relatively insoluble in water and are separated with other components of grease in the filtration procedure involved.

Filtration is considered acceptable practice, since it effectively separates those materials normally referred to as grease and allows low-molecular-weight and soluble materials, which are of no consequence, to escape in the filtrate. Drying of the filtered material removes water so that the solvent can penetrate the sample readily and accomplish separation of the grease in the 4-h extraction period normally provided. It also eliminates the possibility of appreciable amounts of water being carried into the extract and thereby simplifies the drying procedure. A Soxhlet type of extractor which provides intermittent batchwise extraction is used (Fig. 30-1).

Sludges

Sludges are frequently of such consistency and character that they are difficult to filter, and prolonged periods are often required to dry them sufficiently for solvent extraction. The current standard procedure involves the use of a chemical dehydration technique that eliminates the need for filtration and drying. The method consists of weighing a definite amount of sample, acidifying it to release fatty acids as shown in Eq. (30-1), and then adding a quantity of $MgSO_4 \cdot H_2O$ sufficient to combine with all free water by forming higher hydrated forms, $MgSO_4 \cdot 7H_2O$ being the ultimate. With the water in a chemically bound form, the sample is pulverized to facilitate extraction of grease. A Soxhlet extractor, as shown in Fig. 30-1, is used to separate the grease from the $MgSO_4 \cdot 7H_2O$ and the organic matter that is not grease.

30-4 APPLICATIONS OF GREASE DATA

In environmental engineering practice, grease determinations are made more or less routinely for a number of purposes. Many municipalities and other local authorities have ordinances for the regulation of the discharge of grease-bearing industrial wastes to sewer systems or to receiving waters and, of course, use grease and oil determinations for regulatory purposes. Industries that must treat their wastes to remove such materials use the test or some modification of it to determine the effectiveness of treatment units and to keep a record of the grease content of their discharges.

One of the major purposes of waste treatment facilities is to remove unsightly and obnoxious floating matter, of which grease is a major constituent. Grease determinations on raw and settled wastewaters give a measure of the

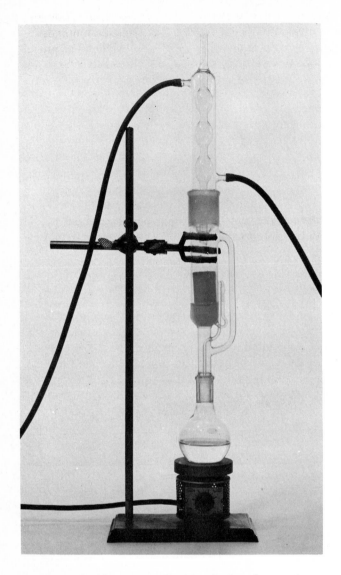

Figure 30-1 Soxhlet apparatus for determination of grease.

effectiveness of primary settling tanks, and determinations on final effluents provide a record of the efficacy of secondary treatment units, as well as the amounts actually discharged to receiving waters. The latter is particularly important where disposal is into recreational areas.

Grease determinations are used extensively in sludge disposal practice. Determinations on the raw and digested sludges, when properly adjusted to volumes involved in a given plant, allow calculation of grease destruction during anaerobic digestion. When scum problems occur in digestion units, determinations of grease often yield information of considerable value. The grease content of sludge is a factor in determining its suitability for use as a fertilizer, and purchasers often require a guarantee that the percentage shall not exceed a certain value.

PROBLEMS

30-1 Name five operating difficulties that are caused or intensified by grease at waste treatment plants.

30-2 List four important sources of grease which occur in domestic wastewater.

30-3 Why are samples of wastewater and sludge acidified prior to the separation of grease?

30-4 Define: (*a*) fats; (*b*) waxes; (*c*) soaps; (*d*) animal and vegetable oils; (*e*) mineral oils.

VOLATILE ACIDS

31-1 GENERAL CONSIDERATIONS

The volatile-acids determination is widely used in the control of anaerobic waste treatment processes. In the biochemical decomposition of organic matter that occurs, saprophytic bacteria of wide variety hydrolyze and convert the complex materials to low-molecular-weight compounds, as discussed in Secs. 7-7 through 7-10. Among the low-molecular-weight compounds formed, the short-chain fatty acids, such as acetic, propionic, butyric, and to a lesser extent valeric, isovaleric, and caproic, are important components. These low-molecular-weight fatty acids are termed *volatile acids* because they can be distilled at atmospheric pressure. An accumulation of volatile acids can have a disastrous effect upon anaerobic digestion if the buffering capacity of the system is exceeded and the pH falls to unfavorable levels.

In anaerobic digestion units that are operating in a stabilized condition, two groups of bacteria work in harmony to accomplish the destruction of organic matter. The saprophytic organisms carry the degradation to the acid stage, and then the methane-forming bacteria complete the conversion into methane and carbon dioxide. When a sufficient population of methane-forming bacteria is present and environmental conditions are favorable, they utilize the end products produced by the saprophytic bacteria as fast as they are formed. As a result, acids do not accumulate beyond the neutralizing ability of the natural buffers present, and the pH remains in a favorable range for the methane bacteria. Under such conditions the volatile-acid content of digesting sludges usually runs in the range of 50 to 250 mg/l, expressed as acetic acid.

Methane-forming bacteria are ubiquitous in nature, and some are always present in domestic wastewater and sludge derived therefrom. Their popula-

494

tion, however, is very small compared with that of the saprophytic bacteria. This disparity in numbers is the reason for the troubles encountered in starting digestion units without benefit of "seeding" sludge. Raw sewage sludge has a relatively low buffering capacity, and when it is allowed to ferment anaerobically, volatile acids are produced so much faster than the few methane bacteria present can consume them that the buffers are soon spent and free acids exist to depress the pH. At pH values below 6.5, methane bacteria are seriously inhibited but the saprophytic bacteria are not until pH levels fall to about 5. Under such unbalanced conditions, the volatile acids concentration continues to increase to levels of 2000 to 6000 mg/l or more, depending upon the solids content of the sludge. Active methane digestion may never develop in such mixtures unless the sludge is diluted or neutralizing agents, such as lime, are added to produce a favorable pH for the methane bacteria. The volatile acids determination, in conjunction with pH measurements, is valuable in control of environmental conditions during the initiation of methane digestion.

Successful operation of anaerobic digestion units depends upon maintaining a satisfactory balance between the methane and saprophytic bacteria. The methane bacteria appear to be the most susceptible to changes in environmental conditions and food load. They are affected much more radically by changes in pH and temperature than the saprophytic bacteria. Inhibitions caused by changes in either of these factors result in a decreased rate of destruction of volatile acids; consequently volatile acids begin to accumulate in the system. The saprophytic bacteria are known to reproduce more rapidly than the methane bacteria. Under increased food loads, volatile acids may be formed faster than the slow-growing methane organisms can take care of them. This discrepancy results in an accumulation of volatile acids in the system. Sludge must be removed or transferred from digestion units on occasion; however, removal of too large an amount will deplete the methane organism population to levels where volatile acids cannot be destroyed as fast as they are formed and accumulations will develop. Volatile acid determinations are extremely important in detecting the presence of unbalanced conditions in digestion units caused by any of the factors mentioned above. The onset of unfavorable conditions can be detected almost immediately, and usually several days in advance of other methods, such as through pH measurements.

31-2 THEORETICAL CONSIDERATIONS

Volatile acids are formed as intermediates during the anaerobic degradation of carbohydrates, proteins, and fats, as discussed in Secs. 7-7 through 7-10. Figure 31-1 shows just a few of the many steps through which a complex waste such as domestic waste sludge must pass during its conversion to methane gas. Propionic acid results as an intermediate mainly from the fermentation of the carbohydrates and proteins present, and about 30 percent of the complex waste is converted to this acid before finally being converted to methane gas. Acetic

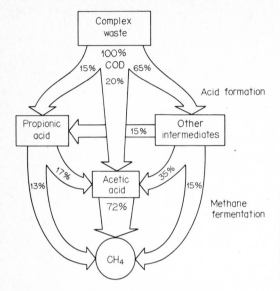

Figure 31-1 Pathways in methane fermentation of complex wastes such as domestic waste sludge. Percentages represent conversion of waste COD by various routes. *(P. L. McCarty, Anaerobic Waste Treatment Fundamentals, Public Works,* **107**, *September 1964.)*

acid is the most abundantly produced volatile acid and is formed as an intermediate during the anaerobic treatment of almost all organic material. With a complex waste such as domestic sludge, about 72 percent of the organic matter is converted to acetic acid before finally being changed into methane gas. The importance of the acetic and propionic acid intermediates is evident when it is considered that about 85 percent of the waste is converted to these two acids before methane gas is formed. A large portion of the remaining 15 percent of the methane results from the fermentation of other volatile acids, such as formic and butyric acids.

The anaerobic biological degradation of wastes thus results in the production of large quantities of organic acids. If these acids are not converted to methane gas as rapidly as they are formed, their concentration will increase, and will lower the pH. The major buffering materials in digesting sludge which tend to prevent a drop in pH are bicarbonates, which in equilibrium with carbonic acid tend to regulate the hydrogen ion concentration:

$$H_2CO_3 \rightleftharpoons H^+ + HCO_3^- \tag{31-1}$$

$$\frac{[H^+][HCO_3^-]}{[H_2CO_3]} = K_1 \tag{31-2}$$

or

$$[H^+] = K_1 \frac{[H_2CO_3]}{[HCO_3^-]} \tag{31-3}$$

On the basis of Henry's law, an equilibrium also exists between the partial pressure of carbon dioxide in the gas phase and the carbonic acid concentration in solution. The effect of this equilibrium and that expressed by Eq. (31-3) on the pH in a digester is illustrated in Fig. 31-2, which indicates that the higher the

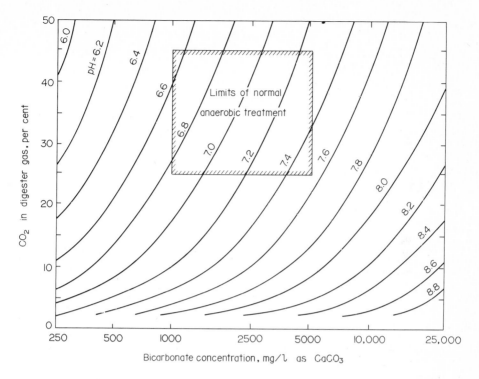

Figure 31-2 Relationship between pH, bicarbonate concentration, and carbon dioxide concentration at 35°C. *(P. L. McCarty, Anaerobic Waste Treatment Fundamentals, Public Works,* **123,** October 1964.)

bicarbonate concentration and the lower the percent carbon dioxide in the digester gas, the higher the pH.

As volatile acids accumulate in a digester, they destroy the bicarbonate buffer and increase the carbon dioxide concentration:

$$R—COOH + HCO_3^- \rightarrow R—COO^- + H_2O + CO_2 \qquad (31\text{-}4)$$

These changes, based on Eq. (31-3) and Fig. 31-2, result in a gradual drop in pH. When the bicarbonate concentration decreases below 1000 mg/l as $CaCO_3$, further accumulation of acid decreases the pH very rapidly. For these reasons it is necessary to maintain buffer capacity by adding basic materials from outside sources or to exercise some control over the rate of volatile acid formation if pH conditions favorable to methane forming bacteria are to be maintained.

31-3 METHODS OF DETERMINING VOLATILE ACIDS

Three methods of determining volatile acids are in current usage. One method uses column-partition chromatography and the other two involve distillation.

Column-Partition Chromatography

The column-partition chromatographic method allows measurement of nearly 100 percent of all volatile acids present and is fairly rapid. Other organic acids such as pyruvic, phthalic, fumaric, lactic, succinic, and oxalic acids are also measured to some degree by this procedure. However, these acids generally do not occur in high concentrations in digesting sludge so that results from the column-partition method are usually about the same as from the distillation procedures. Because acids other than the volatile acids are included in the procedure, "Standard Methods" indicates this method is for "organic acids" rather than for volatile acids.

The chromatographic method is a modification of a procedure originally designed to allow separation and measurement of the individual volatile acids.[1] In partition chromatography two solvents are involved. One is fixed on or in a solid adsorbent, such as silicic acid, and placed in a column. A sample containing the materials to be separated is placed on top of the column. The second or mobile solvent is then added and carries the materials to be separated through the column, where they are continuously partitioned between the immobile solvent and the moving solvent, thus producing a differential migration of the various materials. Materials which are more soluble in the moving phase than in the stationary phase will emerge from the column first. This is the same general principle described for gas chromatography in Sec. 11-4, but with the moving phase in a different physical state.

In routine operation of an anaerobic waste treatment system, the volatile acid analysis is used mainly for control to determine whether or not the system is in balance; thus the total volatile acid concentration rather than the concentration of each individual acid is of most interest. For this purpose, the column-partition chromatography procedure has been simplified, all volatile acids being separated from inorganic acids and other salts in the sample so that a total volatile-acid analysis might be made. The suspended solids are first removed from the sample either by filtration or centrifugation, and the sample is then acidified with a strong inorganic acid, such as sulfuric, to convert the volatile acids from the ionized form to the un-ionized form:

$$R{-}COO^- + H^+ \rightarrow R{-}COOH \tag{31-5}$$

which is highly soluble in the mobile chloroform-butanol solvent used.

The acidified sample is distributed uniformly over a short column containing oven-dried silicic acid, which readily absorbs the sample water, thus forming the stationary solvent phase. A given quantity of the chloroform-butanol mixture is then passed through the column and selectively carries with it the un-ionized volatile acids, which are much more soluble in this mixture than in

---

[1] H. F. Mueller, A. M. Buswell, and T. E. Larson, Chromatographic Determination of Volatile Acids, *Sewage and Ind. Wastes,* **28:** 255 (1956).

the stationary water phase. The sulfuric acid and other ionized salts, however, are more soluble in the water and so are left behind. It is important that sufficient mobile solvent be added to extract all the volatile acids present, but not so much that the inorganic acids are carried through.

The extracted volatile acids are measured by titration with sodium hydroxide to the phenolphthalein end point:

$$R—COOH + NaOH \rightarrow R—COO^- + Na^+ + H_2O \qquad (31\text{-}6)$$

A sodium hydroxide in methanol solution is used for this titration because water is insoluble in the organic solvent. This solution, however, is unstable and must either be standardized frequently or prepared fresh daily. It can be made readily by diluting a proper quantity of 1.00 N aqueous sodium hydroxide with methanol.

Direct Distillation

The direct-distillation method is commonly used for the routine determination of volatile acids. The method is rapid and sufficiently accurate for most practical purposes, since it is not usually necessary to know volatile-acid concentrations more accurately than ±50 mg/l. In this procedure use is made of the fact that all of the low-molecular-weight fatty acids up to octanoic acid have significant vapor pressures at 100°C, the boiling point of water, so that sludge distillates will contain appreciable amounts of these acids. The vapor pressures of the acids are shown in Table 31-1.

Digesting sludges normally have a pH in the range of 6.5 to 7.5, and under such conditions organic acids exist largely in the ionic form and cannot be distilled. By the addition of a strong nonvolatile acid, such as sulfuric, the

Table 31-1 Vapor pressures of volatile acids at 100°C

Acid	Vapor pressure at 100°C, mm Hg*
Formic	753
Acetic	417
Propionic	183 (calc.)
Butyric	70 (calc.)
Valeric	28
Hexanoic	10.6
Heptanoic	4.1
Octanoic	1.6

* Data from "International Critical Tables," McGraw-Hill, New York.

organic acids are converted to the un-ionized form as represented by acetate in the equation

$$\underset{\text{Acetate anion}}{CH_3COO^-} + H^+ \rightarrow \underset{\substack{\text{Un-ionized} \\ \text{acetic acid}}}{CH_3COOH} \tag{31-7}$$

Usually sufficient acid is added to reduce the pH below 1.0.

It would be expected from the vapor-pressure data in Table 31-1 that the low-molecular-weight acids would distill most readily because of their greater vapor pressures. This is not the case, however. Formic, acetic, and propionic acids are the most difficult to separate by distillation in dilute aqueous solution.[2] This is because aqueous solutions are not ideal in character. Molecular association occurs between the acid and water molecules and varies with the particular acid. As a result, binary mixtures of all three classes described in Sec. 3-5 are formed.

Formic acid and water form a Class II mixture, and water is the major component of the vapor phase when dilute solutions are distilled. Acetic acid and water form a Class I mixture, and the vapor phase is predominantly water vapor. Fractionation of the vapor in each case produces a distillate that is essentially pure water and the acids remain in the undistilled liquor. It is for this reason that fractionation of the vapors must be kept at a minimum in the volatile-acids determination.

Propionic, butyric, and other fatty acids that are soluble in water form Class III mixtures. The distillate contains the acid and water in a constant ratio until all the acid is distilled.

It is unnecessary to separate sludge from samples prior to distillation if care is used to prevent excessive heating of the walls of the distillation flask above the level of the mixture in the flask. If this is allowed to happen, some decomposition of sludge may occur in the strong acid environment, which may give rise to the production of organic acids. Also, reduction of sulfuric acid may occur with the production of sulfur dioxide. The latter, being an acid anhydride, will dissolve in the distillate to produce sulfurous acid and give abnormally high results. There is little chance of this happening when electric heaters are used, but it frequently happens if uncontrolled gas flames are used for heat.

The distillation rate is an important consideration in this determination and should be closely controlled. At low rates of distillation considerable fractionation may occur in the neck of the flask and the volatile acids will not carry over in the distillate as desired. Care must be taken to stop the distillation when the appropriate amount of distillate has been formed. The residue remaining in the flask is in contact with the excess sulfuric acid used, and the acid becomes more concentrated as the distillation progresses. If the distillation is carried too far,

---

[2] A. M. Buswell and S. L. Neave, Laboratory Studies on Sludge Digestion, *Illinois State Water Survey Bull.* **30:** 76 (1930).

the sulfuric acid will cause decomposition of the sludge and be reduced as described above.

The distilled acids are measured quantitatively by titration with a standardized NaOH solution to the phenolpthalein end point [Eq. (31-6)]. Approximately 70 percent of the volatile acids contained in sludge samples is distilled from samples treated by the recommended procedure. This matter is taken into account in calculations of the volatile acids present.

Steam Distillation

Many of the limitations of the direct-distillation method can be overcome by use of a steam-distillation procedure. In this method, sludge solids are separated, and the determination of volatile acids is made on the sludge liquor. Steam is generated in a separate unit, and excessive heating of the sample containing sulfuric acid is avoided. Sufficient heat is usually supplied to the distillation flask containing the sample to prevent dilution by condensation, and the accessory heat is controlled in a manner to prevent concentration of the sample. The distillation may be continued as long as desired, and essentially complete recoveries of acid can be obtained in the distillate.

The steam-distillation procedure requires considerable time, and its use is not ordinarily justified except for research purposes.

31-4 APPLICATIONS OF VOLATILE ACID DATA

Volatile-acid determinations have been valuable in providing information concerning the anaerobic degradation of organic matter and the environmental conditions best suited for optimum activity of methane-producing bacteria. Much research on anaerobic processes, particularly in regard to the nature of the substrate undergoing decomposition, remains to be done before anaerobic treatment can be applied intelligently to a wide variety of industrial wastes. The volatile acids determination, because of its ability to detect when methane organisms fail to keep pace with the saprophytic organisms, will continue to be a valuable test in research work.

The value of volatile acid data in routine control of anaerobic digestion units at wastewater treatment plants and other installations has been amply demonstrated. These data are needed especially in the control of units that are operating at or near design capacity. With the development of high-rate digestion processes, the test will become more important in the future, as corrective measures will undoubtedly have to be made more promptly. The volatile acid test offers the most promise of providing pertinent information as quickly as possible and at reasonable expense.

PROBLEMS

31-1 Discuss the significance of the volatile acids determination in anaerobic sludge digestion.

31-2 (a) Define what is meant by volatile acids; (b) indicate what are the most prevalent volatile acids formed during anaerobic treatment; (c) indicate the general classes of organic compounds from which each of the most prevalent volatile acids result.

31-3 What is the purpose of acidification of samples for the volatile acid analysis?

31-4 What is the principle involved in the separation of volatile acids by column-partition chromatography?

31-5 Explain why the lower-molecular-weight volatile acids which have the higher vapor pressure at 100°C have lower fractional recoveries in distillates than the higher-molecular-weight acids.

31-6 What precautions must be exercised in the direct distillation procedure for volatile acids?

THIRTY-TWO

GAS ANALYSIS

32-1 GENERAL CONSIDERATIONS

Anaerobic decomposition of sludges and of some liquid wastes, particularly those with BOD values exceeding 1500 mg/l, is considered the most economical method of treatment when the organic materials involved are amenable to methane fermentation. Anaerobic processes are popular because operating costs are low, although capital investments are usually high.

The gas produced in anaerobic digestion of sludges usually contains 33 to 38 percent carbon dioxide, 55 to 65 percent methane, small amounts of hydrogen, some nitrogen, and traces of hydrogen sulfide; and it has a heating value in the neighborhood of 600 Btu/ft^3. The caloric value of the gas produced is normally in excess of the heat requirements for maintaining proper temperatures in the digestion units. The excess gas has value for producing power for useful work.

The composition of gas produced during anaerobic digestion varies somewhat with the environmental conditions in the digester. It changes rapidly during the period that digestion is being initiated, such as when digesters are first started, and when normal digestion is inhibited. For digesters operating in a routine manner and being fed a given substrate regularly, the composition of the gas produced is fairly uniform; however, the ratio of carbon dioxide to methane varies radically with the character of the substrate undergoing decomposition. Buswell and Boruff[1] have shown that methane fermentation of the three common classes of organic materials produces carbon dioxide and methane in the following ratios:

[1] A. M. Buswell and C. S. Boruff, The Relation between the Chemical Composition of Organic Matter and the Quality and Quantity of Gas Produced during Sludge Digestion, *Sewage Works J.,* **4:** 454 (1932).

	Carbon dioxide	Methane
Carbohydrates	50	50
Low-MW fatty acids	38	62
High-MW fatty acids	28	72
Proteins	{ 24	76
	31	69

Because of the wide variety of organic substances (usually of unknown composition) that are subjected to methane fermentation, it has become common practice to analyze the gases produced to determine their fuel value and, in some cases, to maintain a check on the behavior of the digestion units. The idea has been advanced that the onset of digestion troubles is accompanied by an increase in the carbon dioxide content of the gas produced and that this test can be used in place of the volatile-acids determination to detect such conditions. As a result, a limited analysis of digester gas to determine its carbon dioxide content is becoming common. Experience will demonstrate whether carbon dioxide measurements can substitute for volatile-acids determinations in this capacity. Because of the influence of the substrate on the ratio of carbon dioxide to methane in the gas produced, it is reasonable to assume that interpretations based upon carbon dioxide content alone may lead to faulty conclusions.

32-2 METHODS OF ANALYSIS

Gas analysis can usually be accomplished either by the gasometric (volumetric) procedure or by gas chromatographic analysis, except for the measurement of hydrogen sulfide, which usually occurs in amounts too small to be measured by either of these procedures. In general, gasometric analyses are quite accurate and are suitable for the determination of oxygen, methane, hydrogen, and carbon dioxide. Nitrogen is usually determined in such analyses by an indirect procedure. Gasometric analyses are quite time-consuming. However, the equipment required is relatively simple, needs no calibration before use, and therefore is particularly suitable when analyses are conducted on an infrequent basis.

Gas chromatographic analysis has the distinct advantage of speed; a gas analysis can be completed in only a few minutes. For this advantage to be utilized, however, the instrument must have been previously calibrated for each gas of interest, the oven must have reached a constant temperature, and the detector must be giving a stable response. Because of these requirements, this method of analysis is more suited to routine work where gas analyses are conducted several times each week, making it permissible to keep the instrument always in operating condition. This method is sufficiently accurate for most practical purposes, although not quite as good as gasometric analysis.

Although it is possible to separate hydrogen, oxygen, nitrogen, methane, carbon dioxide, and hydrogen sulfide by gas chromatography, this is not presently practical on a routine basis, using one instrument with a single column and detector. In addition, the quantities of hydrogen or hydrogen sulfide occurring in digester gas are usually too small in concentration to be measured with any accuracy. Gas chromatography is mainly suited to routine analysis to determine the relative proportions of methane, carbon dioxide, and "air" (nitrogen plus oxygen) occurring in a sample of digester gas. Normally, these are the main gaseous components of interest.

32-3 GASOMETRIC ANALYSIS

Early methods of gasometric analysis employed separate measurement of carbon dioxide and oxygen, followed by a slow simultaneous combustion of hydrogen and methane. The analysis was completed by measuring the amount of carbon dioxide produced during the combustion of the methane and then employing a knowledge of Gay-Lussac's law of combining volumes (Sec. 2-9) to determine the amounts of methane and hyrogen present in the mixture.

Apparatus employing a slow-combustion unit, such as the Orsat apparatus shown in Fig. 32-1, is sometimes used in gas analysis. However, its operation in the determination of hydrogen and methane is somewhat hazardous because of the possibility of explosions, and it is therefore not recommended. Other devices, such as the Burrell apparatus shown in Fig. 32-2, provide for separate oxidation of hydrogen and methane. Hydrogen is oxidized by passing the gas through a heated unit charged with cupric oxide, and methane is oxidized in a separate unit by bringing a mixture of it and oxygen in contact with a catalyst at relatively low temperatures. Explosion hazards are completely eliminated.

Carbon Dioxide

Carbon dioxide is measured by bringing a sample of known volume, usually 100 ml, into contact with a solution of potassium hydroxide. The carbon dioxide reacts with the hydroxide to form potassium carbonate, as shown in the equation

$$CO_2 + 2KOH \rightarrow K_2CO_3 + H_2O \qquad (32\text{-}1)$$
$$\text{1 vol.} \qquad\qquad \text{0 vol.}$$

In the reaction, the carbon dioxide disappears from the gaseous phase, and the potassium carbonate formed remains in the liquid phase; therefore the loss in volume of the gas is equal to the carbon dioxide content. Actually, any hydrogen sulfide present in the gas also combines with the potassium hydroxide,

$$H_2S + 2KOH \rightarrow K_2S + 2H_2O \qquad (32\text{-}2)$$
$$\text{1 vol.} \qquad\qquad \text{0 vol.}$$

Figure 32-1 Orsat gas-analysis apparatus equipped with slow-combustion chamber. *(Metcalf and Eddy.)*

but the volume of hydrogen sulfide is usually so small that its effect may be ignored.

Potassium hydroxide is used instead of sodium hydroxide to absorb carbon dioxide because of the greater solubility of potassium carbonate. If sodium hydroxide is used, sodium carbonate tends to precipitate. Some of it usually floats and clogs the capillary passages.

Oxygen

Theoretically there is very little possibility of oxygen being present in the gas produced by anaerobic digestion units. However, small amounts may gain en-

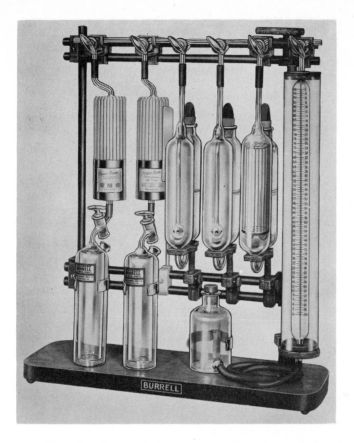

Figure 32-2 Burrell gas-analysis apparatus equipped for separate low-temperature oxidation of hydrogen and catalytic oxidation of methane. *(Burrell Corp.)*

trance to the sample during the sampling procedure and during charging of the gas-analysis apparatus. It is therefore good practice to analyze for oxygen. The presence of more than 0.1 to 0.2 percent usually indicates poor technique in sampling and in transferring the sample to the gas-analysis apparatus.

Oxygen in sludge gas may be measured by the use of any of three reagents: (1) alkaline pyrogallol, a strongly alkaline solution of pyrogallic acid (pyrogallol); (2) acid cuprous chloride solution; and (3) chromous chloride solution.

Under alkaline conditions, pyrogallol (1,2,3-trihydroxybenzene) is oxidized by oxygen. The end products of the reaction have not been clearly defined but are, presumably, carbon dioxide and organic acids, both of which are held as potassium salts in the absorbing solution. If all carbon dioxide has been removed from the sample before bringing it into contact with the alkaline pyrogallol, any decrease in sample volume will be due to removal of oxygen.

Acid solutions of cuprous chloride have been used to determine oxygen.

Their capacity is quite limited, and solutions must be renewed frequently. The reaction involved is

$$4Cu^+ + O_2 + 4H^+ \rightarrow 4Cu^{2+} + 2H_2O \qquad (32\text{-}3)$$

1 vol.   0 vol.

Under the acid conditions prevailing, the oxygen appears as water, which has zero volume as far as the gaseous phase or sample volume is concerned.

Chromous chloride solutions are also used to measure oxygen. They are especially effective and must be carefully protected from the air when units are charged with the reagent. Chromous chloride reacts with oxygen thusly:

$$4Cr^{2+} + O_2 + 4H^+ \rightarrow 4Cr^{3+} + 2H_2O \qquad (32\text{-}4)$$

1 vol.   0 vol.

Hydrogen

Hydrogen may be determined separately in the presence of methane by passing the gas mixture over cupric oxide maintained at a temperature in the range of 290 to 300°C. Under such conditions hydrogen is oxidized to water, but methane is not oxidized. The water vapor formed condenses at the temperatures to which the sample must be reduced for subsequent volume measurements, and therefore loss in volume after contact with heated cupric oxide is a measure of the hydrogen present.

Methane

After removal of hydrogen, methane may be determined in the residual gas by slow combustion or by catalytic oxidation. In either case oxygen is required, but the technique for each is very different.

Catalytic oxidation A mixture of the gas containing methane and oxygen is used in this method, and the oxidation is performed catalytically at temperatures below the kindling point; consequently an explosion does not occur. The volume of oxygen required is determined from the equation

$$CH_4 + 2O_2 \rightarrow CO_2 + 2H_2O \qquad (32\text{-}5)$$

1 vol.   2 vol.   1 vol.   0 vol.

It will be noted that 2 volumes of oxygen are required to oxidize 1 volume of methane and that 1 volume of carbon dioxide is produced. The 2 volumes of water vapor produced are reduced to zero volume at the temperature at which final gas volumes are measured. Since 2 volumes of oxygen are required for each volume of methane, at least $2\frac{1}{2}$ volumes should be used to provide a sufficient excess to carry the reaction to completion.

After the analyses for carbon dioxide, oxygen, and hydrogen are finished, the volume of residual gas containing methane is about 60 to 70 ml if a 100-ml sample is used initially. It is impossible to mix 60 or 70 ml of gas with $2\frac{1}{2}$ times its volume of oxygen in the equipment provided. The maximum-size sample of methane that may be used in a 100-ml buret and be mixed with $2\frac{1}{2}$ times its volume of oxygen is about 28 ml. The usual procedure is to waste sufficient residual gas to leave a sample ranging in size from 20 to 25 ml. Some analysts prefer not to waste the gas but store the excess in the oxygen absorption pipet as a reserve supply in case it is needed. Oxygen is next admitted to the sample in proper amount, and then the mixture is brought into contact with the catalyst.

The methane content of the gas used for the combustion may be determined in two ways. Inspection of Eq. (32-5) will show that 2 volumes of oxygen are used for each volume of methane oxidized and 1 volume of carbon dioxide results. A contraction in gas volume occurs that is equal to the amount of oxygen used. Since this is equal to two times the methane,

$$\text{Volume of methane} = \tfrac{1}{2} \text{ total contraction}$$

The residual gas, after measurement to determine total contraction, still contains carbon dioxide in a volume equal to the methane originally present. It may be measured by bringing it into contact with potassium hydroxide and measuring the residual gas to get the volume absorbed.

Final calculations of methane content must be based upon an adjustment of the values obtained on the test portion of the methane-bearing sample to the total volume. For example, if 65 ml remained after carbon dioxide, oxygen, and hydrogen were removed and 25 ml was used for the methane determination, the values should be multiplied by a factor of 65/25 to obtain the percentage of methane.

Slow combustion The separate determination of methane by the slow-combustion method is not common. The procedure is essentially the same as for a combination of hydrogen and methane. Analysis of the data is exactly as described under catalytic oxidation.

Nitrogen

Nitrogen is a relatively inert gas and remains unchanged at the end of the usual gas-analysis procedure. It is assumed that it is the only component of any significance that behaves in this manner. It is customary to total the carbon dioxide, oxygen, hydrogen, and methane percentages and subtract them from 100. The difference represents inert gases and is reported as nitrogen.

Sources of Error

There are five major sources of error in gas analysis, as follows:

Collection, storage, and handling of samples Unless special care is taken in the collection of samples, contamination by air occurs. Samples should always be collected in glass or metal tubes, particularly if an appreciable time elapses before analysis can be made. Gum-rubber balloons are not suitable because they are pervious to hydrogen and methane. Transfer of gas from the sample tube to the gas-analysis apparatus requires the use of a displacing fluid, and some modification of the sample is apt to occur. Also, some air may gain entrance.

Confining fluid Mercury is the ideal confining fluid because of the insolubility of all gases in it but, because of its great density and cost, it is seldom used, except in precision-type instruments. For ordinary purposes, a high degree of accuracy is not needed and less ideal confining fluids can be used. Water has much too great a solvent power for all the gases involved to serve satisfactorily. However, it has been found that an aqueous solution containing 20 percent sodium sulfate and 5 percent sulfuric acid has markedly reduced solvent powers. This is the mixture normally used in portable equipment but does introduce some error in analysis.

Incomplete combustion of methane During the combustion of methane, a high concentration of oxygen is present at the start of the combustion, but as the combustion proceeds, the oxygen concentration decreases markedly, owing to use and to dilution by the carbon dioxide formed. Unless a volume of oxygen at least $2\frac{1}{2}$ times the size of the gas sample is used, an oxygen deficiency may occur. Incomplete combustion is a common cause of high nitrogen values and of low methane values.

Temperature changes In gas analysis, small changes in temperature can cause serious errors. This is a special problem during the measurement of hydrogen and methane, where the reactions are conducted at high temperatures. There is always a tendency to measure the volume remaining after combustion before the temperature of the gas has returned to the original value. This may lead to positive errors in some instances and negative errors in others.

32-4 GAS CHROMATOGRAPHIC ANALYSIS

Gas chromatography affords a rapid and simple method of gas analysis when used on a routine basis. The principles of this method were outlined in Sec. 11-4. An instrument equipped with a thermal conductivity detector is usually used for digester gas analysis, and helium is normally used as a carrier gas to sweep the gas sample through the instrument. Only about one or two milliliters of a gas sample are required for the analysis, and this is usually introduced into the instrument with a syringe, although some instruments are equipped with a gas sampling port, which simplifies sample measurement and introduction.

One of the most important items in the gas chromatograph is the packing material used in the column for separation of the gaseous components. Several commercially available column packings are listed in "Standard Methods," together with the gases they are particularly suitable for separating. Columns which separate air (oxygen plus nitrogen), methane, and carbon dioxide are necessary for routine analysis of digester gas. Columns which also give separation between the two components of air are more desirable, however. The ability to determine the presence of oxygen is desirable since this is a good check on the adequacy of sampling technique as discussed under the gasometric method.

Most instruments have a sufficiently stable operation so that at a given temperature of operation and gas flow rate a linear relationship will exist between height of a peak on the gas chromatogram and the concentration of a gaseous component. Thus, by using a pure gas standard and injecting different volumes into the gas chromatograph, a curve of peak height versus gas volume may be prepared. The percentage by volume of a given gas in a sample is then equal to its measured volume divided by the total of the sample injected.

Thermal-conductivity detectors tend to lose sensitivity with time, with the result that the peak height for a given volume of gas tends to decrease. This can be corrected by frequent standardization. However, it has been found that the decrease in peak height is proportionally the same for methane, carbon dioxide, and air. Since digester gas is composed almost entirely of these gases, the summation of their respective concentrations should total almost 100 percent. Any significant decrease below this total will indicate a decrease in detector sensitivity, provided that other analytical errors have not been made. If this is true, then the correct percentage for each component can be determined by proportioning each up to give a total of 100 percent. For example, if the measured percentages of the three components total 90 and the measured methane percentage is 54 percent, its actual percentage would be 54(100/90) or 60 percent.

Sources of Error

Errors in collection, storage, and handling of samples will be the same as with the gasometric analysis method. In addition, changes in instrument temperature or carrier gas flow rate, changes in detector sensitivity, degeneration of column packing, and inaccuracies in measurement of volume of injected gas samples will all result in errors. With digester gas the summation of the methane, carbon dioxide, and air percentages should total close to 100 percent. If it does not, then one of the above sources of errors might be the cause.

32-5 HYDROGEN SULFIDE

The measurement of hydrogen sulfide is particularly important where sludge gas is to be used for fuel in gas engines. Most engine manufacturers specify that the gas used should not have more than 50 grains of H_2S/100 ft^3 (1.14 mg/l), in order to prevent harm from corrosion.

Hydrogen sulfide is commonly measured in sludge gas by means of the Tutweiler apparatus shown in Fig. 32-3. The procedure is essentially as follows: A sample of gas is introduced into the apparatus, which contains a small amount of starch indicator. Small amounts of a standard iodine solution are added intermittently, with vigorous shaking between additions of iodine. The iodine reacts with hydrogen sulfide as shown in the equation

$$H_2S + I_2 \rightarrow 2HI + S \tag{32-6}$$

When sufficient iodine has been added to oxidize all the hydrogen sulfide, excess iodine is indicated by the typical blue color produced by starch indicator. The strength of the iodine solution is selected to facilitate calculation of hydrogen sulfide in terms of grains per 100 ft^3.

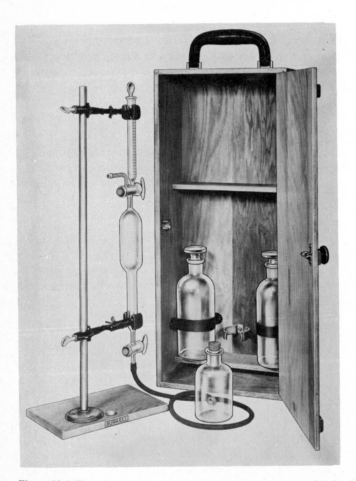

Figure 32-3 Tutweiler apparatus used to determine hydrogen sulfide in sludge gas. *(Burrell Corp.)*

32-6 APPLICATIONS OF GAS-ANALYSIS DATA

In the past, gas analyses have been used largely in research studies and at the larger wastewater treatment plants, where information on the fuel value of gas was important. They will continue to serve in these capacities in the future. In addition, there is a strong possibility that carbon dioxide measurements may be of considerable help in the control of digestion units as a supplement to information provided by volatile-acids and other determinations or as a replacement for some of them.

The determination of hydrogen sulfide will continue to be an important consideration wherever sludge gas is used for fuel in gas engines, particularly in areas where the sulfate content of wastewater is high.

PROBLEMS

32-1 How is gas analysis used in the control of the anaerobic waste treatment process?

32-2 What principle is involved in the separation of gases by chromatography?

32-3 (a) How is the fraction of nitrogen determined in a gas sample using gasometric analysis?
 (b) What sources of error would be of particular concern in determining nitrogen by this method?

THIRTY-THREE

TRACE INORGANICS

33-1 GENERAL CONSIDERATIONS

The matter of trace inorganics in public water supplies in relation to public health has been of concern for nearly a century. The literature shows that lead was the first metal to be brought under scrutiny due to the prevalent use of lead service pipes. The problem was particularly severe in soft water areas, such as New England, because protective coatings did not develop. Zinc and copper were questioned in 1923 and 1926, respectively, because of the use of galvanized services and the growing use of copper salts for algae control. About the same time iodides were found in significant amounts in some supplies, notably in Pennsylvania. In 1931 Churchill showed the correlation between excessive fluorides and mottling of tooth enamel. The literature is replete with references to iron and manganese, largely because of the nuisances they produce rather than threats to the public health.

It was not until the 1962 U.S. Public Health Drinking Water Standards were advanced that limits were set on the levels of inorganics of public health significance other than copper and zinc. These are given in Table 33-1.

The attention of public health and water supply engineers has been refocused upon trace inorganics by four important factors: (1) the outbreak of "Itai-Itai" (Ouch-Ouch) disease that occurred among farmers who drank water containing cadmium from the Jintsu river in Japan; (2) the discovery that metallic mercury escaping from laboratories and industry, mainly chlorine manufacture, was capable of being converted to methyl mercury, concentrated by aqua-

Table 33-1 1962 U.S. Public Health drinking water standards for trace inorganics

Metals	mg/l	Nonmetals	mg/l
Ba	1.0	As	0.01–0.05*
Cd	0.01	CN	0.01–0.2*
Cr	0.05	F	1.0–†
Cu	1.0	Se	0.01
Pb	0.05		
Ag	0.05		
Zn	5.0		

* Permissible if no other source available.

† Depends upon air temperature.

tic life, and passed along through natural food chains to humans, largely through fish; (3) the evidence which exists to indicate that certain forms of the inorganics may be carcinogenic; (4) the realization that purposeful recycling of wastewaters for supplementation of drinking water supplies is imminent in many areas of the country.

The critical nature of the water supply problem has brought Congressional action resulting in Public Law 93-523, known as the Safe Drinking Water Act. It is encumbent upon the Environmental Protection Agency to interpret the law and establish Drinking Water Standards that will apply to all supplies, not just those which supply common carriers, as did the U.S. Public Health Standards.

At this writing, the Environmental Protection Agency has advanced "Interim Primary Drinking Water Standards" as shown in Table 33-2. It will be noted that both copper and zinc have been dropped from the standards. This is largely because the previous standards were to control the aesthetic problem of tastes which these metals cause, rather than from health considerations which is the subject of the newer EPA Primary Drinking Water Standards.

Table 33-2 Environmental Protection Agency interim primary drinking water standards

Metals	mg/l	Nonmetals	mg/l
Ba	1.0	As	0.05
Cd	0.01	CN	0.2
Cr	0.05	F	1.4–2.4*
Pb	0.05	Se	0.01
Hg	0.002		
Ag	0.05		

* Depends upon air temperature.

Of the other metals, mercury has been added to the list and permissible levels of all the ones remaining are the same as expressed in the 1962 U.S. Public Health Standards. For the nonmetals, the limits for arsenic and cyanide have been increased to the former maximum permissible, but the level for selenium has been kept the same.

33-2 TRACE INORGANIC ANALYSIS

The standard procedure for determining levels of arsenic, barium, cadmium, chromium, lead, selenium, and silver is by atomic absorption as discussed in Sec. 11-2. Generally, the graphite furnace is required with atomic absorption in order to measure the low levels of trace inorganics associated with the drinking water standards. The method for mercury represents another modification termed flameless atomic absorption. Here, mercury is reduced to the elemental state and is swept as a vapor from the sample by an air stream into the path of radiation from a mercury cathode-ray tube. The mercury vapor absorbs the radiation in proportion to its concentration, and the resulting reduced radiation reaching a detector is recorded as with the usual atomic absorption procedure.

33-3 SOURCES AND ENVIRONMENTAL SIGNIFICANCE

The principal source of all the trace metals and nonmetals listed in Table 33-2 with the exception of fluoride and selenium is industrial wastes from manufacturing or metal finishing operations. Because of this, municipalities that accept industrial wastes into their sewerage systems are legally responsible for reducing the concentration to acceptable levels before discharge into surface or ground-waters. Usually, treatment at the source is the only practical way of ensuring that the prescribed level of trace inorganics can be met. This poses particular problems in cities where industrial sewer connections have been long established.

Metals

Barium Barium salts are used mainly in the manufacture of paints, linoleum, paper, and drilling muds. Fortunately, the principal form is the sulfate which is highly insoluble. A limit of 1.0 mg/l has been placed on barium because prolonged tests with experimental animals has shown muscular and cardiovascular disorders and kidney damage.

Cadmium Cadmium is used extensively in the manufacture of batteries, paints, and plastics. In addition, it is used to plate iron products, such as nuts and bolts, for corrosion prevention. It is from plating operations that most of the cadmium reaches the water environment. At extreme levels it causes an illness called "Itai-Itai" disease, characterized by brittle bones and intense pain. At low levels of exposure over prolonged periods, it causes high blood pressure, sterility among males, kidney damage, and flu-like disorders. It has recently been discovered that significant amounts are contained in cigarette smoke.

Chromium In the water environment, chromium exists primarily in the form of chromate. Trivalent forms are hydrolyzed completely in natural waters and the chromium precipitates as the hydroxide, leaving minor amounts in solution. Furthermore, there is no evidence to indicate that the trivalent form is detrimental to human health. Chromium is used extensively in industry to make alloys, refractories, catalysts, chromic oxide, and chromate salts. Chromic oxide is used extensively to produce chromic acid in the plating industry. Chromate salts are used in paints and to produce "cleaning solution" in laboratories. Most of the latter eventually reaches the sewer system. Chromate poisoning causes skin disorders and liver damage. There is some reason to believe that chromates are oncogenic (carcinogenic). For this reason, the permissible level in drinking waters has been restricted to 0.05 mg/l.

Copper The restriction of 1.0 mg/l of copper is primarily to avoid the taste which occurs at higher levels. There is no evidence to indicate that copper is detrimental to the public health at levels which are aesthetically acceptable. In surface waters, copper is toxic to aquatic plants at concentrations below 1.0 mg/l and has frequently been used as the sulfate salt to control growth of algae in water supply reservoirs. Concentrations near 1.0 mg/l can be toxic to some fish.

Lead Lead poisoning has been recognized for many years. It has brought about the abandonment of the use of lead service pipes and by many cities, the banning of the use of lead-based paints for interior decoration because of the tendency of some children to gnaw wood and eat the paint. Recognition of the fact that most of the tetraethyllead contained in gasolines is expelled to the atmosphere as lead oxide has assisted in the program of minimizing the amount of leaded gasolines used in internal-combustion engines. Lead has been identified as being a cause of brain and kidney damage. In youngsters it may result in mental retardation and even convulsions in later life.

Mercury Mercury is widely used in amalgams, scientific instruments, batteries, arc lamps, the extraction of gold and silver, and the electrolytic production of chlorine. Its salts are used as fumigants in combating plant diseases and insect pests. It is also used as an antifoulant in ship paints and mildew proofing of canvas tarpaulins and tents. The most spectacular incident of mercury poisoning in humans resulted from the ingestion of sea food taken from Minamata Bay, Japan, during the late 1950s. A chemical plant whose wastewaters discharged into the bay was found to be the source of the mercury. Of a total of

111 cases reported, 43 died. Babies born of afflicted mothers suffered congenital defects.

In the United States and Canada, a potential problem of mercury poisoning was thrust upon us by the discovery of unusual amounts of mercury in fish taken from Lake St. Clair. The major source of the mercury was found to be from chlor-alkali plants employing mercury electrodes. The high concentration of mercury in the fish was found to be due to methylated mercury, CH_3Hg^+ and $(CH_3)_2Hg$, which were shown to be produced by bacterial action in bottom muds under anaerobic conditions.

Most of the recent research concerning the toxicity of mercury has involved the methylated mercury compounds. Because of their extreme effects, the current standard for drinking water has been set at the very low level of 0.002 mg/l or 2 μg/l.

Nickel and cobalt Neither nickel nor cobalt are listed in the current interim primary drinking water standards. Both are used in electroplating, and the rinse waters from these operations constitute the major avenue by which salts of these metals gain access to the aquatic environment.

According to one authority both nickel and cobalt are oncogens (carcinogens). For this reason it is likely that both will be added to the standards in the future, in spite of their rather limited dispersal in nature.

Silver Silver is not a significant pollutant of natural waters. Even in the electroplating industry, losses are minimal due to its high value and efficient recovery systems. Silver poisoning is characterized by a darkening of the skin and eyes. The soluble silver salts are excellent disinfectants. The standard of 0.05 mg/l has been established primarily to discourage use of silver salts for disinfection purposes.

Zinc The toxicity of zinc salts is very low. They gain access to the water environment from mining operations, wastewaters from electroplating, and the corrosion of galvanized piping. The limit of 5.0 mg/l expressed in the 1962 standards was based primarily upon taste considerations.

Nonmetals

Arsenic Arsenic is quite widely distributed in natural waters, occurring at levels of 5 μg/l or more in approximately 5 percent of those tested. The toxicity of arsenic is somewhat erratic. To unacclimated individuals it is quite toxic with dosages of as little as 100 mg, producing serious results. On the other hand, acclimation can occur as evidenced by "arsenic eaters" who consume amounts daily that would be lethal to ordinary persons.

On the basis of the above considerations, there would seem to be little reason to include a standard for arsenic in drinking waters. However, arsenic in certain forms is suspected of being oncogenic; thus a standard is justified.

Arsenic gains access to the water environment through mining operations, the use of arsenical insecticides, and from the combustion of fossil fuels, where part of the fallout occurs on aquatic areas.

Cyanide Cyanides gain access to the water environment through the discharge of rinse waters from plating operations, and refinery and coal coking wastewaters. The cyanide ion has a relatively short half-life because it can serve as a source of energy for aerobic bacteria. For this reason, it should be of little concern where biological treatment systems are employed in the treatment of municipal wastes or where several days detention has occurred in natural waters. The standard of 0.2 mg/l undoubtedly is included to protect against industries with direct discharges to natural waters.

Fluoride See Chap. 27.

Selenium Selenium occurs in natural waters in very limited areas in the United States, notably South Dakota. Selenium is not widely used in industry. Its major use is the manufacture of electrical components: photoelectric cells and rectifiers.

Selenium is an example of an element which may occur in soils in trace amounts but may be accumulated in certain cereals and pasture plants in quantities harmful to animals and humans when ingested. In animals it prevents proper bone formation, and causes "alkali disease" and "blind staggers." The problem is critical with animals because they are dependent upon the local plants for food. Humans obtain their foods from wide geographical areas and thus dilute the effects of selenium from locally grown foods.

Selenium has been reported as being oncogenic but the evidence is very poor. In fact, some report it as being anti-oncogenic. The drinking water standard of 0.01 mg/l is based, undoubtedly, upon the former premise.

REFERENCES

Federal Register, vol. 40, no. 15, March 14, 1975, Environmental Protection Agency, Interim Primary Drinking Water Standards.

"Water Quality and Treatment," 3d ed., American Water Works Association, McGraw-Hill, New York, 1971.

Ehrlich, Ehrlich, and Holden, "Human Ecology," W. H. Freeman, San Francisco, 1973.

Wagner, R. H., "Environment and Man," 2d ed., W. W. Norton, New York, 1974.

"Toxicants Occurring in Natural Foods," National Academy of Sciences, Washington, D.C. 1973.

INDEX